The Cambridge Encyclopedia of Meteorites

Beautifully illustrated with over 150 full color images, *The Cambridge Encyclopedia of Meteorites* provides a thorough guide to these fascinating extraterrestrial rocks.

Meteorites are our only contact with materials from beyond the Earth–Moon system. Using well-known petrologic techniques this book reveals in vivid color their extraordinary external and internal structures. Looking deeper still, right to the atomic level, they begin to tell us of the environment within the solar nebula that existed before the planets accreted. In recent years, meteorites have caught the imagination of scientist and collector alike.

An army of people are now actively searching for them on the hot and cold deserts of Earth. This book is a valuable guide to assist the searchers in the field to recognize the many classes of meteorites. It is further a reference source for students, teachers and scientists, including detailed descriptions of every meteorite type, terrestrial impact crater sites, recent fall and find data, and details of important meteorite collections.

Since earning a degree in astronomy from the University of California, Los Angeles, in 1960, O. RICHARD NORTON has held directorial positions at the Morrison Planetarium, California Academy of Sciences, San Francisco, the Max C. Fleischmann Planetarium, University of Nevada, Reno, and the Grace H. Flandrau Planetarium and Science Center, University of Arizona, Tucson. He has worked as an optical engineer on the design of optical telescopes at the Tinsley Laboratories, Berkeley, California, and the hemispheric projection system for the Desert Research Institute, University of Nevada, the forerunner of the modern Imax system. In 1977 he founded Science Graphics, a company producing science teaching slides in astronomy, space science, meteorites, geology, paleontology and the history of astronomy, used in the science teaching curricula of over 3000 colleges and universities in the United States and further afield. His interest in meteorites extends from his work with Frederick C. Leonard, a renowned early meteoriticist at UCLA. Through the years he has worked to promote the public understanding of science and especially the growing field of meteoritics. He is a fellow of the Meteoritical Society and the author of the best selling book, *Rocks From Space* (1994), and is currently Contributing Editor of the new popular journal, *Meteorite*.

This spectacular painting by Dorothy Sigler Norton shows the violent collision and fragmentation of two main belt asteroids. The larger is a dark, carbonaceous chondrite body; the smaller, a lighter ordinary chondrite body. It is such collisions throughout the history of the Solar System that has spalled off meteoroids and sent them on Earth-crossing orbits to the inner planets, hence to Earth. These meteorites, fragments of other worlds, have much to tell us if we can only learn to read their secrets.

THE CAMBRIDGE ENCYCLOPEDIA OF METEORITES

O. RICHARD NORTON

CAMBRIDGE
UNIVERSITY PRESS

PUBLISHED BY THE PRESS SYNDICATE OF THE UNIVERSITY OF CAMBRIDGE
The Pitt Building, Trumpington Street, Cambridge, United Kingdom

CAMBRIDGE UNIVERSITY PRESS
The Edinburgh Building, Cambridge CB2 2RU, UK
40 West 20th Street, New York, NY 10011-4211, USA
477 Williamstown Road, Port Melbourne, VIC 3207, Australia
Ruiz de Alarcón 13, 28014 Madrid, Spain
Dock House, The Waterfront, Cape Town 8001, South Africa

http://www.cambridge.org

First published 2002

Printed in the United Kingdom at the University Press, Cambridge

Typeface Joanna 10.25/12.5pt. *System* QuarkXPress® [SE]

A catalogue record for this book is available from the British Library

Library of Congress Cataloguing-in-Publication Data

Norton, O. Richard.
The Cambridge encyclopedia of meteorites / O. Richard Norton.
p. cm.
Includes bibliographical references and index.
ISBN 0 521 62143 7
1, Meteorites. I. Title.
QB755.N65 2002
523.5′1′03–dc21 2001035621

ISBN 0 521 62143 7 hardback

To my wife, Dorothy Sigler Norton.
Whose dedication to her husband and his beloved
rocks from space goes well beyond that expected of a loyal
and loving mate. She has over the years shared the joys and pains of
living with an eclectic husband with a hopeless passion for all things celestial.

This striking photograph shows the interior of the 20.2 kg main mass of the Glorieta Mountain pallasite. It is the largest mass yet found in the Glorieta Mountain strewn field north of Santa Fe, New Mexico. Steve Schoner discovered the prized pallasite in 1997 after more than 70 field trips to the site and 15 years of searching. The specimen shows bright yellow olivine with tints of green. The iron network has been etched to show its structure. The slab measures 222 × 187 × 2 mm. (Photograph by Darryl Pitt, curator, the Macovich Meteorite Collection.)

Contents

Preface

Over the past decade public awareness of and interest in meteorites has increased dramatically. Meteorites are no longer just curious, not-so-pretty rocks exhibited in the far-off corner of a geology museum. While they have been the subject of scientific study for over 200 years, only in the final three or four decades of the twentieth century did the study of meteorites gain prominence as a major research field. Perhaps it all began with the 800 pounds of rocks brought back from the Moon in the late 1960s and early 1970s. Someone had to examine these rather dull-looking gray rocks, and who better than that handful of geochemists who had been spending their time studying those rocks that had come to us from space, and free of charge at that. Besides the scientists, there was a smattering of amateur collectors who, for their own reasons, pursued these rare, strange rocks. Slowly meteorites began to appear in rock and mineral shows. They were the common ones: Canyon Diablo irons from Meteor Crater; Odessa irons from the Odessa crater; a few irons from the Henbury craters in central Australia. They were heavy, made of iron and anyone could recognize them as meteorites. How novel to place on your mantelpiece a genuine rock from space, surely the second best thing to owning a Moon rock.

Who could have guessed that 20 years down the road, meteorites would be in great demand, not only by an increasing number of meteorite scientists but by thousands of collectors worldwide racing to get a piece of that new meteorite just found in the Sahara Desert. Almost overnight the Internet became a forum for collectors, providing a place to talk about meteorites as well as a place to buy, sell and trade them. Dozens of meteorite dealers appeared, all trying to get a piece of the action.

As I watched this gain momentum, I had mixed feelings about the whole thing. Why do scientists study meteorites? Why do collectors collect them? Could it be that they have something in common? Meteorites are the oldest materials known, older than Earth itself. To hold in your hand a rock not of this Earth is a unique thought-provoking experience, but to gain access to some of the secrets locked within them is a life-long pursuit shared by both researcher and collector. Every new discovery within these wonderful rocks is announced in papers published in various scientific journals throughout the world. Though in the English language, they might as well have been written in martian. Few amateurs can wade through the technical jargon of the meteoriticist. In fact, only a relative handful of geologists not specifically trained in meteoritics can follow the arguments and methods of the researcher. I know. I've struggled for years to read the journals, with moderate success. For a long time I have felt the need to write a book that could act as a kind of meteoritical Rosetta Stone to help anyone with a keen interest in meteorites, collector or geologist, to follow the trail of the developing science. My hope is that this book will coax the collector to take his meteorites off the display shelf and prepare to go where the action is. This is a book about what's inside those rocks from the asteroid belt. There is another world in there; so stunningly beautiful under the microscope, so revealing of an earlier time when the terrestrial planets were yet to be born. The tantalizing clues of early Solar System history read like a detective story. You can see much of what the researcher sees. In fact, so varied are the meteorites that you are bound to see structures that have not been seen before. This book is a guide to these wonders. It is liberally illustrated with many beautiful pictures never before published in a book about meteorites. Most show the interiors of meteorites illuminated in various ways to reveal unique textures with their curious shapes and forms only seen in extraterrestrial rocks.

I begin this book with the smallest of meteoroids known as micrometeoroids or, better, interplanetary dust particles. These make up by far the greatest amount of extraterrestrial material Earth receives every year. To collect them is a challenge fraught with difficulties, but the scientific rewards are high. These particles have their place among the comets and broken asteroids. There are yet other particles that are perhaps even more intriguing: interstellar dust particles disgorged from red giant stars and exploding supernovae. These particles allow us for the first time to sample "starstuff". They come to us as precious cargo within meteorites that have hosted them since the beginnings of the Solar System. In order for us to sample these early solar and stellar system materials the host meteorites must gain rights of passage to Earth. Succeeding chapters describe the tortuous ablative processes of atmospheric entry. To survive this ordeal is only the beginning. The meteorite must then weather the unique conditions on Earth. Wind and rain, heat and cold team up to destroy them before they can be rescued. How many tons of rare meteoritic material have gone the way of the mountains, eroded to dust and mixed with terrestrial soils? Miraculously, many meteorites make it to Earth each year and they are being found with increasing frequency. On the icy glaciers of Antarctica over 20 000 meteorites have been found in three decades of searching, and the incredible cache from Earth's hot deserts has just begun to appear. A new breed, the meteorite hunter, is seeing to that. Now, at last, there seems to be sufficient extraterrestrial material to sustain the researchers with plenty left over for museum collections and private collectors.

With meteorites well in hand we can begin to explore the secrets we know lie hidden within them. We will cut them, grind them, polish them, etch them, dissect them, photograph them, irradiate them, zap them with electrons, bombard them with neutrons, heat them in ovens and squeeze them into fluorescence – anything necessary to coax them to relinquish their precious store of early Solar System information. And what a story they have to tell. In the remaining chapters of this book I guide the reader through the many meteorite types, both common (if meteorites can really be thought of as common!) and exceedingly rare. Step by step we explore the meteorites with eye, hand lens and microscope, equipment many amateur meteoriticists own today. There is nothing quite as exciting or aesthetically pleasing than to examine a meteorite thin section with a microscope under crossed polarized light. I have tried to impart this excitement on many of the pages of this book. If they produce an "ah" from the reader, then I have succeeded.

We could call thin section analysis the classical approach since it was basically the way researchers in the nineteenth century and up to the first half of the twentieth century studied meteorites. Today, however, there are questions being asked that cannot be answered with these simple tools alone. Fortunately, exotic and wonderful instruments exist that can probe meteorites atom by atom giving us elemental isotopic compositions that lead us in many different directions from determining the composition of the solar nebula to the various ages of meteorites. Herein lies the forefront of meteoritical research where only the highly trained specialist can roam. We depend upon these researchers to help us extend our own knowledge of meteorites beyond that which we can observe on our own. These atom-by-atom observations are intermingled with traditional observations in each chapter, much as they are in meteorite research laboratories. This approach is especially important in the final three chapters in which we consider the role of meteorites and their asteroid parent bodies in the early history of the Solar System.

This is not the first book on meteorites I have written. In 1994, my book *Rocks From Space* was published. It was intended to convey the excitement of meteorites both in history and in the laboratory. It was meant for the general reader with little or no scientific training. It appears to have done its intended job for it met with considerable success. *The Cambridge Encyclopedia of Meteorites* is the next step in the reader's pursuit of meteoritical knowledge. It will hopefully encourage them to look and think deeper. There is much to learn and it is my hope that this book will aid that process.

A book like this is totally dependent upon the work of those who continue to advance the science. I have made liberal use of research papers in such sophisticated journals as *Nature*, *Icarus*, *Geochimica and Cosmochimica*, *Science*, and *Meteoritics and Planetary Science*. All are prestigious journals in which most meteoriticists strive to publish their ongoing research. Membership in the Meteoritical Society earns you a year's subscription to *Meteoritics and Planetary Science*. At the end of each chapter are carefully selected references that I feel the readers can pursue to further their understanding. These journals are found in most college and university reference libraries. Other less demanding journals such as *Sky and Telescope*, *Scientific American*, and *Meteorite* are also included and make for relatively easy reading. The first two are well known among amateurs and professionals. *Meteorite*, a magazine published by Pallasite Press in New Zealand, is a relative newcomer with their first issue appearing in February, 1995. This quarterly journal exists to serve the rapidly growing army of people, from the technically trained to the lay reader, all of whom have at least one thing in common – a fascination with meteorites. It is a place where serious collectors may contribute articles on a variety of meteoritical subjects. Frequently, articles by well-known researchers add to the quality of this fledgling but rapidly growing magazine. I am indebted to the authors of these journal articles for the use of their materials. They have given me quite an education about things meteoritical while writing this book. It is my hope that I have interpreted their work correctly. It seems to be a human trait that errors inevitably enter the most carefully planned and well-written books. This book is probably no different. I must therefore hasten to say that any errors and misinterpretations I might have made are strictly my own and I take full responsibility for them.

A number of people have heeded my call for assistance during the writing of this book. I was determined to use the finest photographs available. Much to my surprise, good thin section photographs of meteorites seemed never to be in color, so the reader must try to follow such caption instructions as "note the olivine (medium gray) mixed with orthopyroxene (white to medium light gray) with individual grains of augite (medium to dark gray) . . .". Just trying to follow the author's descriptions in a world of grays is an exercise in frustration. So I set about the task of photographing meteorite thin sections in color in anticipation that they would be reproduced in color here. Dr Tom Toffoli, an avid meteorite collector and thin section photographer, offered to assist me in this daunting task. For two years we worked together to perfect photographic techniques with minimal equipment but maximal optimism. The beautiful results of this collaboration grace many pages of this book. I can never repay Tom for his dedication, his patience and his skill. As the book is intended to bridge the gap between an introductory text for the lay reader and the much more advanced scientific journals, I asked a trio of very capable scientists to read the manuscript. Dr Joel Schiff, publisher of *Meteorite* magazine and professor of mathematics at the University of Auckland, New Zealand, kindly offered to apply his extraordinary editing talents, which made the book much more readable. His editorial comments and suggestions made me ask many times why I didn't think of that, or worse, why I didn't catch that error in my own multiple rereading. Dr David Mouat, associate research professor at the University of Nevada System's Desert Research Institute, also read the entire manuscript. Though not a meteorite researcher, David is an avid meteorite collector with a strong background in the geosciences. He read the manuscript as a well-informed amateur meteoriticist, exactly the level for which the book was written.

Reading a manuscript for errors is a grueling and thankless task at best and takes many hours of concentrated effort. For this reason, busy research scientists are reluctant to spend inordinate amounts of time correcting other authors' errors. I was especially pleased when Dr Alan Rubin, research geochemist at the Institute of Geophysics and Planetary Physics at UCLA, offered to read the manuscript. Alan is among the finest meteorite researchers in the world. His research over the years has covered a vast area of meteoritics. Although I have known Dr Rubin as a meteoritical research scientist for years through the many scientific papers he has published in the journals, my first opportunity to meet him came by invitation. Shortly after my first book was published, Alan wrote to me and skillfully pointed out some errors I had made in the first edition. There followed two single-spaced pages of criticism. Instead of recoiling from this assault on my work, I was delighted that one of his stature in the field had taken the time to look at it. Thus began a warm relationship that continues to this day. As if to soften the tone of the critical review he invited me to visit UCLA, to meet him and his equally well known coworker, Dr John Wasson, and to view the UCLA collection of meteorites. When it came time to critique *The Cambridge Encyclopedia of Meteorites* I wrote to Alan and asked him to do the honors of "slicing and dicing" the manuscript. This he promptly did, sparing nothing in his desire to "make it right." What agony it was to watch my years of labor taken apart piece by piece. But at the same time what a joy it was to work with him to rebuild it. I only hope that this book will stand as a tribute to his perseverance and insistence to "get it right or you will be criticized." If I have not gotten it all right, then, may the criticisms fall. It will be my fault alone. Thank you, Alan.

A book of such diversity requires the very best illustrations I can acquire or manufacture. My own company, Science Graphics, opened its vast collection of scientific images from planets to meteorites to help illustrate the book. I have been amazed how many people have expressed a desire to help me acquire photographs of meteorites that I did not personally have in the company collection. The same old tired black and white photos have been used for too long in past books. I wanted to include all new photographs. Easy to say; not so easy to do. Yet, people came to my assistance with astonishingly beautiful pictures. Darryl Pitt, curator of the fabulous Macovich collection of meteorites, provided some of the most elegant iron meteorite images I have ever seen. Darryl sees meteorites as art objects, as well he should for his photographs are just that. Michael Casper is adjunct curator of the meteorite collection at Cornell University, Ithaca, New York. At my request he gave me his collection of medium format transparencies of meteorites to use as I wished. Up to nearly the last minute he continued to send me new photographs (the Bilanga diogenite is a good example). Dr Richard Herd, Curator of the National Collections of the Geological Survey of Canada, came to my assistance after he learned of the new book. He suggested that if I would give him a list of needed meteorite photographs he would try to supply them from the Canadian collection. Some of these photographs he made personally, as I could obtain them in no other way.

In other cases, people made their meteorite collections available to me to photograph. Dr Rhian Jones and Dr Adrian Brearley of the Institute of Meteoritics, University of New Mexico, gave me access to their research collection for several days. A few of these photographs were used in my earlier book and others are found in this book also. On the collector end of the spectrum, I was once again given free reign of the Robert A. Haag collection of meteorites, one of the world's finest private collections. I had used many of Bob's meteorites to illustrate my first book. Needless to say, photographs have taken a major role in this book with the philosophy that the written word is not sufficient to describe these fragments of other worlds. Ronald N. Hartman, astronomy instructor at Mt. San Antonio College near Pomona, California, and a colleague and friend (we were college buddies over 40 years ago) introduced me to the use of the dreaded ferric chloride as an etchant for iron meteorites. He and his son James Hartman began experimenting with the agent and encouraged me to give it a try. The technique as they developed it is discussed in Appendix D. Before etching meteorites it is necessary to carefully prepare their surfaces as well as preserve their interiors. Bill Mason, president of Uncommon Conglomerates of St. Paul, Minnesota, introduced me to a process that his company uses to rid iron meteorites of corrosion in preparation for etching as well as preserving them after etching.

Closer to home I was fortunate to have as a good friend, Larry Chitwood, geologist for the Deschutes National Forest here in Central Oregon, which includes some of the most extraordinary volcanic landforms in the world. Larry and I spent many fascinating hours studying meteorite and terrestrial specimens in thin section. His comments especially as applied to terrestrial rocks helped me to bridge the gap between terrestrial and extraterrestrial rock characteristics.

I just happened to be writing this book when Dr Monica Grady of the Museum of Natural History, London, was working on the new *Catalogue of Meteorites, Fifth Edition*. I had been using fall and find statistics from the fourth edition (1985) which was badly out of date. Although the *Catalogue* was not yet published (Cambridge University Press), it was essentially completed. I contacted Dr Grady and boldly requested an advance copy of the crucial first few pages that held the statistics I needed. I expected some reluctance on her part to simply give the statistics to me. After all, she had spent literally years compiling the material and this brazen individual (me) had the nerve to request her work to include in my book *before* publication. Yet, Monica showed not the slightest hesitation. She graciously sent me the material I needed. The book is far better as a result of this charitable act.

Even with the enormous assistance of the *Catalogue of Meteorites* it was more than a year before the manuscript was completed. During that time meteorites continued to be found. To include the very latest fall and find data up to the moment, I called upon Bernd Pauli in Germany. One of his passions is maintaining a complete database of current meteorite falls and finds. I asked him to provide the latest data for the book which he graciously agreed to do up to the time of publication.

I can only briefly mention here the many others who have enthusiastically supported this book since the first words were written more than three years ago. Helen Worth, public affairs officer at Johns Hopkins University's Applied Physics Laboratory, kept me abreast of the progress during the historic NEAR mission to orbit the asteroid 433 Eros. Dr Robert Walker, director of the McDonnell Center for the Space Sciences at Washington University, St. Louis, kindly provided many reprints of their work on presolar interstellar grains as well as photographs of these illusive particles. He had written to me five years earlier to express his dismay that I had not included his beloved interstellar particles in my first book. I vowed not to make that mistake again. The work at the McDonnell Center is summarized in the concluding pages of the first chapter. Carol Schwarz of the Johnson Space Center's Planetary Missions and Materials Branch provided images of interplanetary dust particles and photographs of the high-flying WB57 aircraft with dust collectors. She further provided closeups of impact areas on the Long Duration Exposure Facility. Four scientists in the main stream of meteoritical science kindly provided difficult to obtain photos of thin sections along with their interpretation. Dr Allan H. Treiman of the Lunar and Planetary Institute in Houston provided thin section photos of the Chassigny and the Lafayette nakhlites. Dr Bevan M. French, Research Collaborator, Department of Mineral Sciences, Smithsonian Institution, allowed me free access to materials from his book, *Traces of Catastrophe*, including photos of shocked quartz, Dr Ted E. Bunch, formerly of the Lunar and Planetary Institute, Houston, provided hard to find color images of shock metamorphism showing planar fractures and planar deformation features in quartz. Dr David W. Mittlefehldt of the Johnson Space Center enthusiastically helped me to interpret thin sections of the newest angrite Sahara 99555.

Working with the staff at Cambridge University Press was a unique experience. It all began when I met Dr Simon Mitton, Executive Director, Science and Professional Publishing at Cambridge University Press, at a meeting of the Astronomical Society of the Pacific in Santa Clara, California. He was looking for someone to write a book which he would later title *The Cambridge Encyclopedia of Meteorites*. I gave him a copy of *Rocks From Space* and expressed my interest in writing such a book. He suggested that the new book be written on a higher level and invited me to submit a book proposal, which I did a month or two later. Thus began a seemingly endless cycle of writing, correcting and rewriting. The manuscript in this fluid form did not receive an editor for nearly four years. In this interval the science of meteoritics continued to advance. Exciting discoveries were being made which had to be included. More writing and rewriting. Finally the completed manuscript (an author never considers his manuscript complete) was submitted. It was fortunate that my labor was placed under the care of editor Dr Sally Thomas at CUP. She guided me patiently and gently through the gauntlet of editing. My barrage of questions and concerns were always answered immediately and she was always receptive to my thoughts and ideas. We made a fine team though eight time zones apart, and I will miss her almost daily e-mails. Beverley Lawrence was assigned to be my copy editor. She was the one who would seek out and correct all my composition and grammar errors. She is one of those unsung heros in the publishing business to whom the author, in the strictest of confidence, has to reveal his ignorance of the English language. We worked well together and the result was a readable book written in the Queen's English. There were others I never met who played important roles; book designers, typesetters, graphic artists, far too many to include here but I know you are out there and I thank you and all of the staff at Cambridge University Press for their commitment to this book.

Finally, I reserve the last position of thanks for the person who occupies the first position in my life. My wife Dorothy has suffered with me during ten years of continuous writing and rewriting, reading and rereading, correcting and recorrecting manuscripts and articles about those rocks that incessantly seem to control our lives. Rather than shun them as unwanted infiltrators into our peaceful home, she has accepted them with great gusto. What would the meteorite enthusiast, researcher or collector give to have a spouse who insists upon purchasing that expensive rare specimen for our growing collection? What spouse continues to take the time to educate herself about these enigmatic rocks from space? As if her acceptance and support of my passion with meteorites is not enough, she has given me her considerable skill as a scientific illustrator and fine artist to illustrate both this book and the first. Her talent as a scientific artist is known in meteoritical and paleontological circles around the world. For Dorothy's steadfast devotion to me throughout our lives together, through times both wonderful and difficult, I can only honor her with this book, a symbol of our mutual creative efforts. This book, therefore, I dedicate to her.

O. Richard Norton
Bend, Oregon
September 5, 2001

Foreword

Half a century ago, fewer than 2000 meteorites were known and the number of analytical techniques available to meteorite researchers was limited. Most studies involved little more than bulk chemical analyses, examination of macroscopic structures in polished slabs, and scrutiny of microscopic textures in thin sections. Brian Mason's influential 1962 book *Meteorites* listed only five groups of chondrites (there are now 13 major groups). Among the eight groups of achondrites listed by Mason, there were no lunar meteorites and no recognition that some of the achondrites were of martian origin.

The resurgence of meteorite studies in the past few decades (a fact responsible for the existence of this book) is due to a dramatic increase in the number of meteorites available for study and the number of analytical techniques to study them. Tens of thousands of meteorite samples have been recovered from cold deserts (Antarctica) and hot deserts (mainly Western Australia, the southwestern United States, and the Sahara). Analyses of these samples have led to the identification of many new chondrite groups, the discovery of lunar meteorites, and the recognition that nearly 20 meteorites come from Mars. The number of known meteoritic minerals has shot up from about 40 listed in Mason's book to about 300 today. Modern analytical instruments unavailable half a century ago include mass spectrometers, electron and ion microprobes, and transmission, scanning and analytical electron microscopes. It is now possible to analyze a single mineral grain in a thin section and determine its textural setting, its major-, minor- and trace-element concentrations, and its oxygen-isotopic composition.

To take advantage of modern analytical techniques, researchers have formed an informal world-wide collaborative network. For example, I routinely work with meteorite researchers in Honolulu, Albuquerque, Chicago, Tempe, Houston, London, Berne, and Tokyo. Collaboration in the study of meteorites is not new. In 1802, the English chemist Edward Howard reported his chemical analyses of four ordinary chondrites, two irons and two stony-irons. The significance of this work lay in the discovery of nickel associated with the metallic iron. Nickel is rare in terrestrial rocks and had only been identified in the 1750s. The discovery of meteoritic nickel supported the extraterrestrial provenance of these rocks. Howard collaborated with the French émigré Count Jacques-Louis Bournon who described the mineralogy of Howard's meteorites and first reported the existence of the "globular bodies" now known as chondrules.

Antarctic meteorites began flooding research labs after 1969 when the Enderby Land traverse team of the tenth Japanese Antarctic Research Expedition serendipitously came across nine meteorites near the Yamato Mountains in Queen Maud Land. They found 12 more in the 1973–1974 field season, 663 the following year, and 307 in 1975–1976. In the 1976–1977 field season, a three-man American–Japanese team recovered 11 meteorites on scattered patches of blue ice in the Allan Hills region of Victoria Land, Antarctica, more than 3000 km away from the Yamato Mountains. About 300 Allan Hills meteorites were found the following year. All of these meteorites, and the many thousands found subsequently, were available to researchers; virtually none reached the hands of dealers or collectors.

The situation changed with the recent recovery of more than a thousand meteorites from Northwest Africa. These samples are typically acquired by local entrepreneurs and sold to Western meteorite dealers. The dealers look through the uncut specimens and pull out those that seem most unusual. However, the dealers cannot set prices for their prize specimens until the rocks are properly classified. For example, a rock that looks like a basalt could be a eucrite, an angrite, a shergottite, a lunar basalt, or a terrestrial interloper. The terrestrial basalt is of no commercial value, and the per-gram prices of the other specimens can vary by more than two orders of magnitude. Dealers needed researchers to classify the meteorites; researchers needed samples for scientific study. These complementary needs led to the development of a symbiotic relationship between dealers and researchers. Researchers charge dealers a standard fee for their services – typically 20 g of a stony meteorite (100 g or more of a large iron) or 5% of the total meteorite mass, whichever is the lesser amount. The system works: researchers get scientifically important specimens on the cheap, dealers get to set their prices, and collectors get to pay the bill.

Many collectors are enthusiasts who are as keen to expand their knowledge of meteorites as they are to enhance their collections. However, few collectors have the time or training to pore over hundreds of technical papers loaded with arcane jargon. What was needed was a general reference. This need was filled in part by Richard Norton's first meteorite book, *Rocks From Space*, which has sold more than 20 000 copies in its two editions. This new book, *The Cambridge Encyclopedia of Meteorites*, more completely satisfies the need for a general reference; it is scientifically more rigorous and less anecdotal than *Rocks From Space*, yet it is still accessible to a wide readership.

Some readers may find it odd that a book calling itself an encyclopedia does not have an alphabetic arrangement (e.g., achondrites to Zodiacal light) like those found in *World Book* or *Britannica*. Such an arrangement has been commonly used in encyclopedias since John Harris introduced it in 1704 in his *Lexicon Technicum*. But topically arranged encyclopedias have an even longer history. Aristotle attempted to arrange existing knowledge in a series of topical books in the fourth century BCE. Although Richard Norton has followed Aristotle's encyclopedic arrangement, he breaks ranks with Aristotle's views. In *Meteorologica*, Aristotle described meteors as atmospheric phenomena caused by dry smoky exhalations from crevices in the Earth that rose to the top of the sublunary sphere and burst into flame. Richard Norton corrects Atristotle and sets the record straight. Meteors are light phenomena produced when meteoroids enter Earth's atmosphere and frictionally heat the surrounding air to incandescence. Surviving meteoroids become meteorites – rocks that fall from space. The scientific study of these rocks has yielded fascinating insights into the origin and early history of the solar system. This is their story.

Alan E. Rubin
Institute of Geophysics and Planetary Physics
University of California
Los Angeles
USA

CHAPTER ONE

Cosmic dust:
interplanetary dust particles

The most obvious things that fall to Earth from space usually announce themselves with brilliant fireballs, sonic booms and amazed spectators. Such was the case on October 9, 1992, a few minutes before 8:00 p.m. when a dazzling meteor appeared, traveling in a north-northeasterly direction over the Eastern United States. It was Friday evening and all along its path people watching local high school football games were startled by its sudden appearance and accompanying sonic booms. Thousands of sports fans, their attention captured by the dazzling spectacle, trained video and still cameras skyward. It became the most filmed fireball in history. It took only seconds to travel over Eastern Kentucky, North Carolina, Maryland, and New Jersey, finally ending its 40 s of luminous flight over Peekskill, New York, where it dropped several stony meteorites, the most noteworthy passing through the trunk of a parked car.[1]

The smallest meteoroids – interplanetary dust particles (IDPs)

Most material reaching Earth's surface does so in a much less ostentatious way. Every day, millions of tiny stony bits fall quietly into the atmosphere, burning briefly as meteors, leaving behind the vaporized residue that filters slowly to Earth. Meanwhile, we inhabitants go about our lives unaware that Earth is continuously accreting tons of material from space. Most of this material remains unseen until it encounters the atmosphere. The origin of this material intrigues scientists because most of it is primitive, that is, it contains the first-formed materials to condense out of the solar nebula. Earth and the other planets have been gathering this remnant material for 4.5 billion years. By now, the Solar System should have been swept clean. Yet, Earth collects tons of space debris every day. Meteoroid flux rates depend upon the particle size chosen. Particles one milligram (1 mg) or less in mass, typical IDPs, are collected at the amazing rate of ~110 metric tons every day – and there seems to be no diminution. We can only conclude, then, that they are being manufactured somewhere by something. That "somewhere" and "something" are an important part of the study of interplanetary material.

The zodiacal light

On clear, moonless nights in the Spring or Fall when the ecliptic plane stands high after sunset or before sunrise respectively, a broad, diffuse pyramidal cone of light centered on the ecliptic can be seen. It approaches the brightness of the Milky Way, being brightest near the sunset point on the horizon and gets gradually fainter as it extends upward away from the sunset point nearly to the zenith (Fig. 1.1, right). The light straddles the ecliptic (zodiac) by 15–20° nearest the

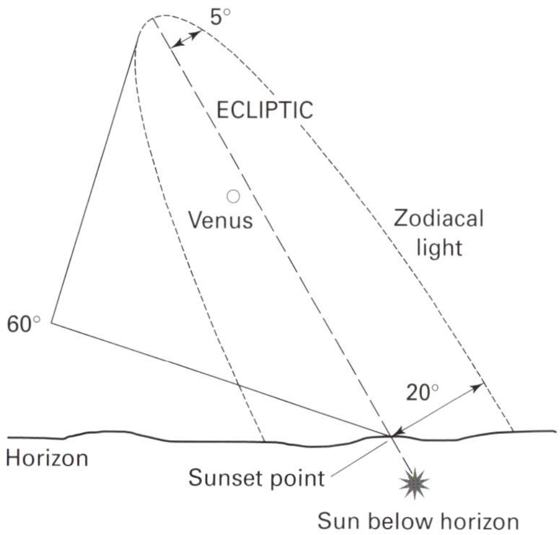

Fig. 1.1. (Right) View of the zodiacal light about one hour after sunset forming a broad wedge extending from the sunset point to an altitude of nearly 60°. The planet Venus shines near the center of the wedge very near the ecliptic plane. (Left) A sketch of the zodiacal light in the photograph showing its Earth-based geometry with respect to the ecliptic plane.

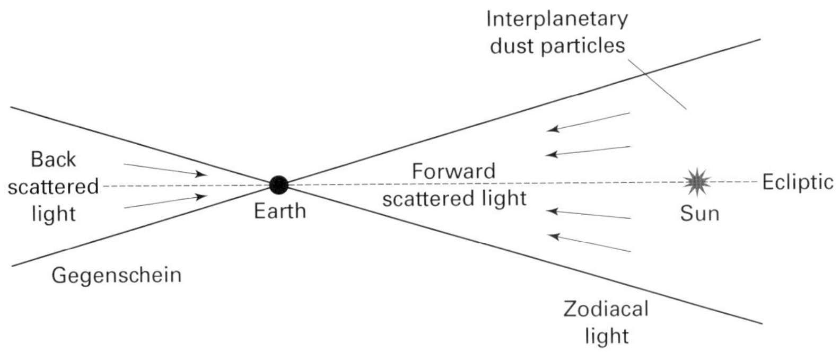

Fig. 1.2. View along the ecliptic plane showing forward-scattering of sunlight off interplanetary dust particles between Earth and Sun producing the zodiacal light; and back-scattering of light from interplanetary dust particles beyond Earth and opposite the Sun producing the gegenschein.

Sun and tapers to a rounded terminus 5–10° at the vertex of the cone. This is the *zodiacal light* and it contributes substantially to the natural light of the night sky (discounting scattered city lights which makes zodiacal light impossible to see from the cities). During Spring and Fall, it can account for as much as 42% of the night skyglow.[2]

The spectrum of the zodiacal light is identical to the absorption spectrum of the Sun, proving it to be sunlight reflected off very tiny micrometeoroids called *interplanetary dust particles* or *IDPs*. The light cone at first seems to be centered along the Solar System plane (ecliptic) but careful measurements show that it is symmetrical around Jupiter's orbital plane which is inclined 1.3° to the ecliptic plane. This suggests that Jupiter's gravity controls the distribution of these particles along the Solar System plane.

There is a much fainter glow opposite the Sun and the zodiacal light, also centered on the ecliptic. This is called the *gegenschein*, a German word meaning "counterglow". It is related to the zodiacal light in that it, too, is sunlight reflecting off interplanetary dust particles, only the geometry is different. Figure 1.2 shows the zodiacal light originating from the scatter and reflection of sunlight by particles between Earth and Sun where the particles are most numerous. Sunlight passes through the cloud of particles and is randomly scattered. This is called *forward-scattering* and is analogous to the scatter of light through a fogged glass window. Gegenschein occurs on the opposite side of the Sun from Earth, the antisolar point. Here, the particles are fully illuminated by the Sun and the light is *back-scattered* or reflected back toward Earth. This occurs where the density of particles is considerably less and the distance over which the light must travel is much greater. Thus, the gegenschein is far fainter than zodiacal light. During exceptionally clear conditions at high altitudes, the zodiacal light has been seen to connect faintly to the gegenschein, forming a nearly continuous light band over most of the sky.

A solar connection

The greatest concentration of particles responsible for the zodiacal light seems to be near the Sun. In fact, there is clear evidence that the particles reach the outermost atmosphere of the Sun. The Sun's outer atmosphere or corona is divided into two components based upon their structure and composition. The inner part, the *K-corona*, is gaseous and shows spectral characteristics suggesting a temperature of greater than 1 000 000 °C. At this temperature atoms readily lose their outer electrons and move freely and at high velocities within the K-corona. Sunlight coming from the photosphere (visible surface) of the Sun is effectively scattered by the electrons giving the K-corona its light. But because the electrons are traveling at high velocities, they Doppler-shift the light all along the absorption spectrum smearing the absorption lines and creating a continuous spectrum. (The K refers to the German word, *Kontinuierlich*, meaning continuous.) This continuous spectrum is observed out to about 3–4 solar radii from the Sun's surface (roughly 2.75 million kilometers). Beyond that point, the continuous spectrum rapidly gives way to the aforementioned original absorption spectrum. This is the beginning of the *F-corona*. (The F stands for *Fraunhofer*, the nineteenth-century astronomer who first found absorption lines in the Sun's spectrum). The absorption spectrum is sunlight reflecting off IDPs, the same dust responsible for the zodiacal light. Thus, the gas of the solar corona gradually grades into the interplanetary medium and the zodiacal light can be thought of as an extension of the Sun's outer corona. Taken as a whole, interplanetary dust particles comprising the zodiacal light can be thought of as a tenuous ring around the Sun, much like the rings of Saturn, extending from the F-corona to beyond the asteroid belt.[3]

Lifetime of interplanetary dust particles

Most of the IDPs range in size from a fraction of a micrometer to several micrometers in diameter. Particles less than 1 μm (one micrometer (μm) = 10^{-3} millimeters (mm)) are subject to the push of the Sun's radiation pressure and the solar wind which applies sufficient continuous radial (outward) force to drive the tiny particles out of the Solar System. Space probes venturing beyond the orbit of Mars have actually detected these small particles leaving the Solar System. Particles more massive than this (several micrometers to a few millimeters in diameter) are only weakly affected by this minuscule outward force. It is more than overcome by the Sun's gravitational force sending them into

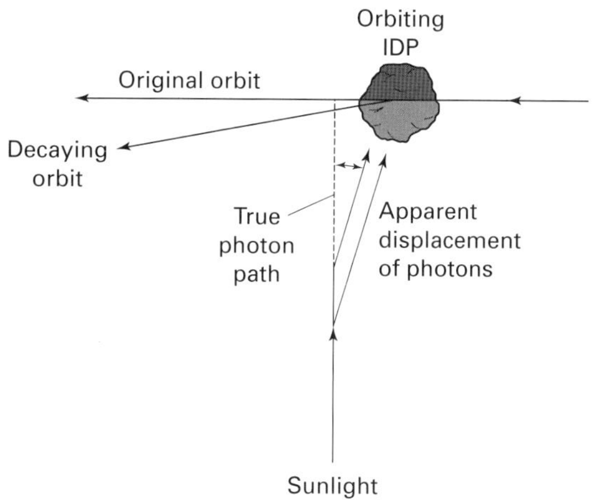

Fig. 1.3. The Poynting–Robertson Effect. Solar photons are displaced by the moving IDP. Striking the IDP at less than 90°, the photons slow the particle causing its orbit to decay slightly. Eventually the particle spirals into the Sun.

elliptical orbits around the Sun. Though too massive to be pushed radially away from the Sun, these particles are affected in a different, more subtle, way. Even though the force of radiation pressure is directed radially away from the Sun, an orbiting particle encounters the photons at *something less than* right angles to its direction of orbital motion. This is analogous to a car driving through a rainstorm where the rain is falling vertically but the car encounters the rain with a horizontal component opposite to the car's motion, causing the rain to hit the windshield. This apparent displacement of the rain drops is due to the forward motion of the car. Likewise, the photons have a small velocity component in a direction opposite to the particle's direction of motion due to the apparent displacement of the photons (Fig. 1.3). The absorption of the photons's momentum produces a tiny but effective drag on the particle. The net result is a gradual slowing of the particle, which first acts to reduce its orbital eccentricity until it becomes circular and then slowly causes it to spiral inward toward the Sun.

The time scale for a particle to spiral into the Sun depends upon its mass and size and its initial perihelion and orbital eccentricity. For example, a particle of 1 mm diameter at a perihelion distance of 1 AU and an orbital eccentricity of 0.7 (a typical comet orbital eccentricity) will spiral into the Sun within about a million years. If the same particle originated in the asteroid belt about 2.8 AU from the Sun, it would spiral into the Sun in about 60 million years. Most of the particles responsible for the zodiacal light lie within the inner Solar System. These particles in the one to ten micrometer range have much shorter lifetimes, being drawn into the Sun in periods as short as 10^4–10^5 years. This phenomenon, called the *Poynting–Robertson Effect*, should therefore clear the Solar System of most micrometeoroid particles within that short time period. Every second, roughly 8 metric tons of these particles spiral into the Sun. At that prodigious rate, the Solar System loses over 250 million metric tons (2.5×10^{11} kg) per year.[4] The fact that the zodiacal light is still observed is strong evidence that it is in equilibrium: the particles are being continually renewed at roughly the same rate.

Density of interplanetary dust particles

Looking at the zodiacal light with the unaided eye (Fig. 1.1) gives one the impression that space is so cluttered with particles that it would present serious drag problems to orbiting spacecraft. The perceived density of IDPs did raise serious questions about the safety of space travel in the early days immediately preceding the first artificial satellites. (As early as 1946, a decade before the first satellites were launched, a respected Harvard University astronomer predicted that one in twenty-five spacecraft traveling to the Moon would be destroyed by orbiting space debris.)[5] The first Earth-orbiting satellite designed specifically to detect interplanetary particles was launched in 1965. Designated *Pegasus*, this satellite had huge wings that could sweep out large areas of space as it orbited, measuring the near-Earth distribution, size and velocity of interplanetary particles. Sensors on the wings could detect the impacts of micrometer-sized micrometeoroids striking the wings at speeds of 10–15 km/s. Three of these satellites were placed in Earth orbit, giving scientists their first sampling of particle densities. The results were happily contrary to earlier dire predictions. Far fewer particles were encountered than anticipated. The data indicated only a few particles per kilometer would be encountered by satellites and Moon-bound spacecraft and these were micrometer-sized, too small to inflict major damage. Since that time, numerous other space probes traveling from Mercury to Mars and beyond have verified these results.

Estimates of the total mass of interplanetary particles composed of micrometer-sized particles to stones several centimeters across varies widely. The value of the mass usually quoted is 2.5×10^{16} kg.[6] This is roughly equivalent to the total mass of a typical comet or perhaps a small asteroid the size of Deimos, Mars' smaller satellite (8 km radius), spread out 10–15° along either side of the ecliptic and extending from the Sun to the asteroid belt.

Comets as sources of interplanetary dust particles

Comets are thought to be a primary source of the particles responsible for over 50% of zodiacal light dust. In addition, infrared data from the IRAS and COBE sensors suggest that collisions of main belt asteroids liberate fine dust that may account for as much as 40%. The remaining few percent comes from two sources: fine debris from the collisions of near-Earth asteroids; and interstellar particles. Some meteorites are hosts to the smallest particles (less than 1 μm) among the IDPs. These have isotopic compositions distinct from the meteorites and probably represent *presolar* or *interstellar* grains. We will look more closely at this source later in this chapter.

As a comet approaches the Sun, its icy components sublimate, being warmed by sunlight. Immense geysers of sublimating water-ice and other volatiles erupt violently

Fig. 1.4. Comet Hale-Bopp photographed on April 2, 1997, seen with its straight blue ion tail (type I) and yellowish dust tail (type II). During its most active phase, Comet Hale-Bopp expelled about 200 times more micron-sized dust particles into space than Comet Halley did in 1986. (Photo by O. Richard Norton and Lynn Carroll.)

from the rock-ice nucleus liberating tons of particles that have been entombed in the ices for billions of years. A dusty tail forms, streaming behind the comet as it is acted upon by the solar wind and radiation pressure (Fig. 1.4). To maintain the density of interplanetary particles making up the zodiacal dust cloud requires a renewal rate of about 2.5×10^{11} kg per year, the amount lost to the Poynting–Robertson Effect each year. An average comet loses $\sim 10^{11}$ kg of solid meteoritic material during every perihelion passage and if about half of the particles remain in the inner Solar System (the smallest particles are quickly removed from the inner Solar System by radiation pressure within a year), the remaining particles would very nearly match the yearly renewal rate requirement. To replace the entire zodiacal dust cloud would require one comet a year for 10^5 years, a rather optimistic outlook. Undoubtedly, other sources such as asteroid dust must lend a helping hand to keep the zodiacal cloud in equilibrium.

Once the comet particles are released, they continue in the comet's highly elliptical orbit about the Sun long after the comet has left the inner Solar System. Tracks of cometary debris crossing the inner Solar System and mingling with the zodiacal light particles have been detected by infrared satellites. Gradually their orbits become more circular as radiation pressure and the solar wind continue to act on the particles, dispersing them within the inner Solar System until they become part of the interplanetary dust that produces the zodiacal light.

Atmospheric collection of interplanetary dust particles

As Earth orbits the Sun, it tunnels its way through the particles, accreting about a hundred metric tons every day. Interplanetary dust particles can filter through Earth's atmosphere without heating and ablating. Their mass to volume ratio is small, meaning they have large surface areas relative to their masses allowing them to be effective radiators of heat energy. They radiate away heat of atmospheric entry faster than they absorb it. Setting the entry velocity at Earth's escape velocity (11.7 km/s), for entry without ablation the critical size limit for such particles is 52 μm diameter.[7] Once their cosmic velocity has been slowed, they become trapped high in the stratosphere where they remain for days or weeks before falling gently to Earth.[I]

The first attempts to collect IDPs were made on the Earth's surface. Since the material falls more or less uniformly all over the Earth, scientists at first searched for IDPs in surface samples. To find these particles among the myriad of terrestrial materials proved to be arduous and fraught with problems. No

[I] Some atmospheric physicists have suggested that these particles may act as condensation nuclei for mysterious high altitude clouds most frequently seen at high northern latitudes in summer. They are referred to as *noctilucent clouds* because at typical altitudes of 80–85 km, they are far above tropospheric clouds and easily capture sunlight for an hour or so after sunset. Collections made from sounding rockets fired through the clouds show the presence of volatile particles (water-ice) with only occasional solid nuclei. If IDPs act as important condensation nuclei for the formation of noctilucent clouds, then these clouds should be more prevalent following annual meteor showers such as the Perseids. The data, however, do not show any such relationship.[8]

Fig. 1.5. The current generation of high-flying IDP-collecting aircraft is the ER-2. Similar to earlier U-2 aircraft, the ER-2 has enormous wings allowing it to reach very high altitudes. It is equipped with large area dust collectors on the wings. It flies out of NASA-Ames. (NASA photo.)

one knew exactly what they should look like. Most of the particles thought to be extraterrestrial turned out to be a mix of natural and manmade terrestrial contaminants. It very quickly became apparent that searches would have to be made in a relatively uncontaminated area, namely, the stratosphere.

The first attempts to collect stratospheric IDPs using balloons and aircraft were made by scientists from the Smithsonian Astrophysical Observatory in the 1950s and met with only limited success. Many particles were collected, most of which were of terrestrial origin. Particles in the micrometer size range were collected but they were too small to be chemically analyzed with any known technique, so there was no way to compare these particles chemically with known meteoritic materials. (This was before the development of the electron and ion microprobes.) Contamination with terrestrial particles continued to be a serious problem using aircraft and high flying balloons through the early 1960s. And to make success even more elusive, it became increasingly apparent that the flux of IDPs through the atmosphere was much smaller than originally estimated.

In the early 1970s, Paul Hodge and Donald Brownlee of the University of Washington's Interplanetary Dust Laboratory began to reinvestigate the use of high flying aircraft. The U-2 airplane (Fig. 1.5) with its enormous wingspan allowing it to reach 20 km altitude and maintain it at relatively slow speeds was nearly ideal for IDP collection. Brownlee designed an efficient collector using collection plates coated with highly viscous silicone oil and mounted them near the wing tips where contamination was minimal. These remained sealed until the aircraft reached collecting altitude. They were then opened to the slipstream. After cruising for several hours, the plates were retracted to avoid lower atmosphere dust contamination. They were then taken to a clean laboratory where the oil was dissolved and any trapped particles released on collection filters. As always, a high percentage of contaminants were picked up but were fortunately easily recognized. The first ten flights collected some 300 IDPs.

The IDPs have diameters between 1 and 50 μm. Many have a fluffy, porous texture made up of loosely consolidated aggregates of small grains ranging in size between 0.1 and 3 μm diameter (Fig. 1.6). Their porous textures are responsible in large measure for their low porosities of between 0.7 and 2.2 g/cm^3. Their mineral compositions are *chondritic*, meaning that they have elemental abundances relative to silicon within a factor of three of those found in carbonaceous chondrite meteorites (see Chapter 7). The minerals include iron-poor pyroxene and olivine, iron-rich sulfides (troilite), layer-lattice silicates (smectite, serpentine), magnetite, and carbonaceous material. Chondritic IDPs are divided into two groups based upon their internal textures: anhydrous (containing no water in chemical combination) silicates and layer-lattice (layered structure in their crystal lattice) silicates. Anhydrous silicates IDPs are the highly porous IDPs mentioned above and illustrated in Fig. 1.6. The most common constituents of anhydrous IDPs are individual mineral grains of olivine and pyroxene with minor amounts of magnetite, kamacite (low-nickel iron), cohenite (iron–nickel carbide) and chromite. Often small crystals of kamacite, olivine and iron sulfide are seen emersed in a carbonaceous matrix. These constituents are collectively called "tar balls". A disordered constituent of carbonaceous matter of unknown composition are frequently found with the tar balls. Layer-lattice silicates IDPs as the name implies, are low porosity, compact structures. They are made of smectite (the montmorillonite group of clay minerals) or serpentine, the former being much more common. This material may have been produced by aqueous alteration of pyroxene or olivine. Like the anhydrous IDPs, carbonaceous matter and tar balls are also found in layer-lattice IDPs.[9]

A most obvious question is how IDPs relate to known meteorites. We have seen similarities between chondritic IDPs and carbonaceous chondrite meteorites but there are many differences that suggests that IDPs are distinct from known meteorites. Although the anhydrous IDPs contain minerals that are common in chondritic meteorites, we have seen other minerals and structures that are unique to IDPs. One striking

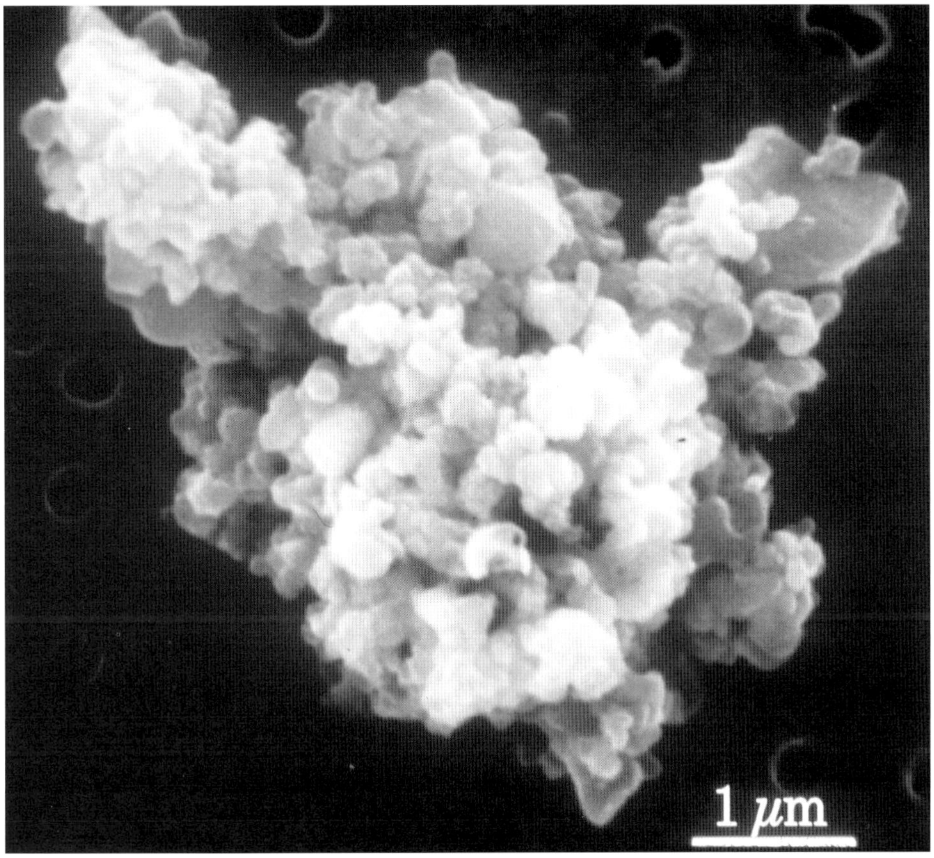

Fig. 1.6. This IDP was collected at 20 km altitude by the ER-2 aircraft. It is composed of loosely consolidated anhydrous silicate mineral particles and carbonaceous material with an elemental composition similar to carbonaceous chondrite meteorites. Note the high porosity of the particle. The voids at one time may have been filled with volatiles. The scale bar is 1 μm. (Johnson Space Center and NASA photo.)

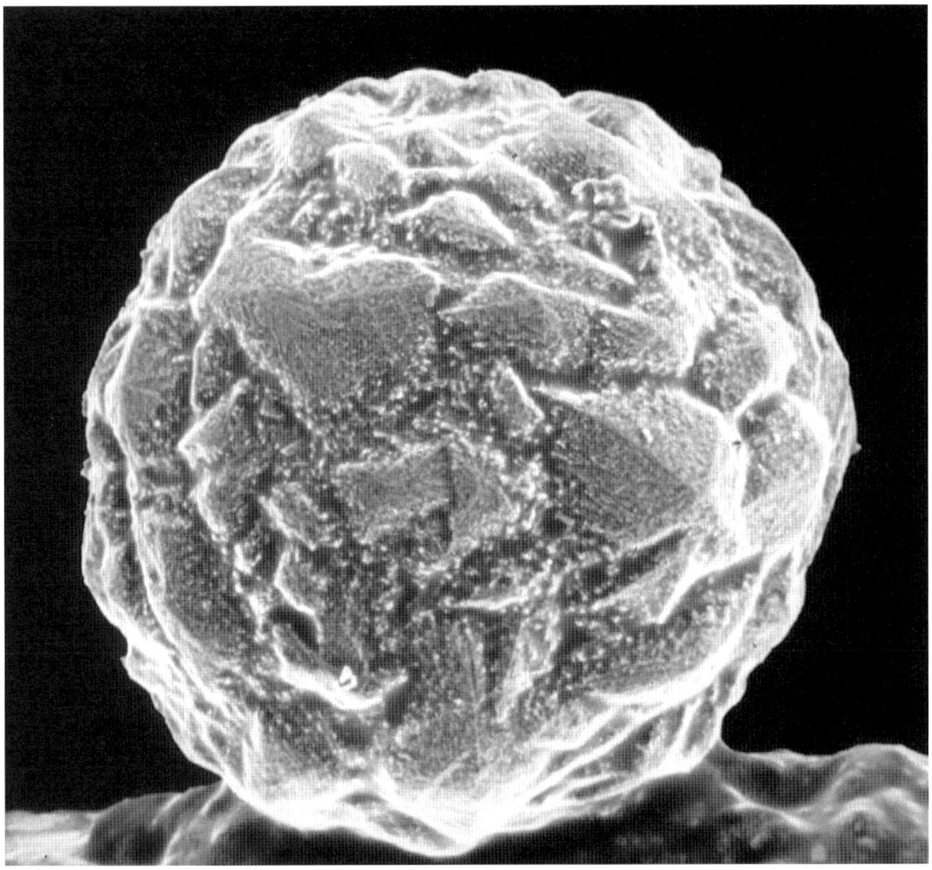

Fig. 1.7. This remarkably spherical 300 μm IDP was collected on the ocean floor by a magnetic collector. Its composition is primarily olivine and magnetite. Such melted and recrystallized spheres are much larger than atmospheric IDPs and may have originally been millimeter-sized particles from a comet. They may be typical of particles that produce meteors. (Johnson Space Center and NASA photo.)

difference is that many IDPs contain as much as 50% carbon, a far higher carbon abundance than in even the most carbon-rich chondrites, the CI chondrites, that contain no more than about 5% carbon. They are therefore considered distinct mineralogically and chemically from meteorites, which strongly suggests that they have a cometary origin.

Collection of deep sea interplanetary dust particles

The deep oceans are obvious collection beds for IDPs that have fallen through the atmosphere. They have been known and collected for a century. Most are spherical, a millimeter or more in diameter, black in color and contain sufficient magnetite to allow them to be retrieved with magnetic collectors dragged along the ocean floor. Most magnetic spheroids are iron-rich with elemental abundances quite similar to iron meteorites. These have been characterized as ablation products from iron meteorites, their parent bodies. The iron occurs in its oxidized phase as magnetite with small amounts of nickel which forms a shell around a core body composed of the iron–nickel alloy, *taenite*. Taenite is a common nickel-rich iron alloy found in iron meteorites, usually composed of between 20 and 40% nickel. This nickel-rich core is found in both metallic and oxidized phases and can be as much as 90% nickel.[10]

More than half of the deep ocean spheroids have silicate compositions. These undoubtedly represent ablation products from chondritic meteorites with compositional similarities to CI, CM and CR chondrites (Fig. 1.7). They are in many ways more interesting because they involve the more complex chemistry of stony meteorites. They differ in appearance from the iron spheroids, having a dull luster and are gray in color. The mineral compositions of the silicate spheroids are what one would expect coming from chondritic meteorites: magnesium-rich olivine and magnetite crystals set in glass. The olivine is recrystallized from the ablation melt and the magnetite is a product of the rapid oxidation of metallic iron present in the meteorite. The magnetite mixes with glass formed during ablation. This composition is similar to the fusion crusts found on freshly fallen stony meteorites. The textures are quite variable; some show randomly arranged relatively large olivine crystals embedded in smaller magnetite crystals, both held together by interstitial glass while others show orderly parallel arrangements of euhedral olivine crystals making them strikingly similar to barred olivine chondrules (see Chapter 6). Some of the spheroids show evidence of relict grains within the spheres; that is, mineral crystals that existed within the original parent body before atmospheric entry. These were originally discovered by Donald Brownlee and his coworkers in 1978.

Impacts close to home – a practical matter

There is a practical side to the study of interplanetary particles, especially near Earth. Of primary concern to engineers and space scientists is an understanding of the space environment where the majority of manned and unmanned spacecraft spend their time – in low Earth orbits. During the Mercury, Gemini, and Apollo programs, every mission encountered hypervelocity particles sufficiently massive to produce tiny craters in the space hardware. Some were energetic enough to penetrate protective coverings surrounding electronic equipment, which could result in damage or complete disabling of operating satellites.

The best data on near-Earth micrometeorites (<50 μm) could only come from retrievable satellites that spent many months or even years in Earth orbit. This opportunity came in a serendipitous but tragic way. In April 1984, the Space Shuttle Challenger placed into low Earth orbit (400 km altitude) the Long Duration Exposure Facility (LDEF). This satellite was a 12-sided cylinder (plus two ends) housing 57 individual scientific experiments (Fig. 1.8). Its mission was to test the space environment near Earth for a nominal period of nine months. Several of the experiments were designed to detect and characterize natural particles (meteoroids and micrometeoroids) and artificial space debris left by human space activities. Launch delays, along with the Challenger disaster (1986), left the LDEF in orbit for 5.7 years. During that time the satellite maintained a stable configuration relative to its velocity vector, providing 5.7 years of exposure to all sides. The large exposed surface area, about 130 m^2, provided a unique opportunity to acquire data never before obtained on the flux of particulate matter and orbital debris including directionality and velocity vectors of particles. NASA officials recognized early in the mission that the instruments on board designed for meteoroid detection would not provide the data needed to accomplish the goals of the mission so they decided to use the entire spacecraft as a meteoroid detector, giving them a much larger area of exposure and data collection.

The LDEF was finally retrieved in January 1990, and brought back to NASA's Langley Research Center in Hampton, Virginia. There, a specially selected team of scientists gathered to examine every square millimeter of the surface. What they found was indeed startling. The satellite was full of tiny impact craters and penetration holes! Over a four month period they optically scanned and photo-documented all craters 0.5 mm in diameter and larger and all penetration holes larger than 0.3 mm across – totaling approximately 4600 individual impact events. The largest craters were 5 mm in diameter. In all, over 20 000 impact features were detected.[II] Figures 1.9 and 1.10 show examples of these craters and holes.

[II] In February 1997, during the second servicing mission of the Hubble Space Telescope, an extensive imaging survey was performed covering about 97% of the HST external surface. The purpose was to identify and characterize micrometeoroid and orbital debris impacts. In all, 788 impact scars to the detectable limits of 0.5 cm were found. In comparing the number of impacts found during the first servicing mission in December 1993 with those found during the second mission, an apparent increase of from five impacts per square meter to 20 impacts per square meter, respectively, was noted. Though some of this increase may be due to better detection equipment and procedures available on the second mission, a change in the micrometeoroid environment in low Earth orbit may also be indicated.[11]

Fig. 1.8. The Long Duration Exposure Facility (LDEF) ready for release from the Space Shuttle in April 1984. In space for over five years, it gathered impact data on near-Earth hypervelocity interplanetary material. (NASA photo.)

Fig. 1.9. Penetration holes (black) in the front side of a thermal blanket mounted on the outside surface of the LDEF and held in place by Velcro strips. Note the large impact feature about 5 mm in diameter right of center that distorts the blanket. Inset, upper right: a closer look at two penetration holes both about 0.5 mm in the thermal blanket material. The surrounding rings are shock waves generated by the impacting particles. (NASA photo.)

Fig. 1.10. A piece of metal cut from an aluminum clamp mounted on the LDEF shows an impact crater. The millimeter scale surrounds the specimen. (NASA photo.)

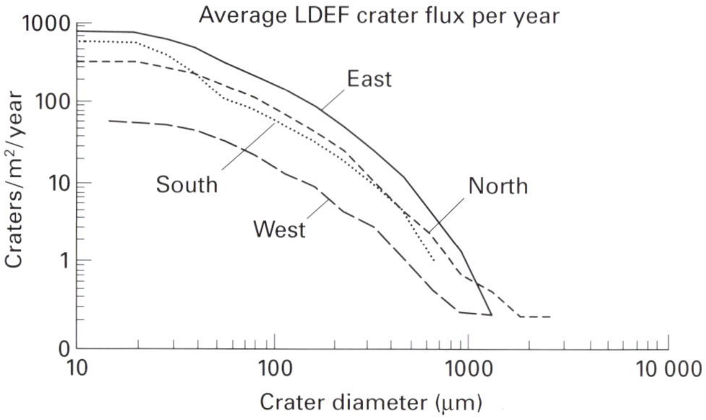

Fig. 1.11. Graph shows the flux of impacting particles per year in terms of crater density vs crater diameter from the LDEF data. The satellite was oriented in a stable configuration while in space. The directions refer to its orientation in space. (Courtesy Johnson Space Center and NASA.)

Most of the craters were the result of impacts at velocities greater than 6 km/s. Surprisingly, there were no obvious physical characteristics allowing a distinction between craters made by meteoroid debris and manmade debris. Both produced craters that were generally symmetrical indentations with shallow rims. Most showed evidence of impact melting in the crater interiors. In about a third of the craters, residue of impactor material was found either within the craters or as ejecta around the craters. Using Secondary Ion Mass Spectrometry techniques, chemical analysis of the residue allowed these craters to be distinguished and classified as the products of either natural or manmade particles. Natural materials were chondritic in composition with relatively high concentrations of Fe, Mg, and Si. Iron–nickel sulfides common in chondrites were also detected.

Because of the difficulty in determining the causative agents of all of the craters, the actual flux of IDPs and millimeter-sized meteoroids cannot be precisely determined from the LDEF data. What can be said is that there is substantial material in low Earth orbit constantly encountered by spacecraft. The number of craters per square meter per year increases dramatically with a decrease in crater diameter suggesting that there are many times more small particles than large in near-Earth space (Fig. 1.11). Much of this material is micrometer-sized, just what one would expect if the smallest particles were IDPs and micrometeoroids.

Meteors

Micrometer-sized particles from the interplanetary medium can find their way to Earth's upper atmosphere or even to Earth's surface without suffering the fate of much larger bodies. They only have to be concerned about the Sun's radiation pressure pushing them out of the Solar System or the Poynting–Robertson Effect which ultimately sends them into the Sun. Millions of much larger particles encounter Earth every day. Some are from cometary debris left behind by passing periodic comets. If the orbits of these objects intersect Earth's orbit they enter Earth's atmosphere at cometary velocities. Earth's atmospheric barrier effectively stops these bodies after a brief flight of a second or two. During their atmospheric flight, they rapidly lose their initial kinetic energy to heat as they encounter air molecules and reach vaporization temperatures about 100 km above Earth's surface. These particles are roughly in the 100 μm to 3 mm diameter range. We become aware of them as they leave brief flashes of light in the night sky. The particle is termed a *meteoroid* and the accompanying light phenomenon, a *meteor*. (Frequently the terms *meteor* and *meteoroid* are incorrectly used as synonyms or the two are confused in the literature in which *meteor* is applied incorrectly to designate the solid object that reaches Earth. A meteoroid that has reached Earth is properly termed a *meteorite*.)

Sporadic meteors

Any astronomy textbook describes two types of meteors: *sporadic* and *shower* meteors. They are distinguished by their origin and predictability. Let's look at the sporadic meteors first. On any clear moonless night, half a dozen or more meteors may be observed within an hour, coming from any point in the sky. These are termed sporadic meteors because of their apparent randomness; the time and place of their appearance is unpredictable. Their velocities with respect to the Sun as they cross Earth's orbit are usually substantially higher than Earth's orbital velocity of 29.9 km/s. This is not surprising since their orbits are highly elongated, with eccentricities on the average of about 0.7, taking them out to Jupiter or beyond. Their *heliocentric* velocities (velocities with respect to the Sun) at Earth's orbit are near the escape velocity of the Solar System or 72 km/s at 1 AU. Depending upon the direction they are traveling with respect to Earth's orbital direction, their *geocentric* velocities (velocities with respect to Earth) can vary enormously. Earth travels in a counterclockwise direction around the Sun, that is, west to east if viewed from above. If the meteoroids are traveling in the direction of Earth's motion then they must catch up to Earth. This occurs any time between noon and midnight for any given location on Earth (Fig. 1.12). A meteoroid's velocity is therefore the vector sum of its heliocentric velocity minus Earth's orbital velocity. A meteoroid approaching Earth at 42 km/s from the west will have a velocity with respect to Earth of only 12.1 km/s. These are referred to as *slow* meteors. Conversely, during the morning hours between midnight and noon, Earth encounters meteoroids head on so that Earth's velocity is added to the meteoroid's velocity resulting in *fast* meteors traveling at over 70 km/s. A meteor's brightness and longevity depends strongly upon its velocity. Prograde meteors (slow meteors) observed in the early evening hours are typically much fainter than retrograde meteors (fast meteors). Any observer counting meteors through the night will easily note that fewer meteors are seen in the early evening hours primarily because they are slow moving and therefore faint, many too faint to be readily noticeable. Conversely, more meteors are observed in the morning hours because they are faster and therefore much brighter.

Origin of sporadic meteors

Traditionally, sporadic meteors are thought to have highly elliptical orbits with no apparent preference to the ecliptic plane. Most astronomy texts accept that without question.

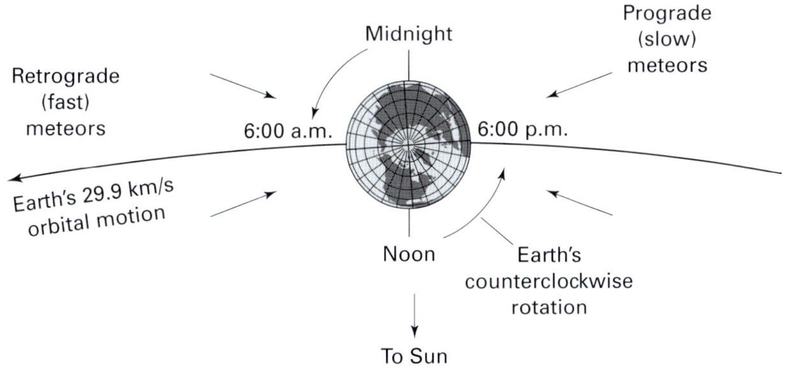

Fig. 1.12. Meteors encounter Earth at different speeds, depending upon the meteors' direction of travel with respect to Earth, that is, either a prograde or retrograde direction with respect to Earth's counterclockwise direction around the Sun. Fast meteors are encountered between midnight and 6:00 a.m.; slow meteors are encountered between noon and 6:00 p.m.

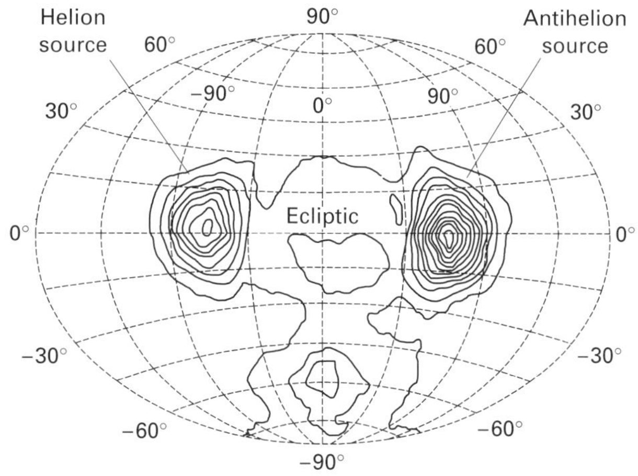

Fig. 1.13. A contour map of the celestial sphere showing the geocentric distribution of radar meteors from the Adelaide surveys. The plot of radiants of meteors show two concentrations on either side of the Sun. The plot uses the ecliptic coordinate system in which the ecliptic is the reference plane. The Sun is located on the ecliptic at celestial longitude −90°, celestial latitude 0°. Two main concentrations are noted 180° opposite each other with the antihelion concentration the greater of the two. (Adapted from *Meteoroid orbits: implications for near-Earth object search programs* by Duncan Steel, *Annals of the New York Academy of Sciences*, **822**.)

This, along with their high velocities, strongly suggests that they are cometary fragments. But their orbits alone cannot be considered conclusive since we have seen that nongravitational forces are at play that change the orbits of small bodies in a surprisingly short time period. It is therefore reasonable to suggest that at least some of the bodies producing sporadic meteors are tiny fragments of asteroids. Given time, such small fragments could be perturbed into elongated elliptical orbits with high geocentric velocities. Duncan Steel with the Spaceguard Foundation, Rome, Italy, conducted two sporadic meteor radar surveys in the 1960s from South Australia (Adelaide) that seemed to suggest just that. In these surveys orbits were determined for particles in the 1–3 mm size range. Most had eccentricities of 0.7 but with orbital inclinations measured from the ecliptic of less than 15°, and semimajor axes (mean distance from the Sun) of between 1.5 and 3.0 AU. This placed them well inside Jupiter's orbit from the middle of the asteroid belt to near Mars. These orbits are similar to Earth-crossing asteroids.[12] The surveys showed two primary concentrations or radiants from which the meteors extended. The centers of both concentrations lie very near the ecliptic and extend nearly 30° on either side (Fig. 1.13). They are positioned toward the Sun (helion position) and 180° away from the Sun (antihelion), the antihelion having by far the strongest radiant. Two other diffuse sources are also present but turn out to be selection effects that drop out when corrections are applied.

Other criteria such as the composition and density of the particles help to distinguish between asteroid and comet particles. The first measurements of the composition of meteoric particles came from the very process that destroys them. As these small bodies hit the atmosphere at cosmic speeds, they are rapidly heated. Their kinetic energy is converted to heat as they encounter air molecules beginning at about 100 km altitude. The heat generated is sufficient to melt and vaporize the small bodies within seconds. Temperatures of thousands of degrees turn the solid body into a glowing gas. Atmospheric molecules surrounding the vaporizing body are also heated to incandescence. Evidence for this vaporization is born out by low dispersion spectra made with wide field cameras equipped with objective prisms. The spectra reveal emission lines of low pressure incandescent gas rather than a continuous spectrum from a solid body heated to incandescence. Spectral lines of hydrogen, nitrogen, and oxygen are present that are probably atmospheric gases but most revealing are lines of ionized metals such as iron, calcium, silicon, magnesium, and sodium. These are the major elemental constituents of chondritic meteorites.

In 1978, Brownlee analyzed 150 micrometeorites ranging in size from 3 to 40 μm collected on U-2 flights. About 60% of the particles were unmelted minerals of CM chondrite composition.[13,14] They were composed of tiny tightly bound aggregates of crystals with interspersed pore spaces that may have held water-ice. Brownlee found elemental abundances of Fe, Mg, Si, S, Ca, and Ni close to abundances in CI and CM carbonaceous chondrites (see Chapter 7). CI chondrites and, to a lesser degree, CM chondrites, contain an abundance of carbon that was not observed in the aggregates but was inferred based upon the abundances of the other elements. (Carbon is a volatile element that may have been lost through heating in the atmosphere. This is certainly the case for meteoric particles.) When the European Space Agency's Halley's Comet probe was sent to photograph its solid nucleus on March 14, 1986, scientists were astonished to find its surface covered with a charcoal gray coating, not the bright white ice they had expected. The dark gray material may be similar in composition to CI or CM chondrites, probably including organic compounds.

The drag upon the particle as it enters the atmosphere depends, among other quantities, upon its density. Its density can be estimated by photographing a meteor's rate of deceleration. High-density meteoroids have greater cohesive properties, tend to stay together longer and decelerate at a slower rate. Most of the IDPs, not surprisingly, were of very low density, less than 1 gm/cm^3. As comets pass through perihelion, they lose tons of volatiles causing a compression of the solid mineral components. Thus, as comets age, they should become more dense. Particles have been recovered with densities of 2.5 or greater and at least one particle with the density of iron, 7.7, has been reported.

Shower meteors

The second type of meteor contrasts with the sporadic meteors in that their time and place of appearance are entirely predictable. Unlike sporadic meteors, there is universal agreement they come from comet dust, small bodies shed by comets and strewn like seeds along comet orbits. Several months during the year are marked by an obvious increase in meteor frequency. This increase usually occurs over a period of a few hours on a specific day or in a few cases over several days of the month. In at least eight of the twelve months of the year, meteor showers are known to occur. Meteor showers result when Earth passes through swarms of particles all moving together in space in a common heliocentric orbit. These particles occupy orbits of known periodic comets whose perihelia (nearest point to the Sun in an elliptical orbit) lie inside Earth's orbit and whose paths cross Earth's orbit. Table 1.1 lists the known prominent meteor showers and their respective comet precursors.

A comet does not deposit dust uniformly along its orbit during every perihelion passage. Instead, it deposits the dust in spurts. Comets are known to form jets like geysers that release dust trapped within sublimating ices. Comets become more active as they approach perihelion. Broad dust tails form that often disconnect and reform as geyser activity subsides and then restarts, resulting in clumps of particles traveling together behind the comet. The particles that produce visible meteors are sufficiently large so they are not subject to radiation pressure. They simply remain in orbit around the Sun following the comet's path. The clumps, however, do not remain together. Instead, gravitational perturbations from planetary bodies that pass near the comet's orbit tend to scatter the particles. Given sufficient time, the particles redistribute more or less uniformly around the orbit. Clumped particles and uniformly distributed particles produce quite different meteor spectacles. The best known and most widely observed meteor shower is produced every year by the Perseid meteors. They begin to appear as early as August 9 and reach a peak around August 12. After midnight when the meteoroids are encountering Earth head-on, as many as 60 meteors per hour may be seen, some exceeding the brightness of the brightest stars. (The Perseid meteor shower was rejuvenated in 1997 due to the reappearance of Comet Swift–Tuttle, the comet responsible for the shower. As many as 125 meteors per hour, twice the normal number, were recorded by observers worldwide.) If one traces the meteor trains backward whence they came, it becomes apparent that their paths seem to diverge from a small area about a degree or two across in the constellation Perseus, seen high in the northeast in the early morning hours in August. This area is called the *radiant*. The meteors seem to radiate from that small area due to perspective. In actuality, the meteoroids are traveling in parallel orbits and therefore hit the top of the atmosphere from the same direction in space. Their apparent paths suffer perspective much as parallel contrails from a jet aircraft seem to converge across the sky in the distance. Since Earth is orbiting the Sun, the radiant does not remain stationary but tends to drift as Earth approaches and recedes from the intersection point of the orbit. This effect is particularly noticeable for meteor showers that occur over several days like the Perseids.

The Leonid meteor storm

Meteor storms occur if Earth happens to encounter a fresh trail of meteoroid particles as they pass across Earth's orbit on

Shower	Maximum activity	Hourly count	Velocity (km/s)	Associated comet
Quadrantids	January 3/4	100	40.2	?
Lyrids	April 21/22	12	45.8	Thatcher (1861 1)
η Aquarids	May 3–5	20	63.2	Halley (outbound)
δ Aquarids	July 29/30	30	39.9	?
Perseids	August 11/12	60	57.3	Swift–Tuttle
Draconids	October 8/9	variable	19.6	Giacobini–Zinner
Orionids	October 20/21	25	64.0	Halley (inbound)
Taurids	November 7/8	12	28.1	Encke
Leonids	November 16–18	variable	68.2	Tempel–Tuttle
Geminids	December 13/14	60	33.1	Asteroid 3200 Phaeton
Ursids	December 22	10	32.2	Mechain–Tuttle

Table 1.1. **Annual meteor showers**

Yearly, Earth passes through clouds of dust particles scattered along orbits of comets known to cross Earth's orbit. This results in annual meteor showers in which meteors appear to originate from a region of the sky occupied by a constellation. These meteor showers are therefore given the names of the constellations in which their radiants are located. Comet Halley's orbit passes through Earth's orbit twice, both inbound and outbound. This produces the Orionids in October and the Eta Aquarids in May. Hourly meteor counts can vary for a given meteor shower from year to year. The Perseids average about 60 per hour but have been as high as 125. The most highly variable meteor shower is the Leonid. Usually the hourly rate is quite low, between 10 and 20, but every 33 years Earth encounters large clouds of particles so that observers positioned in the correct longitude may be treated to a *meteor storm* in which thousands of meteors can be seen per hour.

Fig. 1.14 A fanciful depiction of the Leonid meteor storm occurring on the night of November 12/13, 1833, seen from Niagara Falls, New York. Notice the curious umbrella-like radiant centered on a star. The radiant is actually positioned in the head of Leo near the stars γ and ζ Leonis. This picture appeared in the March, 1864 issue of *Illustrated Astronomy.* (Courtesy J. Gregory Wilson.)

their journey around the Sun. Fresh comet trails contain relatively high densities of meteoroid particles called *meteor swarms* that, given sufficient time, disperse along the orbit of the comet. The most spectacular encounter takes place every 33 years with the Leonid meteor shower. This happened in a big way on the night of November 16/17, 1966, when Earth passed through the trail left by Comet Tempel–Tuttle as it crossed Earth's orbit in 1899. The Leonid meteor shower has a long history, first being recorded in AD 899. Astronomers first became aware of the unusual nature of this shower when an extraordinary display occurred in 1833 (Fig. 1.14). Historical records showed that this shower occurred every 33 years. This was substantiated in 1866 with another spectacular meteor storm. In the same year, astronomers discovered a comet (Comet Tempel–Tuttle) that possessed an orbit almost identical to the meteors. This was the first time a meteor shower was shown to be associated with a comet. The comet and its attendant trail carried the meteor swarm along a highly elliptical orbit taking them out beyond Saturn's orbit. Its perihelion dipped just inside Earth's orbit at 0.983 AU. The orbit was inclined about 162° to the ecliptic plane (similar to Comet Halley) and the comet and trail were moving in a retrograde direction.

It soon became apparent, however, that the particles were spread out along the orbit so that substantial showers could occur one to two years before or after the 33 year period, but an expected shower failed to appear in 1899 and only a weak return was observed in 1901. In 1932/33 the Leonids again failed to appear. Most astronomers believed that the particles had changed their orbit due to perturbations by Jupiter. But in the mid-1950s and early 1960s the meteor count began to pick up again until 1966, when thousands of meteors appeared in the predawn hours, peaking at an incredible 150,000 per hour sustained for nearly an hour. Observers in western North America had a ringside seat (Fig. 1.15). Dawn interrupted the storm on the eastern side of North America and daylight obscured the spectacle for European observers. It appeared that the great Leonid meteor storm had indeed returned.

Predicting the time and intensity of a meteor shower had been fraught with inaccuracies for years until two astronomers, David Asher of the Armagh Observatory in Ireland and Rob McNaught at the Research School of Astronomy and Astrophysics of the Australian National University jointly developed a technique that allowed dense meteoroid swarms in comet trails to be predicted with remarkable accuracy.[15] The technique involves a knowledge of where the dust trails are located as they cross the ecliptic as well as mapping the fine structure in the dust trails within the broad Leonid meteoroid stream. This includes trails going back as much as nine orbits of Comet Tempel–Tuttle, which extends back nearly 300 years. For example, the 1833 meteor storm was produced when Earth passed through the trail left after the comet's passage in 1799. The next strong meteor storm of 1866 occurred when the trail left four orbits previously (1733) was encountered.

An additional piece of the puzzle appeared as a result of the 1998 expected meteor storm. It turned out not to be a "storm", just a good meteor shower with several hundred meteors per hour observed in the Far East within an hour of the predicted time. But something else quite unexpected occurred that sent Asher and his coworkers back to the drawing boards. Some 16 hours ahead of the predicted peak, a fine display of fireballs occurred implying that Earth had run into a swarm of millimeter to centimeter-sized particles that had not been recognized before. Tiny particles usually associated with dust trails tend to spread out along the orbits, eventually becoming evenly distributed in the meteor stream. They discovered that Comet Tempel–Tuttle was in orbital *resonance* with Jupiter (see Chapter 11). Jupiter's great mass had perturbed the comet into a commensurate orbit in which the ratio of their orbital periods was a simple fraction. In this case, the comet makes 5 revolutions of the Sun in 14 revolutions of Jupiter. This had the effect of "herding" the comet's larger particles into a tight arc with an orbital period very nearly that of the comet itself, keeping them from scattering. It was this arc of centimeter-sized particles that Earth had encountered. Apparently there were many such arcs

Fig. 1.15. Leonid meteors photographed over the course of a few minutes from Kitt Peak National Observatory in southern Arizona on the night of November 16/17, 1966. The meteors radiating northward from the head of Leo near the eastern horizon appear to cut across the Big Dipper stars. (National Optical Astronomy Observatory photo.)

creating fine structure in the meteor stream. Searching through the past 42 perihelion passages of Comet Tempel–Tuttle, Asher and his colleagues calculated the motion of the larger particles and found that the particles released in AD 1333 were responsible for the fireballs seen on the night of November 17, 1998. This was the first time that "fine" structure within the meteor stream of a short period comet had been demonstrated.

But 1999 was the real test of the Asher–McNaught theory. The world awaited the Leonids. The two boldly predicted that on the night of November 17/18 a strong peak of about 1000 meteors per hour would occur at 02:08 Greenwich Mean Time as Earth encountered the 1899 trail deposited three revolutions earlier. They further predicted that there would be few if any bright fireballs since the 1333 arc had passed beyond Earth's orbit. As the world watched, observers in the Middle East (Jordan) and in Spain noted that the count began to increase steeply shortly after 01:00 GMT reaching a peak at 02:04 GMT with a zenithal hourly rate (ZHR) above 5000 meteors per hour, five times the predicted rate[III] (Fig. 1.16) and remaining above the 1000 per hour level for nearly 1½ hours. Later, additional observa-

[III] The zenithal hourly rate or ZHR is a standard value astronomers use which takes into account observing conditions and direction from which the meteoroid particles enter the atmosphere. In this case the maximum ZHR was 5400 ± 800. Since this meteor storm included abundant fainter meteors with fewer bright fireballs than expected, the ZHR could rise.

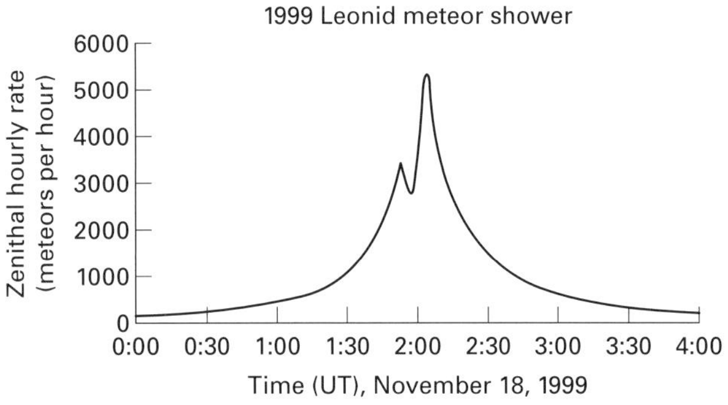

Fig. 1.16. This early graph made immediately after the Leonid shower shows the number of meteors per hour recorded in the Middle East on the night of November 18, 1999. Preliminary data suggested a peak ZHR of over 5000 meteors at 02:04 Universal Time with a secondary ZHR peak of 3500 occurring a few minutes earlier. Later, the addition of more observations resulted in a reduction of the maximum ZHR to 3700 and a disappearance of the secondary peak. Regardless, the remarkable prediciton of Asher and McNaught remained within a few minutes of the observed maximum. (See: Bulletin 15 of the *International Leonid Watch: First Global Analysis of the* 1999 *Leonid Storm* published by WGN, the Journal of the IMO. Vol. 27:6 (December 1999), pp. 286–295.) (Courtesy the International Meteor Organization.)

Fig. 1.17. This spectacular image by Nigel Evans captured more than 70 Leonid meteors radiating from the head of Leo on November 17/18, 1999, from Sharm El Sheikh, Egypt. The image is a digitally combined composite of nine exposures, each of 8 min. The series began at 01:19 and ended over an hour later at 02:38 Universal Time. The composite image beautifully demonstrates the radiant as the meteors appear to diverge out of Leo, making longer trails as they extend away from the radiant point. (Courtesy Nigel Evans.)

tions reduced the prelimary ZHR peak to 3700. And as they had further predicted, few fireballs were seen. Asher and McNaught had confirmed in a spectacular way the validity of their theory. Their predicted peak time was within 5 minutes of the observed peak! For the first time in history, a meteor storm peak had been predicted within minutes; an astonishing accomplishment. An accurate prediction of the hourly rate is much more difficult to predict as it requires a knowledge not only of the path Earth takes through the involved trail but also a knowledge of the fine structure (possibly braided) within the trail.

Asher and McNaught had accurately predicted the best locations on Earth to see the storm. The Middle East, and most of Europe were prime locations (Fig. 1.17). It was not seen at all in North America where Leo was still hours from rising when the peak occurred. The highest count in the United States was a disappointing, but anticipated, 40–50 per hour, though still higher than the normal rate for non-shower years.

Based upon this new theory, Asher and McNaught have released predictions for the years 2000 to 2002. These are listed in Table 1.2. The numbers are exciting but unfortunately the nearly full Moon will interfere with observations in North America in 2002.

The Perseid and Leonid showers were of particular importance in that late eighteenth and early nineteenth century astronomers used observations of these meteors to establish three fundamental properties of meteor showers so obvious to us today. First, meteors occur in the atmosphere. In

Date	Time (UT)	Trail	Estimated ZHR	Moon age	Visibility
1999, Nov.18	02:08	3	500–1000	10 days	Africa, Europe
2000, Nov.18	03:44	8	100?	22	W. Africa, W. Europe, NE S. America
2000, Nov.18	07:51	4	100?	22	N. America, C. America, NW S. America
2001, Nov. 18	10:01	7	2500?	3	N. and Central America
2001, Nov. 18	17:31	9	9000	3	Australia, E. Asia
2001, Nov. 18	18:19	4	15000	3	W. Australia, E., SE., Central Asia
2002, Nov. 19	04:00	7	15000	15	W. Africa, W. Europe, N. Canada, NE S. America
2002, Nov. 19	10:36	4	30000	15	N. America

Table 1.2. **This table summarizes predictions of time, position for best visibility and estimated zenithal hourly rate for the Leonid meteor shower through 2002, based upon estimates by and Robert H. McNaught (Australian National University, Canberra) and David Asher (Armagh Observatory, Northern Ireland)**

The predictions proved accurate to within minutes for the 1999 shower and exceeded the maximum predicted ZHR by a factor of five or better. The position and phase of the Moon is an important constraint to the visibility of the fainter meteors in the shower. Worldwide reports on the November 18, 2000, shower were very poor, as predicted. The November 18, 2001, shower may prove to be the best of the series and is awaited with high expectations. The 2002 shower will be seriously degraded by the full Moon.

November, 1798, the first attempts to measure the altitudes at which meteors were extinguished was made. Although measurements on 22 meteors gave altitudes that varied by a factor of over 20, the average extinction altitude was 89 km, very near the correct value. Second, meteoroid particles originate outside the atmosphere. Although the measured altitudes of meteors was well within Earth's atmosphere, their velocities, greater than Earth's orbital velocity, implied that the particles themselves must originate outside the atmosphere. The radiant was first noted during the November 13, 1833, Leonid meteor storm and the Perseid radiant a year later on August 8, 1834. The periodic return of meteor showers first became evident with these two showers. Third, meteor showers are produced by debris from passing comets. Their association with known comets (Comet Tempel–Tuttle for the Leonids; Comet Swift–Tuttle for the Perseids) was recognized by the mid-nineteenth century. By the century's end periodic comets had been linked with all the recognized periodic meteor showers.

Harvesting interplanetary dust particles in the space environment

Certainly the biggest problem in the study of IDPs through the years has been terrestrial contamination. The only real solution is to collect the particles in space far removed from the terrestrial environment. On February 7, 1999, a new era in the study of IDPs began with the launch of the *Stardust* spacecraft. This is a bold mission, the first designed to collect comet dust and return it to Earth for analysis. Its objectives are twofold: to rendezvous with Comet Wild 2 and collect comet dust released by its nucleus and, on the way to the comet, to collect interstellar dust particles known to be passing through our Solar System. This historic flight is the first ever designed to bring back material from beyond the Earth–Moon system.

Since cometary particles have velocities much higher than orbiting particles from asteroids, their collection poses a difficult technical problem. They will impact the spacecraft collecting surface at such high velocities that they would instantly vaporize on contact. To solve this problem scientists at the Jet Propulsion Laboratory of the California Institute of Technology and NASA developed a material they call *aerogel*, a material with such low density that IDPs will tunnel into the material slowing the particles and trapping them. Aerogel is the lightest, least-dense manmade material ever developed. It has amazing compressional strength for its mass, but more importantly, it has remarkable insulating powers for any kind of energy transfer including the high thermal energies released by impacting interplanetary and interstellar particles. The material's ability to dampen energy transfer is crucial to the success of the mission. After traveling through a stream of interstellar particles it will reach Comet Wild 2 on January 2, 2004. If all goes well, *Stardust* will release its sample return capsule containing the precious cargo as it swings by Earth on January 15, 2006. The capsule will land by parachute in the Utah desert near Salt Lake City.

Fig. 1.18. This is an artist's rendering of the encounter of the *Stardust* spacecraft with Comet Wild 2 scheduled on January 2, 2004. As it passes through the comet's tail it will collect dust streaming from the comet's surface expelled by geysers into its tail. Two years later *Stardust* will release its collector capsule as it swings by Earth, returning to Earth the first comet and interstellar dust collected in space. (Courtesy NASA.)

To touch a star

It is forever the destiny of stellar astronomers to look but never touch those far away places they observe with their telescopes. This began to change about a century ago when astronomers finally realized that meteorites were fragments of asteroids that somehow reached Earth. In the mid-twentieth century Moon rocks were brought back followed by the recognition that rocks from the Moon and Mars, like the other meteorites, had arrived on Earth on their own. We were actually touching fragments of other worlds, something earlier generations of astronomers could never have hoped to do in their lifetimes. Shortly thereafter, Donald Brownlee and his coworkers began to show phenomenal success in capturing interplanetary dust particles, tiny pieces of comet and asteroid debris. The Solar System was becoming more tangible to us. But to the astrophysicists whose primary interests were the stars of the Milky Way Galaxy and galaxies beyond, these nearby objects were not within their domain. Once samples reached Earth they became the property of the geochemists and meteoriticists who could analyze these samples in warm laboratories under controlled conditions. The astrophysicists' laboratory was out in the cold dark of interstellar space. What possible application could these mere Solar System relics have to the Galaxy? Then, about this time IDP researchers began to think in far wider terms. Were not the comets distant messengers that roamed the twilight zone half way to the nearest star (the Oort Cloud of comets)? Had they touched the interstellar medium? Might they have samples of interstellar particles in their primitive makeup? If the interplanetary medium is permeated with comet dust, might not there be interstellar dust grains mixed in with the IDPs? Early in the investigation of IDPs, samples were searched and none were found. This came as no surprise to astrophysical theoreticians who suggested that the early solar nebula was so hot that any presolargrains with their postulated extrasolar isotopic signatures would not have survived.

The search for interstellar grains in meteorites actually began with a search for isotopic anomalies, that is, elemental isotopic differences between terrestrial materials and meteorites. The first searches occurred in the early decades of the twentieth century, all meeting with failure. Then, J.H. Reynolds at Berkeley discovered the first isotopic anomaly in a meteorite in 1960. This appeared in the form of the noble gas isotope xenon-129 and 131–136 through the decay of extinct iodine-129 and plutonium-244.[16] Even though the primary isotopes had short half lives ($^{129}I = 16$ Ma; $^{244}Pu = 82$ Ma) and had become extinct, their signatures remained with their xenon daughter products (see Chapter 10.) At the same time, other noble gas anomalies were found and some researchers proposed that these were signatures of isotopes formed in giant stars and carried by solid grains to the solar nebula and hence to meteorites. These were generally ignored since most astrophysicists at the time believed that presolar grains as carriers of isotopic anomalies could not survive conditions in a hot solar nebula. Attitudes changed when oxygen-16 excesses were discovered in refractory inclusions in CAIs (see Chapter 7) in 1979. These excesses amounted to ~4%, showing that isotopes produced by nucleosynthesis within giant stars could and did survive to be incorporated into meteorites.

Celestial diamonds

Meanwhile, Roy S. Lewis and his coworkers at the Enrico Fermi Institute of the University of Chicago kept looking for interstellar grains. As early as 1973 they had discovered "exotic" oxygen isotopic ratios in carbonaceous chondrites, *exotic* meaning that these ratios were not indigenous to the Solar System, which suggested an interstellar origin. Over the next decade the Chicago group continued their search for other anomalous isotopes. In 1978 they reported finding traces of the noble gas, xenon, within the Murchison CM2 carbonaceous chondrite. Like the exotic oxygen, the xenon isotopic ratios were unlike ratios found in Solar System material. These isotopes are thought to be formed in the interiors of red giant stars. They reasoned that the isotopes needed a carrier grain with a crystal structure that could capture and retain the gas as the grains were expelled into interstellar space by red giant stars ultimately to intermingle with the molecular cloud that spawned the solar nebula. They suspected carbon as the carrier and, in 1983, they announced the detection of two isotopes of carbon, ^{12}C and ^{13}C, with anomalous ratios showing an enrichment in ^{13}C relative to Solar System abundances. So the search for carbon carriers began, using the exotic isotopes as markers. In 1987 they finally presented conclusive proof that carbon was indeed the carrier; the proof was in the form of interstellar diamonds![17] The diamonds were extracted from the Murchison meteorite by dissolving away the entire meteorite and collecting the residue containing the insoluble diamond dust.[IV] Now, these

[IV] Ed Anders, one member of the team of scientists who first extracted the tiny diamonds at Chicago when referring to the technique they used, posed the question, "How do you find a needle in a haystack?" His answer was, "You burn down the haystack."

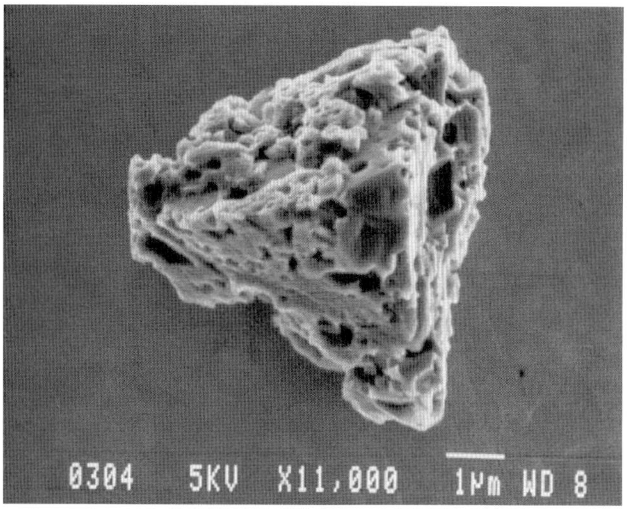

Fig. 1.19. An SEM image of a typical circumstellar SiC grain from an acid residue of the Murchison carbonaceous chondrite (CM2). Scale bar is 1 μm. (Hoppe, P., Amari, S., Zinner, E., Ireland, T. and Lewis, R.S. (1994). Carbon, nitrogen, magnesium, silicon, and titanium isotopic compositions of single interstellar silicon carbide grains from the Murchison carbonaceous chondrite. *Astrophysical Journal* **430**, 870–890.) (Photo courtesy of S. Amari.)

celestial diamonds are not the familiar kind one sees in a jewelry store. They are called *nanodiamonds* because they are exceedingly small, a thousand times smaller than the average IDP or ~1–3 nm (1 nanometer (nm) = 10^{-9} meters (m)). The crystals contain between 60 and 1100 carbon atoms with the mean crystalline size of about 10 angstroms (Å) across (1 Å = 10^{-8} cm). These are the smallest grains studied in meteoritics today and even the latest analytical equipment (the ion microprobe mass spectrometer) cannot analyze them as individual grains, only as bulk samples. Nanodiamonds are by far the most abundant presolar grains in primitive chondrites, amounting to an amazing 6% of the carbon in the Murchison meteorite. Even though they are relatively abundant, approximately 1400 parts per million, they remained elusive for decades because of their small size. With these grains in hand, we can truly say we have "touched" the expelled matter of stars.

A debate continues involving the origin of interstellar diamonds. Infrared astronomical observations have detected interstellar diamonds in dense molecular clouds but they did not form there. They could have condensed within strong stellar winds from red giant *carbon* stars, helium-burning stars that have built up a carbon/oxygen core. Or diamonds may have formed from the violent explosion of a Type II supernova in which the multilayered (including a carbon-rich layer) giant star is disrupted by powerful shock waves traveling through the star during the explosion. Another idea utilizes a Type Ia supernova involving a binary star with a white dwarf and a red giant carbon star. Matter flows onto the white dwarf from the carbon star causing diamonds to condense in the outflow. When the white dwarf star exceeds its theoretical mass limit, a Type Ia supernova explosion occurs creating exotic xenon isotopes that are incorporated into the diamond structure. All of these ideas have merit and account for some of the characteristics of nanodiamonds and their anomalous isotopes, but not all.

Celestial silicon carbide

The year 1987 was a banner year for interstellar grain studies. While still in the glow of the celestial diamonds discovery, researchers identified for the first time interstellar grains of silicon carbide (SiC) in the Murray carbonaceous chondrite. The discovery was made by Thomas Bernatowicz and his team at the McDonnell Center for the Space Sciences, Washington University in St. Louis, Missouri.[18] This carrier showed a 20 000 times enrichment of anomalous isotopic neon and xenon over Solar System values. Silicon carbide turned out to be much rarer than nanodiamonds, only a few parts per million, but the good news was that they were much larger than nanodiamonds, 10^{-3} mm in diameter compared with 10^{-6} mm. They were found as crystalline euhedral grains most averaging between 0.3 and 3 μm in size in either hexagonal or trigonal external shapes, but based upon a cubic lattice (Fig. 1.19). The largest crystals (up to 20 μm) tend to form a different population with anhedral, blocky shapes. Grains over 1 μm are large enough so that once found, they could be analyzed as single grains for their isotopic signatures using the ion microprobe. Up to this point the host meteorite had to be dissolved away with highly corrosive acids to reveal the presolar SiC in the residue but three years after the initial discovery, SiC was found *in situ* in the matrix of both the Murchison and Cold Bokkeveld carbonaceous chondrites (Fig. 1.20).[19]

A large consensus of meteoriticists believe that SiC grains are the product of carbon-rich stars of one solar mass that have entered a second red giant stage called the *asymptotic giant branch* or *AGB*. Stars less than one solar mass to about 1.1 solar masses evolve into their first red giant stage after they have exhausted their core hydrogen through hydrogen to helium fusion, the *proton-proton reaction*. The temperature in the core is insufficient to burn helium so the helium core collapses in an attempt to maintain *hydrostatic equilibrium*, a condition within a stable star in which the internal gas pressure exactly counterbalances internal gravitational forces. This collapse increases the core temperature until the helium core suddenly ignites in an event called the *helium flash*. The star responds by expanding into a red giant, the first of two or more red giant stages. Helium burning proceeds at a much more rapid pace than hydrogen burning and lasts only a few million years, depending upon the star's mass. The helium core is fused to carbon through a fusion reaction called the *triple-alpha process*. The alpha in this case is helium nuclei. Once the core helium has been consumed the star is left with a

Fig. 1.20. *In situ* detection of a circumstellar SiC grain in the Murchison meteorite. The grain was buried below the surface of the polished section. Originally found by "mapping" in Si X-rays, it was subsequently exposed by ion sputtering. Scale bar is 1 μm. (Alexander, C.M.O'D., Swan P. and Walker, R.M. (1990). *In situ* measurement of interstellar silicon carbide in two CM chondrite meteorites. *Nature* **348**, 715–717.)

carbon core surrounded by thin shells of still burning helium and hydrogen, i.e., the star now has a layered internal structure. Once again, thermonuclear reactions temporarily subside in the center of the star causing the carbon core to contract rapidly in response to internal gravitational forces. This increases the temperature and pressure of the carbon core and the surrounding hydrogen and helium burning layers. As a critical temperature is reached the carbon core ignites making the hydrogen and helium layers burn more rapidly. The vast increase in internal pressure at this stage causes the outermost hydrogen layers of the star to expand further, evolving rapidly to an even larger red giant, a *red supergiant*, as its carbon core burns furiously. Concomitant with the expansion of the star is a copious production of carbon dust tied to silicon that is released into the interstellar medium as a strong stellar wind along with its noble gas component.

A small amount of SiC grains have isotopic ratios that cannot be explained through the AGB mechanism. These are attributed instead to supernovae from stars considerably more massive than the Sun that go through a series of red giant stages, building up multiple layers of elements until an iron core end product is created. Stellar iron cores are *endothermic*, that is, they absorb rather than emit energy into the star. With this energy deficit the core begins a final collapse. It reaches a superdense state, rebounds and sends a powerful shock wave through the star. As the shock wave reaches the surface, the star is disrupted by an enormous explosion, a supernova, spewing out vast amounts of newly created isotopes into the interstellar medium.

Interstellar graphite

The last of the trio of presolar grains to be identified was graphite, found again in the Murchison meteorite by S. Amari and coworkers at Washington University in St. Louis in 1990.[20] Graphite turned out to be even rarer than SiC with <1 p.p.m. (part per million) for graphite compared with 7 p.p.m. for SiC and with average diameters between $\sim$1–7 μm. Meteoritic graphite varies morphologically from amorphous carbon to crystalline graphite and takes on a variety of forms from spherules with concentric shells to hexagonal crystals. Only the spherules possess isotopically anomalous $^{12}C/^{13}C$ ratios which clearly distinguish them among the others as interstellar grains (Fig. 1.21). Their low abundance in carbonaceous chondrites makes them difficult to find and isolate. Thus, less work has been done on graphite compared with SiC.

One novel discovery was made when an ultra-thin section of one of the spherules was examined. There, tucked into the concentric layers of carbon were smaller interstellar crystalline particles made of titanium carbide TiC (Fig. 1.22). They had apparently been protected by the spherule during their time in interstellar space. Thus, another presolar grain was added to the list within a year of the original graphite discovery.

To the time of writing (2001) nine presolar minerals have been found in meteorites. These are listed in Table 1.3.[21]

PAHs in graphite

In 1998, working again with circumstellar graphite grains (grains around stars), researchers at Washington University,

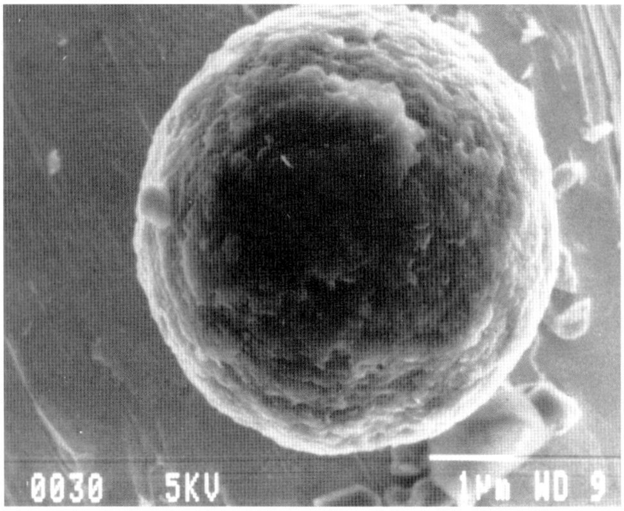

Fig. 1.21. Spherical circumstellar graphite grain from an acid residue of the Murchison carbonaceous (CM2) meteorite. The spherical grain is about 4.5 μm in diameter. (Amari, S., Hoppe, P., Zinner E. and Lewis, R.S. (1993). The isotopic composition and stellar sources of meteorite graphite grains. *Nature* **365**, 806–809.)

Stanford University and The University of Chicago together found yet another molecule attached to the carrier graphite extracted from two meteorites: Murchison and Acfer 094.[22] This time it was an organic molecule: polycyclic aromatic hydrocarbons (PAHs). Roughly 70% of the graphite grains had concentrations of PAHs. This is the same class of organic molecules found in the martian meteorite ALH 84001 that has been the subject of intense research and debate over the past few years (see Chapter 8.) PAHs appear to be indigenous throughout the Milky Way Galaxy. They have been found in chondritic meteorites, in IDPs, in the atmospheres of the outer planets and indirectly through infrared emission from the dust grains of the interstellar medium. Based upon the strength of the infrared emissions in the interstellar medium, PAHs make up between 3% and 15% of the carbon in the Galaxy.[23] Currently astronomical observations of PAHs in interstellar clouds or circumstellar (around stars) regions are limited in that they can detect the broad class of organic molecules but cannot identify specific PAH molecules. The researchers found circumstellar PAHs attached to graphite grains in primitive meteorites and were able to identify specific molecules. Five isotopically anomalous PAH molecules (with $^{12}C/^{13}C$ anomalous ratios) were found with a spread of 178–576 a.m.u. (atomic mass units).

The importance of indigenous large organic molecules in the Galaxy speaks for itself. Apparently these molecules are able to resist the strong ultraviolet radiaton that exists near hot stars where they must have formed. These molecules eventually find a "hiding" place within tiny graphite grains of circumstellar origin. Then they manage to escape their circumstellar birth places, eventually to find a haven at a most critical time in the developing solar nebula. The anomalous

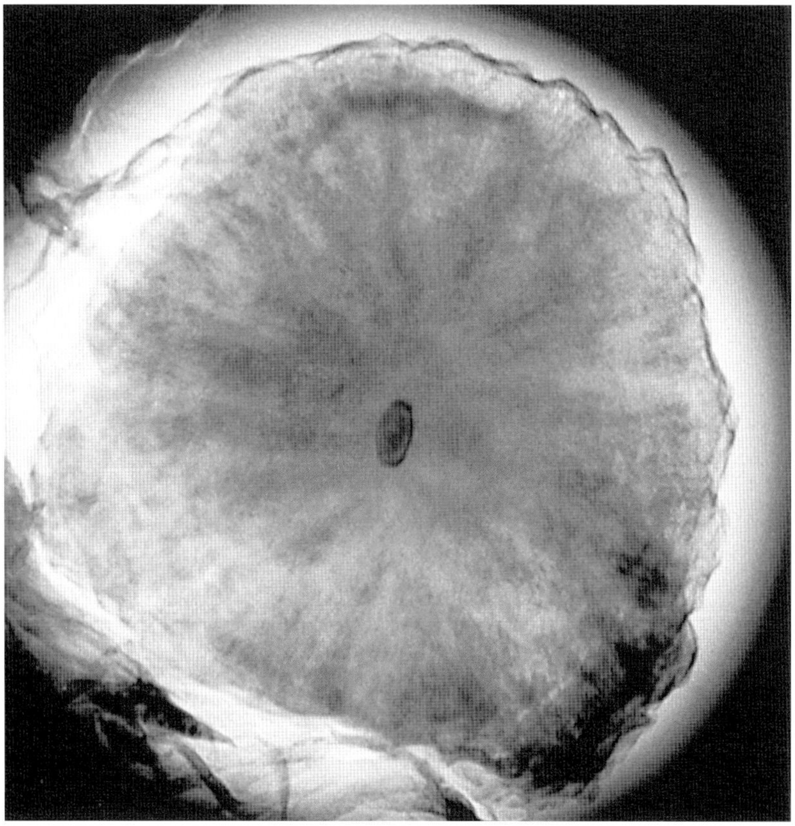

Fig. 1.22. Titanium carbide grain within circumstellar graphite; a 70 nm thick section of a presolar graphite spherule. The crystal in the center of the spherule is titanium carbide, a refractory mineral that formed before the graphite and served as a nucleation center for its growth. The dark, radial spokes in the image are an electron diffraction effect; in reality, the graphite layers are concentric. (Bernatowicz, T. *et al.* (1996). Constraints on stellar grain formation from presolar graphite in the Murchison meteorite. *Astrophysical Journal* **472**, 760–782; Bernatowicz, T. and Walker, R.M. (1997). Ancient stardust in the laboratory. *Physics Today* 26–32.)

corundum (Al_2O_3)	lonsdaleite (C)	*titanium carbide (TiC)
diamond (C)	silicon nitride (Si_3N_4)	*molybdenum carbide (MoC)
graphite (C)	silicon carbide (SiC)	*zirconium carbide (ZrC)

Note: *These three grains are found in graphite and silicon carbide as tiny subgrains that acted as nucleation centers for the host grains.

Table 1.3. **Presolar minerals**

This table is a list of nine presolar (interstellar) minerals found so far in meteorites. Most are either pure carbon or carbides. The single exception is silicon nitride. Stars equal to or more massive than the Sun create these minerals and scatter them through the Galaxy during supernova explosions. In most cases, the presolar minerals are found within carbonaceous chondrites: namely, Murchison CM2 and Cold Bokkeveld CM2.

PAHs tell us that organic molecules of large molecular weight and great complexity are being abiotically but naturally produced in what we on Earth consider the harshest of environments. On Earth life is tenacious, able to exist and even thrive in every environment on the globe. That Earth has been the recipient of organic molecules over the past 4.6 billion years on the wings of carrier grains in carbonaceous meteorites is obvious to us today. What role might they have played in the origin of life on Earth? In the Galaxy? These remain fundamental questions.

A final word

We began this chapter with a local touch and ended with a galaxy of stars. There was a time when early meteoriticists argued for an interstellar origin for meteorites. These were based upon the apparent high velocities and hyperbolic orbits of meteors that strongly suggested that at least some of the stony meteorites came from outside the Solar System. But these high velocities were eventually found to be in error, and the idea of interstellar meteorites faded. Now, the discovery of anomalous noble gas isotopic ratios lacking Solar System signatures in our most primitive meteorites forces us to return to the interstellar medium for a closer look. That reexamination, this time with powerful instruments and techniques undreamt of by earlier generations of scientists, has revealed evidence of stellar material in our precious meteorites. In a way, we have come full circle. Who would ever have thought that the science of meteoritics would lead us far beyond the confines of our Solar System? It has invaded what used to be thought of as astronomers'-only territory. Talk of red giant stars, carbon-rich stars, supernovae, AGB branch supergiants and the like is astronomer jargon. It has permeated our conferences and has dominated many of today's research papers. What is fascinating about all of this is that now meteoritics can offer astronomers a kind of reality check on their ethereal picture of dust-laden circumstellar atmospheres and the processes within stars that result in the nucleosynthesis of heavy elements. And it does so with *real* interstellar dust that can be analyzed on the microscopic level. Now, geochemists and astronomers can join forces to solve fundamental problems in *stellar* (as opposed to planetary) astronomy. Geochemists and meteoriticists have extended their reach to the lofty stars and the dark dust and gas clouds of the Milky Way and at the same time have given astronomers a tantalizing closeup look at their untouchable stars.

References

1. di Cicco, D. (1993). New York's cosmic car conker. *Sky and Telescope* **85**, 26 (Feb).
2. Roach, F.E. and Gordon, J.L. (1973). *The Light of the Night Sky*, D. Reidel Publishing Co., Dordrecht, The Netherlands.
3. Blackwell, D.E. (1960). The zodiacal light. *Scientific American* **203**, 54 (July).
4. Sagan, C. and Druyan, A. (1985). *Comet*, Random House, New York.
5. Gatland, K. (1981). *The Illustrated Encyclopedia of Space Technology*, Harmony Books, New York.
6. Whipple, F. (1967). On maintaining the meteoritic complex, *Zodiacal Light and the Interplanetary Medium*, Weinbert, J. (ed.), NASA SP-150, Washington, D.C.
7. Hodge, P.W. (1981). *Interplanetary Dust*, Gordon and Breach Science Publishers, New York.
8. Gadsden, M. (1986). Noctilucent clouds. *Journal of the Royal Astronomical Society* **27**, 351–366.
9. Bradley, J.P., Sanford, S.A. and Walker, R.M. (1988). Interplanetary dust particles, *Meteorites and the Early Solar System*, Kerridge, J.F. and Matthews, M.S. (eds.), University of Arizona, Tucson, pp. 861–895.
10. Hodge, P.W. (1981). *Interplanetary Dust*, Gordon and Breach Science Publishers, New York.
11. Byrne, G.J., Bretz, D.R., Holly, M.H., Gaunce, M.Y. and Sapp, C.A. (1999). Survey of the Hubble Space Telescope micrometeoroid and orbital debris impacts from Service Mission 2 imagery, *NASA Orbital Debris Quarterly News*, Johnson Space Center.
12. Steel, D. (1997). Meteoroid orbits: implications for near-Earth object search programs. Near-Earth Objects, *Annals of the New York Academy of Sciences* **822**, 31–51.
13. Brownlee, D. and Hodge, P.W. (1978). Chondritic particles from deep sea sediments. *Meteoritics* **13**, 396–399.
14. Brownlee, D. (1978). *Cosmic Dust*, McDonnell, J. (ed.), John Wiley and Sons Inc., New York, p. 295.
15. McNaught, R.H. and Asher, D.J. (1999). Leonid dust trails and meteor storms. *WGN, the Journal of the International Meteor Organization* **27**, 85–102.
16. Reynolds, J.H. (1960). Isotopic composition of primordial xenon. *Physics Review Letters* **4**, 351–354.
17. Lewis, R.E., Tang, M., Waker, J.F., Anders, E. and Steel, E. (1987). Interstellar diamonds in meteorites. *Nature* **326**, 160–162.
18. Bernatowicz, T., Fraundorf, G., Tang, M., Anders, E., Wopenka, B., Zinner, E. and Froundorf, P. (1987). Evidence for interstellar SiC in the Murray carbonaceous chondrite. *Nature* **330**, 728–730.
19. Alexander, C.M.O'D., Swan, P. and Walker, R.M. (1990). *In situ* measurement of interstellar silicon carbide in two CM chondrite meteorites. *Nature* **348**, 715–717.
20. Amari, S., Anders, E., Virag, A. and Zinner, E. (1990). Interstellar graphite in meteorites. *Nature* **345**, 238–240.
21. Rubin, A.E. (1997). Minerology of meteorite groups. *Meteoritics and Planetary Science* **32**, 321–247.
22. Messenger, S., Amari, S., Gao, X., Walker, R.M., Clemmett, S.J., Chillier, X.D.F., Zare, R.N. and Lewis, R.S. (1998). Indigenous polycyclic aromatic hydrocarbons in circumstellar graphite grains from primitive meteorites. *The Astrophysical Journal* **502**, 284–295.
23. Allamandola, L.J., Tielens, A.G.G.M. and Barker, J.R. (1989). Interstellar polycyclic hydrocarbons: the infrared emission bands, the excitation/emission mechanism, and the astrophysical implications. *The Astrophysical Journal, Supplementaries* **71**(1), 733–775.

(a) A large meteoroid enters the Earth's atmosphere and fragments into dozens of smaller pieces 16 km above the Nevada desert floor, the pieces still aglow from its fiery passage through the upper atmosphere. (Painting by Dorothy Sigler Norton. From the Robert A. Haag collection.) (b) On June 13, 1998, after a spectacular daylight fireball accompanied by sonic booms and a corkscrew dust trail, at least 45 individual meteorite fragments landed in Portales Valley, Roosevelt County, New Mexico. Robert Woolard found this 34 kg mass after two hours of searching. It is seen here *in situ* resting in its impact pit before it was collected. The meteorite, an H6 ordinary chondrite, proved to have a unique interior structure (see page 78) unlike any ordinary chondrite known. (Photo by Robert Woolard.)

CHAPTER TWO

The fall of meteorites

Although meteorites do fall during the time of meteor showers, they are coincidental and unrelated to the showers. Of all the meteor showers observed over the past two centuries, not one has been observed to produce a body that survived passage through Earth's atmosphere. This seems to imply that meteor shower particles and meteorites have different origins. Throughout most of the nineteenth and twentieth centuries, there was a clear-cut distinction between meteor shower particles as cometary debris and meteorites as fragments of asteroids. Today the distinction is no longer that clear. Of the many meteor showers known, at least five are suspected of being derived from asteroids: 3200 Phaethon; 2201 Oljato; 944 Hidalgo; 1566 Icarus; and 1937UB Hermes (now lost). Perhaps the best known is 3200 Phaethon whose dust stream produces the Geminid meteor shower every December. All of these asteroids have highly eccentric orbits mimicking those of short period comets. These could be cometary bodies that have expended their volatile elements and now roam the Solar System as dark bodies on comet-like orbits.

Orbits of meteoroids

When a meteoroid large enough to reach Earth's surface plunges into the atmosphere, its luminous phase lasts only seconds. It is only during this all too brief period that the cosmic motion of the body can be observed. The fall phenomena are so startling to observers that little if any attention is paid to the details of the flight. How fast was it moving in degrees per second? What direction was it moving? At what angle was it moving relative to the horizon? How long was the duration of the burn? Answers to these and other questions are needed in order to approximate the atmospheric entry trajectory and initial velocity of the body. The trajectory can then be extended back into space to determine the pre-entry orbit of the meteoroid.

The fireball networks

For nearly two centuries, scientists had to rely upon eyewitness reports of fireballs. Since bright fireballs are unpredictable and occur randomly, scientists could never hope to observe them personally. Gathering data from eyewitness accounts of fireballs was (and still is) typically unreliable and usually cannot be used to determine precise heliocentric orbits or even true atmospheric trajectories. What was needed was two or more automated tracking stations placed many miles apart where precise tracking observations of a given fireball could be made simultaneously by at least two of the stations. Because of the random nature of fireballs, this was technically not feasible and had to await the twentieth century.

All that changed in 1951 when a small photographic program was initiated at the Ondrojov Observatory in the Czech Republic (then Czechoslovakia). It was the first systematic photographic program designed initially to record the passage of faint meteors.[1] On April 7, 1959, with cameras operating, a brilliant fireball was recorded (−19 magnitude) resulting in the recovery of four stony meteorite fragments in Přibram (near Prague), Czechoslovakia. This was the first photographed meteorite fall in history in which meteorites were actually recovered.[I] The primary purpose of the Czech meteor network was not to recover fallen meteorites. Rather, it was designed to study the orbits of relatively faint meteoric particles. This remarkable event was unprecedented in the history of astronomy. It marked the first time a precise pre-terrestrial orbit of an actual recovered meteorite was determined. The aphelion of the Přibram meteorite (4.012 AU) placed it within the outer zone of the asteroid belt and provided the first strong evidence of the connection between meteorites and asteroids (see Fig. 2.8).

The successful recovery of the Přibram meteorite suggested that meteorites could be recovered through a systematic photographic program designed to record bright fireballs, some of which were good candidates for meteorite recovery. The Přibram event provided encouragement for the establishment of two other camera networks, this time specifically with the goal of recovering freshly fallen meteorites. In 1964, the Prairie Camera Network went into operation in the United States, directed by the Smithsonian Astrophysical Observatory. It was composed of 16 widely spaced stations located in the Plains States, each station equipped with four wide-angle cameras that, working simultaneously, covered nearly the entire sky from each location (Fig. 2.1). The

Fig. 2.1. A view of the interior of the Hominy, Oklahoma, Prairie Network Station showing the four surplus aerial cameras converted for meteor photography. The station attendant, a local volunteer, makes his weekly check of the equipment. (Smithsonian Astrophysical Observatory photo.)

[I] Although the photographic data did confirm the location of fall with good precision, there were eyewitnesses in the vicinity of the impact area who almost immediately recovered samples without the aid of the camera data.

Fig. 2.2. Locations of the 16 Prairie Network stations in six Midwestern states. The field headquarters was in Lincoln, Nebraska.

Fig. 2.3. One of 12 camera stations housing five wide-angle meteor cameras of the Canadian MORP Network. (© Geological Survey of Canada.)

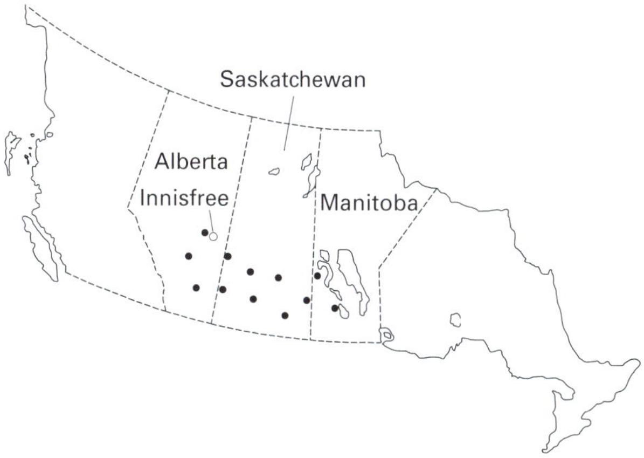

Fig. 2.4. Map of the twelve MORP network camera stations stretching from Alberta to Manitoba. The solid circles are camera locations. Note the location of the Innisfree meteorite site near the Vegreville, Alberta, station. (©Geological Survey of Canada.)

Midwest locations were selected in the event meteorites survived atmospheric passage, for they could be more easily found in the Plains States (where most of the land was cleared and under cultivation. The cameras were arranged over seven states, which effectively covered a total of 1 000 000 km^2 of ground area on which meteorites could fall (Fig.2.2).[2]

The Canadians established the Canadian Meteorite Observation and Recovery Project Network that became fully operational in 1971. It was composed of 12 stations, each equipped with five wide-angle cameras covering 270° of azimuth and from 2 to 55° in altitude (Fig. 2.3). Neighboring stations covered the remaining zenith and azimuth positions. Like the American Prairie Network, the Canadian network was designed with meteorite recovery as a primary goal. The network was therefore set up in a broad relatively flat prairie zone extending from southern Alberta through Saskatchewan and into Manitoba (Fig. 2.4). This extended the effective search area for meteorites to some 700 000 km^2.

Both the US and Canadian networks photographed bright fireballs that led to the recovery of stony meteorites. On January 3, 1970, the Prairie Network photographed a fireball (Fig. 2.5) that dropped four H5 chondrites totaling 17 kg near the small town of Lost City, Oklahoma (Fig. 2.6). The Canadian Network picked up a bright fireball on February 5, 1977, resulting in the fall of a rare LL5 chondrite near Innisfree, Alberta (Fig. 2.7).[3] The Lost City meteorite had an orbit whose aphelion placed it near the sunward edge of the asteroid belt while the Innisfree meteorite aphelion occupied the middle of the belt (Fig. 2.8).[II]

Operation of the US network continued for a decade coming to an end in 1974. The Canadian network continued until 1985. Through the lifetimes of all three camera networks, hundreds of fireballs were photographed, many of

[II] On February 6, 1980, another fireball was observed by the Canadian network cameras over western Canada. This was three years to the day after the Innisfree meteorite fall. The orbits of the two bodies were identical, thus providing the first observational evidence of a multiple fall from the same orbit. The meteorite presumably impacted near the town of Ridgedale, Saskatchewan, but a subsequent search failed to find the meteorite.

Fig. 2.5. Fireball photographed by the Smithsonian Astrophysical Observatory Prairie Network on January 3, 1970. Computer analysis of this and other photos from the camera network enabled Smithsonian researchers to predict a probable impact point near Lost City, Oklahoma, and to recover the 11.8 kg meteorite ten days later. (Smithsonian Astrophysical Observatory photo.)

Fig. 2.6. The 11.8 kg Lost City meteorite. (Smithsonian Astrophysical Observatory photo.)

which should have dropped meteorites. All likely meteorite-dropping fireballs were traced to probable ground locations and subsequent searches made, but no further meteorites were found.

Only the Czech network, now called the European Fireball Network, remains operational. It has been expanded to include several European countries. Today, this network has 12 operating camera stations located in the Czech and Slovak Republics and 22 camera stations dispersed through Germany, Belgium, Austria, and Switzerland.

Fig. 2.7. Fireball photographed by the Canadian MORP camera near Innisfree, Alberta. The camera contained a rotating shutter of known angular speed (4 segments per second) that interrupted the trail allowing the meteor's velocity to be determined. The meteor entered the camera's field of view from the left and traversed the field in 3.82 s. (© Geological Survey of Canada.)

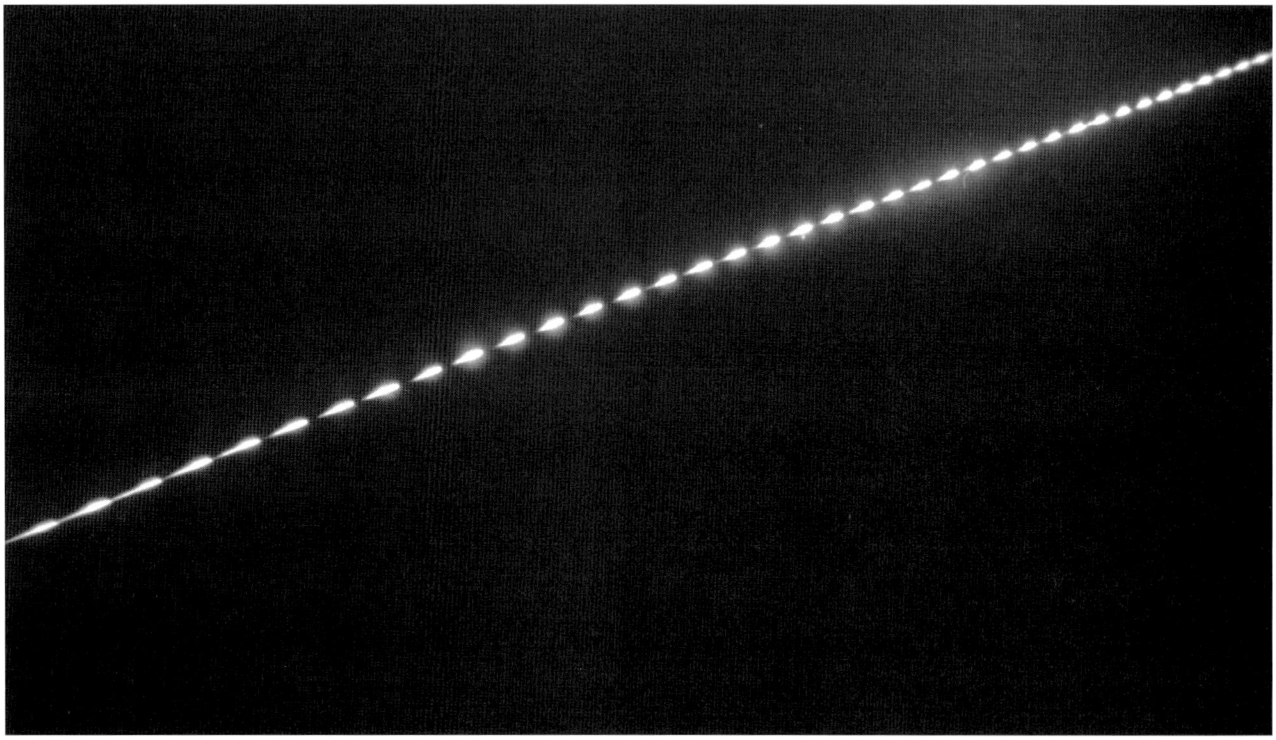

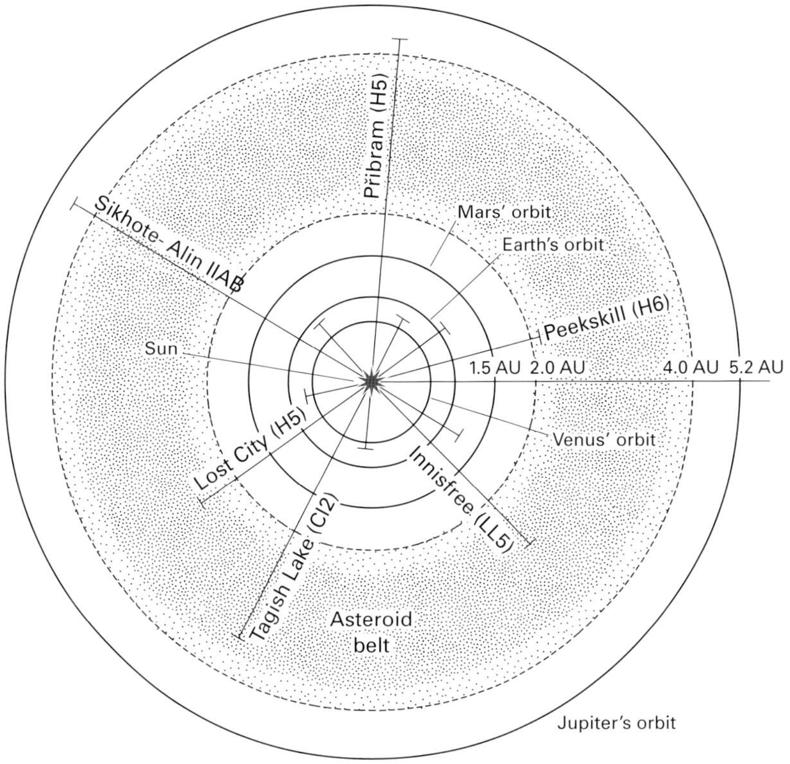

Fig. 2.8. Major axes showing perihelia and aphelia of the Příbram, Lost City and Innisfree meteorite orbits determined from photographic data from the three meteor networks. In addition, three other major axes of recently recovered meteorites are included, as they appear elsewhere in this book: Peekskill, Sikhote-Alin, and Tagish Lake. The designations H, L, CI and IIB are meteorite classifications used in later chapters of this book. The perihelia and aphelia for each orbit are to scale. The major axes only are included for clarity of presentation.

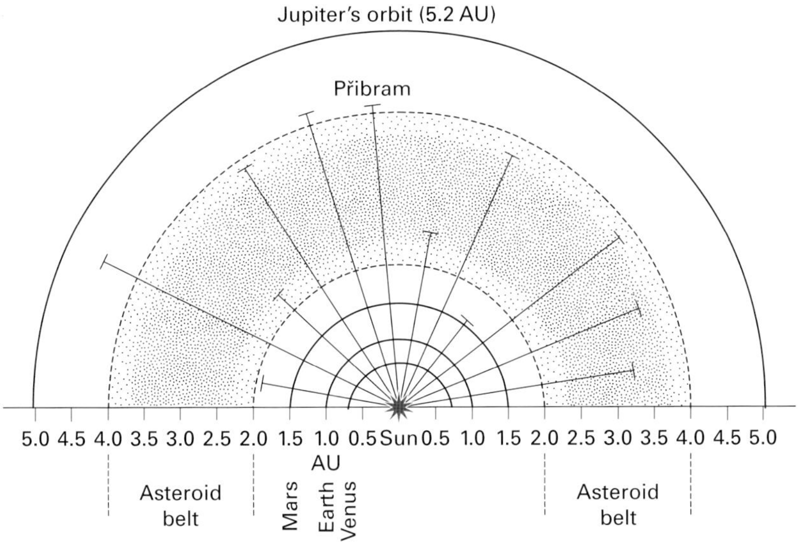

Fig. 2.9. Aphelion distances of twelve bright fireballs photographed by cameras in the European Fireball Network. All have aphelia within the asteroid belt and all have perihelia well inside Earth's orbit (not shown). About half are thought to be asteroidal while the remainder may be low density fragile comet-like bodies. Seven are considered meteorite-dropping candidates but only one has been recovered (Příbram). Aphelia only are included for clarity of presentation. (Data from Ceplecha, Z. (1998). The European Fireball Network: current status and future prospects. *Meteoritics and Planetary Science* **33**, 49–56.)

All three surveys have yielded a plethora of physical data that are still being analyzed. From the trajectories left on film by bright fireballs, scientists have been able to glean a remarkable amount of information. Orbits of hundreds of meteoroid fragments have been calculated. Figure 2.9 shows the major axes of 12 objects that produced exceptionally bright fireballs recorded by the European Fireball Network between 1959 and 1995. About half of these are thought to be bodies with typical stony meteorite densities (3.5 g/cm^3) while the remainder are relatively fragile bodies with densities near that of water ice suggesting they are comet-like. Note that all have aphelia in the asteroid belt although six have orbital inclinations that carry them well above and below the Solar System plane.

Characteristics of a meteoroid's atmospheric passage

A fireball's atmospheric trajectory contains information that can reveal many of the dynamical and physical characteristics of the meteoroid. From the meteoroid's entry velocity and direction of motion relative to Earth's orbital motion, its heliocentric orbital velocity can be found at the Earth–Sun distance (1 AU). This is the first step in determining the meteoroid's orbit around the Sun. The trajectory also contains dynamical information about the effects of the atmosphere on the falling meteoroid and information about its size, shape, pre- and post-atmospheric mass, and density. The meteoroid's atmospheric entry angle and velocity and its initial mass essentially dictate the observed fireball effects.

The true trajectory

The essential data derived from observations should include the altitude and azimuth of the beginning and end of the visible burn phase, the maximum brightness in terms of stellar magnitudes, the angular velocity of the body and the duration of the visible phase. The observers (or cameras) do not see a fireball's true path but only its apparent path projected against the background stars. The third dimension is missing. Another observer several miles away sees the fireball at the same time but its trajectory will appear in a different position with respect to the stars. This displacement is a parallactic shift experienced because the fireball is only a few dozen miles above them and their own positions will affect the observed position of the fireball. Figure 2.10 shows a fireball observed against the celestial sphere from two widely spaced positions on Earth. From each position the observer notes the beginning and end points of the burn phase of the trajectory. The two observations are then combined by projecting lines from each observer's position to the beginning and end points. The true path of the fireball is revealed at the intersection points of these two planes.

Height and angle of descent

Once the true trajectory of the fireball is determined, the heights above the ground of the beginning and end points of the visible trajectory can be found by trigonometric calculation. Figure 2.11 shows the true trajectory of a fireball. We wish to find the angle of descent which can be calculated by solving for the height of the beginning and end points of the trajectory. In this example, the end point of the visible path is selected. This is the point in the fireball's trajectory where the meteoroid has sufficiently slowed its velocity so that heat and light are no longer generated. Two stations marked (A) and (B) positioned a known distance (b) apart simultaneously determine the azimuth (α and β respectively) and angular altitude (θ) for (A) of the end point of the fireball. The height (H) of the end point can be found by solving one of the two triangles AR_pC or BR_pC.

From the Law of Sines:

$$a/\sin\alpha = b/\sin\gamma = c/\sin\beta.$$

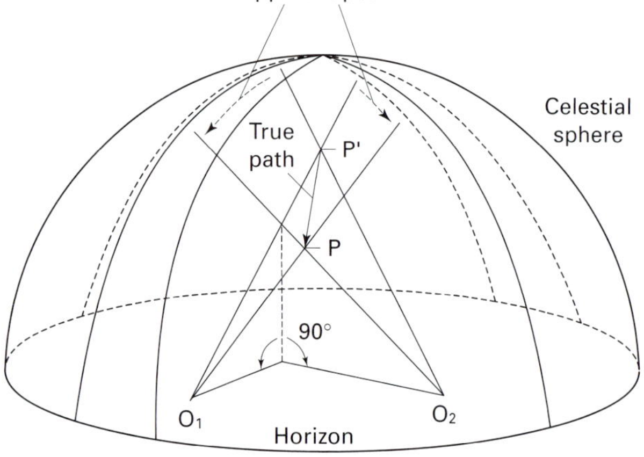

Fig. 2.10. The true path of a fireball (PP') is determined from two or more widely spaced observing positions by noting the apparent path of the fireball against the celestial sphere for each position and combining the observations to reveal the meteor's true path. See text for details. (From *Rocks From Space*, 2nd edition (1988), Mountain Press Publishing Co., Missoula, Montana.)

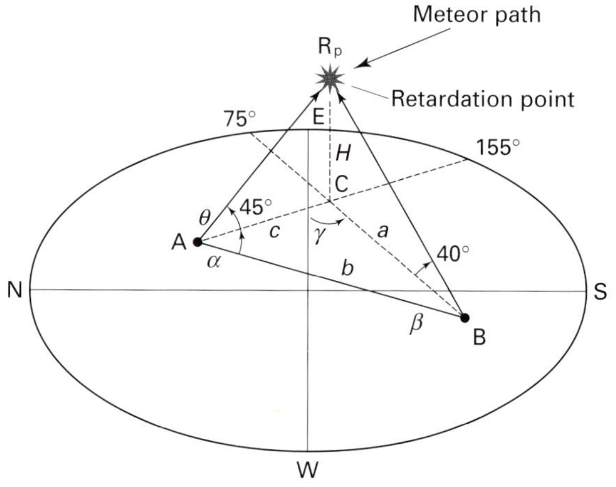

Fig. 2.11. Calculation of the true height of a meteor using observations made from two different positions separated by several kilometers. (Figure adapted from *Rocks From Space*, 2nd edition (1998), Mountain Press Publishing Co., Missoula, Montana.)

Solving for c,

$$c = b \sin \beta / \sin \gamma = b \sin \beta / \sin [180° - (\alpha + \beta)].$$

In triangle AR_pC

$$\tan \theta = H/c, \text{ so } H = c \tan \theta.$$

Substituting c above,

$$H = b \sin \beta \tan \theta / \sin [180° - (\alpha + \beta)]. \quad (2.1)$$

From equation (2.1), the height (H) is found for beginning and end points from which the angle of descent from the vertical is calculated.

The following example using the above equations and Fig. 2.11 illustrates the height calculation from a typical fireball observation. A fireball is seen by two observers from positions (A) and (B). Their positions are separated by 30 km (b). Both observers see the fireball extinguished at the same time. Observer (A) notes its altitude is 45°; observer (B) measures the altitude as 40°. Both observers also note the azimuth direction along the horizon. Using these azimuth readings they construct triangle ABC in which angle (α) and angle (β) are directly measured: 65° and 50° respectively. Then, assuming that the meteor is directly above position C (by definition) and using equation (2.1) derived earlier, the altitude of the meteor is determined. Thus:

$$H = b \sin \beta \tan \theta / \sin [180° - (\alpha + \beta)].$$

$$\text{Hence, } H = 30 \text{ km } (\sin 50°)(\tan 45°) / \sin [180° - (65° + 50°)],$$

So, H = 25.36 km.

In practice, the angular velocity and the rate of change of velocity (deceleration) is measured on the photographic plate. For example, the Canadian Network cameras were equipped with a rotating shutter near the film plane that interrupted the fireball trail four times every second. Thus, the angular velocity in degrees per second can be measured directly on the photographs. The Innisfree fireball pictured in Fig. 2.7 shows the segmented trail with four segments equaling 1 s of flight time. The fireball was observed to travel 37.8° in 3.82 s. Variations in the lengths of the segments demonstrate a change in velocity of the meteoroid as drag forces act to decelerate it.

Meteoroid velocity

Asteroid fragments in Earth-crossing orbits have heliocentric velocities usually greater than Earth's orbital velocity of 29.9 km/s. Orbital data of 213 fireball events from the Canadian network showed heliocentric velocities between 25 and 43 km/s with an average of 38.2 km/s.[2] The geocentric velocity (velocity with respect to Earth) varied considerably from as little as 7.5 km/s to as high as 70.8 km/s with an average of 29.9 km/s. Geocentric velocities higher than 20 km/s either represent comet-like fragments with higher heliocentric velocities or retrograde meteoroids that encounter Earth head on. The initial entry velocity is critical to the survival of the meteoroid. This velocity is higher than the geocentric velocity because it contains an additional component, 11.2 km/s, the acceleration of the meteoroid due to Earth's gravity. A fragment of stony composition entering the atmosphere in excess of 30 km/s will lose 99% of its mass through ablation. Thus, meteoroids that make it to Earth's surface must have entry velocities of less than 30 km/s or geocentric velocities of not much over 18.8 km/s. Of the 213 fireball events recorded by the Canadian network, 46 were selected with high probabilities of atmospheric survival, dropping meteorites of at least 100 g in mass. All of these meteoroids had prograde orbits so that their atmospheric entry velocities were considerably less, averaging 13.3 km/s. The Lost City and Innisfree meteorites had entry velocities of 14.2 km/s respectively and the Přibram meteorite entered a little higher at 20.9 km/s, all well under the survival limits.

Atmospheric drag

The Earth's atmosphere provides an effective shield to incoming meteoroids. At an altitude of about 100 km the density is sufficiently high to produce a significant aerodynamic drag on the body. The drag force on the meteoroid which determines the rate at which its initial entry velocity decreases with altitude varies in a complex way, depending upon a number of factors: the initial entry velocity; the density of the atmosphere; the mass and cross-sectional area of the meteoroid; the drag coefficient (numerically usually between 0 and 1 for meteoroids); the atmospheric entry angle of the meteoroid relative to the Earth's surface;[III] and the constant, g, the acceleration due to gravity, which is added to the final velocity of the meteoroid.[IV] Only the acceleration of gravity

---

[III] The entry angle can vary between 90° to near 0°. Low entry angles allow the fireball to persist since the density of the atmosphere increases more slowly thus reducing the rate at which the meteoroid's velocity decreases with time.

[IV] Meteoriticists use the well known aerodynamic drag equation to calculate the rate at which the velocity of a meteoroid decreases with altitude.[4] Thus:

$$dv/dt = D\rho v^2 A/m + g\cos\theta$$

where dv/dt is the rate of deceleration of the meteoroid with time; D is the drag coefficient (assuming a brick-shaped object at maximum aerodynamic drag, D is between 0 and 1); ρ is the atmospheric density at a given altitude; v^2 is the velocity of the body which is squared here because the drag force varies directly with the square of the velocity; A is the cross-sectional area; m is the mass; g is the acceleration due to gravity which is added to the final velocity of the falling meteoroid; θ is the entry angle relative to the Earth's surface. A thorough mathematical treatment of the physics of meteoroid fall can be found in Buchwald, V.F. (1975). *Iron Meteorites*, University of California Press.

is a constant. All other components vary as the meteoroid passes through the atmosphere. The density of the atmosphere varies exponentially with decreasing altitude; the mass and cross-sectional area of the meteoroid change with time as atmospheric ablation erodes the body; the drag coefficient can vary from 0.4 to 1.4 as the meteoroid's shape changes.

Of all of these characteristics, the initial entry velocity and the angle of descent are the most important in determining the survival of the meteoroid. Aerodynamic drag varies with the square of the velocity for any object traveling faster than sound in the atmosphere. Thus, a meteoroid with a high initial velocity will be more rapidly decelerated than one with a low initial velocity. A high initial entry velocity coupled with a high entry angle (nearly vertical to the ground) is the worst possible scenario for the meteoroid's survival. With a high entry angle, the path length through the atmosphere is shortest and the increase in atmospheric density per unit time is most rapid. Both act together to increase drag forces on the meteoroid, producing high deceleration values that will likely exceed the strength of a typical stony meteorite, especially if it has internal fractures from a past impact history. A low entry angle allows the meteoroid to decelerate more slowly higher in the atmosphere with consequently weaker drag forces on the meteorite. The rate of change of the velocity is not uniform over time but rises to a maximum when the initial entry velocity has been reduced by 39% after which the rate falls.[5]

The atmosphere is very effective in slowing most meteoroids passing through it. Whether or not a meteoroid's entry velocity is totally retarded in the atmosphere depends upon all of the above factors but also upon the meteoroid's initial mass. This is demonstrated in Fig. 2.12 in which the aerodynamic drag equation is used to calculate V_F / V_I, the ratio of the initial entry velocity and the final velocity before impact for meteoroids over a large range of masses. The entry angle (vertical) and initial entry velocity (40 km/s) are identical for each meteoroid. The meteoroids vary in mass from 0.1 tons to 1000 tons. The first thing we notice is that meteoroids of smaller initial mass lose all of their cosmic velocity while still several kilometers above the Earth. At this point the meteoroid's fall path becomes quite steep since its forward momentum has ceased and the body falls by gravity alone which is the *terminal* velocity of the meteoroid. It varies somewhat depending upon the shape of the falling meteoroid creating air resistance around it, but is typically between 125 and 250 m/s when it impacts. The point at which terminal velocity begins occurs at progressively lower altitudes for meteoroids of increasing mass. Finally, for meteoroids of 10 metric tons or more, the drag forces are not sufficient to reduce the initial entry velocities to zero and the meteoroid impacts the ground with some of its cosmic velocity still intact plus an additional velocity due to gravity. For entry angles less than 90°, drag forces are greater resulting in a greater loss of initial velocity with altitude for a given initial mass.

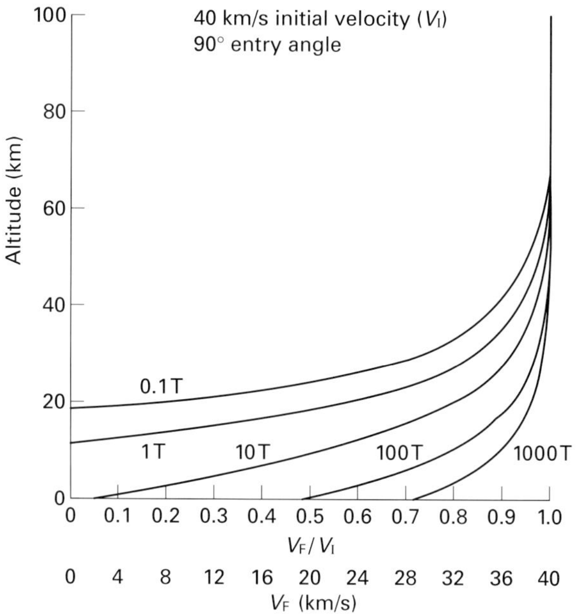

Fig. 2.12. This graph demonstrates how the impact velocity of a meteoroid varies with initial mass, entry velocity, and entry angle. Note that most meteoroids loose all of their cosmic velocity while still several kilometers above Earth. Above an initial mass of 10 metric tons or more, a percentage of the initial cosmic velocity is retained. (Data from Heidi, F. (1964). *Meteorites*, The University of Chicago Press.)

Light phenomena

From the Earth's surface, we can only know of the passage of a sizeable meteoroid because of the light it gives off during the few seconds of maximum dynamic forces upon the mass. A brilliant meteor is certainly one of nature's most sublime deceptions. Virtually everyone who is witness to the phenomenon insists that it was very close, just over the next group of hills, so to speak. The impression is that it was quite close when, in actuality, it may have been many kilometers away. Moreover, their impression of duration of the fireball is usually much longer than it actually was. Concepts of distance and time become highly distorted during the excitement of the event. If the meteor remains visible as it passes behind some distant hills or artificial structures, its line of sight distance approaches 160 km. Even if the meteor appears to go out nearly overhead, it is still many kilometers from the observer.

The time between the beginning and end of a typical fireball's visible trajectory is, on the average, only about 4 s although a few have lasted as long as 20 s or more. The duration of luminosity depends upon entry angle and initial mass. A low entry angle with a gently sloping path allows the fireball to persist longer than a more direct near vertical entry angle. The Jackson Lake, Wyoming, fireball of August 10, 1972, with a near zero entry angle lasted for 101 s before skipping out of the atmosphere. Meteoroids with low entry angles are subject to severe ablation, reducing the chances that they will survive to reach Earth's surface. Longer duration fireballs tend to be more massive. Of some 44 fireball events photographed by the Canadian network, those with durations of less than 4 s had average initial masses of about 2.6 kg while longer durations of 4–6 s or greater had estimated initial masses averaging 6.4 kg.[V]

The height at which the visible trail begins depends upon the meteoroid's entry velocity. Generally, higher entry velocities will produce beginning trails at greater altitudes.[6] For example, from the 44 fireballs mentioned above from which entry velocities of 12.8, 20.6 and 26.5 km/s were selected, the corresponding trail beginnings were 67.0, 78.7 and 91.2 km respectively. The median entry value was found to be 15.2 km/s and a beginning height of 72 km.

A meteoroid in space has considerable kinetic energy defined as:

$$KE = \tfrac{1}{2} mv^2 \qquad (2.2)$$

where m is its mass and v is its cosmic velocity. Unless the meteoroid's mass is thousands of tons, the velocity is the more important quantity since the kinetic energy varies as the square of the velocity and only to the first power of its mass. Thus, two meteoroids with the same mass but varying in initial entry velocity by a factor two will vary in total kinetic energy by a factor of four. The entry velocity therefore has a profound effect on the survival of the body. As the meteoroid is slowed by aerodynamic drag, a large percentage of its kinetic energy is converted to heat and light. The forward end of the meteoroid rapidly reaches the melting point of silicate rock, about 1500 °C, and begins to ablate away as tiny droplets of molten rock. The air rushing rapidly around the meteoroid pushes the liquid material behind the main mass, which then rapidly cools into solid spherical dust specks forming a dark dust trail. Spectra of fireballs and bright meteors show that only a small portion of the light actually comes from the incandescent solid body itself. Its continuous spectrum is weak compared to the light that comes from vaporized meteoritic material. At temperatures beyond 2000 °C, some of the liquefied meteoritic minerals ablating from the main mass vaporize. Disassociation of the mineral molecules into individual elements takes place rapidly and these elements become excited or ionized (lose electrons). This ionized cloud of elements produces an emission spectrum as electrons recombine with the ions. Emission lines of iron, magnesium, silicon, nickel, calcium and other elements that typically make up silicate minerals of stony meteorites are found in the spectrum.

If the total light originated only from the incandescent and vaporized minerals of the meteoroid, the fireball's brightness would be considerably fainter than it actually is. Air molecules violently impacting the invading mass quickly heat and ionize, forming a large plasma ball of charged particles. This produces an additional emission spectrum superimposed upon the meteoroid's spectrum. This ball of superheated air can reach hundreds of meters in diameter. Thus, the fireball seen from the Earth's surface is actually a large ball of glowing gas with a comparatively small nucleus of solid material. As much as 95% of the light from a fireball is from ionized atmospheric gases, primarily nitrogen and oxygen.

Fig. 2.13. Lingering trail of the Tagish Lake (Yukon) fireball. This digital photograph shows the trail glowing orange against the brightening predawn sky. This image was made about a minute after the fireball appeared. The glow was a combination of reflected sunlight and ionization of the gases in the trail. The trail, progressively distorted by high altitude winds, remained visible for more than an hour. (Courtesy Ewald Lemke.)

Behind the fireball a trail of ionized atmospheric gases persist that may continue to glow faintly after the fireball vanishes. This glow, a kind of fluorescence, is caused by remnant ions recombining with free electrons in the atmosphere. This recombination process, if occurring at the appropriate energy levels in oxygen and nitrogen atoms, liberates light energy that gradually fades as the ions are used up, usually in a matter of seconds or at most a few minutes. In one instance, the Pasamonte, New Mexico, fall of March 24, 1933, a glow persisted for 45 minutes! Another extraordinary fireball occurred just before dawn on January 18, 2000. The brilliant fireball was widely observed over the Yukon Territory and northern British Columbia and was detected by satellites in Earth orbit. The ionized glow of the train persisted for nearly a half hour even as the Sun illuminated the distorting cloud (Fig. 2.13). Over 500 meteorites were subsequently recovered on the frozen Tagish Lake in the southwest corner of the Yukon Territory.

Colors are often reported from fireballs. Intense blue-white, emerald green, yellow and red are the most commonly reported hues. When vaporized and ionized, many

[V] These mass values were not directly measured since none were found on the ground. Instead their masses are estimated from the duration and brightness of the fireball trail.

elements give off diagnostic colors. Magnesium burns with a blue-white light; sodium, a yellow light; calcium, orange-red; copper, green. Oxygen can emit a red glow and molecular nitrogen a green glow. The colors often vary over the path of the fireball, usually starting out as an intense blue-white and then turning to an orange or red toward the end of its visible path. Since the composition of the atmosphere and the meteoroid does not vary, the color changes must be the result of the rapid cooling of the solid body as it decelerates.

The maximum brightness of a fireball usually occurs at the point of maximum dynamic force while the meteoroid retains most of its entry velocity. Of the 44 fireballs selected from the Canadian network this occurred at a mean altitude of 47.5 km or about two-thirds of the way along the visible path. The numerical brightness of a fireball is usually given in stellar magnitudes, a concept borrowed from astronomy, in which a difference of one magnitude is numerically equal to 2.512 times in brightness or:

$$\Delta b = 2.512^{(m_1 - m_2)} \quad (2.3)$$

where Δb is the difference in brightness; $(m_1 - m_2)$ is the difference in magnitude between two objects and $(m_1 > m_2)$. Thus, two fireballs of $m = -10$ and another of $m = -8$ have a difference in magnitude of 2 and a difference in brightness of $2.512^2 = 6.3$.[VI] Photographically, the magnitude of the fireball on film is corrected to a brightness as it would be seen on the zenith at a standard distance of 100 km, which takes into account atmospheric extinction. Practically, a fireball's maximum brightness is usually compared to convenient celestial objects such as the planet Venus at −4.7 magnitude; the first quarter Moon at −10, full Moon at −12.5 or the Sun at −26.5. Fireballs that drop meteorites begin at maximum magnitudes of between −8 and −10; the more massive meteoroids attain brighter magnitudes. The Přibram fireball reached a magnitude of −19, one of the brightest fireballs on record. From equation (2.3), we can see that Přibram attained a magnitude very nearly 400 times the brightness of the full Moon!

Figure 2.14 shows how the brightness of a large meteoroid entering the atmosphere changes in time as the meteoroid's velocity changes. The light curve for the fireball begins with a rapid rise to maximum brightness over the first three seconds of travel while its entry velocity remains relatively constant in the rarified upper atmosphere at around 14 km/s. After it attains its peak brightness, the curve flattens out briefly retaining this brightness for about a second or until it reaches about two-thirds of the length of its visible path. It then begins a steep decline as the meteoroid encounters rapidly increasing atmospheric drag. Its velocity decreases rapidly and concomitantly the fireball drops more than 3 magnitudes in brightness during the next 3 s, the final one-third of its visible path.[7]

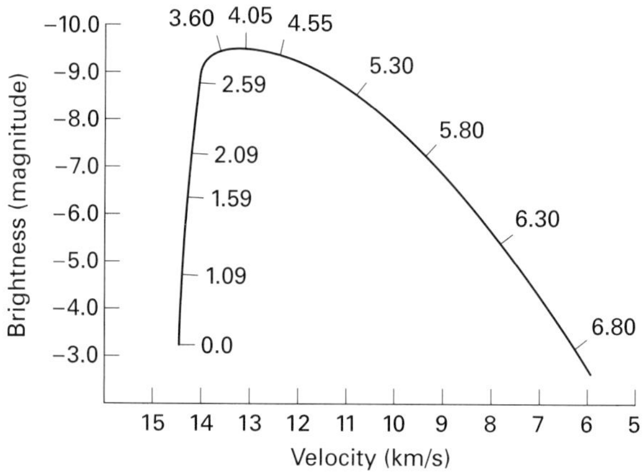

Fig. 2.14. A typical light curve of a fireball. The graph relates the fireball's change in velocity and brightness with time. Seconds of time are noted along the curve. This fireball was photographed by the Canadian camera network (MORP). (Data from Halliday, I. *et al.* (1996). Detailed data for 259 fireballs from the Canadian camera network. *Meteoritics and Planetary Science* **33**, 185–217.)

The altitude at the end of the trail is usually below 30 km and has been measured as low as 5 km. Lost City had an end point of 19.5 km and Přibram, 13 km. At that point its velocity has been reduced to between 3 and 8 km/s.

Once the meteoroid has reached the end of the visible path, the ablation process has ceased and the meteoroid no longer generates light and heat even though it still retains a fraction of its original cosmic velocity. If the meteoroid successfully reaches this point, it stands a good chance of surviving its fall to Earth. It will continue to decelerate due to air resistance until all of its cosmic velocity has been consumed. This is called the *retardation point*. It now falls by gravity alone. Its trajectory becomes essentially a near vertical drop with a speed that is a function of Earth's gravity, the meteoroid's residual mass, and atmospheric drag. Meteoroids under 100 kg impact the ground between 125 and 250 m/s, well under the velocity of sound.

Sound phenomena

On Tuesday, the first of May, at twenty-eight
minutes past twelve o'clock, the people of that

[VI] The magnitude scale in astronomy is based upon the difference in brightness between a first magnitude star (+1) and a 6th magnitude star (+6). A difference of 5 magnitudes corresponds to a brightness ratio of 100:1. Thus, the brightness ratio between two stars with a difference of one magnitude would be $100^{1/5}$ or 2.512. Most stars have positive magnitudes. The smaller the positive number, the brighter the star. There are a few stars brighter than +1 magnitude so the magnitude scale is extended through 0 and into negative numbers with the same mathematical relationship applying. In all cases, $(m_1 - m_2 > 0)$ The brightest star, Sirius, has a magnitude of −1.5. Venus at its brightest is −4.7. Thus, $m_1 - m_2 = 3.2$, and $\Delta b = 2.512^{3.2} = 19.06$ which means that Venus is 19.06 times brighter than Sirius. Since all fireballs have magnitudes greater than 0.0, they are all assigned negative magnitudes.

> vicinity were almost panic stricken by a strange and terrible report in the heavens, which shook the houses for many miles distant. The first report was immediately overhead, and after an interval of a few seconds was followed by similar reports with such increasing rapidity that after the number of twenty-two were counted they were no longer distinct, but became continuous, and died away like the roaring of distant thunder, the course of the reports being from the meridian to the southeast. In one instance three men working in a field, their self-possession being measurably restored from the shock of the more terrible report from above, had their attention attracted by a buzzing noise overhead, and soon observed a large body descending strike the earth at a distance of about one hundred yards. Repairing thither they found a newly-made hole in the ground, from which they extracted an irregular quadrangular stone weighing fifty-one pounds. This stone had buried itself two feet beneath the surface, and when obtained was quite warm.[VII]

Imagine seeing a brilliant fireball blazing a trail across the sky. It's an eerie moment. The sky lights up. Trees cast moving shadows on the ground keeping pace with the alien intruder. Suddenly it explodes, fragmenting into a myriad of pieces but it still maintains its course for a second or two longer. Then it vanishes. All of this occurs in seconds – and in absolute silence. More seconds go by. Still silence prevails. A minute goes by and you have almost recovered from the amazing sight. Suddenly a tremendous series of explosions break the silence lingering like not-so-distant thunder. These are sonic booms caused by a shock or pressure wave generated in the atmosphere by the hypersonic flight of the fireball. They have been likened to sonic booms caused by supersonic aircraft breaking the sound barrier, or a continuous sound like the distant roll of thunder. The sound can be elusive. Some witnesses report hearing sonic booms while others situated along the path hear nothing. Conditions in the atmosphere at the time can effect the propagation of sound waves generated by the shock wave. The atmosphere is heterogeneous in that its temperature, pressure, humidity, and density vary from place to place. Sound waves must pass through these variable conditions which refract the waves by differing amounts along the way. Winds aloft can affect the propagation of sound, sometimes adding to the sound's velocity, sometimes slowing its velocity. The net result is an unpredictable pattern of sound in which some areas on the ground receive highly attenuated waves or none at all while others receive the full intensity of the sound.

Some observers report a series of popping sounds like gun fire from an automatic rifle. I remember a similar series of popping sounds during a Saturn 5 Moon launch at Cape Canaveral, Florida. In that case, each of the five engines expelling exhaust at supersonic velocities produced its own pressure waves in the atmosphere. As the launch vehicle's trajectory changed from vertical to a southeasterly direction, the engine nozzles pointed directly at the viewers back at the Cape, producing a cacophony of popping sounds. A similar situation occurs if a meteorite breaks into pieces while still in supersonic flight. Each individual piece will generate its own pressure wave which together will sound like rapid gun fire. Some observers especially close to the impact point of the meteoroid report hearing a "whomping" sound like that made by the blades of a helicopter. This may be the sound of an irregular-shaped body in rotation.

The light of the fireball and the sound generated by its shock wave obviously cannot occur simultaneously. The light is propagated almost instantaneously while the sound traveling about 330 m/s in the atmosphere lags far behind. Depending upon the distance of the fireball from the observer, the sound may be delayed from 30 s to several minutes after passage of the fireball. This sound delay is commonly experienced with high flying jet aircraft, preceding its own sound waves in the sky.

Electrophonic sounds

Hissing sounds like radio static or the sound of frying bacon, a "swishing" sound, and whistling are the most commonly reported sounds; and fireball witnesses who report them insist that they occurred *simultaneously* with the passage of the fireball. Some reports state quite emphatically that the observers were indoors and were alerted to the passing fireball *before* they actually observed it by looking through a window or stepping outside. The sounds are considerably higher in frequency and much less intense than the sound generated by shock waves. Even more curious, these sounds are selective, that is, they are reported by some observers while others viewing the same fireball at the same time and near the same location report hearing nothing. The average distance over which these sounds are detected is about 100 km from the fireball. Only between 4 and 8% of the witnesses of any given fireball report concomitant light and sound phenomena. Not all fireballs produce the disturbance as they seem to be limited to a brightness of −6 magnitude or greater.

Sounds heard *simultaneously* with the fireball's appearance have been reported for almost 300 years. In 1719, the famous English astronomer, Edmund Halley, recognized the apparent paradox of the reports of sound being transmitted at light speed and dismissed them as "the effect of pure fantasy" in a paper to the Royal Society in which he discussed

[VII] This is from an eyewitness account of the fall of the New Concord, Ohio, stony meteorite in 1860. The account is taken from *Minerals from Earth and Sky*, Part 1, George P. Merrill, *Smithsonian Scientific Series*, Volume 3, 1934.

the observations of a fireball observed over England. In the early nineteenth century scientists finally accepted the idea that meteorites fell from space, but with scientific skepticism on the rise, they continued to reject reports of the simultaneity of light and sound, blaming observers' subjective associations with fireworks and the sounds accompanying them. Even well into the twentieth century such reports were rejected by scientists as psychological rather than real physical phenomena.

But the reports persisted and finally gained the serious attention of a few scientists. By the early twentieth century, some scientists began to suspect that the source of the sounds may be somehow electrical in nature. Thus, in 1917, J.A. Udden, a University of Texas engineering professor, was one of the first to give a hint of the true nature of the phenomenon:

> If these observations are not subjective, the cause of the sound may perhaps be sought in ether waves [electromagnetic waves] that, on meeting the earth or objects attached to the earth, such as plants or artificial structures, are in part dissipated by being transformed into waves of sound in the air. Whether such a transformation of energy is possible, it is not my purpose here to discuss. The suggestion is merely made for what it is worth.[VIII]

Australian physicist Colin S.L. Keay from the University of Newcastle finally solved the anomalous sounds problem in 1980 when he came up with his "magnetic spaghetti" theory. He showed that the ionized trail of a large fireball interacts with the Earth's magnetic field trapping and distorting the field. As the ions quickly recombine with free electrons the magnetic field relaxes and emits several kilowatts of low frequency radio emissions in the 1–10 kHz range.[8] These radio waves travel from the plasma trail to the ground with light speed. If appropriate structures like tall buildings, power poles, metal structures such as towers and antennae, and perhaps even tall trees are near the observer, they could act as natural transducers converting the electromagnetic transmissions into sound waves. The phenomenon has been given the name, *electrophonic sounds*. Plots of the locations where electrophonic sounds have been reported show no preferred direction and may depend upon the location of appropriate structures to act as transducers.

Because fireballs are random and unpredictable, it is nearly impossible to record the radio emissions or resultant sound transmissions. In an article written for a popular astronomy magazine,[9] Keay suggested methods of recording radio emissions and electrophonic sounds but with little hope of success. Yet, almost miraculously, three Japanese observers accomplished this task, making a chart recording of both radio and sound waves from a −7 magnitude Perseid fireball in August, 1988.

Today, electrophonic sounds are not as obscure as they seem. A great fiery meteor enters Earth's atmosphere every few months. It's predictable and its path is well known prior to its entry. I speak here of the Space Shuttle. Its entry corridor lies over populated areas of the United States. Electrophonic sounds have been reported numerous times during its reentry phase.

[VIII] Udden, J.A. (1917). *University of Texas Bulletin*, No. 1772, 46–47.

Meteoroids to meteorites: a lesson in survival

Up to this point, I have described the physical phenomena associated with the fall of a meteoroid through Earth's atmosphere. It is a sobering thought that all meteoroids entering the atmosphere have sufficient kinetic energy such that if it were all converted to heat and light it would completely vaporize them, none surviving to the surface. Fortunately, only a relatively small amount is converted to heat, the remainder consumed by atmospheric drag, formation of a shock wave, loss of material through ablation, atmospheric heating etc. As far as the meteoriticist is concerned, the pyrotechnics of fall is but a fanfare announcing the arrival of a fresh, new meteorite specimen to study. In a way, it is a rite of passage and survival is the goal. Earth with its atmosphere is about the best catchment basin from which to retrieve meteorites from space, bested only by collecting them on the asteroids or comets themselves. Consider this: without Earth's atmosphere to slow their fall, meteorites would hit Earth at cosmic velocity much as they do on the Moon. This would instantly vaporize them. None would survive impact. On the other hand, if Earth's atmosphere was as dense as Venus' atmosphere no meteoroids smaller than many thousands of tons could make it through the atmosphere without total destruction. Venus lacks impact craters smaller than about 3 km in diameter, suggesting that meteorites smaller than around 300 m in diameter do not survive. Thus, given all the other terrestrial planets, Earth (perhaps Mars also) seems the best suited to receive rocks from space.

Ablation

Ablation is a necessary mechanism that enables a meteoroid to survive to the Earth's surface. Earth's atmosphere is sufficiently dense to slow solid bodies of specific mass and size without totally destroying them. Meteorite collections around the world attest to that. In essence, a meteoroid must ablate to survive.

We have already seen that a meteoroid's cosmic velocity is rapidly reduced by aerodynamic forces. Simultaneously, some of its kinetic energy is expended by conversion to thermal energy, which acts to heat the body as well as the surrounding air. The forward end of the meteoroid heats to incandescence, liquefies and begins to slough off the meteoroid. This ablation process acts to cool the body by carrying away the heated portion thereby inhibiting the flow of heat into its interior. A large percentage of the melted material leaves the body and flows into the air stream behind the fireball. The tiny liquid droplets rapidly cool and solidify, building up a dense trail that may persist for hours after the fireball's passage. Ablation is responsible for enormous mass loss in meteoroids. Even though the heating process represents a relatively small percentage of the total kinetic energy of the body, still the mass loss for most meteoroids is over 90%. It was estimated that the "smoke" trail left behind the Sikhote-Alin meteorite carried about 200 tons of ablated meteoritic material from the original iron mass (Fig. 2.15).

Table 2.1 gives some physical data for Příbram, Lost City, Innisfree, Peekskill, and Sikhote-Alin. Only Lost City retained nearly half of its initial mass. The most massive meteoroids in this group retained only about one percent of their original masses. It appears that the amount of mass does not guarantee an increase in survivability. Of greater importance is

Fig. 2.15. A painting of the Sikhote-Alin fireball by Russian artist P.I. Medvedev who witnessed the event in Iman in eastern Siberia on the morning of February 12, 1947. The great iron meteorite shower produced the most massive trail of ablated material ever witnessed in a meteorite fall, estimated at over 200 tons. (Courtesy Dr Michael Peteav, Smithsonian Astrophysical Observatory.)

Meteorite	Initial mass (kg)	Terminal mass (kg)	% Mass loss	Entry velocity (km/s)
Příbram	21 500	*50 (5.6)	98	20.8
Lost City	50	17	66	14
Innisfree	20	3.8	81	14.5
Peekskill	10 000	**11.8	99	10.5
Sikhote-Alin	~205 000	~22 500	89	14.5

Notes:
*50 kg estimated terminal mass but only 5.6 kg actually found.
**Only one specimen found. Undoubtedly others exist but not recovered.

Table 2.1. **This table compares the mass loss of five meteorites whose orbits were calculated through direct observation of their atmospheric trajectory**

A small fraction of all of these meteoroids survived passage through Earth's atmosphere. From their entry velocity and observed trajectory, their initial masses can be estimated and their percent mass loss can be calculated from their terminal masses. Generally, the higher the initial entry velocity, the greater the ablation and mass loss, regardless of initial mass.

the initial entry velocity of the meteoroid. Higher entry velocities mean higher ablation rates.

In summary, Table 2.2 gives data of 20 fireballs taken from a list of 46 from the Canadian camera network that had a high probability of meteorite falls. Note especially the data in column (M_T/M_I) which is the ratio of terminal mass M_T to the initial mass M_I and compare that to the initial entry velocity V_I.

Most meteoroids explode and fragment. If this occurs near the beginning of its visible trajectory, frequently spectacular multiple fireballs result. The Peekskill fireball fragmented early in its flight and a dozen or more fragments formed a multiple fireball (Fig. 2.16). If fragmentation occurs near the end of its visible trajectory, the fireball will appear to explode into a myriad of sparks and quickly become extinguished since fragmentation substantially reduces the cosmic velocity of the body to a point of losing it entirely. For many fireballs, fragmentation and the retardation point are coincident with the end of the visible trajectory after which it falls due to gravity alone.

The distribution ellipse

Meteoroids from the same parent body falling together constitutes a *multiple fall*. The vast majority of stony meteorite falls are multiple. Once the main mass has fragmented, usually explosively during the visible phase of the flight, the pieces continue to travel more or less together during the remainder of the flight since they shared the same velocity during fragmentation. They don't simply fall randomly but instead their fall is dictated by their remaining kinetic energy. The fragments, however, are of different masses which will

Fig. 2.16. Photo of the Peekskill multiple fireball shortly after it fragmented into many pieces over Peekskill, New York, on the evening of October 9, 1992. Fragmentation was early enough so that each fragment had sufficient mass and velocity to produce a fireball and trail. (Photo by Sara Eichmiller. Courtesy *The Altoona Mirror*.)

V_I	V_E	H_B	H_E	D_T	M_I	M_T	M_T/M_I
18.4	5.7	75.5	27.6	5.4	6000	450	0.075
16.3	8.7	81.0	32.6	5.8	1900	1300	0.68
13.1	11.3	70.2	33.7	3.3	940	370	0.39
12.5	8.1	66.5	31.2	3.1	2200	1000	0.45
25.2	7.6	77.6	30.4	3.7	7100	150	0.02
13.0	8.7	61.9	29.5	4.8	16000	13000	0.81
19.7	7.5	72.5	28.9	2.6	6500	160	0.02
17.9	6.5	77.2	25.9	3.6	4300	920	0.21
18.4	7.8	76.6	26.1	3.3	87000	290	0.003
27.1	9.5	77.7	27.1	6.4	240000	8000	0.03
27.9	11.7	92.7	34.2	6.0	4600	260	0.06
13.5	8.9	74.6	39.0	3.6	370	150	0.41
19.1	15.7	66.3	42.9	2.9	1000	290	0.29
23.5	6.5	81.8	24.4	3.3	5600	800	0.14
14.5	2.7	62.4	19.8	4.1	51000	4900	0.10
12.4	4.1	68.6	20.2	6.3	30000	17000	0.57
23.6	14.7	70.5	35.7	2.6	3700	270	0.07
14.1	8.2	67.3	34.9	4.1	3600	150	0.04
21.0	3.8	78.2	22.0	4.6	5700	2200	0.39
12.8	8.5	67.0	30.6	5.3	4300	3200	0.74

Abbreviations: V_I = initial entry velocity; V_E = end velocity; H_B = height at beginning of visible trail; H_E = height at end of visible trail; D_T = duration of photographic trail; M_I = estimated initial mass entering atmosphere; M_T = estimated terminal mass reaching ground.

Table 2.2. **This table gives data on 20 fireballs whose entry into Earth's atmosphere was photographed with the cameras of the Canadian camera network**

These fireballs had a high probability of dropping meteorites, even though subsequent ground searches have not recovered specimens. A study of the data shows again the importance of initial entry velocity to the survival of the meteorite.

ultimately determine where they will fall. Due to their greater kinetic energy, the more massive fragments must travel further before they arch over into a steep near vertical fall. On average, the more massive fragments tend to fall at an angle of about 30° off the vertical while the least massive acquire a 20° drop with all others in between these two values. When they reach the ground they distribute themselves into a *strewn field* which usually defines an elliptical shaped area called a *dispersion or distribution ellipse*. The long axis (major axis) of the ellipse is coincident with the direction of motion of the swarm. The minor axis denotes the lateral scatter on either side of the major axis. The most massive fragments normally fall at the far end of the ellipse while the remainder are dispersed along and on either side of the major axis at positions reflecting their masses. The Homestead, Iowa, fall of 1875 is a good example of a classical distribution ellipse (Fig. 2.17). If the main mass fragmented at high altitude, it is more than likely that winds aloft will effect the fall direction so that the classical distribution is modified. This was the case with the Bruderheim, Alberta, fall of 1960. The strewn field in the distribution ellipse showed a few large specimens mixed with smaller, a more

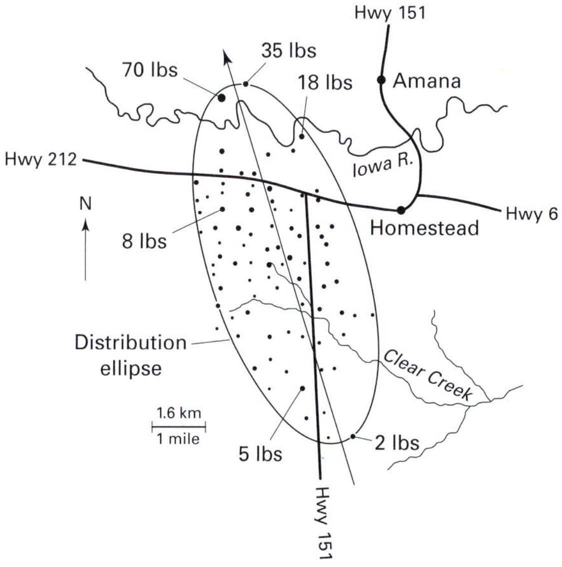

Fig. 2.17. Strewn field of the 1875 fall of the Homestead, Iowa, meteorite shower. This is a classic distribution pattern with the largest masses at the far end of the distribution ellipse. The field measures 10.9 km by 4.5 km. (*Rocks From Space*, 2nd edition (1998), Mountain Press Publishing Co., Missoula, Montana.)

typical distribution. Occasionally, the strewn field does not follow the rules, so to speak. The Johnstown, Colorado, fall of 1924 showed a curious distribution with the larger meteorites at the near end of the ellipse. This might occur if the larger masses fragmented into smaller pieces just before impacting or if the larger pieces shattered upon impact. Often clusters of small meteorites are found in the strewn field, some of which fit together suggesting that they fragmented on impact.

"Fossil" strewn fields

The Earth is sufficiently populated today on most continents so that the fall of meteorites is usually observed by someone and a search often locates the meteorites soon after their fall. But there are still vast areas in Australia, the deserts of Libya and Egypt, vast regions in Saudi Arabia, the Gobi Desert, and even the southwestern deserts of the United States where the population density is minimal and falls could and do occur unobserved. Searches are being conducted in these areas and meteorites are being found almost daily.

Many finds are from prehistoric falls that occurred thousands of years before there was anyone to observe them (Table 2.3). Huge falls of iron meteorites occurred in South America, Australia, and Africa. A recent find in northwest Arizona may represent the oldest stony meteorite strewn field in the world outside Antarctica. The area is within the White Elephant watershed in Mohave County, Arizona, and is known as Gold Basin. For years gold hunters swinging metal detectors kept picking up strong signals as they passed across very ordinary looking brownish rocks scattered on and beneath the surface. They were dismissed as typical "hot rocks" and considered a nuisance by the gold hunters. In November, 1995, one gold hunter, an emeritus professor of Engineering at the University of Arizona, James Kriegh, was less inclined to cast them aside. Suspecting some might be stony meteorites with metal content, he cut them open, found metal and sent samples to the University of Arizona's Lunar and Planetary Laboratory. David Kring and his coworkers identified them as L-type chondrites with a terrestrial age of between 15 000 and 20 000 years. The discovery was kept secret for two years in which time over 2000 meteorites were recovered. The field work continues using many volunteers in an attempt to define the strewn field. After several years of hunting with metal detectors and keeping careful records, a strewn field was sketched, its major axis extending in a north-south line some 13 miles and covering an area of over 50 square miles. The north end passes into the Lake Mead Recreation Area where hunting is severely restricted. The field in this area remains questionable.

There must be many ancient strewn fields scattered across the Earth awaiting discovery. As the Earth's population continues to increase and people become more and more mobile allowing them to conveniently enter remote desert regions, other fossil strewn fields undoubtedly will be discovered.

The temperature myth – hot or cold?

People have many preconceived ideas about the characteristics of freshly fallen meteorites. The number one misconception is they are red hot when they land. This idea is not unreasonable. After all, they race across the sky as intensely hot fireballs and, within seconds, they fall to Earth. There doesn't seem to be time for them to cool off.

> The fall of a stony meteorite at Hessle, Sweden, January 1, 1869 at 12:20 p.m., was accompanied by a sound resembling heavy peals of thunder, followed by a rattling noise as of wagons at a gallop and ending with a sound at first like an

Fall	Found	Location	Number	Type
Brenham	1882	Kiowa Co., Kansas	thousands	Pallasite
Campo del Cielo	1576	Chaco, Argentina	thousands	Iron (IA)
Canyon Diablo	1892	Coconino Co., Arizona	thousands	Iron (IA)
Gibeon	1836	Namibia	thousands	Iron (IVA)
Gold Basin	1996	NW Arizona	thousands	Stone (L4)
Henbury	1931	N. Territory, Australia	thousands	Iron (IIIA)
Imilac	1822	Atacama, Chile	hundreds	Pallasite
Mundrabilla	1911	W. Australia	hundreds	Iron (anomalous)
Plainview	1917	Hale Co., Texas	hundreds	Stone (H5)
Toluca	1776	Mexico	thousands	Iron (IA)
Vaca Muerta	1861	Atacama, Chile	hundreds	Mesosiderite

Table 2.3. **Prehistoric meteorite showers**

This table lists known prehistoric strewn fields in which meteorites continue to be found. In particular, eight of the eleven listed are irons or stony-irons, attesting to their greater resistance to terrestrial weathering.

> organ tone and later like that of hissing. Many small stones fell in this shower. One struck ice close to where a man was fishing and rebounded. He picked it up and found it warm.
>
> At the fall of Knyahinya, Hungary, which took place about 5:00 p.m., January 9, 1866, those nearest the point of fall heard sounds like cannon-shots followed by a noise like the boiling of water and a long roll. At the same time a cloud of smoke appeared in the sky from which stones fell. The observers saw no light, but those at a distance of 10 or 12 miles saw a fireball of the color of white-hot iron with edges of ultramarine blue. Some saw this divide in two. About 1,000 stones, ranging in size from 2 grams to 300 kilograms were precipitated over an area of about 9 miles long by 3 miles wide. The stones picked up immediately after the fall were described as being lukewarm or warm.[IX]

Fig. 2.18. The 11.8 kg main mass of the Peekskill, New York, chondrite. The red color over the black fusion crust is paint from the car it hit and passed through as it landed. (Courtesy Marlin Cilz, Montana Meteorite Laboratory. Photo copyright Colleen K. Smith.)

From the above two accounts, stony meteorites are at most only slightly above ambient temperature when they arrive on Earth. A modern account tells a similar story.

> Eighteen-year-old Michelle Knapp was at home in Peekskill, New York, October 9, 1992, as the fireball was being seen by thousands of high school football fans at games along its path. Suddenly, she heard a loud crash outside, disturbingly close to the house. "It sounded like a three-car crash," Michelle later told reporters. When she stepped outside to investigate, she found the trunk of her car transformed into a twisted mass of metal. A 26-pound stony meteorite had struck the 1980 Chevy Malibu, penetrating the floor of the trunk, barely missing the gas tank, and coming to rest in a shallow impact pit beneath the car [Fig. 2.18]. She reached down and touched the meteorite and noticed it was still warm and smelled of sulfur."[X]

Fortunately, stony meteorites like most terrestrial igneous rocks are poor conductors of heat. Even though the exterior of the meteorite becomes molten and even vaporizes under temperatures of thousands of degrees during the fireball phase, the average duration of incandescence is only 4 s, not sufficient time for heat to be carried to the interior. Moreover, heat is carried away from the body by ablation, a very efficient cooling process. The interior retains the cold of space. Once the meteoroid is past the fireball stage, it begins to cool rapidly. At the average altitude of the end point in the fireball's visible trajectory, the outside air temperature is about −60 °F. This cold air effectively cools the exterior in the minute or two it takes to fall under gravity to Earth. The 150 kg Dhurmsala stone that fell in Himachal Pradesh, India, in 1860 is said to have been found in its impact pit with a layer of frost on its exterior.

Iron meteorites tell a somewhat different story. Iron is a much better conductor of heat and it should penetrate further into the meteorite within the few seconds of heated flight. One would expect that they may retain at least some of their heat upon reaching Earth. The fall of the Mazapil, Mexico, iron meteorite which fell during the Andromaedid meteor shower on November 27, 1886 is especially interesting but is greatly exaggerated.

> . . . When in a few moments we had recovered from our surprise we saw the phosphorescent light disappear, little by little, and when we had brought the light to look for the cause, we found a hole in the ground and in it a ball of fire.
>
> We retired to a distance, fearing it would explode and harm us . . . We returned after a little and found in the hole a hot stone, which we could barely handle, and which on the next day looked like a piece of iron . . .

Particularly interesting is the report of the fall of the Cabin Creek, Arkansas, iron meteorite on March 27, 1886.

> . . . It gave the first indication of its approach to the party who was nearest it, a lady in a house 75 yards away, by a very loud report which caused the dishes in the closet to rattle and was louder than thunder. Running out of the house the lady saw limbs falling from the top of a tall pine tree,

[IX] These two stories were taken from Farrington, O.C. (1915). *Meteorites*, self-published by the author.

[X] Norton, O.R. (1994). *Rocks From Space*, Mountain Press Publishing Co., Missoula, Montana, p. 87.

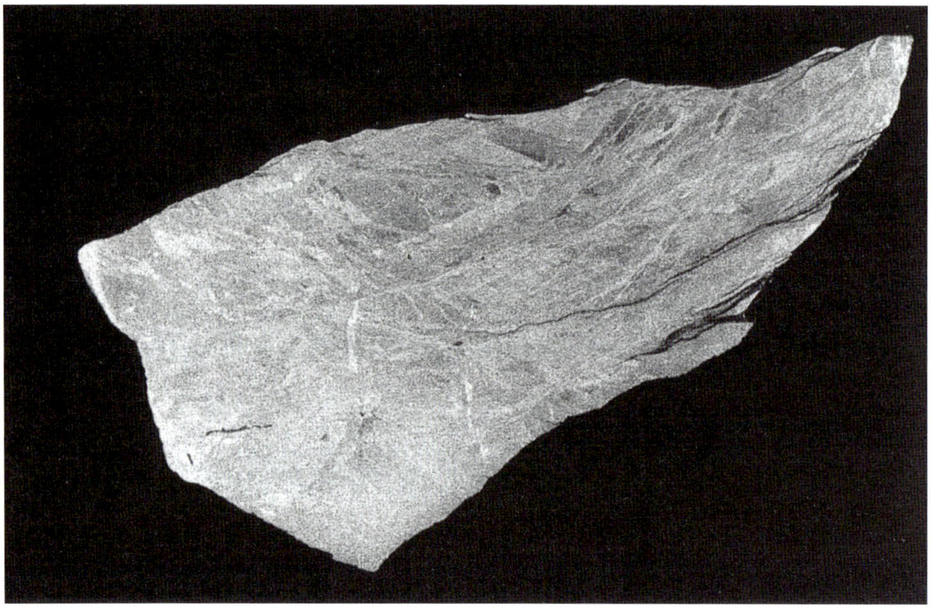

Fig. 2.19. Etched Henbury iron meteorite showing stretched and distorted Widmanstätten structure. The length of the specimen is 132 mm. (Photo by O. Richard Norton.)

> 107 feet high. Three hours later a hole was found near the tree in which an iron meteorite had buried itself to a depth of three feet. The ground was warm and the iron as hot as men could well handle.

It seems from these reports that iron meteorites may be hot enough to be unpleasant to the touch when found soon after they land. That they are glowing red-hot is an obvious exaggeration but in keeping with people's expectations. Under such heated conditions, the meteorite's internal structure would be destroyed. The only iron meteorites found with obvious internal heating have either been involved in a crater-forming process (rim specimens at Meteor Crater) or have been artificially heated by man. A few irons show some heat alteration or even destruction of the crystal structure (Widmanstätten structure – see Chapter 9) around an outer zone that extends a few millimeters into the specimen (see Fig. 3.14).

Mechanical distortion of iron meteorites, especially crater-forming meteorites such as Canyon Diablo and Henbury do often show distortions of the kamacite/taenite bands composing the Widmanstätten structure. Figure 2.19 shows a Henbury iron with the Widmanstätten structure appearing to be stretched and slightly undulating. This might have occurred at the moment of a crater-forming impact when the mechanical stresses on the meteorite were at maximum, or they may be associated with an impact in space. There is often a heat-affected zone around the perimeter of the specimen. Mechanical distortion is frequently encountered in freshly cut Gibeon, Namibia irons. A dramatic example of this is seen in Fig. 2.20 in which the Widmanstätten structure has been bent into a smooth curve and the width of the bands become narrower toward the upper edge. The structure shows some heating effects as the narrowing bands are transformed into a granular zone from the edge to a point about 10 mm inside the specimen. A fracture out of the picture on the left was probably formed simultaneously with and may have been the cause of the distortions.

The final resting place

Meteorites show remarkable variation in the way in which they come to rest on the Earth's surface. Some penetrate the surface to depths of several meters while others have been found lying on the surface with no apparent impact scar whatever. Many variables come into play such as the terminal velocity of the meteorite, its impact angle, and the material's tensional and compressional strength. Stony

Fig. 2.20. Etched Gibeon iron meteorite. The fracture (out of picture on left) probably played a part in the distortion of the Widmanstätten structure. The plates get narrower as they curve and the structure disappears altogether along the edge of the slab, turning into a granular texture. The entire specimen is 223 mm long. (Specimen provided by Edwin Thompson. Photo by O. Richard Norton.)

Fig. 2.21. The 60 ton Hoba iron meteorite in Namibia, South Africa. The specimen's approximate dimensions are 2.743 × 2.743 × 0.975 m or 9 × 9 × 3.2 feet. (Photographed by Dr David Mouat, Desert Research Institute, University of Nevada System.)

meteorites vary considerably in density and friability. Often a meteorite is found in a compacted but shattered condition within a shallow *impact pit* with pieces scattered around the main mass. If the meteorite impacts soft terrain such as a plowed field, it may penetrate to a depth of several meters. This *penetration hole* is usually close to the diameter of the body itself and tilted at an angle from the vertical equal to the fall angle just before impact.

The Norton County, Kansas, fall of February 18, 1948, of a 2360 lb stony meteorite produced a penetration hole 10 feet deep. The meteorite was a fragile achondrite that miraculously held together on impact although several smaller pieces up to 130 lb were found within its distribution ellipse.

Fig. 2.22. The 15.5 ton Willamette, Oregon, iron meteorite. Photograph made by an unknown photographer in 1904. Dimensions are 4.5 feet high, 10.25 feet largest base dimension and 6.5 feet shortest base dimension.

A field party from the Institute of Meteoritics, University of New Mexico, wrapped the huge meteorite in a protective plaster jacket much like paleontologists wrap fragile fossil bones in the field and successfully extricated the mass, which remains the largest intact achondrite ever found. The world's largest chondritic stony meteorite fell March 8, 1976, near Jilin, China, depositing a total of 4 tons of fragments. The largest, weighing 1.9 tons, was excavated from a pit 18 feet deep.

Both Norton County and Jilin were observed falls. One wonders how many large meteorites unobserved to fall or have fallen in prehistoric times around the world remain buried and unknown. The world's largest meteorites are irons. They present something of an enigma. None were seen to fall yet none have been found buried. The world's largest, the 60 ton Hoba iron located on a farm 20 km west of Grootfontain, Namibia, still remains where it was found in 1920. It rests on limestone rock in a shallow depression about 5 feet deep. The depression may have been deeper when it first impacted in prehistoric times but the limestone has weathered away exposing the great iron (Fig. 2.21). The 34 ton Cape York iron was found on the Greenland ice shelf half buried in snow. The bell-shaped 15.5 ton Willamette, Oregon, iron was found in 1902 resting on its side in a conifer forest, sunken only a foot or so into the soft sod (Fig. 2.22). The 1170 kg Goose Lake iron was found on a lava flow with no evidence of an impact pit. Clearly these massive meteorites may not have retained their original positions. They may have been moved by glacial action or flood waters.

References

1. Ceplecha, Z. (1998). The European Fireball Network: current status and future prospects. *Meteoritics and Planetary Science* **33**, 49–56.
2. McCrosky, R.E. (1970). The Lost City meteorite fall. *Sky and Telescope* March, 154–158.
3. Halliday, I., Griffin, A. and Blackwell, A. (1978). The Innisfree meteorite and the Canadian Camera Network. *Journal of the Royal Astronomical Society* **72**, 15–39.
4. Sears, D.W. (1978). *The Nature and Origin of Meteorites*, Adam Hilger Ltd, Bristol.
5. Buchwald, V.B. (1975). *Iron Meteorites*, University of California Press, Berkeley.
6. Halliday I., Griffin A., and Blackwell, A. (1989). The typical meteorite event, based on photographic records of 44 fireballs. *Meteoritics* **24**, 65–72.
7. Halliday, I., Griffin, A. and Blackwell, A. (1996). Detailed data for 259 fireballs from the Canadian camera network and inferences concerning the influx of large meteoroids. *Meteoritics and Planetary Science* **31**, 185–217.
8. Keay, C.S.L. (1992). Electronic sounds from large meteor fireballs. *Meteoritics* **27,** 144–148.
9. Keay, C.S.L. (1985). In quest of meteor sounds. *Sky and Telescope* **70**, 623–625.

Chapter Three

External morphology of meteorites

Ask anyone you know to describe the most general characteristics they expect to see in a meteorite and they invariably give you three: heavy, black, and full of "holes". Although these characteristics are quite true for some meteorites (actual holes are rare in meteorites but many iron meteorites are deeply pitted), they are also true for Earth basalt. Others might say that meteorites are primarily made of iron and are attracted to a magnet. But then so is terrestrial magnetite (frequently mistaken for a meteorite), and, in fact, most stony meteorites. How then can you recognize a meteorite? The recognition and recovery of meteorites in the field is fundamental to everything else that follows. We must first be able to identify a fallen meteorite before we can probe its interior for the secrets it holds. H.H. Nininger, the great meteorite hunter of the twentieth century, in his public lectures and in his writings often lamented the surprising fact that meteorites were not part of the study of rocks in geology classes, either elementary or advanced. Yet, ironically, meteorites, more than terrestrial rocks, tell us much about Earth's origins. They are as old as the Solar System and considerably older than the crustal rocks of the Earth. Except for specialized graduate courses in large universities with planetary science curricula, the situation has changed very little. This chapter (and those to follow), then, is as much for budding geologists (and their teachers) as it is for the meteorite collector.

Some general characteristics

Put in the simplest of terms, meteorites are rocks. Like terrestrial rocks the majority of meteorites are made of aggregates of crystalline minerals. For the most part the minerals of meteorites are common to terrestrial rocks but the way in which they are displayed in meteorites is unique to meteorites. The internal texture of meteorites is an important diagnostic characteristic discussed in detail in the chapters to follow. This chapter deals with the exterior characteristics of meteorites as we find them in the field. No one has ever seen a meteorite as a *meteoroid* in its natural space environment so we do not know what a truly pristine meteoroid looks like. We found in Chapter 2 that meteoroids are modified by heating, mechanical stresses and ablation. What we find on Earth has therefore been terrestrialized to some degree. Fortunately, most of the modifying effects occur on the exterior of the meteorite. The interior remains relatively pristine much as it was in its former space environment provided it doesn't remain undiscovered for too long.

In the most general terms, meteorites are classified into one of three broad categories: *stones*; *irons*; and *stony-irons*. The stones are rocks made up of common silicate minerals. These minerals are usually in a crystalline form small enough to require a magnifier or microscope to see them. Their arrangements within meteorites are quite different than inside most terrestrial rocks which suggests radically different origins for them. The irons are, as the name implies, made primarily of iron alloyed with nickel. Most iron meteorites are at least 90% iron with nickel, cobalt, sulfur and other elements making up the remaining 10%. The stony-irons are a mixture of both silicate minerals and iron–nickel alloy. The ratio of stone to iron can vary considerably from meteorite to meteorite or even within a single meteorite but is roughly two parts stone to one part iron by volume.

Falls and finds

Meteorites are designated as either *falls* or *finds*. The distinction between them is important, statistically. Meteorites classified as falls are those observed to fall and then picked up shortly thereafter. Finds are meteorites that were not seen to fall, remaining undiscovered for a period of years and then subsequently recognized and recovered. If we divide all known meteorites into these two groups, an important conclusion is reached. Of the observed falls, 94% are stones, 5% are irons and 1% are stony-irons. Therefore, we correctly conclude that the vast majority of meteorites falling to Earth are stones. Yet, if we look at the total inventory of meteorites worldwide which includes *both* falls and finds, we find only 69% are stones, 28% are irons, and 3%, stony-irons. Why the contradiction? Simply because stony meteorites, especially terrestrially weathered specimens, are more difficult to distinguish from ordinary terrestrial stones than either irons or stony-irons; the numbers reflect the ease with which iron meteorites are recognized and recovered by the general population. This is strongly substantiated if we look at the totals among the meteorites classified as finds alone. Here, stones drop to 56%, irons increase to 40% and stony-irons increase to 4%. I must add here that scientists seldom search random areas for meteorites (Antarctic and Sahara Desert meteorites excepted) and fewer yet see them actually fall. Most finds are made by ordinary people who work and play out-of-doors: farmers plowing their fields; highway construction crews; and an army of weekend rock and gold hunters scattered across the countryside.

Density

Most meteorites have densities greater than the average terrestrial rock. This means that for a given size, meteorites are heavier than most terrestrial rocks. Most stony meteorites are tightly compacted with little pore space. The most common stony meteorites, the *ordinary chondrites*, have a density range of between 3.5 and 3.8 g/cm^3. By comparison, granite, an igneous rock that makes up much of the Earth's crust, has a density of 2.7 g/cm^3. Meteorite high densities are due primarily to the presence of iron. Some of this iron is locked in iron-bearing silicates but some contain as much as 20% in the metal phase. There is sufficient elemental iron in most of the ordinary chondrites that they are attracted to a strong magnet and metal detectors will detect them (as relic or trash indications!). Rare type chondrites, the so-called *EH chondrites*, can have as much as 35% iron metal. Of the chondritic meteorites the *carbonaceous chondrites* show the greatest variation in density. Of the seven known groups, three contain abundant chemically bound water with little or no metal. The other carbonaceous chondrite groups contain variable amounts of unaltered metal. Thus, the group as a whole vary in density between 2.2 and 2.9 g/cm^3.

The irons, not surprisingly, are the densest of the meteorites. They contain variable amounts of iron alloyed to nickel with accessory metal sulfides, carbides, and phosphides. They have densities twice that of the stones ranging between 7.5 and 7.9 g/cm^3. Their composition resembles no rock type on Earth but they are probably similar to Earth's core material. (Earth's core density is considerably greater than iron meteorites since the core is under a high state of compression. Uncompressed, Earth core material is probably very similar in density to iron meteorites.) There is no mistaking an iron meteorite. Its weight alone attracts attention. The stony-irons, being roughly two-thirds silicate minerals to one-third iron–nickel alloy by volume have densities that lie somewhere between the stones and irons.

Size

Meteorites show an enormous variation in size from micron-sized dust particles filtering slowly through the atmosphere to giants weighing many tons. The average meteorite found on Earth is about the size of a baseball. The size of a meteorite depends upon its impact history in space as well as its fragmentation history in the atmosphere and its further fragmentation when it impacts Earth. The vast majority of meteoroids fragment as they ablate in the atmosphere. It is no surprise that the largest meteorites in the world are irons since they have greater tensional and compressional strength than stones and are better able to withstand the aerodynamic pressures encountered during atmospheric passage as well as the forces at impact. A large iron meteorite has the best chances of remaining in one piece (assuming no pre-entry fractures exist) if it enters the atmosphere at large zenith angles with a velocity of less than 12.3 km/s. A meteorite with a diameter of 5 m entering at a 60° zenith angle and a speed of 10 km/s may impact without fracturing, but impact speeds of more than 2–4 km/s will produce a crater and vaporize the meteorite in the process.[1]

There is therefore a limit to the maximum size of an iron meteorite reaching Earth that depends upon the heat generated at impact. This is determined by the initial mass of the meteoroid before it enters the atmosphere, its initial velocity and its remnant cosmic velocity when it hits the ground. We saw in Chapter 2 that meteorites over about 10 tons retain some of their cosmic velocity right to the Earth's surface. For example, if a 1000 metric ton iron meteorite hits the top of the atmosphere at 20 km/s at a zenith angle of 45°, it retains about 63% of its cosmic velocity and about 90% of its initial mass when it strikes the Earth. Given that it doesn't break up in the atmosphere it retains a residual cosmic velocity of about 12.6 km/s with 900 metric tons of its mass still intact when it impacts. The meteoroid possesses enormous kinetic energy at impact, most of which is converted almost instantly to heat. An iron meteoroid of this mass and speed will generate about 8×10^{13} J of energy or almost 19 000 cal/g of heat on impact. Now, 19 000 cal/g is an enormous amount of heat especially when we compare it to a common value such as 540 cal/g necessary to turn water to a vapor at 100 °C. V.F. Buchwald points out that the element carbon has one of the highest heats of vaporization of any known material, 13 000 cal/g. In the example above, even if the meteorite were pure carbon, it would vaporize on impact. This heat of impact is more than sufficient to vaporize the meteorite on impact even though some of the energy is consumed in producing a shock wave in the ground and a crater a few hundred feet across. Many such craters are known on Earth. Meteor Crater, Arizona, is the classic example and the best preserved impact crater on Earth. It was produced by an iron body estimated to be about 30 m in diameter. Nothing remains of this body today other than fragments off the main mass (Fig. 3.1). Since meteorites heavier than 100 tons retain substantial amounts of their original cosmic velocity and

Fig. 3.1. A 240 g fragment of a Canyon Diablo iron meteorite from Meteor Crater. The hole results from terrestrial weathering within thin-walled ablation cavities. Specimen dimensions are approximately 75 × 67 mm. (Photo by Michael Casper.)

Fig. 3.2. Angularity of a 920 g Gao-Guenie H5 chondrite. Note the flat surfaces nearly at right angles to each other. These are surfaces produced by fragmentation in the atmosphere along preexisting fracture lines acquired while in space. The sharp edges have been rounded by ablation. Specimen dimensions are 113 × 94 mm. (Photo by Michael Casper.)

mass, they tend to self destruct on impact. If the meteorite enters at a zenith angle of 60° to near grazing incidence, its speed might be slowed sufficiently to reach Earth without destroying itself or fragmenting. The fact we see no meteorites larger than the Hoba iron (60 tons) may simply mean that very few large meteorites in the 100+ ton category satisfy these criteria.

Meteorite shapes

Meteorite shapes are dictated primarily by ablative forces and fragmentation. The original shapes are never retained and remain unknown. Most stony meteorites have random shapes since they almost always explosively disrupt and fragment high in the atmosphere. They break along deep, possibly preexisting fractures; and, like crystalline rocks on Earth, when stony meteorites break, sharp edges result. If fragmentation occurs high in the atmosphere at or near the beginning of the ablation process, the sharp edges are rounded by ablation and the rough broken faces often retain a remarkably flat surface smoothed over by the fusion process so that many stony meteorites have obvious angularity (Fig. 3.2). The overall surfaces of stony meteorites are smooth and slightly undulating, suggesting uneven ablation processes. Sometimes stones develop flattened depressions with raised edges. Superimposed over these depressions may be small pits where turbulent flow of superheated gas has ablated specific areas (Fig. 3.3). They range in shape from nearly circular to highly elliptical. Hand specimens frequently show pits the size and shape of a human thumb print. These "thumb prints" are called *regmaglypts* or sometimes *piezoglypts*. Their shape and size relate to their location on the meteorite and to areas of differing ablation.

Iron meteorites have irregular shapes. If they are associated with impact craters, they tend to be flattened and fragmented like shrapnel from an exploded bomb. Meteorites from the Sikhote-Alin fall that were produced by large individuals exploding on impact are typically jagged and distorted. Meteorites from the same fall that did not shatter on impact or were not involved in explosions are more rounded and show remarkable deep pitting and flow structures. Some specimens show characteristics of both types. Figure 3.4 illustrates the two opposing sides of a Sikhote-Alin iron meteorite. The left side shows shallow smoothed-over

Fig. 3.3. This 800 g Holbrook L6 chondrite shows a slightly concave face (right) containing shallow cavities or regmaglypts produced during the ablative process. The face measures 300 mm in the longest dimension. The largest Holbrook specimen recovered weighs 6.6 kg and resides in the collection of the Center for Meteorite Studies at Arizona State University. (Photo by Michael Casper.)

Fig. 3.4. This is an iron meteorite from the Sikhote-Alin fall showing opposing sides. The left side shows a heavily ablated and melted face where the ablation pits have been smoothed over. The right side shows deep pitting on a fragmented side that suffered less ablation and melting. The specimen is 12 cm in length. (Photos by O. Richard Norton.)

pitting, many with elliptical shapes where melting has taken place after the pitting formed. The right side shows deep, subcircular pitting on an obvious fragmented side. This meteorite was in the early stages of ablation when fragmentation took place off the main mass. The raw fragmented side apparently turned, orienting itself toward the forward-proceeding side, producing deep pits and all but erasing the fragmented nature of the meteorite. What was once the side of a large mass became oriented away from the forward end. Material accumulated on this now rear-facing side, flattened and broadened the pits.

Unlike stony meteorites that are composed of minerals each with individual crystal structures, irons are constructed around a single cubic crystal form, either octahedral or hexahedral and they tend to break along natural crystal boundaries. Figure 3.5 shows two sides of a Sikhote-Alin specimen. On the left is a surface with many ablation pits. Near the end of its ablation period, the meteorite suffered fragmentation off the main mass. The right picture shows the fragmented side and a careful examination will reveal triangular-shaped structures that are expressions of the octahedral dipyramids around which the nickel–iron alloy is structured.

Pits on iron meteorites are much deeper than on stones, suggesting differential melting of softer minerals. They tend to be edged by sharp ridges that in some cases terminate in sharp points, a quality much valued by the collector. Commonly, nearly spherical inclusions or nodules of iron sulfide (troilite) are found in the interiors of many iron meteorites that have a substantially lower melting point than the surrounding iron–nickel alloy. If these inclusions are near the surface, they can rapidly ablate away leaving round cavities. A most extraordinary case of pitting is found in the Goose Lake iron. Within this meteorite some of the pits have larger diameters than their openings. Just how ablation could have formed them is something of a mystery. Many believe these pits were produced by terrestrial weathering of less resistant materials (Fig. 3.6).

Fig. 3.5. These two photos show front and back sides of a Sikhote-Alin iron meteorite. The left image shows an ablated smooth surface with extensive pitting. The right image shows the opposing side where fragmentation took place near the end of the ablation period revealing parallel-running linear features typical of octahedrite meteorites. The specimen measures 11.8 × 9.5 cm. (Specimen provided by Dr David Mouat, Desert Research Institute, University of Nevada System. Photos by O. Richard Norton.)

Fig. 3.6. View of a surface section of the 1169.5 kg Goose Lake iron meteorite showing deep, rounded cavities. The largest cavity shown is over 50 mm in diameter. (Photo by O. Richard Norton.)

Oriented meteorites

During their atmospheric flight, most meteoroids tumble randomly. This tends to smooth out irregular meteorites into roughly elliptical or subspherical shapes. Many mimic the appearance of rounded riverbed stones. About 5% of all stony meteorites exhibit a symmetry that suggests they have somehow oriented themselves as they ablated. Such meteorites are referred to as *oriented*, usually having the appearance of a steep-sided cone or flattened shield. These shapes strongly suggest that the meteoroid has maintained stabilized flight through most of its visible trajectory. Meteoroids tend to orient themselves in a configuration that produces maximum drag forces. This means that a meteoroid with a flattened shape or a large flat area on an otherwise random shape will orient the flat side perpendicular to the direction of maximum drag (Fig. 3.7). This stabilizes the meteoroid and prevents tumbling. To produce the oriented shape requires that the stone remain stationary as it ablates or that it rotate about an axis perpendicular to the flat face and along its direction of motion. Once the meteoroid has attained its orientation, the front end (end facing the direction of travel) begins to experience the greatest ablation, being sculpted by the air currents flowing at hypersonic velocity around the front end. Here the material melts rapidly and flows toward the rear of the meteoroid where it accumulates, building a broad apron. At the same time, the front end or *brustseite*[1] becomes narrow, forming a symmetrical cone or in some cases, a flattened shield shape.

Orientation is more common in iron meteorites if they don't disrupt in flight. About 28% show some signs of orientation and many are nearly perfect cones or shields. The great Willamette meteorite is remarkably bell-shaped (Fig. 2.22). The stony meteorite from Adamana, Arizona, is nearly a

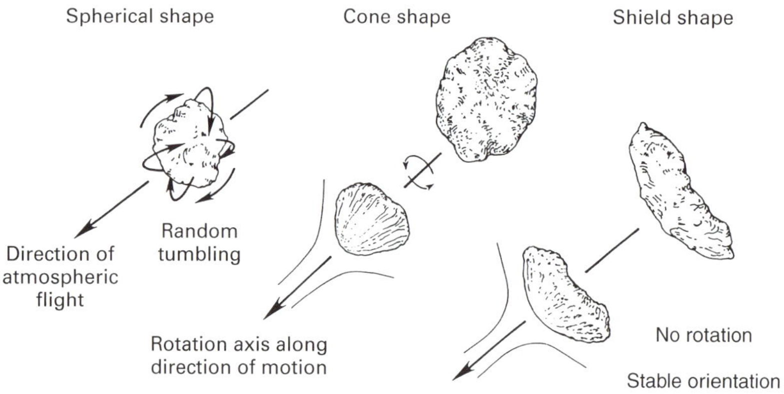

Fig. 3.7. A meteorite that stabilizes during atmospheric flight may become oriented with respect to its direction of motion. Here the flat side experiences maximum drag, causing it to orient so that it spins around its direction of travel with maximum ablation on the forward-facing side. Ablation soon forms a cone or shield shape.

[1] The word *brustseite* is German literally translated as *breast side*, or more appropriately, *front side*. It is a seldom used term in meteoritics today but still useful when referring to oriented meteorites which in many cases mimic the shape of the human female breast.

Fig. 3.8. Nearly perfect steep cone-shaped stony meteorite from Adamana, Arizona. Note the elongated ablation pits positioned radially from the smooth vertex or *brustseite* of the cone. Specimen measures about 9 cm high with a 12 cm diameter base. (Robert A. Haag collection. Photo by O. Richard Norton.)

perfect steep-sided cone (Fig. 3.8) and the Cabin Creek, Arkansas, iron (Fig. 3.9) is a perfect shield shape with elongated radial pits.[II] Usually the brustseite is smooth and free of pitting. Pitting begins to appear beneath the nose and extends in a radial fashion toward the rear. This radial pitting can be quite spectacular with the pits becoming progressively elongated until they are grooves lining the sides of the meteorite. The two meteorites above are exceptional examples of this radial texture. From these illustrations it is evident that pitting is a characteristic of stable flight.

Along with pits and running parallel to them are flow lines of once melted material flowing in rivulets toward the rear of the cone or shield. Many stony meteorites from the Gao-Guenie fall (1960), Upper Volta, Africa, show very fine flow structures best seen under low incident light (Fig. 3.10, left). These often terminate in a curved lip along the edge of the shield (Fig. 3.10, right). One of the most perfectly oriented meteorites known was discovered in a rock and mineral collection at Purdue University in 1931. This meteorite, found in Lafayette, Indiana, has a perfect shield shape accentuated by radial flow lines extending from the brustseite to the rear of the specimen. According to H.H. Nininger, this superb specimen was thought to be an ordinary rock marked by glacial scratches. The meteorite itself, even without its perfect orientation, is a priceless scientific specimen. It is a very rare achondritic meteorite called a nakhlite, a meteorite from Mars!

Fusion crust

All iron-bearing meteorites develop a dark coating over their surfaces when they pass rapidly through Earth's atmosphere during the fireball stage. More than its weight and density, the external coating of the meteorite is the most important

Fig. 3.9. One of the most remarkable shield-shaped oriented meteorites is the Cabin Creek, Arkansas, iron located at the Natural History Museum in Vienna. A forward-facing (left) and side-facing view (right) shows the shield shape to best advantage. The meteorite weighs 47.4 kg and stands 44 cm high. (Photos by O. Richard Norton.)

[III] Oriented meteorites are similar to the shapes of early manned spacecraft. This is no coincidence. As early as the 1950s, H.H. Nininger suggested a cone for the shape of nose cones of rockets and reentry vehicles. Later, rocket engineers undoubtedly studied the shapes of meteorites that had survived atmospheric passage and saw the advantage of the cone as an ideal shape for reentry vehicles.

Fig. 3.10. Flow lines of once molten material on the surface of a 156 g oriented Gao-Guenie stony meteorite (left). Flow lines terminate in a lip that curves around to the rear side of the meteorite (right). The specimen is shield-shaped, 6.3 cm in diameter. (Photos by O. Richard Norton.)

identifying criterion in the field – *if the meteorite is fresh and unweathered*. This coating is called a *fusion crust*.

Both stones and irons develop fusion crusts but they are distinctly different. Let's look at stony meteorite crusts first. A fusion crust results when the surface of a stony meteoroid melts under the 1800 °C fusion temperature within the fireball.[III] The meteoroid constantly sheds molten rocky material through the visible trajectory phase when temperatures are highest. As long as this shedding continues, a fusion crust cannot form. Only during the last second or so of the fireball stage when ablation is reduced and cooling begins does a fusion crust form. Its formation can be thought of as a healing of the molten surface. Stony meteorites are composed of iron and magnesium-bearing silicate minerals; that is, minerals containing silicon and oxygen forming a tetrahedron-shaped atomic structure that provides for attachments of iron, magnesium and other metals. Under melting temperatures, these minerals lose their normal crystal structures and flow freely much like a magma does during volcanic eruptions. The elements that compose these minerals are still within the melt but the high temperatures do not allow the minerals to recrystallize. Also distributed more or less uniformly within the body of the meteoroid are grains of elemental iron. This iron also melts and mixes with the other elements. During the dark phase of atmospheric flight, the meteoroid cools rapidly, too rapidly to recrystallize the melted minerals on the exterior. Instead, a structureless glass forms with the composition of the original minerals. The magnesium–iron silicates, olivine and bronzite (an orthopyroxene), produce a light brown glass. At the same time the elemental iron rapidly oxidizes in the heated oxygen-rich environment forming the black iron oxide, magnetite. Magnetite (and iron sulfide) mixes with the glass to form a crust that ranges from dark brown to black.

The color, texture and luster of the fusion crust can vary greatly, even among meteorites from the same parent body. Elemental iron is not always distributed uniformly in a chondritic meteorite, so it is no surprise that fusion crusts within a given fall can vary from black to dark brown. Not all stony meteorites contain free iron. Certain meteorites called iron-poor achondrites (enstatite achondrites or aubrites) contain so little total iron (metal plus iron in silicates) that little magnetite forms. The crust instead takes on a light beige to almost clear tone, the color of the glass formed by melted iron-poor minerals. Some achondrites (igneous meteorites without chondrules) have an abundance of minerals with calcium-rich compositions. These include calcium-rich plagioclase, its glassy form, maskelynite, and the calcium-rich pyroxene, augite. These are found in eucrites (basaltic achondrites) and shergottites and nakhlites (both Mars meteorites). The calcium-bearing minerals mixed with magnetite produce a crust with an unmistakable glossy black luster. The vast majority of stony meteorites are composed of densely packed refractory minerals with high melting points. These typically produce very dull, porous black fusion crusts. So the color, texture and luster of the fusion crust reveals much about the nature of the meteorite wearing it. Figure 3.11a–d shows a variety of fusion crusts.

Microscopically, the crust usually encloses small silicate crystals of incompletely fused minerals along with iron particles and iron sulfide. Often thicker portions of crust have a frothy texture filled with bubbles. Three zones are sometimes recognized under the microscope and especially in thin sections (Fig. 3.12). The outermost zone is very thin, usually not over 0.1 mm thick. It is composed of nearly pure glass that can be opaque, brown or clear with few or no inclusions. Beneath is a much thicker zone in which the glass has been forced between the crystalline grains of the matrix. Surprisingly little if any changes due to heating are observed in the grains.

[III] Laboratory studies show that typical chondritic material begins melting at approximately 1300 °C.

Fig. 3.11 (a) Black satin-luster fusion crust typical of ordinary chondrites. Kitchener L6 chondrite. (Photo by Dr Richard Herd, Curator, National Collections, Geological Survey of Canada.) (b) Dull, black and porous fusion crust typical of carbonaceous chondrites. Murchison CM2. Specimen approximately 11.4 cm largest dimension. (Institute of Meteoritics, University of New Mexico. Photo by O. Richard Norton.) (c) Shiny, black fusion crust typical of calcium-rich eucrites and Mars meteorites. This specimen is a 29.4 g Millbillillie eucrite from West Australia. Horizontal dimension is approximately 35 mm. (Institute of Meteoritics, University of New Mexico. Photo by O. Richard Norton.) (d) Light cream-colored fusion crust of iron-poor achondrites. Norton County aubrite. Specimen weighs 894 g and is 102 mm largest dimension across the fusion crust area. (Robert A. Haag collection. Photo by O. Richard Norton.)

Crusts of stony meteorites vary considerably in thickness. On average they are less than 1 mm thick. On oriented specimens, crusts can vary from about 0.1 mm on the front side to several millimeters toward the rear where fused materials have accumulated forming an apron. Typically, a fusion crust covers the entire meteoroid, even in the depressions and pits. This is called the *primary crust*. When the meteoroid fragments, fresh interior is exposed to the high temperatures and a *secondary crust* begins to develop that is considerably thinner than the primary crust. The secondary crust usually barely coats the fragmented faces before the ablation process ceases. The fresh interior is not completely fused at this stage and, unlike the primary crust that overlies a relative smooth ablated surface, the secondary crust lies over bumps and pits accentuating the interior texture (Fig. 3.13). By noting the number and maturity of crusts on a fragmented meteorite, something of the history of its passage through the atmosphere can be surmised. Incidentally, the interior of many stony meteorites are usually light gray in color. This is in stark contrast to the black crust and provides a useful identification tool. No terrestrial stone is known to develop such a crust with a melting history although there are terrestrial stones that take on a dark patina of manganese as a weathering product that can superficially mimic a fusion crust.

Wouldn't it be instructive to take a terrestrial rock of known composition outside the atmosphere and then launch it toward Earth to see what an Earth "meteorite" would look like? Just such an experiment, the first of its kind, called "Flying Stones" was attempted in 1999. On September 9, a Russian Soyuz-U rocket launched a Foton-12 spacecraft into a low Earth orbit. Scientists from the United Kingdom installed three types of Earth rock on the exterior heat shield of spacecraft. These included an igneous rock, basalt; a sedimentary rock, dolomite (limestone); and a simulated clod of soil composed of 80% crushed basalt and 20% gypsum cemented together by carbonates and sulfates. The purpose of the experiment was to gain information on how terrestrial rocks were modified during the heat of entry through Earth's atmosphere. The rocks were selected to be as close as possible

Fig. 3.12 (a) Cross section of a fresh fusion crust on a Leedey, Oklahoma, chondrite. The black area is a glass and magnetite mix. The lower edge is irregular where heating and melting has penetrated into the meteorite. Note that the refractory minerals in the crust have not melted. The crust averages about 0.35 mm thick. (b) Cross section of the fusion crust of the Lake Labyrinth chondrite. The minerals trapped within the crust have not melted but show alteration from weathering (reddish grains). The thickest section is 0.45 mm. (Photo by Tom Toffoli.)

to martian rocks and soils. The goal was to study the characteristics of ablated rocks in an effort to improve identification and recovery of different Mars rocks (different than those already found) that must exist on Earth at favored meteorite collecting sites in Antarctica and the Sahara Desert. Planetary scientists believe that there must be large areas on Mars with sedimentary rock so there must be samples of Mars sedimentary rocks on Earth. To date, however, all of the Mars meteorites recovered have been igneous rocks. So, where are the Mars sedimentary rocks? Perhaps we do not recognize sedimentary meteorites or maybe they are too fragile to survive atmospheric passage.

The spacecraft successfully landed on September 24, 1999, 133 km northwest of Orenburg, Kazakhstan. The stones were retrieved and are now being studied by Colin Pillinger and his coworkers at the Planetary Sciences Research Institute, The Open University, in the UK. Unfortunately, the fixture holding the basalt, the control sample, became dislodged during reentry and was lost. The surviving sedimentary rocks did not produce a typical black fusion crust. Besides lacking iron, sedimentary rocks release carbon dioxide and sulfur dioxide that explosively erupt onto the rock surface. Since about 40% of the dolomite survived, the researchers believe that Mars sedimentary rocks could survive atmospheric passage and land safely on Earth. Without a typical fusion crust, they would be far more difficult to identify, however.

Iron meteorites also develop a fusion crust but because the

Fig. 3.13. A secondary fusion crust formed over a fracture in this meteorite. The primary crust (left) is thicker and coarser. The secondary crust (right), overlying the undulating fractured surface, appears thinner and is partially removed by weathering. The meteorite is Sahara 97001. The fractured face is approximately 100 mm in its longest dimension. (Photo by Michael Casper.)

irons contain little if any silicate inclusions, glass is not produced. The crust is composed of a heat-affected zone up to 2 mm thick (Fig. 3.14). The upper part of this zone is totally oxidized iron, namely magnetite and wüstite (FeO). Below that is a complex mixture of minerals of different melting points that respond differently to heating. Kamacite has the highest melting point of all the iron minerals (~1500 °C) and is found throughout the heat-affected zone. It is usually converted to the α_2 – iron nickel phase, martensite, appearing uniformly granular where the Widmanstätten structure is subdued or absent. High-nickel taenite melts a few tens of degrees under the melting point of kamacite depending upon the nickel content. Troilite (iron sulfide) with a melting point of about 1190 °C is usually found as subspherical inclusions and often burns out during the ablation process leaving behind small cavities on which remnant recrystallized troilite is found coated with magnetite–wüstite. Schreibersite (iron phosphide) is most affected at a melting temperature of ~1000 °C and is usually completely melted and removed from the affected zone or sometimes recrystallized.

A fresh magnetite surface looks very similar to the bluish surface of a freshly welded piece of iron. This outside veneer is very thin, usually not much thicker than 50 μm. And it is very fragile. A few months of exposure in a humid environment is all that is necessary to destroy the original patina. Years of chemical weathering can completely destroy the 2 mm heat-affected zone. What one usually finds is an iron meteorite coated with a scaly brown rusty material (limonite) which has completely replaced the fusion crust. If the rust is removed, a black equally scaly coating is revealed. This is not the original crust but a replacement composed of magnetite. If the meteorite is cut, the heat-affected zone is no longer present. Recently, large numbers of beautifully shaped Sikhote-Alin iron meteorites have been unearthed from the strewn field in eastern Siberia and have been sold to collectors in the world's meteorite markets. These have been beautifully cleaned, revealing a satin smooth black coating looking like gun bluing. This is probably not the original magnetite–wüstite layer but only a carefully smoothed terrestrially deposited magnetite coating. Amazingly, the delicate thread-like flow marks of once molten metal are still evident on these specimens, attesting to the resistance these meteorites have to terrestrial weathering and, equally important, the care that was taken during the cleaning process.

Fig. 3.14. This 290 g slab from a IIIAB iron meteorite observed to fall in Kayakent, Turkey, in 1961 shows the effects of heat alteration. It has been etched to reveal the internal structure typical of many iron meteorites. Note the destruction of the structure in a remarkably uniform band completely surrounding the meteorite. The medium octahedrite structure has been erased to a depth of about 2 mm inside the edge. A centimeter rule gives the scale. (Photographed by Dr Richard Herd, Curator, National Collections, Geological Survey of Canada.)

Weathering of meteorites

Meteorites are alien to Earth. They survived billions of years in the near vacuum of space far removed from the damaging effects of water and oxygen. Under normal terrestrial conditions they cannot survive long without measures taken to preserve them. Even under the protective environment of the museum and laboratory, meteorites continue their slow terrestrialization. Meteorites are subject to the same mechanical and chemical weathering that all terrestrial rocks are. Unfortunately, they are more fragile than most terrestrial rocks because they contain minerals that rapidly undergo destructive changes as weathering proceeds.

Mechanical weathering

Without question, the most destructive moment in a meteorite's brief existence on Earth is when it fragments. Fragmentation is a mechanical process that begins the weathering process in meteorites. We saw earlier that most meteorites suffer one or more episodes of fragmentation while still in the atmosphere. They can and often do further fragment when they impact, traveling at several hundred kilometers per hour. This immediately exposes fresh interior to the weathering elements that begin to work to break down its minerals. The glassy fusion crust provides some level of temporary protection on those sides that have not fragmented on impact. Even if the stony meteorite makes it to Earth without fragmenting and is completely covered with fusion crust, it may still be subject to mechanical weathering. Fusion crusts form in the final second or two of the ablation process after which the crust rapidly cools. Being a magnetite impregnated glass, contraction cracks often form, looking much like the crazing on pottery (Fig. 3.15). These cracks are tiny avenues allowing water into the meteorite's interior where chemical weathering can begin. If the meteorite has landed in an area that undergoes rapid fluctuations in temperature or a daily freeze/thaw cycle, this could be sufficient to further widen the tiny cracks through ice wedging.

Many very weathered stony meteorites have a thick rind that seems to have replaced the fusion crust. This rind can be peeled off as though the meteorite has suffered exfoliation, probably the result of both chemical reactions that have expanded the crust and mechanical forces that have lifted the layer. Figure 3.16 (left) shows a weathered 240 g L4 chondrite from the Gold Basin, Arizona, strewn field. It was found by the author about 6 inches beneath a rocky desert pavement surface. A thick rind envelopes the surface in place of a fusion crust. Expansion cracks, possibly induced chemically, are resulting in the slow removal of the surface rind. Partially buried meteorites show more chemical weathering on below ground portions than on those portions exposed above ground. Another Gold Basin specimen (Fig. 3.16, right) shows a remnant of the black crust still intact on the side remaining above ground.

Chemical weathering

Mechanical weathering leads to fragmentation and inadvertently prepares the meteorite for the far more destructive *chemical weathering*. This involves a series of chemical reactions when exposed to water and oxygen that changes its minerals to more stable forms within Earth's environment. Ultimately, chemical weathering destroys the meteorite's cosmic characteristics and results in its complete terrestrialization.

Fig. 3.15. Contraction cracks in the crust of the Pasamonte eucrite. The specimen is about 62 mm long. (Institute of Meteoritics, University of New Mexico. Photo by O. Richard Norton.)

Fig. 3.16. Two Gold Basin, Arizona, L4 stony meteorites demonstrate the effects of weathering on the exterior surface. (Left) A 240 g specimen found below ground level with its fusion crust completely replaced by a thick rind. The specimen measures about 8 cm longest dimension. (Right) A 53 g Gold Basin meteorite showing a thin, black remnant fusion crust on the portion of the exterior surface that was positioned above ground level. The specimen is 3.5 cm longest dimension (Photos by O. Richard Norton.)

Let's look at some typical chemical reactions. Chemical weathering only takes place in the presence of water, oxygen and carbonic acid. Without water, chemical weathering ceases. Carbonic acid forms by the chemical combination of water and carbon dioxide (3.1).

$$H_2O + CO_2 \rightleftharpoons H_2CO_3 \qquad (3.1)$$

The carbonic acid quickly ionizes producing hydrogen ions (H^+) and bicarbonate ions (HCO_3^-). The atmosphere contains only about 0.03% carbon dioxide so little carbonic acid is actually produced in the atmosphere. Of that, only about 0.001% of the carbonic acid actually ionizes. Thus, it is a relatively weak acid. Within the soil, however, pore spaces can contain as much as 10% or more CO_2 concentrating the forming acid. For this reason, meteorite specimens remaining on the surface should last longer. This is particularly true in desert environments where dry conditions prevail most of the year on the surface.

Among the common meteoritic minerals affected by water are plagioclase feldspar, olivine, pyroxene, and, of course, iron–nickel. Plagioclase, a *silicic* mineral (minerals with a high percentage of silica in the form of silicon–oxygen tetrahedra) is usually found as a mixture of both sodium (albite) and calcium (anorthite) plagioclase in terrestrial rock such as a granodiorite. Plagioclase is actually a continuous series of minerals called a *solid solution* with sodium and calcium plagioclase as end members.[IV] In chondritic meteorites the plagioclase is primarily sodic and in stony irons (mesosiderites), primarily calcic. Both minerals are altered in the following way:

$$CaAl_2Si_2O_8 \cdot 2NaAlSi_3O_8 + 4H_2CO_3 + 2(nH_2O) \rightarrow \qquad (3.2)$$
$$Ca(HCO_3)_2 + 2NaHCO_3 + 2Al_2(OH)_2Si_4O_{10} \cdot nH_2O$$

Sodium and calcium plagioclase both react with carbonic acid and water to yield their respective bicarbonates and a clay mineral with bound water (3.2). This clay is highly stable and represents an end product of chemical weathering.

Plagioclase is only an accessory mineral in ordinary chondrites but an important mineral in some achondrites and stony-irons. Of major importance in ordinary chondrites are the *mafic* minerals (dark colored minerals rich in iron and magnesium), olivine and pyroxene. Both minerals are actually a family of minerals determined by their ratio of magnesium to iron. This can vary from all magnesium and no iron to all iron and no magnesium. Like the plagioclase series, olivine and pyroxene form as individual solid solution series in which there is an ionic exchange of magnesium and iron; the atoms of both elements are nearly identical in size and can occupy the same place in the crystal structure of both minerals. In meteorites, a mixture of olivine and pyroxene is seen and is chemically designated, $(Mg,Fe)_2SiO_4$ for olivine and $(Mg,Fe)SiO_3$ for pyroxene. This is developed in more detail in Chapter 5.

The weathering of both olivine and pyroxene are essentially the same. In the example below we look at the weathering of the iron end member of the olivine solid solution series, the mineral *fayalite* – Fe_2SiO_4. I have picked iron instead of the magnesium end member because of the more interesting and varied weathering products involving iron that are more easily seen in heavily weathered meteorites. The reaction involves water and atmospheric oxygen (3.3).

$$2Fe_2SiO_4 + 2H_2O + O_2 \rightarrow 4FeO(OH) + 2SiO_2 \qquad (3.3)$$

Here, the reaction includes both oxidation and hydration of the iron. First, the olivine dissolves with both SiO_2 and Fe^{2+}

[IV] Although plagioclase does have a solid solution, the complete solid solution between the end members is a *coupled substitution*. That is, $Na^+Si^{4+} \Leftrightarrow Ca^{2+}Al^{3+}$. For each Ca^{2+} that replaces Na^+, one Si^{4+} is replaced by Al^{3+}. Note that the electrical charges on both sides are identical and the structure remains neutral.

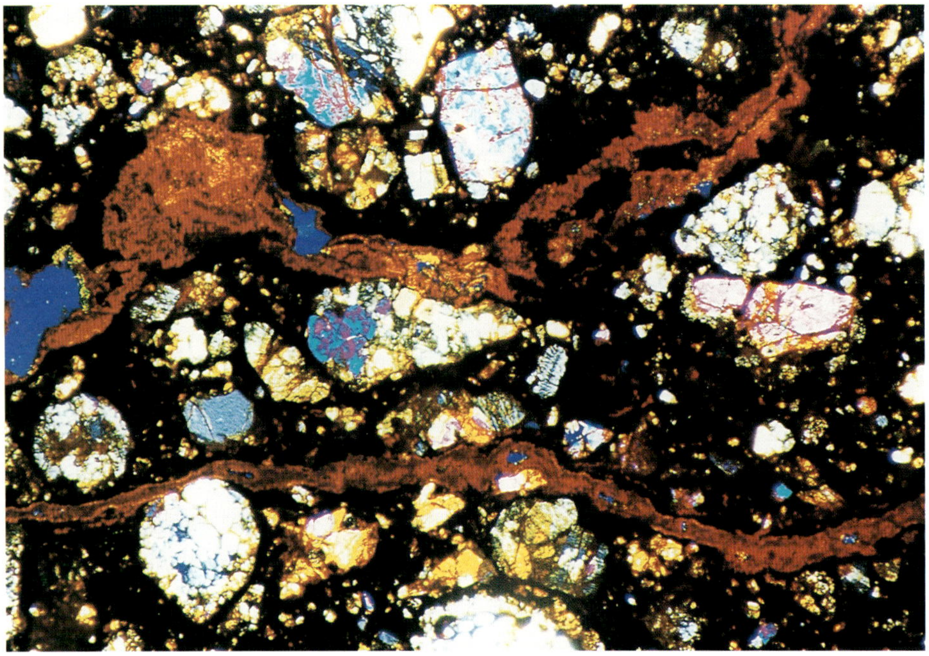

Fig. 3.17. A thin section of the Brownfield H3 chondrite seen under crossed Polaroids shows the bright red hydrated iron oxide, iddingsite, invading the interior of the meteorite along fine cracks. Some of the olivine grains also show signs of alteration to iddingsite in their interiors along fine cracks. The horizontal field of view is 3.4 mm. (Photo by O. Richard Norton and Tom Toffoli.)

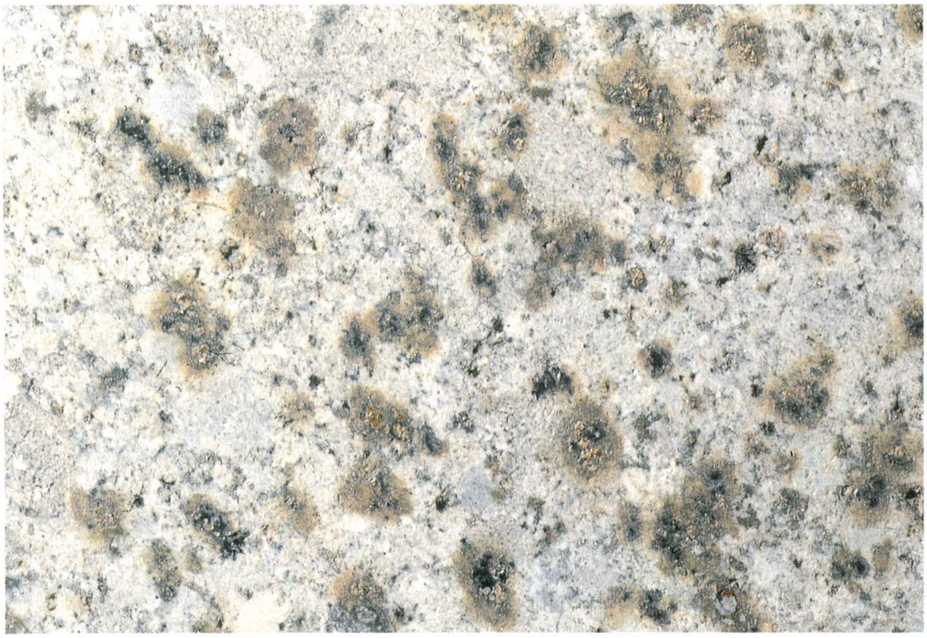

Fig. 3.18. A photomicrograph of the cut surface of a Bruderheim L6 chondrite showing limonite staining of the matrix as it forms around included iron–nickel grains. The horizontal field of view is 35 mm. (Photo by O. Richard Norton.)

(ferrous iron), going into solution. The Fe^{2+} is then oxidized by atmospheric oxygen to Fe^{3+} (ferric iron) which combines with water and precipitates as a yellow-brown substance commonly referred to as *limonite* but which is actually a mixture of two hydrated iron oxide minerals of the same composition but with different crystal structures: *goethite* (named after the German poet, Goethe) and *lepidocrocite*. Goethite usually dominates. It is this reaction that slowly changes the meteorite's exterior color from black or dark brown to a reddish-brown. Exposed interiors also turn reddish-brown. In thin sections of weathered chondrites seen under crossed polarized light, a bright red substance often invades the interior along fractures extending from the fusion crust (Fig. 3.17). This material is *iddingsite*, a hydrated iron oxide mixed with clay. It is associated with the aqueous alteration of iron-bearing olivine as it reacts with water. The presence of iddingsite is a sure sign that water has entered the meteorite. Carbonic acid can also become involved in the decomposition of the mafic minerals resulting in iron and magnesium carbonates both of which are soluble in water and tend to leach out of the meteorite in humid climates.

The iron–nickel metal in ordinary chondrites is, of course, attacked by oxygen. It can be transformed into different oxides of iron depending upon conditions. In cut specimens, frequently a corona of rust-colored material forms around the iron grains and stains adjacent minerals as the limonite

Fig. 3.19. Thin section of a badly weathered chondritic meteorite, Dar al Gani 521. Most of the interior is stained brown with iron oxide and clay. Only a few chondrules (white) appear to lack staining although microscopically alteration of the chondrule grains is evident. The section is about 15 mm across. (Specimen provided by Dr David Mouat, Desert Research Institute, University of Nevada System. Photo by O. Richard Norton.)

diffuses through the matrix (Fig. 3.18). This is limonite formed as atmospheric water reacts with the iron grains. Nickel is carried by the iron into the reaction with water and shows up with the standard chemical nickel test[V] applied to the limonite. Since nickel tends to leach out of the limonite in time, a test for nickel may give a negative result.[2]

A severely weathered stony meteorite lacks textural details. The interior takes on a dark brown uniformity with little left to suggest its meteoritic nature. Under the microscope, the field appears a uniform brown-yellow color as iron oxide and clay minerals stain the matrix and included grains (Fig. 3.19). Olivine is attacked first so that under crossed Polaroids little olivine appears in its usual vivid birefringent colors. At this stage of weathering the meteorite has lost most of its primary characteristics and is of limited scientific value.

The myth of lawrencite in iron meteorites

Iron meteorites quickly loose their thin fusion crusts to weathering and a glossy black terrestrially produced magnetite replaces it. The majority of iron meteorites are octahedrites, in which the crystal structure is cubic, centered around an octahedral dipyramid structure. Iron–nickel alloy is grown as thin plates or lamellae, called *Widmanstätten* structures, along the faces of the dipyramid (see Chapter 9). The minute spaces between individual plates are avenues through which water can slowly penetrate, especially if the meteorite is buried.

At this point the long-accepted myth begins. Nineteenth-century meteoriticists were all too aware of a corrosive chemical agent that seemed to pervade many iron meteorites, reacting with the iron and, if unchecked, resulting in its eventual destruction. As early as 1802, the noted French chemist Jacques Louis Bournon noticed that the Krasnojarsk pallasite seemed to attract water, since it was covered with liquid drops. At almost the same time the English chemist Edward Charles Howard also noticed drops of liquid on the pallasite. Howard washed the pallasite in distilled water, removing the drops, and then analyzed the water. In it he found iron chloride. Curiously, both chemists noted their observations in prepared manuscripts but then chose to delete them before publication. In 1838, C.T. Jackson first reported the presence of chlorine (Cl) in a yellow-green semi-liquid material exuding from between the kamacite plates of a heavily corroded iron meteorite (Lime Creek iron, Monroe County, Alabama, found 1834). Jackson claimed that the chlorine was indigenous to the iron, although this report was questioned by other workers.[3] Jackson and his coworker, A.A. Hayes, countered with further arguments in 1844 that seemed to settle the matter.[4] About a decade later, J. Lawrence Smith reported his discovery of ferrous chloride ($FeCl_2$) in two iron meteorites. In honor of the discovery, the new mineral was named *lawrencite*. Incredibly, without any further confirmation, the idea of indigenous chlorine-bearing lawrencite in meteoritic iron was universally accepted.

Taking the presence of lawrencite in meteoritic iron as a given, I will characterize the reactions of lawrencite with the iron as it was understood by these nineteenth-century chemists.

[V] The test for nickel is considered diagnostic for meteoritic metal. Iron meteorites can contain between 4% and 20% nickel, and sometimes higher. Nickel in the metal in ordinary chondrites is usually less than 10%, more likely around 4–5%. It is a function of the chondrite group; i.e., H, L and LL metal have different nickel contents.

Fig. 3.20. Ferrous chloride oozes out from between the Widmanstätten plates of this Canyon Diablo iron meteorite and forms an intricate pattern progressing along plate boundaries. The section is about 10 mm across (Photo by O. Richard Norton.)

A ferrous chloride mineral, lawrencite ($FeCl_2$), is often found within iron and stony-iron meteorites that acts as a "cancer" which slowly destroys the Widmanstätten structure. The destructive reactions are exacerbated by heat, high humidity and etching processes.

$$6Fe^{2+}Cl_2 + 3H_2O + 3O_2 = Fe_2O_6H_6 + 4Fe^{3+}Cl_3 \quad (3.4)$$

Ferrous chloride, beginning in the solid form within the meteorite, becomes a liquid as it absorbs water and is usually first observed on the surface of the meteorite as a green-yellow substance forming tiny droplets along plate boundaries (Fig. 3.20). Within hours, the ferrous chloride is oxidized as it changes to hydrated ferric hydroxide and solid ferric chloride. The two new minerals are earthy brown and give the impression of rust outlining the Widmanstätten plates. In the presence of iron, the oxidation state of the iron in ferric chloride is reduced,[VI] forming ferrous chloride again.

$$4Fe^{3+}Cl_3 + 2Fe = 6Fe^{2+}Cl_2 \quad (3.5)$$

This means that the process will continue until all of the chlorine is consumed in the reaction. In practical terms, the meteorite will continue to rust regardless of the care taken in preparation. The amount of lawrencite in meteorites of the same fall can vary widely. In Canyon Diablo, Odessa, and Toluca irons, some show no signs of the mineral after cutting, polishing and etching while others rust to pieces with time, showing a remarkable difference in the amount of lawrencite within a particular fall. Another consequence of these reactions is that space is consumed by the forming minerals causing an internal pressure that can lift the Widmanstätten plates above the level of the slab or even completely out of the specimen! Likewise, when lawrencite reacts with the reticulated iron–nickel structure in pallasites, the build up of pressure behind the olivine crystals is sufficient to pop out the olivine leaving an unattractive rusty void.

As if these reactions are not serious enough, there is yet another reaction that may even be worse on the meteorite. If the ferric chloride is not reduced to ferrous chloride, it can react with water to form rust (limonite) and hydrochloric acid. Hydrochloric acid easily dissolves away the iron.[5]

$$4FeCl_3 + 6H_2O = Fe_2O_6H_6 + 6HCl + 2FeCl_3 \quad (3.6)$$

Although the above reactions and observations are valid with the results easily observed, what appears not to be true is the indigenous nature of meteoritic lawrencite. In the mid-1980s, Roy S. Clarke Jr. of the Smithsonian Institution and Vagn F. Buchwald of the Technical University, Denmark, undertook a study of Antarctic iron meteorites in various states of corrosion.[6] Four major corrosion products were found using electron microprobe analysis: *akaganèite* (β-FeO(OH)); *goethite* (α-FeO(OH)); *lepidocrocite* (γ-FeO(OH)); and *maghemite* (γ-Fe_2O_3). What was conspicuous by its absence was chlorine-bearing lawrencite. Of the four minerals akaganèite[VII] is the primary carrier of chlorine varying from a low of 0.3 wt% to 5.4 wt% Cl. The Cl is *not* indigenous to the meteoritic iron but comes from the terrestrial environment. The researchers found Cl-laced akaganèite in cracks along the low iron–nickel alloy kamacite plate boundaries, the Cl content being highest in contact with fresh uncorroded metal. Here, the kamacite is converted directly to akaganèite. If the corroding meteorite experiences moderate warming and humidity changes through the year, the OH^- ion can exchange with the Cl^- ion releasing the Cl to migrate

[VI] This is a typical oxidation–reduction reaction involving iron. Ferric iron is more oxidized than ferrous iron in that it contains fewer electrons. Oxidation in chemical reactions is defined as a net loss of electrons and, regardless of the name of the process, may or may not involve oxygen. In this case, however, oxygen is involved in the reaction. Reduction is the net gain of electrons. In this case, ferrous iron is reduced by gaining one additional electron from another chlorine ion.

[VII] Akaganèite was first discovered as a corrosion mineral in the Akaganè limonite mine in Japan in the early 1960s and first described on an iron meteorite by Ursula Marvin in 1963.

to other sites or it could be flushed out of the meteorite. If this occurs, the mineral becomes unstable and decomposes converting to goethite and maghemite (a hematite-like mineral).

With all this said, however, it is academic whether lawrencite or akaganèite is responsible for the corrosion of the iron in meteorites. The net result is basically the same – great harm to the meteorites. Many methods have been used to attempt to prevent the above chemical reactions. After etching the meteorite slab with dilute nitric acid to bring out the Widmanstätten structure, the specimen is neutralized, washed with water, oven dried and then soaked in 100% alcohol in an attempt to dry out the specimen. It is then protected from the outside air by coating with an acrylic finish. If the meteorite still contains chlorine, it will most likely begin the cycle anew. The author has found that placing the etched side tightly against a flat piece of glass helps reduce if not eliminate the problem by squeezing out the air layer between the specimen and the glass. An experiment was tried in which a cube of Odessa iron was etched on all sides and then placed on glass. None of the faces were coated. All sides *except* the side against the glass badly rusted from the chlorine reaction. One good piece of news is that the chlorine is not indigenous to the meteorite which means that it should be used up relatively rapidly after the meteorite is taken from the chlorine-infested environment. When a buried iron is found it has rust over its entire surface. This is limonite and is easily removed with an electric wire brush. Beneath is a thin terrestrially deposited magnetite layer. In unstable irons this magnetite coating can oxidize further to limonite as a yet deeper layer of magnetite forms. The result is a "peeling" of the surface in time.

The ease or resistance of a mineral to weathering depends primarily upon the mineral itself. Igneous rock minerals on Earth are most stable when they are near the environment in which they originally formed. The surface environment is as alien to terrestrial rocks as it is to meteorites. All igneous rock minerals form in a melt under specific temperatures and pressures. In a slowly cooling melt of ferromagnesian composition, olivine is the first to crystallize out at the highest temperature. Next, at a lower temperature the olivine reacts with the magma forming a pyroxene. The last formed mineral is quartz, crystallizing at the lowest temperature in the remaining magma which has been depleted of iron, magnesium, calcium and other metals. On the Earth's surface, the first formed minerals being furthest from their environment of formation will react more quickly to the surface environment than the last formed minerals. Thus, quartz (only rarely found in meteorites) is highly resistant to chemical weathering whereas olivine, pyroxene, and calcic plagioclase rapidly weather. Unfortunately, the major minerals making up ordinary chondrites are the first to chemically weather.

A weathering scale[7]

F. Wlotzka of the Max Planck Institute in Mainz, Germany, developed a weathering scale for ordinary chondrites in 1993. He detailed six weathering grades as seen on polished sections and in thin sections. He designated seven progressive weathering states labeled W0–W6 with the following characteristics.

- W0 No visible oxidation of the iron–nickel metal grains or sulfides. Fresh falls that are recovered within days show this grade.
- W1 Minor oxide rims around metal grains and troilite; some minor oxide veins.
- W2 Moderate oxidation of metal grains and iron sulfide with 20–60% involvement.
- W3 Heavy oxidation of metal grains and troilite with 60–95% replacement.
- W4 Complete oxidation of metal grains and troilite. No alteration of silicates.
- W5 Beginning chemical alteration of magnesium/iron-bearing (mafic) silicates, primarily along cracks. Thin section shows olivine to be affected first. Alteration begins within the grains, not on the rims of the grains.
- W6 Extensive replacement of mafic silicates by iron oxides and clay minerals. Interstitial feldspars in the matrix not affected.

Meteorites were collected from Roosevelt County, New Mexico, a typical southwestern dry hot desert environment. Terrestrial dates for stages W2–W6 were correlated in an attempt to determine the length of time needed to develop each weathering state. These are as follows.

W2 – 5000–15 000 years
W3 – 15 000–30 000 years
W4 – 20 000–35 000 years
W5 and W6 – 30 000 > 45 000 years

Meteorites from other desert environments showed a similar weathering/terrestrial age correlation. Meteorites from Antarctica, however, showed far less weathering over much greater terrestrial ages. This is undoubtedly because they remained locked frozen within ice for thousands of years before being exhumed through natural processes. Under these frigid conditions, chemical reactions proceed very slowly if at all. Weathering only begins after they have been released from the ice and lie on the surface. For Antarctic meteorites a simpler classification is used.[8] Like the Wlotzka system, it depends upon the degree of rustiness since this characteristic is most readily observed. Only three classes are recognized.

A. Minor rustiness; rust haloes on metal particles and rust stains along fractures are minor.
B. Moderate rustiness; large rust haloes occur on metal particles and rust stains on internal fractures are extensive.
C. Severe rustiness; metal particles have been mostly, if not totally, converted to rust and the specimen is stained by rust throughout.

Recently the Antarctic scale has been modified to include a lower case "e" after the A, B, C designation if the meteorite contains observable surface evaporite deposits.

Both the Wlotzka and Antarctic systems are strictly qualitative and subjective. Thus, different observers may classify the weathering of a particular meteorite differently. These classifications work reasonably well for iron-rich meteorites but fail for iron-poor meteorites. For those meteorites that contain no free metal, the value of the scale is diminished. For these, the weathering grade is determined by an "overall rustiness" where the iron in the silicate minerals has been reduced by weathering of the minerals.[7]

Terrestrial lifetimes of meteorites

Imagine the thousands of tons of meteoritic material that have successfully passed the most severe test, atmospheric passage, only to succumb to the slow ravages of mechanical and chemical weathering. How much time do we have to find and preserve these meteorites? Assuming a desert environment with an annual precipitation of between 6 and 12 inches of rain, the average lifetime of a chondritic meteorite is between 50 000 and 100 000 years. In wetter, more humid climates the lifetime falls to a few thousand years for ordinary chondrites and even less for many carbonaceous chondrites and certain achondrites. In Gold Basin, an area in northwest Arizona, old, very weathered stony meteorites were found in a strewn field in 1996. The terrestrial age of these meteorites, determined using radioisotopic dating methods is between 15 000 and 20 000 years. The meteorites, though showing extensive weathering (W3), still retain their internal cosmic textures. Rain in the area averages about 8 inches annually.

Of the roughly 35 000 tons of meteoritic debris that fall to Earth annually (most of which is meteoritic dust), roughly 46 000 meteorites weighing from a few grams to 1000 kg or more survive and are scattered around the world, including the oceans.[9] Assuming a uniform distribution worldwide, and that 71% of the Earth's surface is covered by water, only about 13 000 meteorites fall on the remaining 127.5 million square kilometers of land surface. At this rate, about 100 meteorites will land on every square kilometer of land surface every million years or roughly one meteorite per square kilometer every 10 000 years. Now assuming that they have average lifetimes of 100 000 years (optimistic for some areas, pessimistic for others), 10 would have fallen during that period and would continue to do so assuming the flux remained constant. This means that there should be at least 10 meteorites to be found over every square kilometer of Earth's land surface in some stage of the weathering process. This further means that if one were to select carefully an area where the rainfall is only a few inches or less per year (the American southwestern deserts, 3–4 inches; the Sahara Deserts of Libya and Egypt, $<<1$ inch; the Atacama Desert of Chile, 0.25 inch, for example), and a detailed search conducted, there should be at least that many meteorites recovered for every square kilometer.

Just such an experiment was conducted beginning in 1957 by I.E. Wilson, an associate of H.H. Nininger. Wilson selected a desert area in Roosevelt County, New Mexico, and searched every square foot for surface specimens. After five years in which he searched four square miles (10.24 km^2), he recovered 159 meteorites.[10] This is a far greater number than the estimated flux and average lifetime of meteorites predict. This suggests that either the average lifetime of meteorites in very dry climates such as that found in the dry, hot deserts of the world must be greater than 100 000 years or that the influx of meteorites per year is greater than 50 000 per year worldwide. Using the Wilson field study and others conducted at the American Meteorite Laboratory in Denver, Colorado, G.I. Huss estimated that the influx was over 200 000 per year. If this is the case, then one would expect to find meteorites representing several falls in a single strewn field. Such multiple falls were actually found in the Allende strewn field in Chihuahua, Mexico. Over the past 30 years, meteorites from at least 13 separate falls have been located in the Allende field. In 1999, a new chondrite was found in the Gold Basin field (Fig. 3.21). This meteorite shows a fresh, unweathered fusion crust and interior unlike the original Gold Basin find, suggesting relative youth. High meteorite abundance in desert climates may be due, in part, to deflation, the removal of the thin desert soil by surface winds leaving behind a residue of rocks forming a pavement and exposing old falls among the pavement stones. Indeed, Gold Basin meteorites are found in such deflated areas.

Over the past several years interest among meteorite enthusiasts (this includes collectors, commercial dealers and researchers) has turned to the Earth's driest, hottest deserts

Fig. 3.21. A new Gold Basin meteorite find made in 1999. Notice the fresh-looking fusion crust, unlike the majority of Gold Basin meteorites. This meteorite is a relatively recent fall within the Gold Basin strewn field. It was found February 16, 1999, by a codiscoverer of the Gold Basin strewn field, Twink Monrad of Oro Valley, Arizona, and named after a hill within the field called Golden Rule Peak. It is an L5 chondrite measuring 100 × 89 × 82 mm and weighs 797.6 g. (Courtesy Twink Monrad, Tucson, Arizona.)

and the above predictions seem to be reaching fruition. Using four wheel drive vehicles to traverse the enormous distances in the Egyptian and Libyan deserts, hundreds of meteorites have been found lying on the surface. Here the thin veneer of soil has been removed by the winds, exposing hard desert pavement with meteorites resting on top. Most show the effects of severe weathering, suggesting great age. Others are in near mint condition and must be freshly fallen. In 1998, a lunar meteorite, an anorthositic breccia labeled DaG 400 weighing 1.425 kg, was recovered. This remains the largest lunar meteorite found on Earth.

Without question, the best area for the preservation of meteorites in the natural environment has proven to be Antarctica. The first Antarctic meteorite was found in Adelie Land in 1912. A few meteorites (Thiel Mountains) were recovered in 1962 but it wasn't until meteorites were accidentally rediscovered there by Japanese glaciologists in 1969, that there began the recovery of many thousands of meteorites in various states of preservation, most showing considerable weathering, a few nearly pristine.[VIII] Before their release they had been locked in ice for hundreds of thousands of years which protected them from weathering. Antarctic meteorites have proven to have the greatest terrestrial age of any known. Two iron meteorites with ages of 3.2 and 5 million years currently hold the record to date from the ice fields.[11] This suggests that given the right conditions, meteorites can last many times longer on Earth than previously thought. Antarctic chondritic meteorites tend not to be as old. An L-chondrite, Lewis Cliff 86360, was found to have an age of 2.35 million years,[11] followed by an Antarctic H-chondrite, Allan Hills 88019, with a terrestrial age of around 2 million years.[12]

The message here is clear. Although in most regions of the Earth the environment is not conducive to the preservation of meteorites, there are places where conditions continue to be favorable for meteorite survival. Moreover, the influx of meteorites may be as much as 10 times earlier estimates, i.e. 100 per square kilometer. Taken together, I can only conclude that the Earth abounds with meteorites and the diligent and clever searcher will succeed in finding them.

[VIII] They were found on the surface of blue ice after natural processes had exhumed them from the ice pushing up against the mountains ringing the edge of the Antarctic continent.

"Fossil" meteorites

From the cratering record on Earth and on the Moon we know that meteorites have been raining down on Earth throughout all of geologic time. The first granitic crust that began to form continents was very mobile. Plate tectonics operating some 4 billion years ago destroyed and recycled much of this first continental crust so that no rock older than this is found on Earth. Meteorites reaching Earth at that time did so unchecked by the thick atmosphere we see today. Most of these would have been destroyed as they arrived carrying much of their cosmic velocity at impact. Those that didn't were ultimately destroyed by plate movements. About 3.5 billion years ago, Earth's atmosphere had thickened through volcanic outgassing and became oxygenated by the oxygen-producing blue-green algae in the primitive seas. Oxygen is a highly reactive gas leading to oxidation reactions that quickly destroy fossils, rocks, and meteorites. Thus, it seems at first thought highly unlikely that meteorites could survive as "fossils" dating back many millions of years.

The Brunflo "paleo" fossils

At the beginning of the Paleozoic Era about 670 million years ago, shallow warm seas were teeming with life. In the Cambrian Period much of this life had developed hard calcareous shells that protected the soft parts of the animals. When these animals died their hard shells drifted to the ocean floor, mixed with limy deposits and formed a calcareous sediment that hardened to a fossiliferous limestone. In the Ordovician, the second period of the Paleozoic Era, beginning about 460 million years ago, a group of chondritic meteorites fell into a shallow sea near what is today Brunflo, Sweden. They survived their plunge into the sea and buried themselves in the limy sediments. In 1980, in the Rödbrottet limestone quarry near Brunflo, a small rock was discovered embedded in the limestone. Entombed with it only a few centimeters away was a fossil orthocone, a primitive straight nautiloid gastropod. The minerals of the "paleo" chondrite had reacted with the surrounding sediments leaving a stain in the limestone surrounding the meteorite. All of the meteoritic minerals had been replaced except chromite.[13] Like the fossils in the sediments, there was a major replacement by calcite, apatite and barite along with a complex assortment of accessory, minor and trace minerals.[14] What was amazing was a replacement of a radial pyroxene chondrule still showing excentroradial structure though now a *pseudomorph*. In other words, though the primary meteoritic minerals had been replaced, its chondritic structure was still recognizable in many areas of the meteorite, so much so that it has been classified a petrographic type H4–5. A second Ordovician paleochondrite was discovered in another limestone quarry near Österplana, Sweden, in 1988. This one was also a chondrite with replacement minerals similar to Brunflo. To date, 13 paleometeorites, all chondrites, have been recovered from the two quarries.

The oldest "paleo" iron?

There is some evidence for the discovery of an iron meteorite located during well-drilling operations in 1930 in Texas. It was found at a depth of 465 m in Eocene epoch deposits. It is said that it passed the nickel test and showed Widmanstätten structure. Unfortunately, and for reasons that are not clear, no material from this meteorite was preserved.[15]

The most remarkable preservation of a fossil meteorite found to date may be the Lake Murray, Oklahoma, iron. This meteorite was discovered in 1933 after being struck by a plow. The rock was found in a shallow gully emersed in but partially weathered out of Lower Cretaceous Antler sandstone. Undisturbed Middle and Upper Cretaceous sedimentary beds were above the meteorite. Fossils in the sandstone were used to date the beds and the terrestrial age of the meteorite, about 110 million years. The meteorite was recognized *in situ* in 1952 by paleontologist Allan Graffham. It was taken to the Institute of Meteoritics at the University of New Mexico where it was sawn in half. There were six to eight inches of rusty rind around the meteorite that undoubtedly served to protect the core. Its original weight was estimated by Institute director, Lincoln LaPaz, at over a ton. Amazingly, a core about two feet across remained unaltered. Etching of the meteorite core showed it to be a IIB coarsest octahedrite with 6.3% nickel (Fig. 3.22). Lake Murray may be the oldest and best preserved paleoiron meteorite yet discovered but further investigation needs to be done before Lake Murray can be definitively labeled a "fossil" meteorite.[16]

A K–T boundary meteorite?

I can't end this section without mentioning the discovery of a 5 mm fragment brought up in a deep-sea drill core in 1996. The core sediments marked precisely the Cretaceous–Tertiary boundary. The fragment consists of hematite mixed with clays and tiny grains of iron–nickel and

Fig. 3.22. Lake Murray, Oklahoma, iron meteorite. This is the core of an iron meteorite that may have fallen into a shallow sea in the Lower Cretaceous Period. A thick shell of limonite rust formed around the 275 kg core mass that apparently protected it from further chemical weathering. The slab has been etched to show its coarse Widmanstätten structure. The slab is 23 cm long. (Photo by O. Richard Norton.)

iron sulfide. This may be a small remnant of the 10 km diameter asteroid that impacted the Yucatàn region 65 million years ago bringing to an end the Cretaceous period with the extinction of the dinosaurs and 65% of the world's animal and plant species.[17]

References

1. Hills, J.G. and Goda, M.P. (1997). The largest mass of nickel–iron meteorites. *Meteoritics and Planetary Science* **32** (Supplement), A60–61.
2. Nininger, H.H. (1952). *Out of the Sky*, Dover Publications, Inc., New York, pp. 173–280.
3. Shepard, C.U. (1842). Analysis of a meteoritic iron from Cocke County, Tennessee, with some remarks on chlorine in meteoritic iron masses. *American Journal of Science* **43**, 354–363.
4. Jackson, C.T. and Hayes, A.A. (1844). Remarks on the Alabama meteoritic iron, with a chemical analysis of the drops of green liquid which exude from it. *American Journal of Science* **48**, 145–156.
5. Farrington, O.C. (1915). *Meteorites*. Self-published, Chicago.
6. Buchwald, V.F. and Clarke, R.S., Jr. (1989). Corrosion of Fe–Ni alloys by Cl-containing akaganèite (β-FeO(OH)): the Antarctic meteorite case. *American Mineralogist* **74**, 656–667.
7. Wlotzka, F. (1993). A weathering scale for the ordinary chondrites. *Meteoritics* **28**, 460.
8. Goodin, J.L. (1989). Significance of terrestrial weathering effects in Antarctic meteorites. *Smithsonian Contributions to the Earth Sciences* **28**, 93–98.
9. Hughe, D.W. (1981) Meteorite falls and finds: some statistics. *Meteoritics* **16**, 269–281.
10. Maurette, M. (1993). *Hunting for Stars*, McGraw-Hill, Inc., New York.
11. Welten, K.C. *et al.* (1997). Lewis Cliff 86360: an Antarctic L-chondrite with a terrestrial age of 2.35 million years. *Meteoritics and Planetary Science* **32**, 775–780.
12. Scherer, P. *et al.* (1997). Allan Hills 88019: an Antarctic H-chondrite with a very long terrestrial age. *Meteoritics and Planetary Science* **32**, 769–773.
13. Thorslund, P. and Wickman, F.E. (1981). Middle Ordovician chondrite in fossiliferous limestone from Brunflo, central Sweden. *Nature* **289**, 285–286.
14. Thorslund, P., Wickman, F.E. and Nyström, J.O. (1984). The Ordovician chondrite from Brunflo, central Sweden. *Lithos* **17**, 87–100.
15. Lovering, J.F. (1959). Frequency of meteorite falls throughout the ages. *Nature* **183**, 1664–1665.
16. Norton, O.R. (1999). The Lake Murray "fossil" meteorite. *Meteorite!* **5**, No.4, 22–23.
17. Kyte, F.T. (1996). A piece of the K/T bolide? (abstract). *Lunar and Planetary Science* **27**, 717–718.

CHAPTER FOUR

Classification of meteorites: a historical viewpoint

Scientists seek order in the universe. They look for sequences and logical progressions in nature. In virtually every scientific discipline scientists have attempted to arrange the subjects of their study into logical sequences since often such sequences reveal important relationships not otherwise apparent. The classification of animals and plants into various groups based upon anatomical characteristics immediately suggested biological relationships among the species that pointed to an evolution over time from simple to complex. Mineralogists long ago developed a useful classification of minerals, dividing the over 2000 known into 32 classes based upon their mineralogy, crystal structure, and chemical composition. At the same time, geologists looked at rocks and saw important relationships among their mineral compositions, textures and modes of origin and evolution. Even in a science as remote as astronomy, spectroscopic studies of stars revealed many different spectral types that suggested an evolutionary sequence.

The science of meteoritics is no different in this respect. Meteorites have been divided into distinct groups and subgroups that show similar chemical, mineral and structural relationships. The recognized groups probably represent uniquely different parent bodies that formed under a variety of conditions and environments early in Solar System history. Some have remained much as they first formed, nearly pristine and unchanged. Others have gone through periods of metamorphism that have changed their internal structures and compositions. The rocks that come to Earth therefore represent a moment in time after they accreted from the solar nebula, some remaining primitive, others differentiating and most forming before the terrestrial planets.

Any classification is only as valid as the criteria used to develop the classification. These criteria are still developing for meteorites. *Meteoritics* is a multidisciplinary science, that is, an intermix of several scientific disciplines: mineralogy, petrology, geochemistry, geophysics and astronomy. Meteoritics borrows heavily from all of them. Thus, it was necessary to develop these sciences before a workable meteorite classification could appear.

(Opposite) Ernst Florenz Friedrich Chladni (1756–1827). Chladni was the first to propose the meteorites and fireballs were related; that meteorites were stones tfat fell from the sky; that meteorites were celestial bodies that came from interstellar space. Chladni saw his first two postulates generally accepted in his lifetime but the third, that meteorites came from interstellar space, was never accepted by nineteenth-century scientists and was finally discredited by the mid-twentieth century.

The early years

Even though there were many reports of stones falling from the sky in the seventeenth and eighteenth centuries, these reports were looked upon with considerable skepticism by the scientists of the day. In the last decade of the eighteenth century, Europe experienced several well-placed fireballs that dropped stones in France and Italy, producing many eyewitness accounts that scientists could not easily ignore. These were called thunderstones, since most scientists insisted that they originated from "igneous clouds" that somehow condensed them. The igneous minerals, they said, were concentrated in these clouds, perhaps originating from ash deposited in the atmosphere by volcanoes such as Vesuvius. During violent thunderstorms, these terrestrial rocks suspended in the clouds were struck by lightning, blackening them and sending them hurtling to Earth. The argument was not so much the reality of the stones or that they were actually falling from the sky – there were plenty of witnesses who would swear to that – rather it was their origin that was in question. As absurd as this may sound to us today, it seemed far more believable that these were terrestrial stones falling from huge clouds rather than that they were actually coming from beyond the Earth. No less an authority than Isaac Newton had said that the spaces between the planets were essentially empty. The rain of stones in the 1790s encouraged a lone German scientist, Ernst Friederick Chladni (1756–1827), to write a small book in 1794 in which he summarized his investigations of numerous well-witnessed falls. From this he drew the following conclusions: that stones and irons did indeed fall from the sky; that fireballs were attended by meteorites and dropped by them; and that meteorites originated beyond the confines of Earth or Sun. The first two conclusions met with considerable resistance. The third conclusion was firmly rejected by scientists. Chladni considered that meteorites might be remnant interplanetary material but rejected this idea since the velocities seemed far too high to assign them to the gravitational field of Earth or Sun. He believed instead that they came from interstellar space, an idea that persisted well into the twentieth century.[1]

If meteorites had terrestrial origins, their compositions, like all terrestrial stones, should vary widely depending upon their point and mode of origin. In 1802, the English chemist Edward Charles Howard (1774–1816) was one of the first to chemically analyze four stone meteorites that fell in widely separated localities across Europe and India. He found them to be remarkably similar in composition and not at all related to the terrestrial rocks among which they were found. He could only conclude that these meteorites must have come from a common source, but a source beyond Earth.[2]

The first rudiments of a classification came before meteorites were recognized as extraterrestrial objects. It began simply enough. There seemed to be two basic types: stones and irons.[1] Howard analyzed the metal found in both types and found nickel alloyed with the iron suggesting that these distinctly different meteorites were indeed related. Through the nineteenth century the primary problem was to develop criteria that would help distinguish terrestrial stones from meteorites. The presence of nickel alloyed with iron in all but a few meteorite types was one of the first criteria discovered and still remains an important test for meteorite confirmation. With this developing criteria, increasing numbers of meteorites were recovered that gave nineteenth century scientists sufficient representative material from which to build a classification. A concerted effort to identify minerals in meteorites began to reveal important similarities and differences that could be used in a classification scheme. Karl Ludwig von Reichenback (1788–1869) was the first to study the minerals and textures of meteorites with a microscope in 1857; and in 1861 Nevil Story-Maskelyne (1823–1911) designed a polarizing microscope that was fabricated in 1863 in which he first studied meteorite thin sections in crossed-polarized light. With only these basic tools of classical mineralogy and with chemical analysis still in its infancy, nineteenth century scientists laid out the first significant classification.[3]

---

[1] It wasn't until 1863 that a third major type, the stony-irons, was added by Nevil Story-Maskelyne, keeper of minerals at the British Museum. He referred to the three classes as aerolites (stones), siderites (irons) and mesosiderites (stony-irons). These names were used well into the twentieth century with only the term mesosiderites changed: aerolites, siderites, siderolites.

The Rose–Tschermak–Brezina classification

Gustav Rose (1798–1873) at the Mineralogical Museum of the University of Berlin devised a classification in 1863 based upon mineral composition and texture.[4] He named the spherical bodies in stony meteorites, *chondrules*, and called meteorites containing them *chondrites*. There were two basic meteorite types: stones and irons. He divided the stones into seven groups: *chondrites*; *howardites* (named in honor of Edward Howard); *chladnites* named in honor of E.F.F. Chladni (Chladnites are enstatite achondrites and are called *aubrites* today); *chassignites* (an SNC meteorite); *shalkites*, named after the Shalka meteorite, called a *diogenite* today; *carbonaceous chondrites*; and *eucrites*. These were distinguished by their mineralogical differences. The irons were divided into three groups: the "pure" nickel-irons; *pallasites* (named in honor of Peter Simon Pallas who first described this type in 1772 after examining the Krasnojarsk stony-iron); and *mesosiderites*. The value of Rose's classification today is in the terminology he established. Many of the meteorite names used in our modern classification extend from the Rose system.

Gustav Tschermak (1836–1927), director of the Mineralogical and Petrographical Institute at the University of Vienna, modified and added to Rose's classification over a period of eleven years beginning in 1872 with further modifications in 1883. Tschermak was a highly skilled mineral microscopist. His classification system was based primarily upon petrographic observations in which he studied *in situ* properties of meteorite minerals, determining their crystal form, twinning and cleavage, and optical properties. Some chemical analysis was done on individual mineral crystals that Tschermak managed to pry from the stony mass. Tschermak's microscopic petrographic studies of stony meteorites were recorded on film, one of the first scientists to do so. By the time of his death in 1927, he had accumulated and described the largest and most comprehensive collection of meteorite photomicrographs in history.[II]

Tschermak separated all meteorites into five main classes: three stone classes and two iron classes. The chondrites were the largest class which he subdivided into nine groups based upon their mineralogy and texture. These all contained olivine, orthopyroxene (bronzite) and iron–nickel clasts. He further included the carbonaceous chondrites into this group. Another stone group contained olivine and pyroxene but no chondrules. This class was probably metamorphosed chondrites in which the original chondritic texture had been destroyed through recrystallization. He did not divide the stones into chondritic and achondritic as we do today but he certainly recognized their fundamental differences. In the third stone class he included achondritic meteorites such as eucrites, howardites and aubrites (the enstatite achondrite Rose called a chladnite, Tschermak called a *bustite*) and he renamed Rose's Shalka group, *amphoterites*. (Today amphoterites are called LL chondrites and placed with the chondrites since they have chondritic composition, iron–nickel metal and contain chondrules.) Finally, meteorites containing the iron rich orthopyroxene, hypersthene, he called a *diogenite*.[III]

The iron class Tschermak separated into three subclasses based upon their texture after acid etching: *hexahedrites*; *octahedrites*; and *ataxites*. The hexahedrites showed a hexahedral cleavage with the crystals crossed by parallel lines which Rose had earlier termed *Neumann lines*. The largest iron subclass comprised the octahedrites which upon etching showed a profusion of lamellae (Widmanstätten structures) lying along the faces of an octahedral dipyramid. Tschermak subdivided them according to the width of the lamellae into fine, medium and coarse. One additional form he described as curvilinear. This is not a true form but only distorted lamellae due to mechanical stresses probably caused by collision while in space. The ataxites showed no structure when etched but only a granular appearance.

The second iron class contained stony components. Tschermak subdivided them into four subclasses: *pallasites*; *mesosiderites*; *siderophyres*, and *grahamites*. Only the first two, included by Rose, survive as separate groups today. Grahamites were subsumed into the mesosiderites. Siderophyres were represented by only one specimen, the Steinbach iron (1724), and is not considered a true group. Rather, it is considered an *anomalous* iron of subgroup IVA with silicate inclusions.

Aristides Brezina (1848–1909) of the Natural History Museum in Vienna was the third author of the classification system. He substantially expanded the Rose–Tschermak system, especially for chondrites. These he divided into 32

[II] In 1885, Tschermak published *Die mikroskopische Beschaffenheit der Meteoriten* (The Microscopic Properties of Meteorites) which is a summary of his work on meteorites. His meteorite classification is described in detail and is accompanied by 100 thin section photographs. Very few copies of the original work published by E. Schweizerbart'sche Verlagshandlung of Stuttgart survive today. In 1964, it was translated into English by meteoriticist John A. Wood and E. Mathilde Wood and published by the Smithsonian Institution under *Contributions to Astrophysics*, Vol. 4, No. 6.

[III] The term *diogenite* refers to the Greek philosopher Diogenes of Appollonia who Tschermak believed was the first to teach that meteorites came from beyond Earth.

subclasses, differentiating these groups on the basis of texture and color as well as chemical and mineral constituents. His terms *black chondrite* and *white chondrite* simply referred to the general darkness or lightness of the interior. Although color is no longer considered an important criterion, the nomenclature still exists but with different meaning. Thus, it is common in the literature today to encounter the term, black chondrites, which is used as a shock indicator. The term white chondrite is not used. The presence of veins and brecciated texture were also used as distinguishing criteria but like color, these criteria fell into disuse because later meteoriticists considered these only secondary features produced by impact events which only acted to mask the true fundamental properties of meteorites. The octahedrites were subdivided into eleven subgroups by quantitatively defining the octahedrite lamellae with limiting widths in millimeters for each subgroup. Brezina introduced the term *achondrites* for those meteorites without chondrules. After his modifications, the final classification had a rather cumbersome 76 subclasses.[IV]

[IV] For a complete listing of the Rose–Tschermak–Brezina classification, the reader should refer to: Mason, B. (1962). *Meteorites*, John Wiley and Sons, Inc., New York, pp. 47–50.

The Prior classification

Although the Rose–Tschermak–Brezina classification was adopted and used well into the mid-twentieth century, it proved to be awkward to use, overly complex and error prone. The use of secondary features such as color, brecciation and veins proved to be poor distinguishing criteria and were deleted in the Prior classification. The problem now was to simplify the system into a more useable form. In 1916, George T. Prior (1862–1936), Keeper of Minerals at the British Museum, recognized the extraordinary similarity in mineral and chemical composition among the chondrites. As distinguishing features, he used the variations in iron–nickel metal as well as variations in the iron incorporated in the two main silicates, olivine and pyroxene. These chemical and mineral relationships were the basis of his classification of the chondrites and are known specifically as Prior's rules.[5] They can be stated as follows.

1. *The less iron–nickel metal in a chondrite, the richer the metallic iron is in nickel*. This is usually shown in terms of the ratio Ni/Fe which increases with decreasing iron. In chondrites, the nickel remains relatively constant and all of it is found alloyed with iron. It is usually not found in the silicates. If the iron is oxidized and incorporated into the ferromagnesian minerals, the remaining iron will be richer in nickel.
2. *The less iron–nickel metal in a chondrite, the richer in iron are the magnesium silicates*. This is simply an extension of the first rule. Chondrites with less metal have more iron locked in their silicates. This is usually shown as the ratio Fe/(Fe + Mg).

Using the above criteria, Prior divided the chondrites into five groups based upon the ratio of oxidized iron to iron metal or their Fe/FeO bulk content. In applying names to these groups, he used the dominant orthopyroxene (enstatite, bronzite, hypersthene) along with olivine. The orthopyroxene is a solid solution series and as listed here increases in iron from enstatite to hypersthene. Table 4.1 shows the Prior classification in its entirety.

In all cases, the various groups are distinguished by their mineralogy. Note that the achondrites are divided into calcium-rich and calcium-poor subgroups. The stony-irons are distinguished by their silicate minerals and the irons are divided using structure and nickel content. Until the mid-twentieth century, the Prior classification was used much as it is seen here with some modifications by Brian Mason of the American Museum of Natural History in 1967.[6] That year Klaus Keil and Kurt Fredriksson at the University of California in San Diego performed the first microanalysis of meteorites using the electron microprobe. This new analytical tool was a giant leap forward and enabled meteoriticists to determine the elemental composition of meteorites with an accuracy never before attained. In particular, measurements of trace elements in iron meteorites led to a completely new classification no longer based upon structure and nickel content alone. In the chapters to follow we will develop the most recent classification for each major meteorite class. A summary of the currently used classification of meteorites is found in Appendix A.

Group	Class	Major minerals
Chondrites	Enstatite	Enstatite, iron–nickel
	Olivine–bronzite	Olivine, bronzite, iron–nickel
	Olivine–hypersthene	Olivine, hypersthene, iron–nickel
	Olivine–pigeonite	Olivine, pigeonite
	Carbonaceous	Serpentine
Achondrites (Ca-poor)	Aubrites	Enstatite
	Diogenites	Hypersthene
	Chassignites	Olivine
	Ureilites	Olivine, pigeonite, iron–nickel
Achondrites (Ca-rich)	Angrite	Augite
	Nakhlites	Olivine, diopside
	Eucrites	Pigeonite, plagioclase
	Howardites	Hypersthene, plagioclase
Stony-irons	Pallasites	Olivine, iron–nickel
	Siderophyre	Bronzite, iron–nickel
	Lodranites	Bronzite, olivine, iron–nickel
	Mesosiderites	Pyroxene, plagioclase, iron–nickel
Irons	Hexahedrites	Iron–nickel alloy (kamacite)
	Octahedrites	Iron–nickel alloy (kamacite, taenite)
	Ataxites	Iron–nickel alloy (nickel-rich taenite)

Table 4.1. **The Prior meteorite classification**

This table shows the Prior classification of meteorites which replaced the earlier Rose–Tschermak–Brezina classification. George T. Prior considerably simplified the classification, especially the chondrites, relating them to the amount of iron–nickel metal in the meteorites vs the combined iron in the minerals. The chondrite names are related to the principal mineral(s) found in the meteorite. The achondrites related to their mineral compositions, the stony-irons to the primary silicate mineral and the irons to their internal structure and nickel content.

References

1. Chladni, E.F.F. (1794). *On the Origin of the Mass of Iron Discovered by Pallas and Others Similar to It, and on Some Natural Phenomena Related to Them*, J.F. Hartknoch, Riga. 63 pp.
2. Howard, E.C., DeBournon, J.L. and Williams, J.L. (1802). Experiments and observations on certain stony and metalline substances, which at different times are said to have fallen on the Earth; also on various kind of native iron. *Philosophical Transactions of the Royal Society London* **92**, 168–212.
3. Burke, J.G. (1986). *Cosmic Debris, Meteorites in History*, University of California Press, Berkeley, California. 441 pp.
4. Rose, G. (1863). Systematisches Verzeichnis der Meteoriten in den mineralogischen Museum der Universität zu Berlin. *Annalen der Physik* **120**, 419–423.
5. Prior, G.T. (1916). On the genetic relationship and classification of meteorites. *The Mineralogical Magazine* **18**, 26–44.
6. Mason, B.H. (1962). The classification of chondritic meteorites. *American Museum Novitiates* **2085**, 1–20.

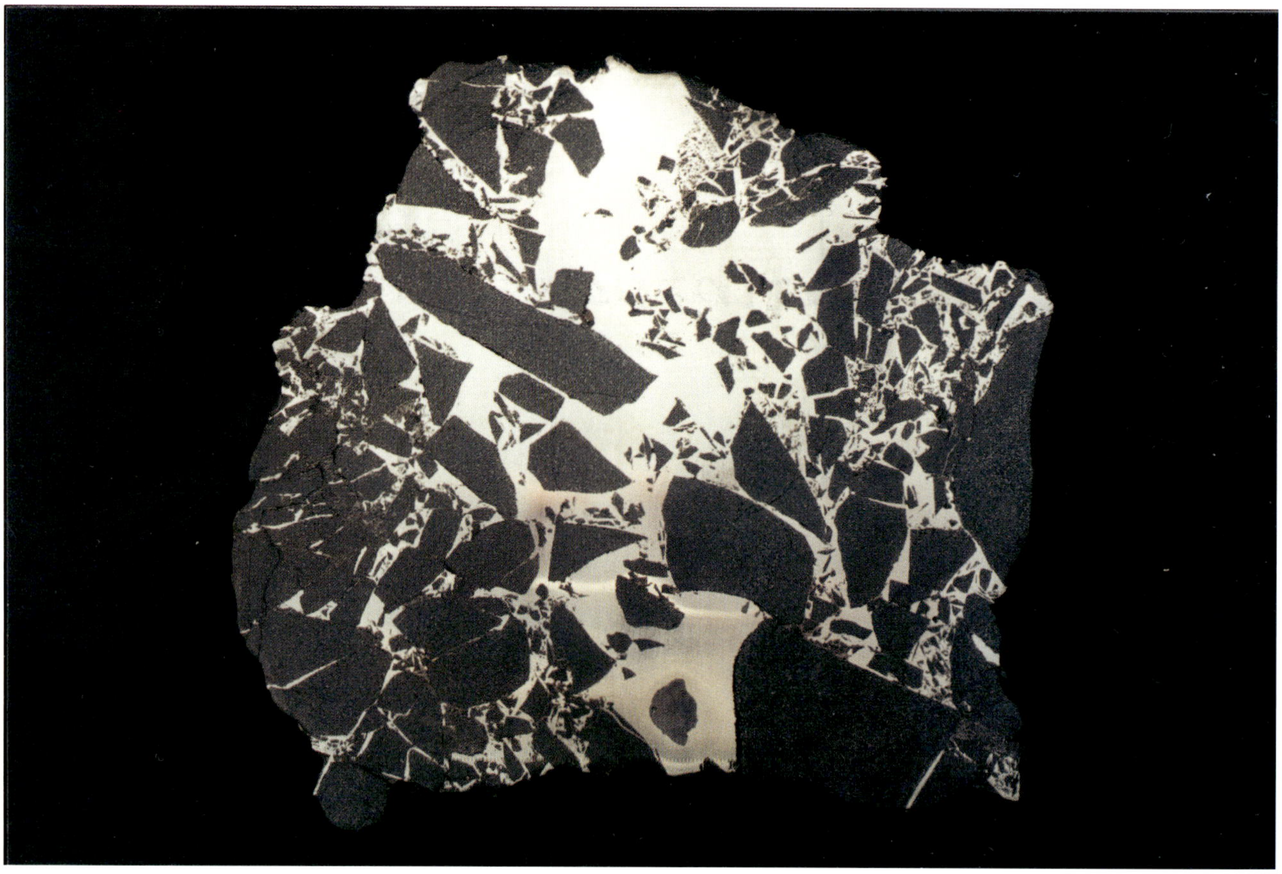

A unique ordinary chondrite. This is a 2115 g slab cut off the 34 kg main mass of the Portales Valley H6 chondrite. The slab measures 295 × 285 × 7 mm. Comparing this texture to other ordinary chondrites illustrated in Chapters 5 and 6 demonstrate its uniqueness. The silicate matrix has major fragmentation by the intrusion of wedges of metal into fractures within the silicate matrix. Many of the angular fragments show obvious signs of displacement. So much metal is present that early investigators suggested it was a silicated iron or a stony-iron or, lacking a better name, an anomalous iron (with chondritic xenoliths). The silicate matrix proved to be a very common H6 petrographic type but the large abundance of iron makes this otherwise ordinary chondritic meteorite an enigma. (Photo by Robert Woolard.)

CHAPTER FIVE

Primitive meteorites: the chondrites

When a new meteorite is observed to fall to Earth and subsequently recovered, there is an 82% chance that it will be a chondrite. Using data from the *Catalogue of Meteorites*[1] the total number of meteorites *witnessed* to fall to the year 2000 was 1005. Of these, 821 (82%) were chondrites, making them by far the most common of all the meteorites. Today we recognize many different kinds of chondrites, and to distinguish them from each other requires that their interiors be studied. The study of meteorites is basically the study of rocks. It doesn't matter that these rocks are not of this Earth. The tools and methods used are the same and they would certainly be familiar to any geologist. Before we look at the chondrites it may be helpful to briefly introduce the scientific disciplines within the broad area called petrology that geoscientists use to study rock samples with the goal of determining their mineralogy, their classification, and something of their history. From this point forward we will make continuous use of the techniques, observations and discoveries stemming from three essential disciplines: petrology, mineralogy, and geochemistry.

Petrology is a broadly based branch of geology that looks at the natural history of rocks which includes their origin, their past history, their present occurrence, and their structure. It is primarily a study of igneous and metamorphic rocks. Petrology can be divided into two main branches: petrography, and petrogenesis. *Petrography* is more restrictive than petrology in that its primary goal is to describe the structure of a rock, and thence determine its place in the rock classification scheme. Petrography is the microscopic study of rocks usually through the use of thin sections. *Lithology* is included in petrography in that its goal is similar – to determine the physical characteristics of a rock – but is restricted to a megascopic (naked eye) study, or at most, a low magnification (10× magnifier) study of hand specimens in the field. Lithology is what is practiced when a suspected meteorite is first examined in the field or laboratory. Both disciplines are very important in the identification and classification of meteorites. These lead to the second branch of petrology, *petrogenesis*, which deals with the origin of rocks, in this case, usually igneous rocks. The determination of the origin of meteorites is the ultimate goal of meteoritical science and the classification of meteorites tends to suggest their origins or at least points in the most logical direction. In this chapter, we will look at each of the chondrite groups, note their characteristic lithology and petrography and relate them to the current classification of meteorites.

Up to about the mid-twentieth century, meteorites were studied principally through the microscopic study of rock thin sections using a petrographic or polarizing microscope. This yielded their basic mineral compositions and textures. Chemical analysis was much slower to develop. The mineral samples had to be extracted from the meteorites in order to use traditional "wet" chemical analysis which was difficult and tedious at best and subject to error. It wasn't until the development of microanalytical techniques in the 1960s using powerful new instruments and technologies such as electron microprobe microanalysis and neutron activation technology, that chemistry took on added importance. For example, it became possible to analyze individual mineral grains *in situ* in chondrites to determine their elemental compositions. These are noninvasive techniques that allows the meteorite to remain intact with little or no destruction of material other than surface preparation. Differences in

[1] Grady, M.M. (2000). *Catalogue of Meteorites*, 5th edition, Cambridge University Press, Cambridge. 720pp.

mineral composition among the grains or even within a single grain became an important criterion in the classification of chondrites. Trace elements became especially important in the classification of the irons.

As we develop the classification scheme in this and the following chapters, it will become apparent that the different meteorite classes and groups are discontinuous in that their properties vary significantly from group to group. It is precisely their differences rather than their similarities that lead us inexorably to consider the different physical and chemical conditions that must have existed at different locations across the solar nebula when their parent bodies accreted ~4.5 billion years ago. Although for convenience, I have used the four main categories of meteorites (chondrites, achondrites, irons, stony-irons) in discussions so far, in today's classification this distinction is inadequate and says little about the relationships between them. Today, there are two broad divisions based upon their petrology and mineral chemistry. Meteorites are classed as either *primitive* or *differentiated*. All meteorites fall into one of these two categories. *Primitive* meteorites show characteristics that suggest they have never been melted since their parent bodies formed. Though many have experiencd some reheating, they have remained essentially as they formed in the early solar nebula. These are the meteorites of most interest to meteoriticists considering the petrogenesis of the Solar System's first rocky bodies. *Differentiated* meteorites, though similar in age to the primitive meteorites, possess properties that, without question, demonstrate that their parent bodies were either partially or entirely melted, leading to a differentiation of their masses into cores, mantles, and crusts much like the differentiated planets of the Solar System (Mercury to Mars).

Petrologic–chemical classification of the chondrites

The name *chondrite* is derived from the texture of these meteorites. In all but one group,[II] their interiors show a texture unlike any found in terrestrial rocks. Dispersed more or less uniformly throughout the rock are spherical, subspherical and sometimes ellipsoidal structures called *chondrules*. These range in size from about 0.1 to 4 mm diameter with a few reaching centimeter size. Their abundance within a given chondrite can vary enormously from only a few percent of the total volume of the meteorite to as much as 70% with fine-grained matrix material dispersed between the chondrules (Fig. 5.1). The single group (CI chondrites) that contains no chondrules at all is chemically related to the other chondrites therefore qualifying it as a chondrite without chondrules.

Several clans divide the chondrite class. These are (1) the *ordinary chondrites*, designated OC; (2) the *enstatite chondrites*, designated E; and (3) a group of clans comprising the *carbonaceous* chondrite "set", designated C. These clans are in turn divided into a total of 13 groups based upon their mineral and chemical properties. These chondrite groups are introduced in this chapter but are the subject of the next two chapters.

Fig. 5.1. Abundant spherical and subspherical chondrules are evident in this thin section of the Parnallee LL3.6 chondrite made in transmitted light. The chondrules are set in a dark matrix of similar composition. The large black object on the left is an opaque iron–nickel inclusion. A variety of textures within the chondrules is evident. The largest chondrule in the lower right has metal blebs within it and is 3 mm in its longest dimension. (Photo by O. Richard Norton.)

[II] The nomenclature in the classification of chondrites is still developing. At the present time there is no accepted complete hierarchical classification of chondrites but there is a nomenclature in common use. In the chondrite "class", the meteorites are divided into *clans* and *groups*. A clan may be defined as one or more groups that formed in a relatively restricted region of the solar nebula. A group is defined as being comprised of meteorites from a single asteroid or parent body. The carbonaceous chondrites are composed of several clans but taken as a set, they will be called the carbonaceous chondrite "class" in this book.

Cosmic abundances

Meteoriticists are fond of pointing out that chondrites are the most primitive of all the meteorite classes. Their elemental compositions are very nearly the composition of the Sun – minus the gaseous and highly volatile components, most of which have not been retained by the meteorites. These include hydrogen, helium, carbon, nitrogen, oxygen and neon. By mass, the Sun is composed of 73.4% hydrogen and 25% helium, which leaves only 1.6% for the remaining elements. The elemental composition of the most primitive chondritic meteorites, the CI carbonaceous chondrites, is usually compared to solar abundances.[III] Figure 5.2 shows the solar elemental abundance plotted against the elemental abundance of CI chondrites. In this diagram it is the ratios of the nonvolatile elements such as Fe, Si, Al and Ca to Si that we wish to compare. Measuring absolute abundances of nonvolatile (refractory) elements in a CI carbonaceous chondrite is relatively easy to do but determining absolute abundances of any element in the solar photosphere is much more difficult and the results are not nearly as accurate. Moreover, absolute values are meaningless since the Sun contains many orders of magnitude more of every element than that in any meteorite. What can be done is to compare *ratios* of refractory elements found in both. This is accomplished by comparing each element to an arbitrarily selected "standard" element found abundantly in meteorites and in the Sun. This procedure, called *normalizing*, is used in petrology for determining the chemical (elemental) composition of a rock with respect to an abundant standard mineral or element. The normative element in this case is silicon. The elemental abundances in Fig. 5.2 are compared to one million atoms of silicon: element/10^6 Si. If the elemental abundances of the meteorite and the Sun were identical, all elements (except the most volatile) would plot exactly along a line with a 45° slope. As you can see, the chemical abundances are remarkably close to this line, showing that CI chondrites have compositions very close to the Sun. This is interpreted to mean that CI chondrites have retained the original composition of the solar nebula and are therefore quite primitive. The elemental abundances of CI chondrites are often referred to as *cosmic abundances* since they are thought to represent mean elemental abundances found throughout the Galaxy. The volatile elements C, O, and N are displaced to the upper left, showing that these elements have escaped the meteorite but are found prominently in the Sun. One element, lithium (Li), is depleted in the Sun compared to the CI chondrite. During nucleosynthesis within the core of the Sun, lithium is destroyed.

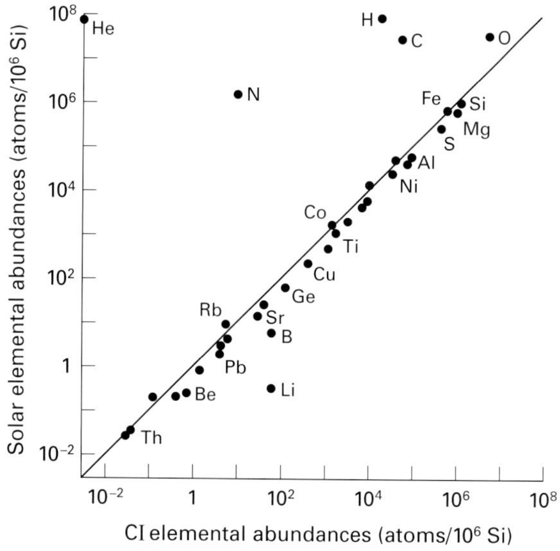

Fig. 5.2. This graph compares the elemental abundances of the Sun with the elemental abundances in CI chondrites, both normalized to 10^6 silicon atoms (see text). The solid diagonal line is the solar abundances of nonvolatile elements. CI abundances of nonvolatile elements are scattered tightly around this line, demonstrating that primitive chondrites closely match the Sun's chemical composition. Most stony meteorites are depleted in the volatile elements C, H, O, and N relative to solar abundances; the graph shows them displaced to the upper part of the graph.

[III] The CI chondrite, Orgueil, has been studied more than any other CI making this meteorite the "type" specimen usually used for this chemical comparison with the Sun. Even though CI chondrites show considerable low-temperature aqueous alteration, they are selected because they contain low-temperature minerals suggesting that they have retained volatile elements lost by the ordinary and enstatite chondrites.They are therefore the most primitive meteorite and thus the best candidate for comparison with solar abundances.

Chemical classification of the chondrites

Of all the meteorites, the chondrites show the greatest similarities in composition, so much so that it took substantial advances in analytical techniques to discover subtle chemical differences between them. These differences form the basis of the current classification of chondrites. Beginning in the 1950s, geochemists determined elemental ratios of certain refractory elements to silicon found within the silicate minerals of the meteorites. The lithophile elements (elements with a strong affinity for oxygen that tend to concentrate in the silicate phase) magnesium and calcium, showed the most distinct divisions among the chondrites. Figure 5.3 is a histogram plot of Mg/Si and Ca/Si abundances in the chondrite groups. Here we see in both cases an obvious distinction between the chondrite groups. The atomic ratios show three distinctive groupings: E or *enstatite chondrites* which exhibit the lowest element/Si ratio while the C or *carbonaceous chondrites* are clustered among the highest ratios. The *ordinary chondrites* fall in a tight cluster between the two with evidence that they actually represent at least three groups making a total of five. An even more striking distinction among the chondrites is evident when oxidized iron is plotted against iron in the metal phase and iron sulfide seen in Fig. 5.4 (data from Mason, 1962). To maintain the true iron metal to oxidized iron ratios, Mason pointed out that only fresh, relatively unweathered meteorites were used in the analysis to insure that the iron was not contaminated (oxidized) by terrestrial weathering. Once again, the plot shows a clear distinction between the three clans of chondrites where each group forms a tight cluster with the E chondrites showing little oxidation and the carbonaceous chondrites showing the greatest oxidation of its iron. Again, the ordinary chondrites fall in between. The ordinary chondrites can now be classified according to their range of FeO/(FeO + MgO)[IV] molecular

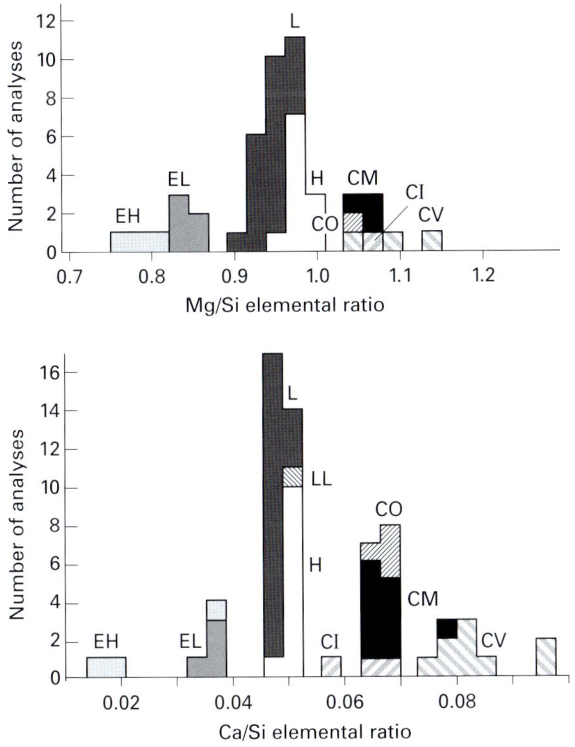

Fig. 5.3. Two histograms showing the Mg/Si and Ca/Si composition of the chondrites. The atomic ratios differ significantly so that three distinct divisions or clans of the chondrites appear: the enstatite chondrites; ordinary chondrites; and the carbonaceous chondrites. The data even allows each clan to be resolved into groups: enstatite chondrites into EH and EL; ordinary chondrites into H, L, LL; and carbonaceous chondrites into CI, CM, CV, CO. (Data and figure after Von Michaelis, H., Ahrens, I.H. and Willis, J.P. (1969). The compositions of stony meteorites – II. The analytical data and an assessment of their quality. *Earth and Planetary Scientific Letters* **5**, 387–394; Van Schmus, W.R. and Hayes, J.M. (1974). Chemical and petrographic correlations among carbonaceous chondrites. *Geochimica Cosmochimica Acta* **38**, 47–64.)

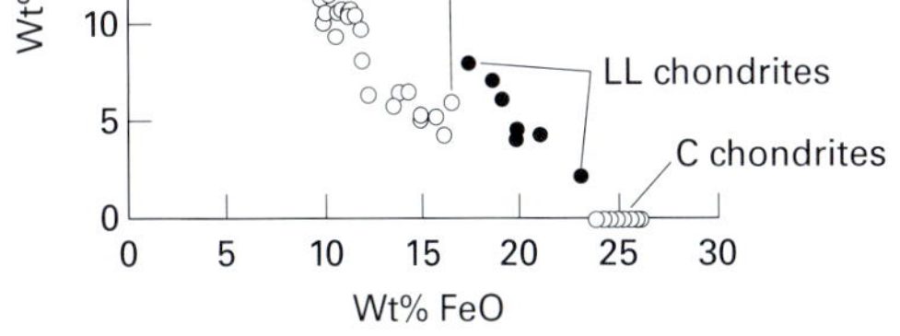

Fig. 5.4. This graph plots the weight percent oxidized iron against the weight percent of iron metal plus FeS in ordinary chondrites observed to fall and recovered shortly thereafter. A clear division of the three clans is obvious and the ordinary chondrite clan is divided into its three groups, H, L, LL. (Data from Mason, B. (1962). *Meteorites*, John Wiley and Son Inc., New York.)

[IV] By convention, elemental compositions are usually designated as oxides. Thus, FeO, MgO, etc. For the olivines and pyroxenes in meteorites, mole percent iron is usually given as a ratio, written FeO/(FeO + MgO). This compares the oxidized iron (FeO) with (FeO + MgO), the total composition of the olivine; that is, iron **plus** magnesium.

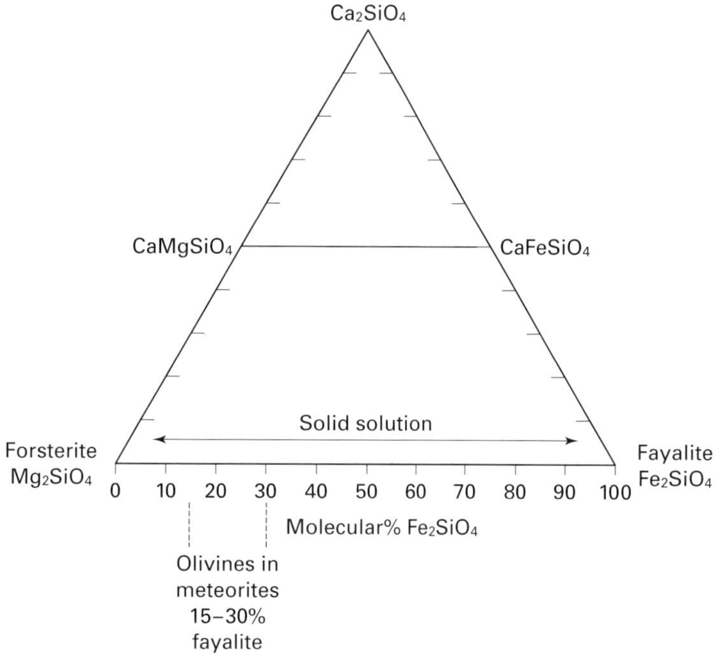

Fig. 5.5. Composition diagram for olivine in the system Ca_2SiO_4–Mg_2SiO_4–Fe_2SiO_4. Most important in meteorites is the forsterite–fayalite solid solution series in which the fayalite to forsterite ratio usually extends from 15 to 30% fayalite. (Adapted from *Rocks From Space*, 2nd edition (1998), Mountain Press Publishing Company, Missoula, Montana.)

percentages in their two most common ferromagnesian minerals, olivine and pyroxene.

Olivine, generally thought of as one mineral, is actually a whole family of minerals with an orthorhombic crystal structure but with varying composition. Its general formula is $(Mg,Fe)_2SiO_4$. Both the magnesium and iron have similar atomic sizes and can readily substitute for each other in the crystal lattice. This is known as a *solid solution* and is illustrated in the composition diagram in Fig. 5.5. As the diagram shows, olivine can vary from 100% magnesium and 0% iron (*forsterite* – Mg_2SiO_4) to 100% iron and 0% magnesium (*fayalite* – Fe_2SiO_4). These are the end members of the solid solution series. The olivine composition in meteorites is usually given as the mole percent fayalite, which is a statement of the oxidization state of the iron or how much iron is bound up in the olivine. For example, a typical ordinary chondrite in group H has a fayalite composition of 18%, written: Fa_{18}. The ratio FeO/(FeO + MgO) therefore yields the fayalite composition. For meteorites, the fayalite composition most commonly lies between 15 and 30%.

Pyroxene, like olivine, is not one mineral but many with similar structures and compositions that vary between magnesium and iron. Like olivine, it crystallizes in the orthorhombic system. Its general formula is $(Mg,Fe)SiO_3$. Its solid solution series is shown in Fig. 5.6. Most of the pyroxene is orthopyroxene which varies from 100% magnesium ($MgSiO_3$ – *enstatite*) to 100% iron ($FeSiO_3$ – *ferrosilite*) as end members. In the past, meteorite nomenclature often included the name of the dominant orthopyroxene. Thus, there are *enstatite* chondrites or olivine–*pigeonite* chondrites. This nomenclature is no longer used in the literature but it is useful nevertheless since it is a statement of the dominant pyroxene helpful for remembering the type of chondrite. Like the olivine, the orthopyroxene composition in meteorites is given as the mole percent of the iron-bearing end member, ferrosilite (Fs_{22}, as an example). This would be a typical composition for the L group of ordinary chondrites. Some of the pyroxenes, especially in some of the achondrites (see Chapter 8) contain calcium. These are clinopyroxenes and crystallize in the monoclinic crystal system. Pyroxenes such as pigeonite, diopside, and augite are examples. All of these are found in meteorites.

The five groups of chondrites in Table 5.1 opposite are distinguished by their total iron, both oxidized iron and metal. The table gives the chemical classes of the chondrites with the normal variations found in their metal phase, total iron, fayalite and ferrosilite content.

The three ordinary chondrite groups are labeled H, L, and LL. The H means high total iron (both metal and oxidized), L stands for low total iron, and LL stands for low total iron and low metal. This table is an excellent demonstration of Prior's rules. Prior (and later Mason, 1962) used the chemical names of the prominent orthopyroxene found in the three groups of ordinary chondrites: *olivine–bronzite* and *olivine–hypersthene*. Today, they are simply designated type H, L, and LL. LL chondrites were not recognized as a separate group in Prior's day and he listed them with the achondrites. Mason added an *olivine–pigeonite* group in 1962. Pigeonite is a low calcium pyroxene (a clinopyroxene) which Prior did not recognize in 1920. This group contained a small amount of carbonaceous compounds and most of their iron was in the oxidized state. In 1967, Mason deleted the olivine–pigeonite group and added it correctly to the carbonaceous chondrite class. Finally, in 1964, the first electron microprobe analysis was made by Klaus Keil and Kurt Fredriksson on 95 equilibrated chondrites. They found that the iron oxidation within

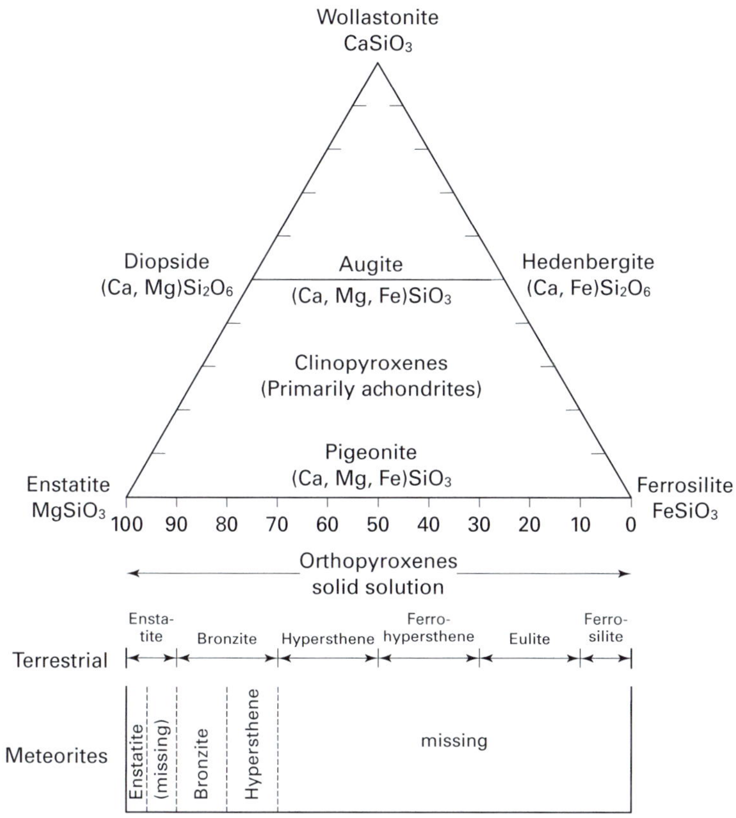

Fig. 5.6. Composition diagram for pyroxene in the system $CaSiO_3$–$MgSiO_3$–$FeSiO_3$. Most important in meteorites is the orthopyroxene solid solution series enstatite–ferrosilite. In meteorites the orthopyroxene does not form a complete solid solution. The range of orthopyroxene in meteorites extends from pure $MgSiO_3$ (enstatite) to iron-rich bronzite referred to as hypersthene. Hypersthene in terrestrial orthopyroxenes usually extends to higher iron content. Clinopyroxenes (augite, diopside) in which calcium appears with magnesium and iron occurs commonly in achondrites. (Adapted from *Rocks From Space*, 2nd edition (1998), Mountain Press Publishing Company, Missoula, Montana.)

Clan	Group	Metal (wt%)	Total iron (wt%)	Fa (mole%)	Fs (mole%)
E	H & L	17–23	22–33	<1	0
OC	H	15–19	25–30	16–20	14–20
	L	1–10	20–23	21–25	20–30
	LL	1–3	19–22	26–32	32–40

Table 5.1. **The enstatite chondrites (EH, EL) and the ordinary chondrites (H, L, LL) are classified by their total iron content**

In this table the H, L, and LL symbolize the chemical abundance of iron found in each, both as metal (unoxidized) and combined iron (oxidized). The H and L symbols represent high total and low total iron respectively. The LL means low total iron and low metal. Note that the fayalite content of the olivine and the ferrosilite content of the pyroxene are both distinguishing indicators of each group.

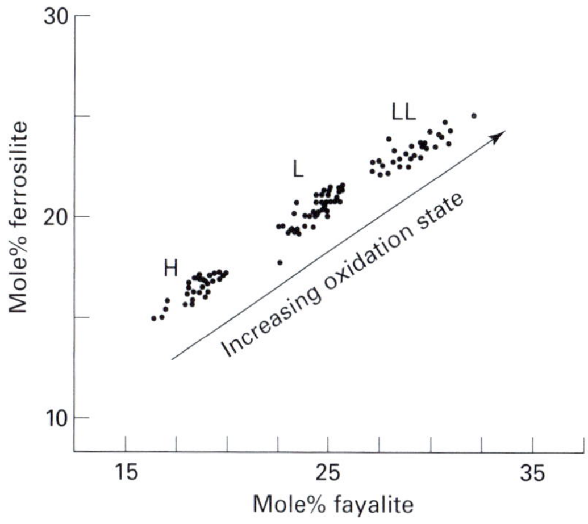

Fig. 5.7. A plot of the fayalite (Fa) content of olivine against the ferrosilite (Fs) content of orthopyroxene in equilibrated ordinary chondrites clearly reveals the existence of three oxidation groups: H, L, LL. (Adapted from Fredriksson, K. and Keil, K. (1964). The iron, magnesium and calcium distribution in coexisting olivine and rhombic pyroxenes in chondrites. *Journal of Geophysical Research* **69**, 3487–3515.)

the major silicate minerals revealed differences among the chondrites that clearly suggested not two but three groups: the H and L group and also a new small but significant group, the LL chondrites. If the mole percent fayalite in the olivine (Fa) and mole percent ferrosilite (Fs) in the pyroxene are plotted against each other the three ordinary chondrite groups clearly appear (Fig. 5.7). Once again, this is a statement of Prior's rules that *the more iron in these minerals as fayalite and ferrosilite, the less iron as metal in the bulk composition*. At the same time, it is also a statement of the increase in oxygen within the mineral as the oxidized iron increases. Thus, the H chondrites are the least oxidized and the LL chondrites the most oxidized of the ordinary chondrites.

The Van Schmus–Wood criteria for the petrologic[V] – chemical classification of chondrites

The classification of chondrites based upon chemical/mineralogical criteria is considered a *primary* classification in that the bulk chemistry is thought to be a primary characteristic given to the meteorites or their parent bodies while forming in the solar nebula. It was clear, however, that meteorites within a particular chemical group, the three groups from the ordinary chondrite clan, for example, had remarkably similar bulk compositions, but under hand lens and microscope there were striking petrographic differences. Meteoriticists wanted a classification system that would also take into account these petrographic differences, thus allowing a meteorite to be at least roughly classified by visual inspection. This required that *secondary* properties be considered, that is, properties that formed from processes which modified the original primary petrographic characteristics. An effective classification of chondrites could take into account *both* their petrographic properties as well as their chemical differences. Petrographic differences can be used to subdivide and further define the chemical groups.

In 1967, W. Randall Van Schmus and John A. Wood published a landmark paper in which they presented a comprehensive classification for the chondrites which, with some modification, is still being used today.[1] Table 5.2 is a chart giving ten criteria Van Schmus and Wood used to determine the petrographic type of each chondrite group. Of the ten, most involve precise chemical and mineral analysis. Fortunately, among the criteria (numbers 3, 4, 7and 8) there are well-defined properties that are readily observable through microscopic study of thin sections so that one can visually estimate the petrographic type with some confidence without chemical analysis. Criteria numbers 7 and 8 are presented first. Though out of order historically, they establish criteria needed to understand the classification of petrographic type upon which the remaining criteria refer. Historically, the order in which the 10 criteria were originally presented is noted in parentheses. The original order is presented in Table 5.2. We now look at these criteria in some detail.

1. Chondrule texture (7)

Of the 10 criteria, chondrule texture and density is the most easily observed. Petrographic types range from 1 to 6. Type 1 chondrites are absent of chondrules. This is a very rare C chondrite of which only five are known (see Table 7.1). Type 2 contains distinct chondrules but they are rather sparsely distributed within a matrix that constitutes nearly 50% of the meteorite by volume. The Murchison CM2 chondrite is a good example. Type 3–6 show progressive stages of thermal metamorphism (Fig. 5.8 a–d). The chondrule boundaries become progressively indistinct as solid state recrystallization occurs. This causes alteration of the original chondrule boundaries due to the intergrowth of chondrule and matrix material. This recrystallization does not represent heating to the point of fusion but only sufficient heating to allow migration and recombination of the mineral elements into new minerals. This is called solid state recrystallization and occurs between 400 and 950 °C.[VI]

Ordinary chondrites show petrographic types from 3 to 6. Often Type 5 and 6 ordinary chondrites show brecciated textures, composed of light clasts set against a dark matrix. It is not unusual to see more than one petrographic type in these breccias. Typically, the clasts show Type 5 or 6 while the matrix shows Type 3 or 4. In this case the entire petrographic range is designated: Type 3–6.

2. Matrix texture (8)

This criterion is easily observed in thin sections. Matrix textures in Type 1 and 2 chondrites are opaque (black) and very fine-grained with scattered recognizable crystal fragments. Type 2 show small chondrules, enclosing only about 12% of the meteorite by volume. Type 3 is still unequilibrated and its matrix is still dark but chondrules have increased in number and take up 30% or more of the volume. From Type 4 to 6, increasing thermal metamorphism in ordinary chondrites produce recrystallization of the matrix in which the crystals grow from cryptocrystalline to near naked-eye

[V] Most books and papers on meteorites use the term *petrologic* when they really mean *petrographic*. John F. Kerridge and Mildred Shapley Matthews point out that the correct term is *petrographic* since the meteorite types described in this classification are based upon microscopic studies of their texture and mineralogy. This is only part of the *petrologic* study of meteorites. (Kerridge, J.F. and Matthews, M.S. (eds.) (1988). *Meteorites and the Early Solar System*, University of Arizona Press, Tucson, p. 25.)

[VI] There have been a few ordinary chondrites and at least one E chondrite that appears to be even more recrystallized than would be expected of a Type 6. Thus, some researchers designate a Type 7 in which all vestiges of chondrule texture have disappeared. All of these meteorites show intense thermal metamorphism beyond solid state recrystallization to the point of melting and may be better termed *impact melt breccias*. The melting may be due to shock on the parent bodies as a result of impact. Type 6 is the most metamorphosed state beyond which melting occurs. There is, therefore, no need for petrographic Type 7.

Criteria	Petrographic Types 1	2	3	4	5	6
(1) Homogeneity of olivine and pyroxene compositions	—	Mean deviation of pyroxene ≥5%, olivine ≥5%		<5% mean deviation to uniform	Uniform ferromagnesian minerals	
(2) Structural state of low-Ca pyroxene	—	Predominantly monoclinic crystals		Monoclinic crystals >20%	<20%	Orthorhombic crystals
(3) Degree of development of secondary feldspar	—	Absent		<2 μm grains	<50 μm grains	Clear interstitial glass; >50 μm grains
(4) Igneous glass in chondrules	—	Clear and isotropic primary glass; variable abundance		Turbid if present	Absent	
(5) Metallic minerals (maximum wt% Ni)	—	Taenite absent or very minor (Ni <200 mg/g)	Kamacite and taenite present (>20%)			
(6) Sulfide minerals (average Ni content)	—	>5 mg/g	<0.5%			
(7) Chondrule texture	No chondrules	Very sharply defined chondrules		Well-defined chondrules	Chondrules readily distinguished	Chondrules poorly defined
(8) Matrix texture	All fine-grained, opaque	Much opaque matrix	Opaque matrix	Transparent microcrystalline matrix	Recrystallized matrix	
(9) Bulk carbon (wt%)	3–5%	1.5–2.8%	0.1–1.1%	<0.2%		
(10) Bulk water content (wt%)	18–22%	3–11%	<2%			

Table 5.2. **Criteria for distinguishing petrographic types in chondrites (R. Van Schmus and J. Wood, 1967)**

In 1967 Van Schmus and Wood devised a now widely used scheme to distinguish between one of six petrographic types for E, OC and C chondrite clans. This table displays the ten criteria used in this distinction as they were originally devised. The broken lines are intended to reflect the lack of sharpness of the boundaries between two petrographic types. A few meteorites show higher states of metamorphism and have been given a petrographic Type 7. This type is still in dispute. These criteria are detailed in the text. (Adapted with modifications from Van Schmus and Wood, 1967.)

Fig. 5.8. These four ordinary chondrites of increasing petrographic type from Type 3 to Type 6 demonstrate the effects of progressive stages of thermal metamorphism. (a) Sahara 95035 shows well defined and densely packed chondrules and fine-grained dark matrix typical of an L3; (b) Julesberg L3.6 shows less densely packed chondrules and less defined boundaries with some recrystallization of the matrix ; (c) Cole Creek H5 shows considerable recrystallization with fewer chondrules and less defined compared to Type 4; (d) Tenham L6 has few chondrules and a recrystallized matrix. Solid state recrystallization between the matrix and chondrules blurs the chondrule boundaries. All photographs were made in crossed-polarized light which helps to distinguish the individual crystal grains within the chondrules and matrix. All specimens were photographed to the same scale. The horizontal field of view is 10 mm. (Cole Creek H5 thin section kindly furnished by Michael Blood. Photos by O. Richard Norton.)

Fig. 5.9. (Top) The Saratov L4 chondrite shows a dark interior with a microcrystalline matrix texture. (Bottom) The Mbale L6 chondrite is completely recrystallized showing a lighter interior and a much coarser matrix. Both specimens were photographed to the same scale. The horizontal field of view is 8 mm. (Photos by O. Richard Norton.)

visibility. This turns the matrix transparent giving the interior of the chondrite a white appearance (Fig. 5.9). Groups EH, EL and CK chondrites also show thermal metamorphism with Type 3–5, Type 3–6, and Type 4–6 respectively.

3. Homogeneity of olivine and pyroxene compositions (1)

From the textures of the ordinary chondrites, we can assume for the moment that they all began in a relatively unmetamorphosed state designated Type 3. At this point, the chondritic body is chemically *unequilibrated*; that is, its mineral composition is *heterogenous*, showing wide variations in chemical composition within each mineral. For example, the two most common minerals in chondrites, olivine and pyroxene, show wide variations in their Mg/Fe compositions. The chondritic body apparently cooled too rapidly to allow the various chemical reactions to proceed to completion. The minerals in Type 3 ordinary chondrites are therefore not in chemical equilibrium with their surroundings. When the parent body is subsequently reheated, the individual minerals became progressively more homogeneous as chemical reactions renew and continue. In unequilibrated Type 3 ordinary chondrites, the iron composition in olivine and orthopyroxene vary from grain to grain by more than 5%. This variation is progressively reduced through Type 4 until it reaches nearly a singular composition at Type 5 where both have become more ferrous. All the olivine and all the orthopyroxene now have similar iron

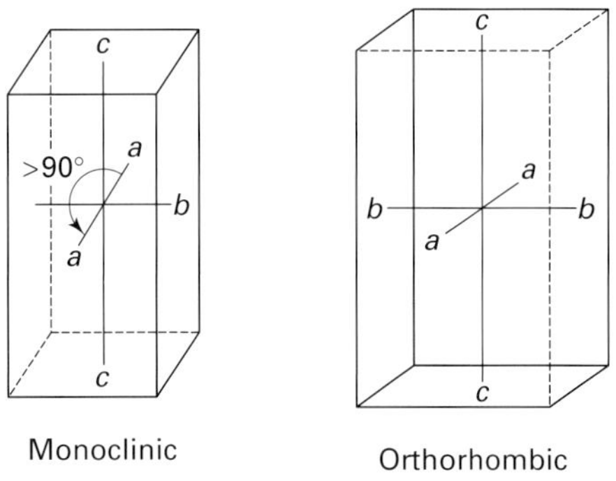

Fig. 5.10. (Left) A monoclinic crystal showing internal axes. (Right) An orthorhombic crystal showing internal axes. Low-calcium pyroxene has a monoclinic crystal form in unequilibrated chondrites but converts to an orthorhombic crystal form in equilibrated chondrites (see text).

compositions. Type 5 and 6 chondrites are both *homogenous* and *equilibrated*.

4. Structural states of low-calcium pyroxene (2)

Low-calcium pyroxene in unequilibrated chondrites (Types 2 and 3) is usually found as monoclinic crystals – three unequal axes, two of which are obliquely inclined to each other with the third axis perpendicular to the plane of the other two (Fig. 5.10). In this crystal system the pyroxene is metastable, that is, it retains this crystal form only if the conditions of formation remain stable. If the chondrite is heated, the pyroxene recrystallizes into orthorhombic crystals – three mutually perpendicular axes all of different lengths. The ratio of monoclinic to orthorhombic crystals is an indicator of the petrographic type. In Types 2 and 3, monoclinic pyroxene predominates. Type 4 is transitional with both monoclinic and orthorhombic crystals coexisting. Types 5 and 6 are primarily orthorhombic. These different crystal systems can often be observed with a high power microscope in well formed euhedral crystals.

5. Degree of development of secondary feldspar (3)

Other effects of metamorphism in ordinary chondrites can be seen on a microscopic level. In Type 3 chondrites, olivine and pyroxene crystals in the matrix are often seen suspended in a clear glass. The amorphous glass, lacking a crystalline structure, is optically homogenous in all directions (isotropic) and goes to extinction under crossed-polarized light at any optical rotation angle. Typically, glass is metastable and begins to devitrify (change to a crystalline state) when conditions change. In Type 4 chondrites, the glass begins to show devitrification becoming more turbid, appearing ("feathery") as tiny sub-micrometer-sized crystals begin to form and scatter light. By Types 5 and 6, the glass has disappeared, replaced by crystals. A glassy mesostasis of feldspathic composition is present in the chondrules of Type 3 ordinary chondrites. This is igneous glass formed during melting associated with chondrule formation. This glass tends to devitrify becoming feldspar grains (usually the sodic feldspar, *oligoclase*) of progressively increasing size with higher petrographic types. Feldspar grains within the matrix is a clear sign of an equilibrated Type 6 chondrite. It can be recognized under the microscope in reflected light as dark gray crystals 50–100 μm across, sometimes showing twinning under crossed-polarized light. The feldspar is referred to as a "secondary" mineral because it is not present in Type 3 but only appears as a consequence of recrystallization.

6. Igneous glass in chondrules (4)

A glassy mesostasis appears within chondrules of Types 2 and 3. This is igneous glass formed during melting associated with chondrule formation. It is a result of rapid cooling of the chondrule before crystals can form. It is called igneous because it formed through a rapid melting of silicate minerals to form a liquid droplet which is then quenched (see Chapter 10). The composition is usually feldspathic. It is commonly seen in Types 2 and 3 within barred olivine chondrules as a clear, isotropic glass (Fig. 5.11). If present in Type 4, it is usually translucent or cloudy (turbid) because it has begun to recrystallize into tiny micro-crystals that effectively scatter light. It is generally not seen in Type 5 and may appear in Type 6 as recrystallized plagioclase. Criteria (3) and (4) are obviously closely related.

7. Metallic minerals (maximum weight percent nickel in FeNi) (5)

Metallic iron alloyed with nickel is found as small grains in most chondrites. Unequilibrated Type 2 chondrites contain the low nickel mineral, kamacite. High nickel taenite along with kamacite is found in Type 3–6 with higher nickel content trending to higher petrographic types. This is because much of the iron metal has been oxidized within the silicates leaving nickel alloyed with the remaining metal (Prior's rules).

8. Sulfide minerals (average nickel in FeS) (6)

These include troilite (FeS) and pentlandite $(Fe, Ni)_9 S_8$. Those chondrites showing no metamorphism contain the above iron sulfides that are rich in nickel. In equilibrated chondrites, Type 4–6, the nickel content of the sulfides decreases with increasing petrographic type.

9. Weight percent carbon (9)

Type 1 to Type 3 include the carbonaceous chondrites. The carbon content is greatest in Type 1 with 3–5% and decreases through Type 2 to Type 3, which can have as little as 0.2%. Ordinary chondrites from Type 4 to Type 6 have the least carbon with less than 0.2 wt%.

Fig. 5.11. (Top) Thin section of a 1 mm diameter barred olivine chondrule from a Brownfield, Texas, H3 chondrite. The olivine bars are embedded in igneous glass that is isotropic and goes to extinction (black) in this image photographed in crossed polarized light. (Bottom) This image in plane polarized light reveals the interstitial glass in a translucent brown color. (Photos by O. Richard Norton and Tom Toffoli.)

10. Water content (10)

Groups CI, CM, and CR carbonaceous chondrites contain the greatest weight percent of water of all the chondrites. Since carbon and water are volatile components, they would have been easily drawn out of these meteorites under the high temperatures experienced by the ordinary chondrites. The fact that they contain as much as 22 wt% water (Type 1), suggests that they were never heated but instead show aqueous alteration of their minerals. Water content decreases with increasing petrographic type. The ordinary chondrites contain less than 0.2 wt% water.[VII]

[VII] An extraordinary fall of two H5 chondrites in the city limits of Monahans, Texas, occurred on March 22, 1998. They were examined by Everett Gibson, M. Zolensky and G. Lofgren of NASA's Johnson Space Center in Houston within hours of the fall. Besides the normal ferromagnesian minerals expected in the matrix of an H5 OC, the team found crystals of halite (sodium chloride or common table salt) and sylvite (KCl), some up to 3 mm across. They were purple in color as a result of exposure to UV radiation in space. Trapped within the crystals were found tiny inclusions of briny water. This is the first time that liquid water has been found in a meteorite. An H3–6 chondrite called Zag fell in the Moroccan Sahara desert five months later that also had halite crystals with water inclusions.

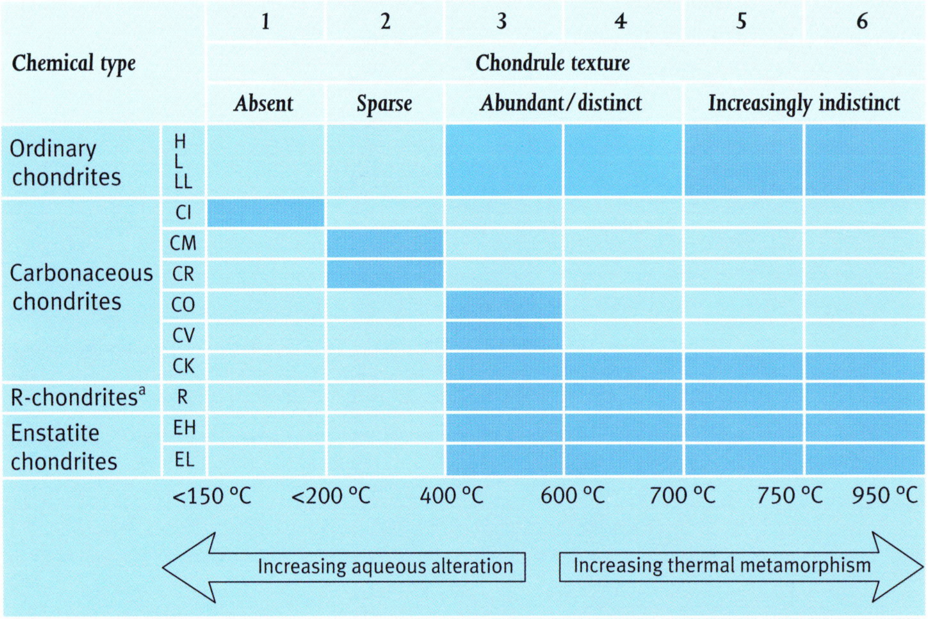

Note: [a] R-chondrites have brecciated textures with light clasts and dark matrix. Typically, the clasts show high petrographic types (5–6), and the matrix shows low petrographic types (3–4).

Table 5.3. **Petrographic types**

This chart shows the grouping of all chondrites into chemical and petrographic types. The chemical types are thought to represent different asteroid parent bodies while the petrographic types refer to various states of thermal metamorphism or aqueous alteration occurring on or within the parent bodies. The ordinary chondrites show thermal metamorphism, while the carbonaceous chondrites can be divided into those that show aqueous alteration and those that show thermal metamorphism. The blank boxes indicate the combinations that either do not exist or have yet to be found.

The Van Schmus–Wood petrographic types added a new designation to meteorites. Today a meteorite is classified using both chemical and petrographic designations. For example, a chemical type H chondrite can have a petrographic type between 3 and 6 or H3 . . . H6. Ordinary chondrites have ranges from H, L, or LL3 to 6. Taken together, the carbonaceous chondrites vary from C1–6 and the E chondrites 3–6. Table 5.3 gives a summary of the known petrographic types for the chondrite groups.[VIII]

[VIII] Type 3 ordinary and carbonaceous chondrites are an exception. These have been subtyped from 3.0 to 3.9 using a different set of criteria. This was found necessary because Type 3 ordinary chondrites have gone through an unusually large range of thermal metamorphism, more so than the other types. Among the new criteria is thermoluminescence sensitivity (tendency to emit light or infrared energy upon heating), percent matrix recrystallization, variation of the element cobalt in the low nickel kamacite, variations of the fayalite in the olivine, and the FeO/(FeO + MgO) ratio in the matrix. CO chondrites are likewise subtyped from CO 3.0 to CO 3.9.

Shock metamorphism and shock stage classification[2]

Among the processes that change the primordial characteristics of ordinary chondrites is dynamic or shock metamorphism due to low- and high-velocity impacts during and following the accretion stage of the parent bodies.[IX] Most ordinary chondrites show some evidence of impact shock. The most obvious shock-produced properties seen easily with the naked eye are brecciation and veining. Such properties were used in early classification schemes (Brezina) but were later disregarded because it was thought that such properties represented secondary changes that only acted to mask the more fundamental properties. In 1991, D. Stöffler, K. Keil, and E.R.D. Scott pointed out that shock metamorphism occurring very early in parent body formation was in essence part of the process that affected the fundamental characteristics of the body and should therefore be considered an important factor in the classification of chondrites.

Since olivine is the primary mineral used for determining shock classification, it can be used with any olivine-bearing meteorite including C chondrites and some achondrites (ureilites). E chondrites are composed of enstatite with less than 1% olivine. Thus, the shock scale using olivine alone does not include these rare meteorites. However, in 1997, A.E. Rubin and E.R.D. Scott extended the scheme to include orthopyroxene and measured the shock classification of enstatite chondrites.[3]

The Stöffler–Keil–Scott shock classification system is divided into six progressive stages (S1–S6) and relies upon shock effects observed in olivine and plagioclase (oligoclase). Plagioclase usually does not reveal itself until the chondrite is equilibrated (petrographic Types 5 and 6) and the crystals are very small making undulose extinction and mosaic deformation textures difficult or impossible to observe. Thus, the usefulness of plagioclase to determine shock stage is limited. Pyroxene shows similar effects but is not used in the classification because, unlike olivine and plagioclase, the effects occur over such a wide range of pressures that it is difficult to designate the shock stage with any confidence. Fortunately, the identification of shock features requires only the preparation of a thin section and examination under a polarizing petrographic microscope. Magnifications of several hundred is necessary to see the effects.

The following is a summary of the criteria used to classify the intensity of shock metamorphism.

The Stöffler–Keil–Scott shock classification system

S1 – *unshocked*

The olivine and plagioclase show sharp optical extinction as the thin section is rotated. There may be some irregular fractures which is typically seen in olivine and plagioclase in terrestrial rocks and unshocked chondrites. The pressure should be less than 5 GPa.[X]

S2 – *very weakly shocked*

Both olivine and feldspar show undulose extinction, that is, the extinction is not uniform but instead occurs across the mineral as a wave of darkening. Undulose extinction is commonly found in stressed quartz grains in igneous and metamorphic rocks but is produced by static stress (crystallization or tectonism) rather than through impact shock. Irregular fractures are common (Fig. 5.12). The pressure is 5–10 GPa.

S3 – *weakly shocked*

Olivine shows sets of planar fractures, *shock lamellae*, in which there are three or more parallel fractures in the set or two or more sets of parallel fractures intersecting each other (Fig. 5.13). These appear as closely spaced fractures distinct from cleavage plains but oriented parallel to planes in the crystal lattice. Shock lamellae are the most important shock indicators at this stage. Undulose extinction is still present in the olivine. Feldspar does not show planar fractures but still displays undulose extinction. Opaque melt veins begin to appear along with pockets of melted material. This occurs at pressures of 15–20 GPa.

S4 – *moderately shocked*

Planar fractures still exist in this shock stage and a weak mosaic pattern forms on the olivine crystals. This is the most

[IX] In the Van Schmus–Wood petrographic classification, several of the criteria involve thermal metamorphism. This is *static* metamorphism since the heating takes place over an extended period of time within a parent body but does not result in melting. Metamorphism due to impact shock is *dynamic* metamorphism and the heating occurs almost instantaneously upon impact, with the most energetic impacts producing melting.

[X] The pascal (Pa) is the standard for pressure measurements relating to impact shock. Its magnitude is small compared to standard atmospheric pressure (bar): 1 bar ≅ 10^5 Pa; 1 GPa = 10^9 Pa = 10 000 atmospheres.

Fig. 5.12. This sequence of photographs demonstrate undulose extinction. An unshocked grain goes through complete instantaneous extinction as the specimen is rotated. The white grain in the center has been fractured, displaced and shocked. As the thin section is rotated, extinction begins at the upper edge of the grain, moves from left to right across the fracture and down from the top, finally concentrating in the center of the grain. The field is 3 mm across. (Photo by O. Richard Norton and Tom Toffoli.)

Fig. 5.13. Sets of parallel-running shock lamellae at angles to each other traverse a shocked olivine grain within a chondrule in a Gold Basin L4 chondrite. This is shock grade S3. The olivine grain is about 1.5 mm across. (Photos by O. Richard Norton and Tom Toffoli.)

important criteria for the S4 shock stage. Plagioclase displays undulose extinction and shows planar deformation structures that appear partially isotropic under polarized light. This is the beginning of the phase change of oligoclase into maskelynite glass but without flowing (*diaplectic glass*). Diaplectic glass is a pseudomorph after the original crystal and does not exhibit evidence of flow. Melt pockets with interconnecting veins become more prevalent. The shock pressure ranges from 30–35 GPa.

Fig. 5.14. Multiple shock veins running through a slab of the Etter, Texas, L5 chondrite. Large grains or melt pockets are scattered along the veins. Longest dimension 12 cm. (Photo by O. Richard Norton.)

S5 – *strongly shocked*

Mosaic textures in olivine appear very strong along with planar fractures and planar deformation features. Oligoclase is completely transformed to maskelynite glass. This is the distinguishing feature of the S5 stage. Melt pockets and connecting melt veins spread throughout the meteorite. Opaque shock veins made of a mix of crystalline material formed by melting and intergrowth of various silicate minerals, troilite and metal (Fig. 5.14). These increase with increasing shock stages and frequently enlarge into melt pockets. Shock pressure reaches 45–55 GPa.

S6 – *very strongly shocked*

Olivine and pyroxene recrystallizes in the solid state. Melting of olivine and pyroxene along the crystal edges takes place adjacent to melt pockets. Some olivine changes to ringwoodite, a mineral with the same chemistry as olivine but with a different crystal structure – a dimorph of olivine with a spinel (octahedron) structure. Maskelynite melts and converts to a normal glass in melt pockets. Shock pressure at 75–90 GPa.

Shock pressures higher than S6 will melt the meteorite. The chondritic structure is destroyed and the meteorite will take on an igneous rock texture. These meteorites are termed *impact-melt breccias* and go beyond the shock scale. The shock stage of a chondritic meteorite is usually given along with its chemical and petrographic designation. Thus, an L chondrite of petrographic Type 6 and a shock stage 4 is designated: L6(S4).

Brecciated chondrites

Shock studies tell us that virtually all chondrites have been subject to both low- and high-velocity impacts throughout their history. They all show some degree of shock. Perhaps the most obvious signs of this impact history is in their macro textures. Brecciated chondrites with moderate shock (>S3) show dark, glassy veins of chondritic composition running through their interiors. Many are visible without optical aid (Fig. 5.14). The veins appear as filaments branching off larger main veins. They form through local impact melting of the rock and injection of this melt into preexisting fractures. Shock veins of iron metal and troilite are commonly found in chondritic meteorites, especially those with relatively high metal content. In some cases these metal veins run the length of the meteorite or end in large nodules of metal. Large irregular blebs of metal are other signs of shock melting of metal. Dark veins, metal veins and metal nodules require impact pressures greater than 25 GPa.

Chondrites subjected to such high pressures are almost always brecciated, that is, they are not homogenous solid bodies but are composed of angular rock fragments cemented together into a coherent rock by fine-grained interstitial host material. The lithification process can occur anytime after brecciation and even may require a second impact event thousands or millions of years later. Shock-melting of silicate grains in the matrix of the meteorite furnishes the cementing agent. If the fragments all share the same clastic material among themselves and are of the same composition as the matrix, the meteorite is a *monomict breccia*. These meteorites have been impacted and crushed but their material has remained generally intact and the pieces tend to fit like a jigsaw puzzle but are separated by dark shock-melted matrix material (Fig. 5.15). A more interesting type are breccias composed of two or more clastic fragments of different texture and composition from the host interstitial rock. These are called *polymict breccias* and often show fragments of light and dark clasts. The meteorite, Cumberland Falls, an aubrite achondrite (see Chapter 8), shows light clasts which are part of the original enstatite material and dark fragments of a chondritic composition. Here, material from both impacting bodies are intermingled into one rock (Fig. 5.16). Occasionally, a chondritic breccia is found with clasts of similar composition but of different metamorphic grades (petrographic types). These are called *genomict breccias*. If these meteorites came from the same parent body, they suggest that parent bodies consist of rock of different metamorphic grades probably arranged with the more highly metamorphosed rock more deeply seated within the parent body. High-velocity impact excavated this rock from the deep interior and mixed it with lightly metamorphosed rock near the surface where further impacts lithified the mix into a genomict breccia. These are most easily recognized in thin section under the microscope. This is the so called *onion shell model*. This model suggests that the parent body is broken apart and then reassembled as a *rubble pile*. Both onion shell and rubble pile models will be looked at again in Chapter 11.

Regolith breccias

All of the above breccias still retain the characteristic textures of ordinary chondrites and probably represent fragments of the earliest-formed primitive chondritic parent bodies released either during or shortly after formation.There are other chondritic meteorites that show evidence of having been part of the surface or near surface of a parent body long

Fig. 5.15. A brecciated and shocked L6 chondrite from St. Michel, Mikkeli, Finland. Its white interior shows clasts of the same composition neatly fitted together into a *monomict* breccia, welded together by dark glassy shock-melted matrix material. The specimen is 52 mm across. (Specimen furnished by Blaine Reed Meteorites. Photo by O. Richard Norton.)

Fig. 5.16. The Cumberland Falls achondrite (aubrite) shows a classic *polymict* breccia texture. The white clasts are magnesium-rich enstatite composing the original pre-shocked parent body and the dark xenoliths are of anomalous chondritic composition intermediate between E and H groups. This specimen weighs 49.53 g and measures 60 mm across the longest dimension. (Courtesy Institute of Meteoritics, University of New Mexico. Photo by O. Richard Norton.)

after formation. This shallow layer was subjected to impacts throughout most of its history. This created an unconsolidated layer of fragments and dust resting on solid bedrock similar to that found on the lunar surface. Such a surface is called a *regolith*. Here the surface material has been overturned or *gardened* by impacts. Then the local rock rubble is mixed with transported rock from distant impacts. The fragments and dust eventually compact (through shock lithification) to form a *regolith breccia*. Recent high-resolution photographs of chondritic asteroid surfaces show a characteristic surface permeated with old impact craters covered with dust and rocky debris, alongside of fresh appearing impact craters with sharp rims and surrounding ejecta blankets, all typical of regolith formation (see Chapter 11).

The smaller impacts produce local impact-melted glass which bonds fragments of minerals and rocks into an assemblages called *agglutinates*. These were first discovered in lunar material. The glass is usually highly vesicular indicating high gas content of the original unshocked material and is usually found adhering to or containing other regolith fragments.

About 10% of the ordinary chondrites are regolith breccias. They are easily recognized by a light–dark structure; light clasts and dark matrix (Fig. 5.17). The darkness of the matrix is attributed to a variety of causes. The fine-grained matrix material is surficial and receives a maximum amount of irradiation. Radiation tends to turn the crystals dark. The matrix is dust material composed of the broken down remnants of crustal rock that has been continually crushed and ground down to smaller and smaller sizes through most of the asteroid's impact history. This comminuted state has produced dust-size crystals that are more efficient at trapping incident light, making them appear darker than the fresher light-colored clasts. Finally, the dark matrix material has been heavily shocked which has dispersed fine-grained metal and troilite particles through the silicate minerals making them less transparent and consequently darker. This is especially true of regolith breccias of chondritic composition that contain an abundance of uncombined iron–nickel metal and troilite.

Regolith breccias' greatest distinction comes from their noble gas content. Surface material on small bodies are subjected to constant bombardment from the solar wind. Among the solar wind particles are isotopes of helium, argon and neon. Without an atmosphere on the parent body to impede and absorb them, these isotopes penetrate those minerals directly exposed on the surface. Penetration levels vary from the surface to about 10^{-5} cm. This creates an excess of these noble gases within the breccias detectable with a mass spectrometer.[XI] One curiosity is that the excess is almost

Fig. 5.17. This LL6 chondrite from Beeler, Kansas, shows the typical light–dark structure of a regolith breccia. The dark, finely comminuted material contains solar flare tracks (see text). The slab measures 9.2 cm largest dimension. (Photo by O. Richard Norton.)

[XI] The first relatively accurate measurement of helium abundance within a chondritic meteorite was made by Frederick Adolf Paneth (1887–1958) using an early mass spectrometer of his own design at Durham University in England in 1931. Paneth was trying to determine the age of meteorites by measuring the helium isotope content as a decay product of radioisotopes occurring naturally within the meteorites. Unusually high concentrations of noble gases were first recognized in some meteorites (regolith breccias) by Gerling and Levski in 1956. Suess suggested in 1964 that most of the excess probably originated from the solar wind. This was confirmed by H. Wänke in 1964 and P. Eberhart in 1965. (Suess, H.E., Wänke, H. and Wlotzka, F. (1964). On the origin of gas-rich meteorites. *Geochimica Cosmochimica Acta* **28**, 595–607.)

Fig. 5.18. This large slab about 75 cm across shows the texture of the L5 Rio Limay, Argentina, impact-melt breccia. Note the two distinct fields. The light area is a typical chondritic texture with large chondrules visible. The dark area is an impact melted field lacking chondritic texture. Dark shock veins run through this area. (Photo by Edwin Thompson.)

invariably found in the dark matrix with much lower levels in the lighter clasts. This can be explained by considering the accretionary growth of an asteroid parent body. As accretion proceeds, the surface is quickly covered with fragmented rock which is gradually broken down into a comminuted dust layer or regolith. This layer is exposed to the solar wind. Occasional large impacts excavate deep layers of metamorphosed chondritic rock (H5 or H6) bringing it to the surface as an ejecta blanket. This deeply buried rock has been protected from solar wind and solar flare particles. These rocks, now on the surface, are eroded by micrometeoroid impacts producing chips of light-colored metamorphic rocks that become mixed with the dark regolith debris. There the loose mix of fine-grained and brecciated debris remain until a strong shock wave from a nearby impact causes minor melting along grain boundaries sufficient to cement the regolith debris and light-colored clasts into a strong coherent rock. Lithification of the dark dust surface layer and rock debris forms an H-chondrite regolith breccia with different noble gas contents.

The most energetic solar wind particles are produced by solar flares. They leave tiny tracks along their paths as they penetrate the outer 1 mm of the mineral crystals. These solar flare tracks were first found in meteoritic regolith breccias but later discovered in regolith breccias brought back from the Moon. In the dark matrix between 2 and 20% of the grains showed tracks.

Impact-melt breccias

Crater-producing impact events on asteroids frequently involve shock pressures in excess of 90 GPa. This is sufficiently large to produce wide-spread melting of rock. In this case, the floor of the newly produced crater instantly melts upon impact, forming a melt lens as much as 1 km beneath the crater floor. This rock loses its original texture and, upon cooling, shows more of an igneous texture. Surviving impactor rock, usually a chondrite, loses much of its chondritic texture as it is partially melted in this lens. Usually, however, the melting is incomplete in that the resultant rock is a mix of melted rock and unmelted clasts retaining some of their pre-impact texture. These meteorites are termed *impact-melt breccias* (Fig. 5.18).

These breccias often preserve a record of multiple impact events. This is especially characterized in an extraordinary meteorite found on a hiking path on the south end of the Tucson Mountains near Tucson, Arizona, in the early 1980s. In 1993 the rock was taken to the University of Arizona's Lunar and Planetary Laboratory where David Kring identified the 2.7 kg rock as a rare impact-melt breccia with chondritic clasts showing an L5 texture (Fig. 5.19). The meteorite was

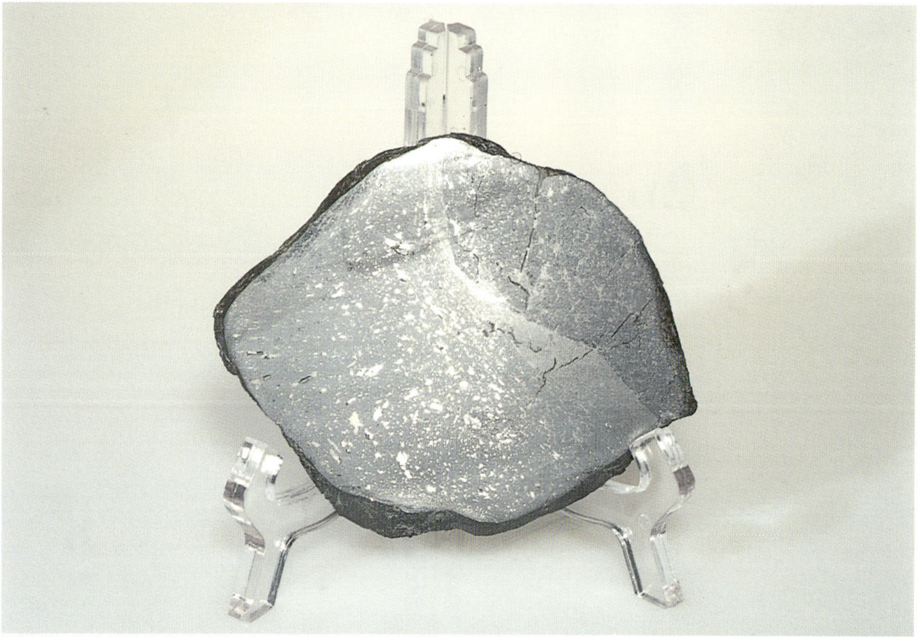

Fig. 5.19. Slab of the Cat Mountain, Arizona, impact melt breccia. The lighter gray areas have clasts with an L5 texture. Metal veins and pools of metal attest to shock events. The L5 texture was destroyed by partial melting in the dark gray fields. The specimen weighs 120 g. Its largest dimension is approximately 97 mm. (Michael Casper photo.)

given the name Cat Mountain, for a prominent southernmost peak in the Tucson Mountains. Subsequent research by Kring and his coworkers led to some extraordinary conclusions. Cat Mountain has yielded one of the most complete records of the collisional history of a primitive chondritic body within the asteroid belt. It is a complex history of multiple impact events concluding with its arrival on Earth only a day before being collected. The original parent body accreted as an L5 chondrite about 4.550 billion years ago. Then, sometime after it formed, one or more impacts created a regolith surface breaking up the original chondritic body to a depth of several meters. Shock veining in the L5 chondritic fragments of the regolith breccia record these events. Then, about 880 million years ago, a major impact took place on the regolith surface creating a crater at least 1 km or more in diameter, producing shock pressures of greater than 80 GPa, melting the floor of the crater and embedding chondritic clasts in a shock-melted matrix. There, the melt cooled, trapping the clasts and forming an impact-melt breccia. Another large impact splintered off a large fragment of the crater floor sending it into space along a trajectory that eventually became Earth-crossing. Other impacts reduced the large fragment to several meter-sized bodies. There they remained in an Earth-crossing orbit for another 20 million years (the cosmic ray exposure age) after which at least one of these fragments collided with Earth.[4] Thus, impact-melt breccias reveal violent impact events from the origin of parent bodies in the asteroid belt to arrival on Earth of their fragments.

References

1. Van Schmus, W.R. and Wood, J.A. (1967). A chemical–petrologic classification for the chondritic meteorites. *Geochimica Cosmochimica Acta* **31**, 747–765.
2. Stöffler, D., Keil, K. and Scott, E.R.D. (1991). Shock metamorphism of ordinary chondrites. *Geochimica Cosmochimica Acta* **55**, 3845–3867.
3. Rubin, A.E. and Scott, E.R.D. (1997). Shock metamorphism of enstatite chondrites. *Geochimica Cosmochimica Acta* **61**, 847–858.
4. Kring, D.A., Swindle, T.D., Britt, D.T. and Grier, J.A. (1996). Cat Mountain: a meteorite sample of an impact-melted asteroid regolith. *Journal of Geophysical Research* **101**, No. E12, 29, 353–29, 371.

Chapter Six

Chondrites: a closer look

Chondrites have characteristics of the three basic terrestrial rock types: igneous, sedimentary and metamorphic. Chondrules, which are the major textural component of chondrites, have igneous properties in that they formed from a molten or partially molten state. Their spherical, subspherical and sometimes ellipsoidal shapes suggest that they formed through an unknown heating mechanism that melted weakly consolidated solid precursor particles that, through condensation from nebular gas (from gas to solid state), were forming and accumulating along the plane of the solar disk. As they reached their liquidus point, they responded to tensional forces on their surfaces and formed spherical droplets in which mineral crystals began to form before being quenched. We can imagine myriads of these now solid spherical bodies attracted to the solar disk plane where they concentrated, gently impacting each other, compacting together with fine-grained matrix material, slowly constructing a chondritic parent body (Fig. 6.1). In this case, gravitational forces controlled the accretion process.

This is not unlike the accretion processes occurring on Earth as small clastic material accumulated, destined in time to become sedimentary rock. Many of these allogenic particles (particles derived from preexisting rocks) had an igneous origin. Here, within a water medium they accumulated by gravity; compacted, desiccated, sometimes recrystallized, and ultimately cemented together. A look at sedimentary rock, say a quartz sandstone, under the microscope shows a well-sorted texture with rounded particles in contact with each other like stacked marbles. In the interstices between the grains, a cementing agent like calcium carbonate or hematite has slowly deposited, acting as a binding agent to lithify the rock.

There are other similarities between the two sedimentary processes. Rounded quartz particles in sandstones have a

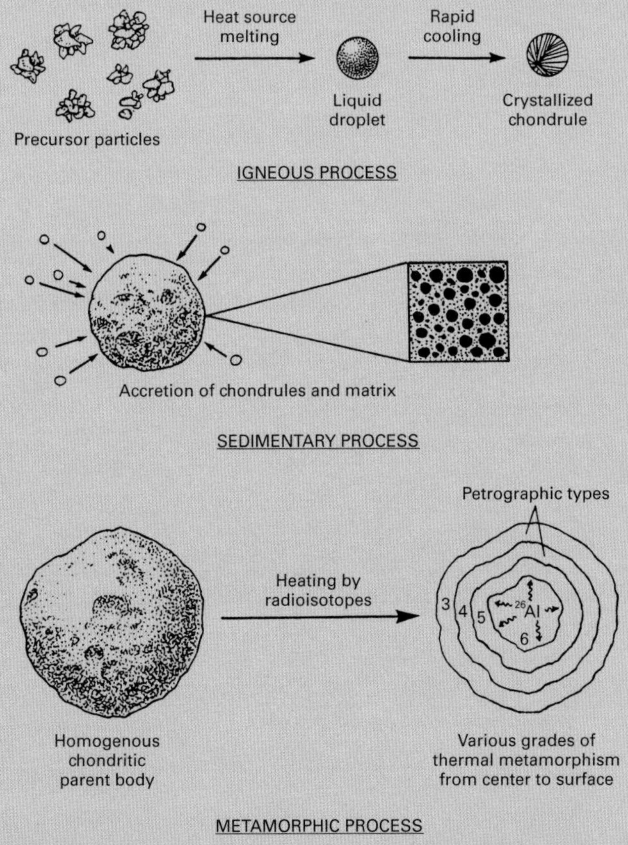

Fig. 6.1. This schematic illustrates the building of a chondritic parent body. During the building phases it passes through igneous, sedimentary and metamorphic processes. An igneous process involves melting of chondritic precursor grains forming spherical chondrules that crystallize upon cooling. The sedimentary phase occurs during the accretion of chondrules and matrix material onto a growing homogeneous parent body. Thermal metamorphism occurs as short-lived radioisotopes (^{26}Al) heat the body creating a heterogeneous, layered structure with metamorphism increasing from surface to center.

range of size from fine sand at about 125 μm to very coarse sand between 1 and 2 mm. Wind and water tend to size sort them into a narrow range of size for a given sandstone rock. Chondrule sizes vary among the different chondrite groups and each group shows variations in chondrule sizes over a range that seems fairly consistent for each group. This suggests that there may have been some form of size sorting within the local environment where each group formed. Finally, after the chondritic parent bodies had formed through accretion of matrix, chondrules and inclusions, they experienced thermal metamorphism deep in their interiors that changed the original pristine chondritic textures throughout the bodies. The degree of metamorphism depended upon the depth of burial at various radii with the least metamorphism near the surface and the greatest in the center.

Chondrite density and porosity

In H3, L3 and LL3 chondrites, the chondrules are very sharply defined and closely packed (Fig. 6.2). If a chondrite had uniform, perfectly spherical chondrules and if they were packed with the tightest possible geometrical arrangement for maximum density – geologists refer to this as *close packing* – the chondrules would form a rock with 26% porosity. So even stacking perfect little spheres as tightly together as possible, an unlikely natural arrangement, still leaves small spaces between each sphere. In actuality, chondrules show a random combination of packing arrangements, sometimes uniform if the chondrules are of similar size and spherical, but much more commonly arranged in a haphazard fashion. This is called *chance packing*. Even if the chondrules were all the same size in this arrangement, the meteorite's porosity would increase to about 40%. In the pore spaces of chondrites, matrix material accumulates. This fine-grained material may be the dusty precursor material that originally formed around the chondrules, helping bind them together, reducing the meteorite's porosity and increasing its density.

Typical sandstone rock has a porosity of between 15 and 30%. It varies widely depending upon particle size and shape, compaction, and cementation. By comparison, ordinary chondrites also have widely varying porosities, ranging from less than 1% to a high of 25% with most lying between 8 and 12%. The L5 chondrite Baszkówka (Fig. 6.3) is exceptionally porous, about 20%, making its density only 2.9 g/cm^3. This is unusual since the density of ordinary chondrites varies between 3.4 and 3.8 g/cm^3. Another example of a highly porous meteorite is the L/LL4 chondrite Bjurböle, with a porosity of 16.7%. This meteorite is not well consolidated and is so friable that its chondrules can be easily picked out of a hand specimen (Fig. 6.4).

Fig. 6.2. The Marlow, Oklahoma, L4 chondrite exhibits sharply defined densely packed chondrules. The chondrules vary greatly in size and shape, many being more ellipsoidal than spheroidal. They are far from a *close packed* geometrical arrangement, rather the arrangement here is *chance packing*. The pore space between chondrules is filled with dark matrix material. The horizontal field of view is 15 mm. (Photo by O. Richard Norton.)

Fig. 6.3. The Baszkówka L5 chondrite from Poland has a 20% porosity. Note the large voids in the interior. This meteorite lacks the compaction of most chondrites. The specimen measures about 8 cm in the longest dimension. (Photo by Andrzej Pilski.)

Fig. 6.4. Bjurböle L/LL4 hand specimen. Its extreme friability makes it subject to crumbling. The chondrules frequently fall out of the matrix leaving obvious cavities. The specimen is 5.3 cm in the largest dimension. (Photo by O. Richard Norton.)

Meteorite porosity values vary widely in the literature. The values given above are bulk measurements utilizing the entire meteorite. Some researchers prefer to measure random matrix values; others measure both. Usually the matrix is less porous. For example, measuring the matrix of ordinary chondrites shows porosities over a narrower range of between 2 and 10%. Porosity also depends upon the weathering grade of the meteorite. Terrestrial chemical weathering can act to fill in fractures and pore spaces with weathering products, decreasing the bulk porosity of the meteorite. For such meteorites, a weathering correction factor must be applied.

Carbonaceous chondrites generally have greater variations in porosity and higher average porosities compared to ordinary chondrites. The highest porosities are, surprisingly, among the CV3 chondrites. They can vary between 1 and 30%. Allende, for example has a matrix porosity of 25% with a bulk porosity of 20%. CM chondrites have a similar range but their modal value is only 6%. One would think the highest porosity would be with the CI chondrites since they have an average density of only 2.2 g/cm^3. Yet, of two that were measured (there are only five known) their matrix porosities were between 4 and 5%. Taken together, all chondrites of Type 1 to 3 show porosity ranges from 2 to 30% with a sharp peak at 2%.[1]

One would think that chondrite porosities would decrease with greater states of metamorphism (higher petrographic type) since high metamorphic states require deep burial where confining gravitational pressures would reduce the pore spaces. This does not appear to be the case, however. Terrestrial sandstones become progressively more compacted with increasing pressures of progressively deeper burial. With gravitational pressures of about 0.4 GPa, the particles are reduced to smaller sizes forcing them into the pore spaces. Such gravitational pressures do not exist within asteroids due to their limited size, mass and internal gravity. On the other hand, porosity of ordinary chondrites seems to vary with shock state. Compaction of the forming parent body was by impact processes. High velocity collisions that produce higher shock stages would be expected to produce a denser, less porous body. Thus, a lightly shocked ordinary chondrite (shock stage S2) could have a porosity of as great as 30%. Highly shocked ordinary chondrites (shock stage S6) have porosities of 10% or less.[2] Carbonaceous chondrites suggest a similar correlation between porosity and shock. Olivine grains in CI, CM, CO, CV, and CR show no significant shock effects, and they are predictably among the more porous of the chondrites.

Chondrules[3]

There are many structures in chondrites that are unrelated to chondrules. To qualify as a chondrule, the object must conform to rather specific criteria. The most important is that they must show clear evidence of having been partially or completely melted. This means that they must show the presence of glass, a texture showing quenching, and a spherical or ellipsoidal or partially spherical/ellipsoidal form which is evidence of the original liquid droplet having formed in a space environment. Some chondrules appear as fragments because of incomplete melting; but like their cousins, the completely melted droplet forms, they must have a ferromagnesian composition.

Chondrule sizes[4] and abundances

Chondrules vary considerably in size within each chondrite group. For the ordinary chondrites, they range from a few hundred microns to as large as a centimeter. For the H, L, LL groups, the mean sizes are 300 μm, 500 μm, and 600 μm respectively, with a mean size for all three of about 450 μm diameter. Group R falls on the OC chondrule average at 400 μm. The mean diameter also tends to vary with the type of chondrule. Droplet chondrules with barred olivine and radial pyroxene textures (see below) tend to be larger than the average. The E chondrites vary significantly among the two groups: 220 μm for EH and 550 μm for EL. Carbonaceous chondrites show the greatest variation in size. In order of increasing diameters they are: 150 μm (CO); 270 μm (CM); 700μm (CR); 800 μm (CK); and 1000 μm (CV). CV chondrites, especially Allende, are known for their larger than average chondrule sizes. Figure 6.5 shows a complete, remarkably spherical unbroken chondrule still within but protruding from the matrix of an Allende CV3 carbonaceous chondrite. The chondrule measures over 6 mm across.

The chondrule density within a chondrite is usually given as a volume percent of the entire meteorite. The volume percent varies considerably for the different chondrite groups. Taken as a whole, the ordinary chondrites contain the greatest number of chondrules, 65–75 vol%, followed by the CV3 carbonaceous chondrites with 35–45 vol% (Table 6.1).

Chondrule textures and compositions

There is perhaps no other study within the chondrite class that is more pleasing aesthetically than the microscopic study of chondrules. Under crossed-polarized light their crystals display interference colors that rival the artist's palette. Chondrules consist of ferromagnesian minerals in various states of crystallization, sometimes emersed in a glass, and in some cases, bounded by a rim of similar mineralogy. If you think of chondrules as spherical bodies (we see cross sections of the chondrules in thin sections) then the rim is an outer envelope that completely confines the chondrule. Chondrule bodies are embedded in a matrix of fine crystals of similar

Fig. 6.5. An unusually large complete spherical chondrule over 6 mm in diameter projects through the crust of an Allende CV3.2 carbonaceous chondrite. Other smaller chondrules can be seen on the surface. (Photo by O. Richard Norton.)

	Type	Texture	Abundance (%)
Group 1 (porphyritic)	PO	porphyritic olivine	23
	PP	porphyritic pyroxene	10
	POP	porphyritic olivine–pyroxene	48
Group 2 (nonporphyritic)	RP	radial pyroxene	7
	BO	barred olivine	4
	C	cryptocrystalline	5
Group 3	GOP	granular olivine–pyroxene	3

Table 6.1. **Classification of chondrules**

Most chondrules of all three chondrite clans show remarkable mineral consistency, being composed of olivine, pyroxene, or both. Although they seem to show a bewildering array of textural forms, close examination reveals only seven basic textural types with variation within each type. These are divided into either porphyritic, nonporphyritic or granular types. The different textures imply different states of formation that tell us something of the environment in which they formed. This table is based upon the work of J.L. Gooding and K. Keil (1981).[5]

bulk mineralogy. The most sharply defined and most densely packed chondrules occur in Type 3 ordinary and carbonaceous chondrites. Figure 6.6 shows a slab from the Axtell CV3.0 chondrite with well-defined, spherical chondrules ranging from about 3 mm to 0.1 mm diameter. Of all the chondrite petrographic types, Type 3 is thought to have somehow escaped the severe thermal metamorphism and other secondary processes that have clearly changed ordinary chondrites of petrographic Type 4–6. On the other hand, petrographic Type 2 show far fewer chondrules and also alteration due to the presence of water. Type 3 chondrites have retained their original primary properties so that their texture and composition are relatively unaltered. Studies of Type 3 chondrules are therefore most likely to reveal the conditions in the early solar nebula.

Chondrules show a bewildering variety of structures and textures but a careful study also reveals a remarkable consistency of textures, so much so that the most common types have been classified on the basis of texture. The texture of a chondrule depends upon its mineral composition, cooling rates, degree of melting, and secondary reheating. J.L. Gooding and K. Keil (1981) proposed a classification of olivine/pyroxene-rich chondrules having seven categories[5] (Table 6.1).

They first divided the chondrules into two broad categories labeled *porphyritic* and *nonporphyritic*. This terminology is borrowed from igneous petrology used to describe the texture of igneous rocks. It is used here with similar meaning: relatively large crystals (phenocrysts) within a fine-grained groundmass. In igneous rock this texture implies that there were two cooling stages of the magma. The first was relatively slow allowing large crystals to form in the magma chamber. This was followed by a relatively rapid cooling period (a lava flow) in which glass or very small crystals formed in the remaining melt enclosing the larger phenocrysts.

Chondrules are studied by preparing a thin section of the meteorite. A thin slice is removed from the meteorite, cemented to a microscope glass slide and ground to a thickness of 0.03 mm. At that thickness most minerals become transparent (except metal, FeS and some metal oxides). The section is then polished to a bright luster. Representative

Fig. 6.6. Axtell CV3.0 carbonaceous chondrite. Weathering of the meteorite has altered some of the chondrules turning them an orange color. The slab stands 10 cm high and the largest chondrules are about 2 mm in diameter. (Michael Casper photo.)

samples of many types of chondrules and their variations seen in transmitted, reflected, and crossed-polarized light, either singly or in combination, liberally grace the pages of this book.

Porphyritic chondrules

Among the porphyritic chondrules are *porphyritic olivine* (PO), *porphyritic pyroxene* (PP) and *porphyritic olivine–pyroxene* (POP). These are named according to the dominant mineral composing them. Porphyritic chondrules are the most common types, totaling 81% of all chondrules in ordinary chondrites. They also show the greatest variations in texture. Porphyritic chondrules are thought to have been only partially melted during formation. They probably formed around tiny grains, acting as condensation nuclei within the solar nebula. In unequilibrated ordinary chondrites (Type 3), porphyritic chondrules come in two different varieties: FeO-poor, Type-I; and FeO-rich, Type-II. They can be recognized by the iron content in their silicates: their fayalite (olivine) and ferrosilite (pyroxene) contents. In ordinary chondrites, the FeO-rich Type-II varieties are more commonly seen.

Porphyritic olivine (PO) chondrules usually contain well-formed olivine crystals in which flat crystal faces are often seen. The crystals can be euhedral (a fully developed crystal showing well-formed crystal faces) to anhedral (a crystal not bounded by its crystal faces) and are often embayed (crystals penetrated by another crystal, glass, or metal). Chondrules with large euhedral to subhedral (a crystal only partially bounded by its own crystal faces) crystals set in a dark microcrystalline mesostasis (remnant glassy interstitial material) are typical of FeO-rich Type-II (Fig. 6.7), while PO

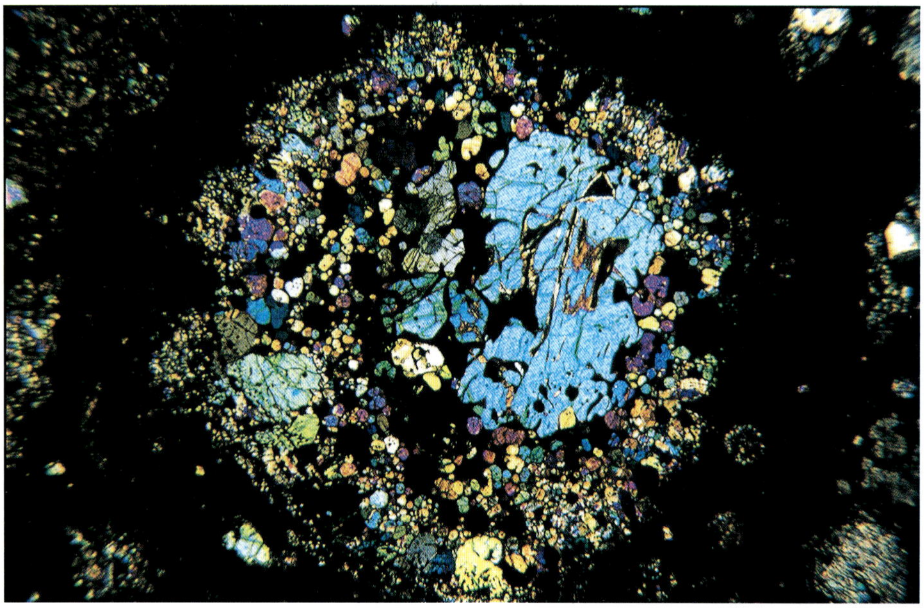

Fig. 6.7. (Top) An FeO-rich (Type-II) porphyritic olivine (PO) chondrule with large euhedral and subhedral olivine grains in bright second and third order interference colors seen under crossed-polarized light. The grains are set in a glassy matrix appearing in extinction (black) under crossed-polarized light. The somewhat elliptical chondrule from a Gold Basin L4 specimen is 2 mm in its largest dimension. (Bottom) An FeO-poor (Type-I) porphyritic olivine (PO) chondrule with small anhedral olivine grains showing bright interference colors. A large deeply embayed blue olivine grain occupies half of the interior of the chondrule, possibly a relict precursor grain formed before the chondrule. This exceptionally large chondrule in an Allende CV3.2 chondrite measures 4 mm in diameter. (Photos by O. Richard Norton and Tom Toffoli.)

Fig. 6.8. This section showing zoning in a clinopyroxene crystal. Note the dark band around the edges of the crystal viewed under crossed Polaroids. This condition called zoning denotes a change in the mineral composition of the pryoxene, usually becoming more iron-rich toward the edge of the grain. This causes a change in birefringence or optical properties from the core to the outer edge. The horizontal field of view is 3.3 mm. (Photo by O. Richard Norton and Tom Toffoli.)

chondrules with small anhedral crystals set in clear glass are the FeO-poor Type-I (Fig. 6.7). Under crossed-polarized light, olivine crystals are easily recognized by their vivid birefringent second and third order colors. Sometimes reaction rims or zoning can be seen along the edges of the crystals (both olivine and pyroxene) where the crystal composition has been altered through reaction with the medium, usually changing to a more FeO-rich composition from the center to the edge (Fig. 6.8). Olivine shows no cleavage or twinning planes but is often permeated with fine random cracks. Alteration is often seen along the cracks. The olivine is set in clear glass, appearing dark (isotropic) under crossed-polarized light or a light brown in plane polarized. The glass usually has the composition of oligoclase feldspar and in this vitrified state is referred to as *mesostasis*. In petrographic Type 4, the glass has devitrified into a microcrystalline texture and becomes anisotropic.

Porphyritic pyroxene (PP) chondrule grains in Type 3 chondrites are usually low-Ca pyroxene or *clinoenstatite*. They are easily recognized by their low birefringent colors usually in shades of gray viewed in crossed-polarized light. The crystals are often crossed by multiple parallel twinning planes called polysynthetic twinning planes (Fig. 6.9). Sometimes tiny colorful crystals of olivine can be seen enclosed within the pyroxene grains. This texture is called *poikilitic* and is common among Type-I PP chondrules (Fig. 6.10). In Type-II PP chondrules, the pyroxene crystals are often much larger than the olivine crystals and are tightly packed within the chondrule (Fig. 6.11). Undulatory extinction due to shock is commonly seen in these crystals when rotated with polars crossed

Fig. 6.9. A single clinoenstatite pyroxene grain showing low birefringence colors and polysynthetic twinning planes, dark parallel lines running the length of the crystal. From an H5 Faith, South Dakota, specimen. Magnification: 156×. (Photo by O. Richard Norton.)

Fig. 6.10. A Type-I porphyritic pyroxene (PP) chondrule (white rim area) with brightly colored olivine grains scattered poikilitically within the pyroxene grains seen in crossed polarized light. Thin section from a Saratov L4 specimen. The horizontal field of view is 2.0 mm. (Photo by O. Richard Norton and Tom Toffoli.)

6.11. A Type-II porphyritic pyroxene (PP) chondrule in crossed-polarized light showing large tightly packed clinoenstatite grains, many showing polysynthetic twinning planes under crossed-polarized light. Thin section from the Semarkona LL3.0 chondrite. The chondrule is about 1 mm in diameter. (Courtesy Institute of Meteoritics, University of New Mexico; photo by Dr Rhian Jones.)

because most chondrites have been shocked to Stage 2 or above.

Porphyritic olivine–pyroxene (POP) chondrules are the most common type of porphyritic chondrule, accounting for 48%. These contain mixtures of both olivine and clinoenstatite. Often the interior of the chondrule is occupied by small olivine grains with the rim composed of larger pyroxene crystals (Fig. 6.12). It is important to keep in mind that the classification presented here is somewhat artificial. There is an almost infinite variation seen among the porphyritic textures. Transitional types exist between all of the major types. For example, the POP is quite similar to the GOP type (see below), differing only in grain size. Where one type ends and the other begins is a matter of interpretation. In later petrographic types, thermal metamorphism produces yet other variations.

Granular olivine–pyroxene chondrules (GOP)

GOP chondrules are the rarest type comprising only about 3% of the total. They are usually a tightly packed mix of olivine and pyroxene anhedral grains of uniform but very small size set in a glassy mesostasis. The small grain size requires high magnifications to see them well. Often they are surrounded by much coarser grains of the same minerals forming a rim (Fig. 6.13). This small group is sometimes classified with the nonporphyritic chondrules.

Nonporphyritic or droplet chondrules

These chondrules are thought to have formed from fully molten droplets. They condensed out of the solar nebula as "fluffy" snowflake-like masses of sub-micron-sized mineral

Fig. 6.12. A porphyritic olivine–pyroxene (POP) chondrule under crossed-polarized light shows small colorful euhedral to subhedral olivine grains with interspersed large lath-shaped white clinoenstatite grains. Thin section from an Allende CV3.2 chondrite. The horizontal field of view is 3.1 mm. (Photo by O. Richard Norton and Tom Toffoli.)

Fig. 6.13. A granular olivine–pyroxene (GOP) chondrule composed of minute grains of pyroxene and olivine with a rim of large pyroxene grains. Viewed under crossed-polarized light. Thin section from a Nuevo Mercurio H5 chondrite. The chondrule is 0.5 mm in diameter. (Photo by O. Richard Norton and Tom Toffoli.)

crystals and then were rapidly heated to a liquid state forming liquid spheres, crystallizing as spherical chondrules before being rapidly quenched. They are also known as *quenched droplets*. These include *radial pyroxene* (RP), *barred olivine* (BO), *cryptocrystalline* (C). All of these are texturally distinctive and easy to recognize.

Radial pyroxene (RP) chondrules are the most spherical of the chondrules and represent 44% of the droplet forms. They are composed of extremely thin fibers or *laths* of orthopyroxene (bronzite) or low-calcium pyroxene arranged in a radiating fan-like texture (Fig. 6.14). These fibers extend from one or more nucleation points eccentric to the chondrule. For that reason they are often called *excentroradial* or *exocentric* radial chondrules. In many cases the laths are so fine, 1–10 μm thick, that they cannot easily be resolved with a light microscope. Sometimes they do not extend across the chondrule field and it is not unusual to find several groups of laths extending from an equal number of nucleation sites. Under crossed-polarized light, the pyroxene usually shows low birefringence and extinction moves across the chondrule as the thin section is rotated. Often, especially in petrographic Type 4 and 5, the chondrules frequently are not round but appear scalloped like a sea shell (Fig. 6.15a). In these cases, the chondrules have suffered a corrosive chemical alteration. Related to the fibrous RP are chondrules with exocentric structure but with relatively large blades of bronzite radiating from an eccentric nucleation point. These are coarse enough to be seen with a hand magnifier or even the naked eye (Fig. 6.15b).

Barred olivine (BO) chondrules account for some 25% of the droplet chondrules. Their distinctive texture is unmistakable under the microscope. The name comes from its easily

Fig. 6.14. A radial pyroxene (RP) chondrule in crossed polarized light showing thin pyroxene laths radiating from an eccentric nucleation point. A smaller second RP chondrule appears enclosed within the larger chondrule and shares the same nucleation point. The horizontal field of view is 1.9 mm. Thin section from a Faith H5 specimen. (Photo by O. Richard Norton and Tom Toffoli.)

Fig. 6.15.(a) A radial pyroxene chondrule with scalloped borders suggesting chemical erosion of the rim. The pyroxene laths have sub-micrometer thicknesses making them too narrow to be seen individually but different domains in the chondrule go to extinction at different times as the thin section is rotated demonstrating that the chondrule is composed of several sets of laths. Thin section from a Gao-Guenie H5 chondrite. The chondrule diameter is 0.6 mm. (b) This radial pyroxene chondrule under crossed-polarized light shows large blades of clinoenstatite radiating from a nucleation point. Scattered poikilitically within the pyroxene are brightly colored olivine grains with yellow reaction rims. The field of view is 3.4 mm. Thin section from an Allende CV3.2 chondrite. (Photos by O. Richard Norton and Tom Toffoli.)

Fig. 6.16. An ellipsoidal barred olivine (BO) chondrule with parallel bars of olivine and an attached thick olivine rim. Both bars and rim share the same interference colors under crossed-polarized light and go to extinction simultaneously showing the entire structure is a single crystal in optical continuity. Thin section from a Mbale L5/6 chondrite. The major axis of the chondrule is 1.7 mm long. (Photo by O. Richard Norton and Tom Toffoli.)

Fig. 6.17. A polysomatic barred olivine chondrule showing several sets of parallel olivine bars oriented at different angles to each other resulting in different interference colors and extinction points. A thick rim encloses the polysomatic bars except at the upper edge where the rim has been disrupted. Thin section from an Allende CV3 chondrite. The horizontal field of view is 2.1 mm. (Photo by O. Richard Norton and Tom Toffoli.)

recognized morphology – a series of parallel bars or plates of olivine set in glass. Under crossed-polarized light, they often feature a singular interference color and when rotated go to extinction simultaneously, demonstrating that they are in optical continuity *or monosomatic*, with all the plates oriented identically. The plates are usually surrounded by a rim of the same mineral which also displays the same bright interference colors and the same optical continuity, going to extinction simultaneously with the interior plates (Fig. 6.16). Frequently an outer rim also exists usually much thicker than the inner rim and containing tiny opaque droplets of troilite.

There are many different transitional forms of BO chondrules. Some show several sets of parallel prisms that are optically distinct. They show different orientations, interference colors and go to extinction separately. These are *polysomatic* BO chondrules (Fig. 6.17). In rare cases, these chondrules can take on remarkable symmetry. Figure 6.18 shows a polysomatic chondrule with four sets of parallel olivine plates arranged in four quadrants set close to 90° to each other. Each quadrant shows separate optical continuity.

Cryptocrystalline (C) chondrules have a nearly featureless morphology. They are composed of microcrystalline orthopyroxene grains so fine-grained that they cannot be easily seen under the light microscope.[1] Under crossed-polarized

[1] A.E. Rubin at the University of California, Los Angeles, and J.N. Grossman of the US Geological Survey have defined cryptocrystalline chondrules as containing grains smaller than 2 μm and having multiple extinction domains.

Fig. 6.18. A polysomatic barred olivine chondrule with sets of parallel olivine bars arranged near 90° angles to each other. Thin section from a Forrest 002, L6 chondrite. The chondrule is 2.1 mm across. (Photo by O. Richard Norton and Tom Toffoli.)

Fig. 6.19. A cryptocrystalline chondrule viewed in crossed-polarized light revealing differently oriented multiple domains of pyroxene grains that go to extinction at various times as the section is rotated. From a Gao-Guenie H5 chondrite. The chondrule is 0.8 mm in diameter. (Photo by O. Richard Norton and Tom Toffoli.)

light different parts of the chondrule go to extinction at different times (Fig. 6.19). This multiple extinction demonstrates that there are domains where the grains are oriented differently. They take on a patchy appearance as the thin section is rotated. C chondrules are probably related to RP chondrules in that they were both formed as molten drops that quenched rapidly before crystal growth could proceed very far.

Glassy chondrules are relatively rare being composed of almost pure sodium-rich feldspathic glass. They are more or less featureless with the exception of occasional feathery needles of olivine or pyroxene that were crystallizing out of the glass while still molten. Glass chondrules are found only in unequilibrated chondrites of Type 2 and 3.

There seems to be no distinguishing chondrule textural types among the three chemical groups of ordinary chondrites (H, L, LL). Their distribution and textural variations are essentially identical in all three. Most noteworthy is that PO chondrules decrease and PP chondrules increase preceding through the carbonaceous chondrites, ordinary chondrites to the enstatite chondrites. E chondrites being composed of nearly pure enstatite pyroxene, contain no olivine chondrules. (Occasionally, rare POP chondrules are found in E chondrites, however.) FeO-poor porphyritic chondrules are more abundant than FeO-rich chondrules in all chondrite groups, especially in the carbonaceous chondrites and E chondrites. Although there are far more FeO-rich chondrules in ordinary chondrites than in carbonaceous chondrites, there are still more FeO-poor than FeO-rich chondrules in ordinary chondrites.

The other chondrites

The chondrites comprise more than 90% of all the stony meteorites that fall to Earth. Of these, 86% are ordinary chondrites with the remainder divided unequally among three distinctly different chondrites: the *C chondrites*; *E chondrites*; and *R chondrites*. The C or carbonaceous chondrites comprise the largest of the remaining chondrite classes amounting to 4% of the chondrites. These are of such a unique character that a separate chapter is given to them (see Chapter 7). The remainder, the E or enstatite chondrites comprise only 1.4% of the chondrites and the R chondrites even less at about 0.5%. Both of these chondrites make an interesting pair for comparison as they are almost diametrically opposed mineralogically, chemically and petrographically. In essence, they occupy the far ends of the chondrites with the ordinary and carbonaceous chondrites in between (Fig. 6.20).

Enstatite chondrites (E)

E chondrites are named for their primary silicate mineral, enstatite. Enstatite, if you recall, is the magnesium-rich end member of the orthopyroxene solid solution series. It is nearly pure magnesium silicate and comprises 60–80 vol% of the meteorite. Iron oxidation is rare. There is ~0.03 wt% FeO in the pyroxene of EL6 and some EL3 and EH3 chondrites contain pyroxene grains that range up to Fs_{12}. Virtually all of the iron is in the metal phase (13–28 vol%) or as sulfide (5–17 vol%). From this, it is evident that E chondrites are the most reduced of the chondrites with very little iron in the silicate phase, i.e., nearly all of the iron is in the metallic state or combined with sulfur. Lithophile elements such as Ca, Mg, K, Na, and Mn that normally have an affinity for the silicate phase are found partly in sulfides, further reflecting the highly reduced state of E chondrites. A low bulk Mg/Si ratio further distinguishes them from the ordinary chondrites. The result is an excess of silica that prevents the formation of olivine ($Fa_{<1}$). Reflecting this silica excess is the presence of polymorphs of silica such as *tridymite* and *cristobalite* and it is the only meteorite class that contains small quantities of quartz. Even the iron metal contains some dissolved silicon.

E chondrites top the list for chondrites containing more metal than any other stony meteorite class (until recently).[II] Their rich metal component is particularly evident looking at a polished slab or thin section in reflected light (Fig. 6.21). Total iron varies between 22 and 33 wt%. The metal, between 17 and 23 wt%, is entirely kamacite, the nickel-poor iron alloy found in all iron meteorites. The remaining metal is found as the iron sulfide, troilite, also nickel-poor. Like the ordinary chondrites, the E chondrites are divided into two groups, EH and EL, depending upon the total iron and metal. EH chondrites average about 30% total iron of which about 5% is sulfide; the EL chondrites have about 25% total iron with 3.5% sulfide.[6]

Texturally, enstatite chondrites contain chondrules in various states of metamorphism. EH chondrites contain 15–20 vol% chondrules in petrographic Types 3–5. These are quite small with average diameters of about 0.2 mm. Since E chondrites contain very little olivine, nearly all of the chondrules are orthopyroxene, most with porphyritic textures. They are Type-I (FeO-poor) varieties as would be expected of highly reduced chondrules. Of the nonporphyritic chondrules, most have radial pyroxene textures. Until recently it was thought that all EL chondrites were petrographic Type 6. New meteorites recovered from Antarctica and the hot deserts have revealed EL chondrites with petrographic Types 3, 4 and 5. Type 6 is still much more common.

The origin of enstatite chondrite parent bodies is something of a mystery. The paucity of oxygen suggests that they

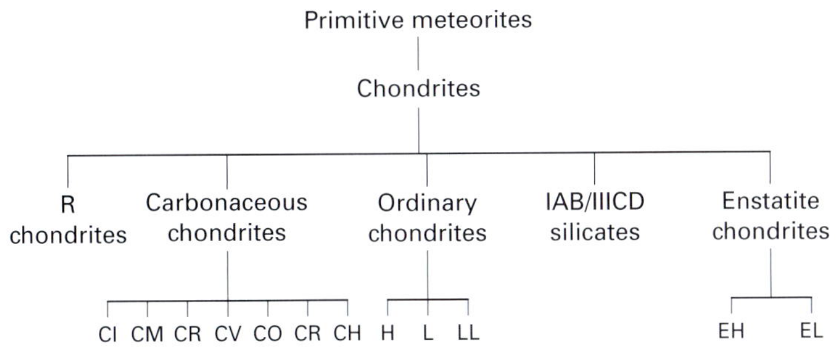

Fig. 6.20. A flow chart showing the four classes of chondrites and their groups. They are arranged by their iron oxidation or Fe_{metal}/Fe_{total}. For example, the E chondrites are the most reduced of the chondrites as opposed to the R chondrites that are the most oxidized. The C chondrites are closer to the R chondrites and the OC chondrites are closer to the E chondrites in terms of their metal to total iron.

[II] This was true until recently when a new group was added to the carbonaceous chondrite class. The CH chondrites are richer still in metallic iron.

Fig. 6.21. An EH3 chondrite seen under reflected light plus crossed-polarized light. Metal appear as gray irregular grains. An exceptionally large radial pyroxene chondrule is shown among small chondrules, pieces of chondrules and individual mineral grains (yellow). No olivine is present. Thin section from Sahara 98096. The horizontal field of view is 4.95 mm. (Photo by O. Richard Norton and Tom Toffoli.)

may have formed in a reducing environment probably close to the Sun; perhaps even inside the orbit of Mercury. There may be a genetic connection between the enstatite achondrites or *aubrites* and enstatite chondrites. Aubrites are almost pure iron-poor enstatite but they contain little or no metal. We will look further at the origin of E chondrites and aubrites and their possible relationship again in Chapter 8.

"Rumuruti" chondrites (R)[7]

R chondrites represent the newest chondrite group. Members of this group have been known since 1977 when a single highly weathered 49.5 g meteorite was discovered near Carlisle Lakes on the northern edge of the Nullarbor Plain, Western Australia. Two others with similar characteristics were found in Antarctica forming a "grouplet" called the Carlisle Lakes type.[8] These specimens were highly weathered to a point where their important distinguishing characteristics were obscured. Though they appeared to have characteristics distinct from ordinary chondrites, meteoriticists were understandably reluctant to designate these meteorites a new chondrite group until a fresh, unaltered specimen could be found and examined.

Forty years earlier, in 1934, a witnessed meteorite shower occurred near Rumuruti in southwestern Kenya scattering meteorites over a square kilometer area. A number of stones were collected. One of these stones, a 67 g specimen was purchased by the Berlin Museum of Natural History where it remained unstudied and unrecognized until 1993. Since this was a witnessed fall in which collection had been rapidly made, the meteorites acquired no weathering. Once recognized, this specimen became the type specimen for the new chondrite group, the *Rumurutiites* or *R chondrites*. Nineteen R chondrites are known: Carlisle Lakes, Rumuruti, eleven from Antarctica, five from the Sahara Desert and one from Australia. Two others, one from the Libyan Sahara Desert and one from the western United States are being classified at the time of writing.

All R chondrites are brecciated (Carlisle Lakes is the only unbrecciated) with light equilibrated clasts set against a dark fine-grained clastic matrix. At least three of the nine appear to be regolith breccias (Acfer 217, Rumuruti, and Pecora Escarpment 91002) in that they have light/dark structure and contain solar wind noble gases typical of regolith breccias. The clasts show petrographic Type 5–6 while the matrix is less metamorphosed to Type 3–4. They are therefore genomict breccias. All show light to moderate shock metamorphism (S3–S4) and local impact melting.

The most abundant silicate, about 70 vol%, is FeO-rich olivine, mostly Fa_{38-41}. Plagioclase (14 vol%), Ca-rich pyroxene (5 vol%), pyrrhotite (4.4 vol%), and pentlandite (3.6 vol%) make up the composition of these unusual meteorites. They have by far the highest iron oxidation of any chondritic meteorite. There is almost no free iron metal (a few grains here and there) and most of the iron occurs as pyrrhotite (FeS), pentlandite $(Fe, Ni)_9S_8$) or in the olivine.[9]

Summary

This chapter summarizes the important characteristics of ordinary chondrites and their cousins the E and R chondrites. These characteristics (texture, mineralogy, chemistry) tell us a great deal about conditions that existed early in the history of the Solar System when crystallized mineral assemblages began to accrete to form the first parent bodies. There seems to be little question that each chondrite group represents an individual parent body that must have suffered impact and fragmentation somewhere in the asteroid belt, eventually sending representative meteorite samples to Earth. Ordinary chondrites, especially unequilibrated Type 3 are being actively studied by meteoriticists worldwide. They are confident that these primitive meteorites hold answers to the processes that resulted in the evolution of the Solar System bodies from comets and asteroids to the terrestrial and Jovian planets and their satellites. We will return again to the ordinary chondrites in Chapter 10 when we consider the origin of chondrules and the formation of chondritic parent bodies.

References

1. Corrigan, C.M. *et al.* (1997). The porosity and permeability of chondritic meteorites and interplanetary dust particles. *Meteoritics and Planetary Science* **32**, No. 4, 509–515.
2. Consolmagno, G.J., Britt, D.T. and Stoll, C.P. (1998). Metamorphism, shock, and porosity: why are there meteorites? *Meteoritics and Planetary Science* **33**, No. 4, (Supplement), A34.
3. Grossman, J.N., Rubin, A.E., Nagahara, H. and King, E.A. (1988). Properties of chondrules. *Meteorites and the Early Solar System*, University of Arizona Press, Tucson, pp. 619–659.
4. Rubin, A.E. (2000). Petrologic, geochemical and experimental constraints on models of chondrule formation. *Earth Science Reviews* **50**, 3–27.
5. Gooding, J.L. and Keil, K. (1981). Relative abundances of chondrule primary textural types in ordinary chondrites and their bearing on conditions of chondrule formation. *Meteoritics* **16**, 17–43.
6. Heide, F. and Wlotzka, F. (1994). *Meteorites: Messengers from Space*, Springer Verlag, New York, p. 126.
7. Kallemeyn, G.W., Rubin, A.E. and Wasson, J.T. (1996). The compositional classification of chondrites: the R chondrite group. *Geochimica et Cosmochimica Acta* **60**, 2243–2256.
8. Rubin, A.E. and Kallemeyn, G.W. (1989). Carlisle Lakes and Allan Hills 85151: members of a new chondrite grouplet. *Geochimica et Cosmochimica Acta* **53**, 3035–3044.
9. Schulze, H. *et al.* (1994). Mineralogy and chemistry of Rumuruti: the first meteorite fall of the new R chondrite group. *Meteoritics and Planetary Science* **29**, 275–286.

CHAPTER SEVEN

Primitive meteorites: the carbonaceous chondrites

One's first impression upon seeing carbonaceous chondrite hand specimens is that they are plain, rather nondescript and ordinary-looking; not the kind of rock that would compel you to pick it up. An accurate description is that many of them look deceptively like charcoal briquets. Back in the 1960s only seventeen were known in the world. Most of them were quite small; nine of the seventeen were well under 1 kg total mass and only two had total masses of 10 kg or more. The first recognized fall (Alais) occurred in 1806 in Gard, France. A 4 kg mass landed in Saint Etienne and a 2 kg specimen in Valence separated by a distance of about 65 km. Had they not been witnessed falls, they probably would not have been recognized as meteorites. Today, many new C chondrites are being recovered from the Antarctic and the world's hot, dry deserts so that the inventory is now 36 witnessed falls and 525 finds[1] and others are not fully classified.

Carbonaceous, or C chondrites, are so distinctive that I have chosen to devote an entire chapter to them even though they are technically classified with the chondrites described in Chapter 6. Relative to the ordinary chondrites, they are quite rare. Only about 3.8% of the chondrites fall into this classification. Before 1969 roughly half of these were witnessed falls and were recovered very soon after fall. This is fortunate since many of the C chondrites are very friable, weakly consolidated, highly porous and loaded with soluble minerals. A few months or so in a wet climate would cause them to disintegrate. This is, in part, the reason why they were so rare. But on February 8, 1969, between 1:05 and 1:10 a.m. CST all that changed and the science of meteoritics took a giant leap forward. At that early hour an unprecedented fall of over 2 tons of a rare carbonaceous chondrite occurred in the Mexican state of Chihuahua, 200 km south of the city of the same name. The strewn field was enormous,[1] a roughly elliptical shape extending in a northeast-southwest direction for 50 km and covering about 300 km^2. This was the largest stony meteorite strewn field ever investigated.

Numbering in the thousands of individuals, it produced the greatest total weight ever recovered in a stony meteorite strewn field up to that time. The first specimen was recovered in the northern section of the field in the small town of Pueblito de Allende located in the obscure Valley of the Rio de Allende. This made available to researchers worldwide for the first time, a plethora of a relatively rare CV3 carbonaceous chondrite. This fall and the fall of the Murchison CM2 carbonaceous chondrite in Victoria, Australia, in September of the same year marked the beginning of intensive study of this class of meteorite, which has continued unabated to this day.

Carbonaceous chondrites are the most complex and heterogenous of all meteorites. At the same time, they are the most primitive and contain organic compounds, a few of which are found in life forms on Earth. These characteristics tantalize us with hope of discovering within them clues to the origin of life as life-dependent precursor materials formed in space. Ultimately they may lead us to the origin of life on Earth.

[1] Grady, M. M. (2000). *Catalogue of Meteorites*, 5th edition, Cambridge University Press, Cambridge, 720pp.

Ordinary vs carbonaceous chondrites

A brief comparison of C chondrites with ordinary chondrites will serve to introduce these most sought after meteorites. The ordinary chondrites we have seen are closely related, having remarkably similar elemental and mineral compositions. Ordinary chondrites are essentially igneous in character in their chondrules and quasi-sedimentary in their matrices. The chondrules are composed of refractory ferromagnesian minerals while their matrices are a collection of dust grains and chondrule fragments in between the chondrules. Their textures vary as a function of the thermal metamorphism they experienced. Thus, they cover a range of petrographic types from Type 3 to 6, with their minerals in matrix and chondrules showing increasing homogeneity. C chondrites are heterogeneous in composition, and they differ in composition among their various groups. This is, in part, because they have experienced little thermal metamorphism (except CK chondrites, which are almost all Type 4–6.) On the contrary, all C chondrites show various states of aqueous alteration. Either during or immediately after their parent body formation, water must have percolated through these rocks at temperatures from above the freezing point to well above the boiling point, reacting with the original refractory silicates to form hydrated silicate minerals. Aqueous or low-temperature chemical alteration is a hallmark of many of the C chondrite groups, especially CI, CM and CR groups. Aqueous alteration in CV, CK, and CO groups is lower and is not yet well established.

Ordinary chondrites on the average contain less than 1 wt% water. C chondrites contain various amounts of water from as high as 17–22 wt% (CI) to less than 1 wt%. All of the water is in the combined state. It is therefore not surprising that they contain water-bearing minerals similar to terrestrial phyllosilicates. Even further, some groups contain hydrous sulphates of calcium and magnesium that appear to have been deposited in veins running through the meteorites much as we find in water-bearing rocks on Earth.

Some C chondrites contain substantially more volatile elements than any other meteorite type. The ordinary chondrites are depleted in volatile elements telling us that they formed in a relatively high temperature environment. The C chondrites may have formed farther from the Sun where temperatures were low enough to retain these volatiles. Yet, all but one group, the CI chondrites, contain chondrules with igneous textures and aggregates of highly refractory minerals that necessitated formation temperatures well beyond 1500 K or higher. Even the CI chondrites contain isolated grains that are probably chondrule and CAI fragments.

Some C chondrites show high oxidation states with little uncombined metal whereas ordinary chondrites vary considerably in oxidation state showing, as we saw in Chapter 5, various amount of elemental iron and sulfide. But like the ordinary chondrites, other C chondrite groups contain abundant metal. These include reduced CV, unaltered COs, and even CMs have some metal. The newest C chondrite group, the CH chondrites, contains the most metal of all.

Typically, the olivines and pyroxenes in the matrices of C chondrites show moderately higher iron content than the ordinary chondrites. This implies that the C chondrites formed in a more oxidizing environment further from the Sun while the ordinary chondrites formed perhaps closer to the Sun under less oxidizing conditions. The chondrules, on the other hand, are magnesium-rich, quite similar to the ordinary chondrites implying lower oxidizing conditions for their formation. These contradictory characteristics make C chondrites the most enigmatic but also the most profoundly interesting of all the meteorites.

Finally, as the "C" implies, C chondrites are carbon-rich, containing various amounts of carbon in the form of carbonates and complex organic compounds including amino acids (see page 140). Although other meteorites contain as much or even more carbon,[II] the C chondrites are the only meteorites containing organic compounds that could be considered precursors of life.

[II] The name *carbonaceous chondrite* was the result of an early mistaken belief that these chondrites contained more carbon than other chondrites. While this is generally true, there are some carbonaceous chondrites that are carbon-poor, containing less carbon than some ordinary chondrites (Type 3). CI and CM chondrites typically contain >20 mg/g while the CV and CO chondrites contain <10 mg/g. Carbon is not a particularly rare element in meteorites in general. At least one achondrite (i.e., ureilites) is relatively abundant in carbon and iron meteorites commonly have inclusions of carbon in the form of carbides, graphite and diamond. The criterion for classifying a chondrite into the C class is the relative abundance of lithophile elements over that found in the most primitive C chondrites, CI. C chondrites are presently defined as having CI-normalized refractory lithophile abundances greater than or equal to 1.0 and O isotopic compositions below the terrestrial fractionation line (except for CI.) What is unusual and significant about the carbon in carbonaceous chondrites is that it is a major constituent of organic compounds found only in this meteorite class.

Carbonaceous chondrite groups

From the above comparison, it seems that we are dealing with a different type of stony meteorite but one that shares many characteristics with the ordinary chondrites. In fact, many researchers feel that carbonaceous chondrites are not that different from ordinary chondrites as we will see. The Van Schmus–Wood petrographic chart shows the C chondrites varying from C1 to C3 with a single exception, the group CK. Petrographic Type 3 seems to be a natural subdivision of the meteorites with those to the left showing increasing aqueous alteration and those to the right increasing thermal alteration (see Table 5.3). But even the Type 3 chondrites cannot be considered truly pristine as most show evidence of moderate amounts of hydrothermal alteration. Of the Type 3 chondrites, the reduced CV3.0 and CO3.0 chondrites come closest to being pristine.

Up to the early 1980s meteoriticists recognized four different groups of carbonaceous chondrites. Today, with the impressive meteorite discoveries in Antarctica the number has risen to six with a seventh recently acquiring group status (the CH group). The capital letter "C" designates the major class followed by a letter corresponding to the prototype meteorite whose characteristics were used to define the group. These are usually given the names of the geographic locations or towns in which they fell (all were witnessed falls): *CI*, where the "I" designates Ivuna, a town in Tanzania where in 1938 a 704 gram meteorite fell; *CM*, with "M" standing for the 8 kg Mighei meteorite which fell in the Ukraine in 1889; *CV*, the "V" standing for the town of Vigarano, in Emilia, Italy, in which two stones weighing a total of 16 kg fell in 1910; *CO*, the "O" designating the town of Ornans in Doubs, France where a 6 kg stone fell in 1868; *CR*, with the "R" standing for Renazzo, near the town of Ferrara, in Emilia, Italy, where 10 kg of meteorites fell in 1824; and *CK*, where "K" stands for Karoonda, a small town in South Australia where about 50 kg of stones fell in 1930. Still evident in the literature is an earlier designation for some of these groups. These are noted in parentheses next to the currently used group designation below.

CI chondrites (C1)

Without question, the CI carbonaceous chondrites are the most primitive in terms of solar elemental abundances and highest content of volatiles, being significantly higher in volatiles than the CM chondrites, the type most closely related to the CI group. It is the meteorite type with which all others are compared since it is closest to solar composition (see Chapter 5).[III] Only a scant five are known, barely enough material to do meaningful research. Even with the great harvest of C chondrites from Antarctica and the hot, dry deserts over the past 30 years, not one new CI has been recovered. Table 7.1 lists the specimens, giving date and location of each fall (all falls, no finds), number of individuals and total weight in grams.

From Table 7.1, it is not surprising that most of the work on CI chondrites has been on the most abundant fall, Orgueil.

Name	Location	Date of Fall	Number of individuals	Total mass (g)
Alais	France	Mar. 15, 1806	2	6000
Ivuna	Tanzania	Dec. 16, 1938	2–3	705
Orgueil	France	May 14, 1864	20	10000
Tonk	India	Jan. 22, 1911	Several (?)	7.7
Revelstoke	B.C., Canada	Mar. 31, 1965	2	1

Table 7.1. **The five known CI chondrites**

These five meteorites represent the world's collection of CI chondrites. All of them were seen to fall and collected shortly thereafter. It is no wonder that none are in the "find" category since they are extremely fragile and easily and quickly disintegrate under terrestrial conditions. Two other meteorites have been recovered that may add to the group of five. During the Apollo 12 mission to the Moon in December, 1969, a carbonaceous chondrite was inadvertently collected in a soil sample found near Bench Crater in Oceanus Procellarum. It has a matrix similar to a CI chondrite. It was later classified as a shocked CM1 chondrite. The "1" means that, like the CI class, it does not have chondrules. The second specimen fell in Tagish Lake, Yukon Territory on January 18, 2000, and was recovered almost immediately (see page 122). This proved to be an intermediate CI type now classified as a CI2, that is, a CI with chondrules.

[III] A comprehensive compilation of solar abundances normalized to 10^6 Si atoms and elemental abundances found in CI chondrites may be found in the text: Wasson, John T. (1985). *Meteorites: Their Record of Early Solar System History*, W.H. Freeman & Co., New York.

Fig. 7.1. A CI, Type I carbonaceous chondrite, Orgueil, from Mountauban, Tarn-et-Garonne, France. The black fusion crust blends with the black magnetite-laced interior of clay-like minerals. The white areas are hydrated carbonate and sulphate minerals. The specimen measures 20 mm and weighs 2.39 g. (Courtesy of The Institute of Meteoritics, University of New Mexico. Photo by O. Richard Norton.)

Recovered CI chondrites are never large. Their friable nature assures their disruption during atmospheric flight often leaving only small pebbles to survive. A case in point is the Revelstoke fall which made an impressionable fireball but left only two tiny recoverable stones, the larger weighing under 1 g. CI fusion crusts have a mat black luster matching the interior to such a degree that one has to look carefully to distinguish it from the matrix. In Orgueil (Or `gay), white hydrous minerals in the matrix help to distinguish the crust.

The hallmark of the CI chondrites is their lack of chondrules, aggregates and inclusions. This may seem a contradiction, a chondrite without chondrules, but they are clearly related to the other C chondrite groups. Their compositions are very closely allied to the CM chondrites, a more important comparative quality than their textures. Nevertheless, it remains a mystery why CI chondrites lack chondrules. They do contain isolated olivine and pyroxene grains and olivine/pyroxene intergrowths that are probably chondrule fragments. They are classified petrographic Type 1 (Fig. 7.1). These meteorites are essentially all matrix with a uniformity unlike any other chondrite. In thin section the matrix is black and opaque due to very fine-grained magnetite (about 120 mg/g), small amounts of pyrrhotite, and carbonaceous matter mixed with water-bearing silicates. These silicates are clay-like minerals usually grouped together in textbooks under the label "serpentine", which is actually a group of hydrous minerals. Some analyses show the presence of montmorillonite clay which has an affinity for water, swelling in its presence. It is derived from the aqueous alteration of magnesium-rich olivines and pyroxenes.[2] Also found are isolated olivine, pyroxene and refractory mineral grains, a small number of disrupted chondrules and CAIs. These high-temperature minerals are present in the matrix as tiny, nearly equant idiomorphic crystals (equidimensional crystals bounded on all faces) revealed under high magnification. Both forsterite and fayalite (Fa_{10-20}) are present along with orthopyroxenes and clinopyroxenes. It is a puzzle that no signs of alteration of the ferromagnesian minerals have been found in the matrix.[3] Characteristic forms of magnetite are found in CI. These include framboids (microscopic aggregates of spherical grains), spherulites, and platelets, all best seen with a scanning electron microscope.

CI chondrites contain between 17–22 wt% water, which is easily extracted by heating in a closed vessel. The high porosity (~30%) of these meteorites must have aided in the retention of some of this water though liquid water has not been found. All the water is locked in the hydrous silicates. The average density of these meteorites is only 2.2 g/cm^3. When these meteorites are cut, the matrix may show a mottled or speckled appearance due to the presence of hydrated sulfates and carbonates. Microscopic veins running through the matrix appear to be lined with epsomite ($MgSO_4 \cdot 7H_2O$). This strongly suggests that liquid water at low temperatures (probably between 20 and 125 °C) existed near the surface of the parent body and that it permeated these meteorites through fractures in the parent body material. Aqueous alternation of olivine, sulfides and clay-like minerals produced sulfates and carbonates that precipitated within the veins as the water circulated.[4]

A new CI chondrite?

As this book was being written an extraordinary meteorite fall occurred in Canada. On January 18, 2000, a brilliant fireball with accompanying loud detonations was observed over the Yukon Territory and northern British Columbia. The fall was widely observed, and on January 25 and 26 several dozen small meteorites totaling about a kilogram were recovered on the Taku Arm of Tagish Lake near the town of Whitehorse in the southern Yukon. Fortunately, the lake was completely frozen over at the time, allowing collection to be made with minimal terrestrial contamination. The meteorites were kept frozen as they were transported to laboratories in Canada, the United Kingdom, and the United States.[IV] This was exceptionally fortunate, for these meteorites proved to be a rare, fragile carbonaceous chondrite. Additional meteorites were found three months later but by that time some were altered by water from melting ice. In total, over 500 individual specimens were recovered.

The pristine first-collected specimens were examined with microscope and electron microprobe and it soon became apparent that this was no ordinary C chondrite. Petrographically, the Tagish Lake meteorites resembled CM chondrites in that they contained sparcely distributed chondrules, altered calcium–aluminium inclusions (CAIs) and individual olivine grains not seen in CI chondrites. By contrast, the composition was more like CI chondrites. The high bulk carbon content (5.4 wt%) was more typical of CI chondrites. There was abundant carbonates, magnetite and iron sulfide. The abundance of magnetite was more typical of CI than CM chondrites. But unlike both, the carbonates showed compositional variations not seen in either CI or CM chondrites. These and other petrographic and compositional details convinced the researchers not to classify Tagish Lake meteorites as a CI. So where does it fit into the present classification scheme? Tagish Lake is obviously a new C chondrite for which there is no convenient slot to place it in the current C chondrite hierarchy. For the time being, the investigators decided to call Tagish Lake a CI2, essentially a CI with chondrules, but this may change as investigations continue.

CM chondrites (C2)

Before the discovery of the fabulous meteorite fields in Antarctica in the late 1960s, CM chondrites were the most plentiful of the carbonaceous chondrite groups, constituting about half of all known C chondrites. In the 1985 *Catalogue of Meteorites*, the total known C chondrites was given as 67 with 33 constituting group CM. By the year 2000, 561 carbonaceous chondrites were known. Of these 161 were CM chondrites amounting to 29% of the total.[V] This reflects the incredible increase in finds from Antarctica and the hot deserts. By far the largest fall occurred on September 28, 1969. It was a clear Sunday morning when they fell over Victoria, Australia, centering on the town of Murchison about 130 km north of Melbourne. People on their way to church heard sonic booms and "hissing" (electrophonic) sounds and saw a fireball explode. Over 700 stones rained out of the sky onto the edge of town, over fields, in front yards and on the streets. Murchison had been blessed with the fall of over 100 kg of fresh, rare CM chondrites, the largest weighing 7 kg. Many showed orientation forming cone-shaped specimens with extraordinary flow features (Fig. 7.2). The strewn field covered some 8 km^2. Most of the recent research on CM chondrites has been done on Murchison meteorites because they are plentiful, freshly fallen, and in a

Fig. 7.2. A beautiful oriented CM2 individual that fell in Murchison, Victoria, Australia in 1969. The dark fusion crust is lined with flow structures and a system of cracks can be seen along the top. The specimen weighs 107 g and measures 58 mm at the widest point. (Michael Casper photo.)

[IV] Tagish Lake meteorites investigators were: M.E. Zolensky (SN2 NASA Johnson Space Center, Houston, TX 77058 USA); M.M. Grady (Natural History Museum, London, United Kingdom); R.N. Clayton and T.K. Mayeda (University of Chicago, Chicago, IL, 60637, USA); A.R. Hildebrand (University of Calgary, Calgary, AB T2N 1N4, Canada); P.G. Brown (University of Western Ontario, London, Ontario, N6A 3K7, Canada); C.F. Roots (Canadian Geological Survey, Whitehorse, Yukon, YIA 2C6, Canada).

[V] Grady, M.M. (2000). *Catalogue of Meteorites*, 5th edition, Cambridge University Press, Cambridge, 720 pp.

Fig. 7.3. A low-magnification photograph in transmitted light of the Murchison CM2 interior shows a paucity of complete chondrules. A rare, relatively large porphyritic olivine chondrule is in the upper right. Pieces of chondrules, individual olivine crystals and small calcium–aluminum inclusions (CAIs) are scattered in the matrix. The thin section field of view is 21 mm. (Photo by O. Richard Norton.)

relatively uncontaminated state – all highly desirable qualities especially for specimens containing organic compounds.

CM chondrites, earlier designated C2, are now classified as CM2, the "2" designating the presence of well-formed but sparsely distributed small chondrules in a matrix chemically very similar to the CI chondrites (Fig. 7.3). They are more compact than CI chondrites with a density ranging between 2.5 and 2.9 g/cm^3. The high-temperature phases include chondrules and chondrule fragments, crystal aggregates, and individual crystals, altogether constituting about 50 wt% of the meteorite. This contrasts sharply with the CI chondrites, which are composed of 99% matrix.

Of the high-temperature phases, the chondrules are a relatively minor constituent, amounting to only about 12 vol% of the meteorite. Most are small, averaging less than 0.5 mm diameter. The smallest chondrules are nearly spherical with a granular texture made up of very small pyroxene and olivine crystals. The largest chondrules tend to be irregular in shape, have no rims and are composed of relatively large euhedral to subhedral olivine crystals (Fig. 7.4). The classic chondrule textures (barred olivines, radial pyroxenes, porphyritic olivine pyroxene, etc.) are present but scarce compared to ordinary chondrites. FeO-poor porphyritic chondrules are abundant in CM. Chondrule fragments and individual olivine crystals as large as many chondrules are much more common. Individual olivine crystals make up as much as 20 vol% or more of the CM chondrites. Also abundant are irregular-shaped loose aggregates of very small olivine crystals of forsterite composition set in a glassy mesostasis (Fig. 7.5) suggesting that they have been melted like the larger chondrules. These aggregates constitute as much as 18 vol%. Grains of metallic iron are frequently seen in the chondrules and aggregates. Small amounts of highly refractory calcium–aluminum–titanium minerals are present in the aggregates. These minerals take on major importance in other groups of C chondrites but are minor constituents in CM2s.

The matrix is opaque and black, composing about 48 vol% of the meteorite. Its mineralogy is closely related to that found in CI chondrites but with much less magnetite ($\leq$8 mg/g). The major constituents are phyllosilicates. CM meteorites contain 3–11 wt% water, roughly half as much as CI chondrites. Complex hydrocarbons and other organic compounds are present (see page 140). Interestingly, they are found almost exclusively in meteorites with phyllosilicates composing their matrices (CI and CM chondrites).

Fig. 7.4. A relatively rare large rimless porphyritic olivine chondrule 0.9 mm across in a Murchison CM2 chondrite. Crystal sizes vary greatly as do their structure from small anhedral to large euhedral shapes. An amoeboid olivine aggregate appears in the upper left. (Photo by O. Richard Norton and Tom Toffoli.)

Fig. 7.5. A loose aggregate of olivine crystals 0.4 mm across set in glass is common in this Murchison CM2 chondrite. This is analogous to amoeboid olivine aggregates in CV3 chondrites. (Photo by O. Richard Norton and Tom Toffoli.)

Are CI and CM chondrites related to comets?

Up to this point in our discussion of stony meteorites we have assumed with good reason that the chondritic meteorites are simply fragments of chondritic asteroids that have somehow escaped the confines of the asteroid belt and have come to Earth. Although, in the most general terms, Earth and its sister worlds in the inner Solar System have chondritic compositions in that their silicate chemistry is similar, they differ substantially in their more volatile elements. For example, Earth and Moon are heavily depleted in potassium (K). Ratios of K to refractory elements are much higher in chondritic meteorites than in the Earth. This shows that the meteorites arriving on Earth today are not the same bodies that accreted to form the inner planets. This could mean that the solar nebula was not homogenous across its radius and that the inner planets were in a zone of depleted volatiles; or it may be that the Sun experienced a short period of increased, intense activity that removed the more volatile components of the nebula in the zone of the future terrestrial planets.

CI and CM chondrites are high in volatiles including water, suggesting that they formed in the outer regions of the asteroid belt, 4 AU or beyond. At that distance under conditions thought to exist in the solar nebula, water-ice condenses at 160 K (sometimes referred to as the *snow line*). Beyond 4 AU volatile primordial ices remained solid and were preserved. Indeed C chondrites appear quite similar in composition to some of the icy satellites of the outermost planets. When the parent bodies of the CI and CM chondrites formed near the fringes of the asteroid belt, they accreted silicate minerals, water-ice and other volatiles, and organic compounds, sharing these materials with other icy bodies that formed by the millions in the outer parts of the Solar System: comets.

In March 1986, an international armada of spacecraft visited Comet Halley when it was closest to Earth. Video cameras on board gave real time images of the approaching nucleus, never before seen (Fig. 7.6a,b). Mass spectrometers designed to measure the mass of impacting particles determined the composition of the solid particles jetting from geysers on the icy surface. This material formed a dusty cloud around the nucleus that extended into its dust tail. Among the lighter dust components were ices composed of the most volatile elements: carbon (C), hydrogen (H), oxygen (O), and nitrogen (N). Other components were dust particles rich in refractory materials composed of silicon, magnesium, and iron, common silicates making up chondritic meteorites. Some of the silicate grains appeared to be coated with carbon (organic) compounds, probably derivatives of the CHON ices.

Released in enormous quantities from the jet-like geysers was water which sublimated to a gas and quickly dissociated into H^+ and hydroxyl $(OH)^-$ ions as well as free neutral oxygen and hydrogen. These components were measured from the ground. Assuming these ions and atoms came from original water locked as ice in the nucleus, experimenters were able to estimate the total amount of water issuing from the nucleus per unit time. At the most active period one week on either side of its perihelion, Comet Halley was disgorging between 33 and 56 tons of water every second.[5]

The first hint that comet nuclei were coated with dark carbonaceous material came in February 1985, when ground-based observations in the infrared showed the color of Halley's nucleus to be similar to C-type asteroids from which carbonaceous chondrites originate.[6] These asteroids have albedos of around 4%. Since most of the ice composing Halley's nucleus was water-ice, it was evident that only a small portion of the icy nucleus was exposed. Meanwhile,

Fig. 7.6 (a) Comet Halley's nucleus photographed by the European Space Agency's Giotto spacecraft from a distance of 18 270 km. Two dusty jets vent from the sunward side. The icy nucleus is 16 km long. (b) The closest picture made by Giotto during it brief flyby of Comet Halley's nucleus made from a distance of 4500 km on March 14, 1986. The resolution is about 50 m. Two active jets release tons of dust while a third in the dark area is just turning off. The surface is covered by a dark material of carbonaceous composition. Crater-like features are present on the surface. (Courtesy Max Planck Institut für Aeronomie, Lindau/Herz, Germany.)

the European Space Agency's Giotto spacecraft was rapidly approaching Halley with instructions to photograph the brightest area of the nucleus, a decision that almost proved disastrous. The ground-based observations were correct. Halley was covered by a layer as black as charcoal and the cameras zeroed in on the white geysers on the sunlit side almost missing the blackened shape. The dark material was carbon-based dust released from the water ice nucleus during geyser activity. Infrared emission measurements at 3.4 μm made aboard the Russian VEGA spacecraft as it passed by Comet Halley in 1986 demonstrated the presence of organic compounds on the surface and as coatings on silicate dust of the inner coma and dust tail. These are presumed to be aliphatic and aromatic hydrocarbons, though they have not been specifically identified. We saw in Chapter 1 that interplanetary dust particles presumed to originate from the tails of comets have been collected high in the Earth's stratosphere. These particles, most less than 10 μm across are composed of loosely interlocking chondritic minerals with a coating of carbonaceous matter similar to that of CI and CM chondrites (see Fig. 1.6).

It appears from the above observations that comets and water-bearing C chondrites may be closely related. Comets have densities very close to that of water or even less, telling us that they are composed of CHON ices with relatively small amounts of solid silicate and carbonaceous material trapped within the ices, a kind of a "dirty iceberg", as comet authority Fred Whipple proposed in 1963.[7] As periodic comets return to the Sun, they continue to shed water and other volatiles along with solid particles. These comets eventually lose most of their volatiles leaving behind dark silicate and carbon-rich cores of higher density in which only a fraction of their water and other volatiles remain trapped. Two observations seem to confirm that comet nuclei do indeed contain substantial silicates: (1) Comet Enke is known to contain sizeable silicate bodies within its icy nucleus; and (2) when the pieces of Comet Shoemaker/Levy impacted Jupiter, water was not detected (they may have completely lost their volatiles by that time) but elemental analysis of spectra indicated the presence of Fe, Mg, and Si, the all-important elements of meteorite compositions and just what would be expected to compose a comet core.

These "dead" comets, still retaining their elongated orbits, are free to cross the asteroid belt inviting impacts with the main belt asteroids. If disrupted by impact with, among others, ordinary chondrite asteroids, they would release carbon-rich fragments which would become embedded in the impacted asteroid (Fig. 7.7a). It seems clear that collisions between high-density (ordinary chondrites) OC and much more fragile comet cores would shatter the cores but leave the OC parent bodies well intact. Pieces of the comet core would embed themselves into the matricies of OC parent bodies. More than 60% of the clasts found in chondritic meteorites are carbonaceous clasts of CI and CM-like

Fig. 7.7 (a). A carbonaceous inclusion found in the Zag H3–6 chondrite. It measures about 4 mm across. Little textural detail can be seen. This may be a CI clast. (b) A clast of carbonaceous matter about 7 mm across in a Plainview H5 chondrite regolith breccia photographed in reflected light. The white inclusions in the clast demonstrate it to be a CM2 carbonaceous chondrite. (Plainview specimen provided by Dr David Mouat, Desert Research Institute, University of Nevada System. Photos by O. Richard Norton.)

material. They are commonly found as inclusions in regolith breccias of ordinary chondrites (Fig. 7.7b). The usual explanation for these inclusions is that early in Solar System history the ordinary chondrite parent bodies were embedded in a large population of C-type asteroids in which collisions were quite frequent. It is true that about 75% of the known main belt asteroids are C-type but since the C-type asteroids are found primarily from the middle to the outer part of the asteroid belt, far removed from the inner belt where most ordinary chondrite asteroids are thought to reside, it is difficult to understand how such frequent collisions could occur. Moreover, we almost never see the reverse; that is, OC-like clasts in C chondrites. This suggests either that ordinary chondrite asteroids are rare or, more probably, that OC parent bodies remain intact during collisions with C asteroids while the C asteroids will more likely fragment. Thus, far fewer OC fragments are likely to form and embed in C chondrite parent bodies.

In recent years, the division between comets and asteroids has become increasingly blurred. We know there are large nonvolatile bodies in orbits with aphelia near Jupiter and perihelia near 1 AU. These are Earth-crossing "dead" comets that have the potential of colliding with Earth. Meteoriticists correctly point out that comets are moving far faster in their elongated orbits than do near-Earth asteroids. Cometary bodies or pieces of them, they point out, are not strong enough to survive shock pressures and ablative processes in Earth's atmosphere. But, if by chance they did, they would be small remnants of their preatmospheric passage mass. At most, only a handful of small meteorites from spent comets could be expected to have reached Earth intact. The paucity of known CI chondrites (Table 7.1) seems to support the above constraints. Even if cometary "meteorites" do not make it to Earth on their own, is it possible that they have been surviving as fragments within the protective envelope of ordinary chondrites? The relationship, if any, between water-bearing C chondrites and spent comets must await a comet sample-return mission for a definitive answer.

CV chondrites[VI]

With the Allende fall in 1969, more CV chondrites became available for study and distribution than all other C chondrites combined. Allende is probably the most studied carbonaceous chondrite in meteoritical history (Fig. 7.8). Individual CV chondrite falls are far fewer in number than CM chondrites, comprising only about 20% of the C chondrites. Compared to the CI and CM chondrites, CV chondrites have much higher densities (~3.5 g/cm^3), essentially OC densities. This is, in part, because the matrix is much more compact – there is less pore space – and the water content is less than 2 wt%. The interior is altogether different than the CI and CM chondrites. The matrix is medium gray in tone rather than black and contains an abundance of large, well-defined chondrules giving CV chondrites a petrographic Type 3; i.e., CV3.[VII] Also embedded in the matrix are large irregular-shaped, white, refractory inclusions high in calcium, aluminum, and titanium, designated CAIs. Using Allende as the "type" specimen, we now look at the three major components of CV3 chondrites: matrix; chondrules; and inclusions (Fig. 7.9).

Matrix

CV3 chondrites contain on the average, about 42 vol% matrix. Allende has a higher than average value of about 60 vol%. Examination of matrix material in thin section show it to be black and opaque. High magnification of ultrathin sections show the dominant mineral to be olivine, seen as a uniform distribution of tiny, euhedral crystals dispersed

Fig. 7.8. An Allende CV3.2 carbonaceous chondrite with a partial fusion crust and lead-gray interior. The white specks are calcium–aluminum inclusions. The specimen measures about 127 mm horizontal dimension. (Photo by Michael Casper.)

[VI] Most of the descriptive material in this section comes from studies of the Allende CV3.2 carbonaceous chondrite. Numerous papers have been published on Allende since its recovery in 1969. An excellent treatise on Allende may be found in the *Smithsonian Contributions To The Earth Sciences*, 1970, The Allende, Mexico, Meteorite Shower. This work gives colorful historical accounts of the fall and recovery of the Allende meteorites as well as summary descriptions of the external and internal characteristics of the meteorites.

[VII] Earlier classifications distinguish three petrographic type CV chondrites classified CV2, CV3 and CV4. CV2 and CV4 have since been reclassified as distinctly different groups, now CR and CK chondrites respectively. These additional groups are described later in this chapter.

Fig. 7.9. A typical Allende field showing a bewildering array of components. Among them are chondrules of many different structures and mineralogies, olivine aggregates, dark inclusions, individual mineral grains and convoluted calcium-aluminum inclusions, all set in a black matrix of opaque minerals. The field is 22 mm on the long side. The elliptical chondrule below center right is 3 mm in its longest dimension. (Photo by O. Richard Norton.)

Fig. 7.10. A POP chondrule in an Allende CV3.2 thin section viewed in crossed Polaroids with equant crystals of olivine and long blades of orthopyroxene. The chondrule measures 1.8 mm across. (Photo by O. Richard Norton and Tom Toffoli.)

in opaque material. Like the matrix of ordinary chondrites, the olivine is iron-rich with an average fayalite content of 50% (Fa_{50}). The opaque components consist of pentlandite $(Fe, Ni)_9S_8$, troilite (FeS) and minor grains of nickel-rich metal, *awaruite* (Ni_3Fe). Most of the metal in Allende (~0.5 wt%) is awaruite, an alloy similar to taenite. Low-nickel kamacite may have existed in the early parent body but subsequent oxidation probably destroyed it. It can still be found in a few rare chondrules. Contributing to the opacity of the matrix is evenly distributed carbon-based material that thinly coats much of the olivine. CV3 chondrites are carbon-poor compared to the water-bearing CI and CM chondrites, averaging less than 1 wt% carbon. (Allende contains only 0.29 wt% carbon). Hydrous silicates are rare and water-deposited sulfates and carbonates are absent.

Chondrules

The most obvious distinction between CV3 chondrites and all other C groups is the texture of the chondrule fields. Chondrules cover an average of 44 vol%. (Allende is about 30 vol%.) The chondrules are much larger on the average than those in the other groups, typically between 0.5 and 2 mm or more. Occasionally, especially in Allende, much larger chondrules are found, some exceeding 4 or 5 mm diameter with a record being a whopping 25 mm! The most common are porphyritic olivine chondrules made up of small tightly packed euhedral and subhedral olivine grains remarkably uniform in size. Some contain elongated prisms of clinoenstatite showing polysynthetic twinning (Fig. 7.10). Commonly seen within these chondrules are opaque grains

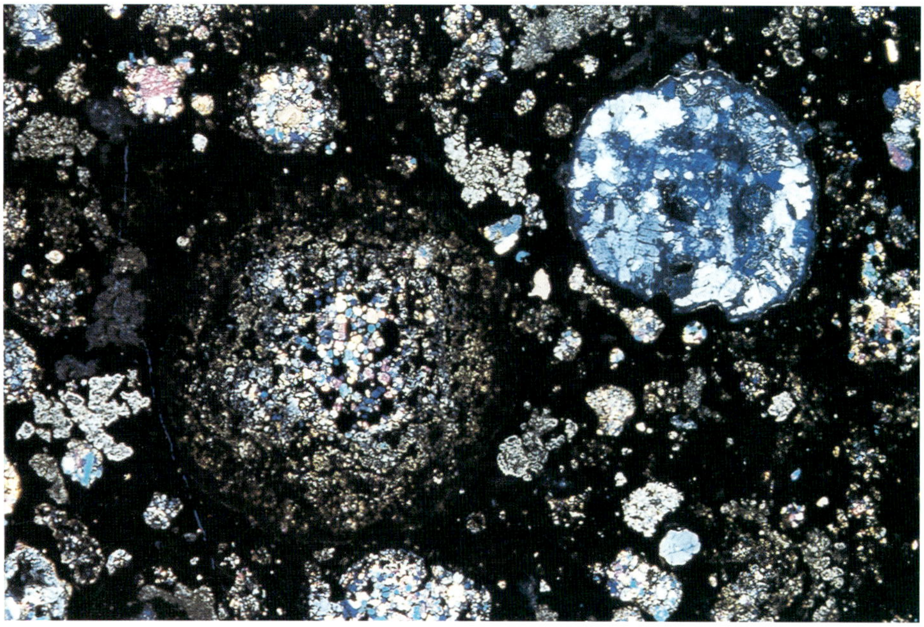

Fig. 7.11. A large granular olivine chondrule with tiny anhedral olivine crystals. The entire chondrule is surrounded by a brown serpentinite produced by the hydrous alteration of olivine; diameter is 3.7 mm. The blue chondrule to the upper right is a PP chondrule. Viewed in crossed-polarized light. From an Allende CV3.2. (Photo by O. Richard Norton and Tom Toffoli.)

7.12. A BO chondrule from an Allende CV3.2 with bars and double rim in optical continuity seen under crossed-polarized light. The dark object below, interrupting the thick rim, is a missing section with the mounting glass in extinction. Several circular blebs of metal appearing black are in the rim above. The field of view is 1.9 mm in the longest dimension. (Photo by O. Richard Norton and Tom Toffoli.)

of metal, iron sulfide, and magnetite. Surrounding the chondrules are halos of dark brown material composed of serpentinite, a product of aqueous alteration of olivine followed by dehydration (Fig. 7.11). Mixed with the serpentinite is opaque FeS and $NiFe_2S$. The olivine within many of the chondrules is almost pure forsterite averaging (Fa_6), but larger crystals frequently show a zoning with an increasing iron content toward their edges and may be caused by reaction with the iron-rich matrix. The zoning is often noticed under crossed-polarized light as a change in birefringence across the crystal.

Barred olivine chondrules are rare in CV3 chondrites compared to the phorphyritic chondrules but no less striking under crossed-polarized light. The bars may be contiguous with the rim material as one crystal, or several sets of parallel prisms with different orientations and different birefringent colors may compose the chondrule. All are set in a glassy mesostasis (Fig. 7.12). Often surrounding BO chondrules is a "corona" of loosely packed tiny olivine crystals that is in optical continuity with the chondrule. This is simply a spherical shell that forms a "rim" around the chondrule. The common optical continuity between the chondrule and shell demonstrates their common origin. This is not unique with CV3 chondrites. They can occur in any chondrite group that contains BO chondrules but they are more obvious in CV3s.

Porphyritic pyroxene chondrules are even rarer than BO chondrules. Their individual crystals are usually large, tightly grouped and lack rim material. Excentroradial pyroxene chondrules are very rare, amounting to less than 0.1 vol%.

All of the chondrules so far described are texturally similar to chondrules found in ordinary chondrites. There are chondrules in CV3 chondrites that are extremely rare in other chondrite groups but rather common in CV3s. In particular, in thin sections in transmitted light some remarkably round

Fig. 7.13. An anorthite–forsterite–spinel chondrule from an Allende CV3.2. The large white crystals along the edge are plagioclase. Spinel is scattered through the chondrule as very small opaque crystals. The darker blades are forsterite. Note that the forsterite and plagioclase are radially arranged with respect to the center. A sparse olivine aggregate is in the lower left and upper right corners. Viewed in crossed-polarized light. The chondrule is 1.6 mm diameter. (Photo by O. Richard Norton and Tom Toffoli.)

forms ranging in size from 0.1 mm to about 2 mm stand out. They have an overall dark gray color with white needle-like crystals arranged around the perimeter roughly aligned radially with the center. These are *anorthite–forsterite–spinel* chondrules. The white crystals are Ca-rich plagioclase or anorthite prisms usually arranged around a fine-grained core with the prisms getting progressively smaller from edge to center (Fig. 7.13). Spinel ($MgAl_2O_4$) is found as tiny dark, opaque crystals throughout the chondrule and within the anorthite prisms with forsterite occurring between the prisms.

Olivine aggregates, dark inclusions

Scattered more or less uniformly in the matrix are aggregates of micrometer-sized crystals composed primarily of olivine with lesser amounts of pyroxene and feldspathoids.[VIII] In the least-metamorphosed C chondrites, the olivine aggregates consist of forsteritic olivine with minor diopside and anorthite and in some cases spinel. With increasing alteration/metamorphism, the olivine and spinel (where present) become increasingly FeO-rich.

Olivine aggregates can be associated with olivine-rich chondrules, forming a surrounding halo or they can appear amoeboid-like as singular, irregular crystalline olivine masses with projecting arms (Fig. 7.14). Like the olivine chondrules, the olivine in aggregates is nearly pure forsterite. Often enclosed within the aggregates are spherical nodules of high temperature oxides and silicates (melilite). The aggregates do not show characteristics of having been melted but accreted around the nodules, consolidating into aggregates within the matrix.

Dark inclusions are relatively common in CV3 chondrites and they remain perhaps the most enigmatic of the inclusions. They are usually revealed in cut slabs as angular in shape and much larger than the chondrules. To the eye, they appear fine-grained and featureless (Fig. 7.15a), but under the microscope, some show very small, roughly equant chondrules averaging about 0.1 mm in diameter or less, superficially resembling a CO3 chondrite fragment (Fig. 7.15b). The matrix around the fragments sometimes show signs of alteration probably related to parent body aqueous alteration. The composition is very close to the CV3 matrix.

Calcium–aluminum inclusions (CAIs)

A very different inclusion is prominent in CV3 chondrites. They are rich in calcium, aluminum and titanium giving their name calcium–aluminum inclusions, or CAIs. They stand out conspicuously on broken faces of Allende, Vigarano and Axtell appearing as whitish or pinkish irregular masses contrasting against the black matrix (Fig. 7.16). They are larger than most of the chondrules. Although they have been known for many years, they became the subject of intense research after the fall of Allende. While all CV3 (and CO) chondrites contain these white inclusions, Allende seems to possess more than any other with about 5–10 vol%. There are two types based upon texture: fine-grained and coarse-grained. The fine-grained variety is composed of crystals less than 1 μm in size, too small to see clearly with a light microscope. Their irregular shapes terminate in lobate forms (Fig. 7.17a). Under crossed-polarized light they have very low or no birefringence, remaining a gray-blue-silvery tone. The coarse-grained types have much larger crystals and some appear to have reaction rims (Fig. 7.17b).

[VIII] A feldspathoid is a mineral chemically similar to feldspar but with a deficiency of SiO_2. In C chondrites these minerals would include nepheline and sodalite, also found in CAIs.

Fig. 7.14. An amoeboid olivine aggregate from Allende. Note the "pseudopod-like" arms projecting into the matrix. The aggregate is 0.8 mm at its narrowest. Viewed in crossed-polarized light. (Photo by O. Richard Norton and Tom Toffoli.)

Fig. 7.15 (a) A xenolithic dark inclusion about 32 mm long in an Allende slab. (b) Photomicrograph of the xenolithic inclusion of (a) showing tiny equant chondrules in a dark matrix. Around the edge of the inclusion is an altered zone that separates the CV3 texture from the inclusion texture. This mimics a fragment of a CO3 chrondrite. (Photo by O. Richard Norton.)

Fig. 7.16. A 13.9 g slab of the Axtell CV3.0 chondrite showing a large calcium–aluminum inclusion 9 mm long in the center. (Courtesy Institute of Meteoritics, University of New Mexico. Photo by O. Richard Norton.)

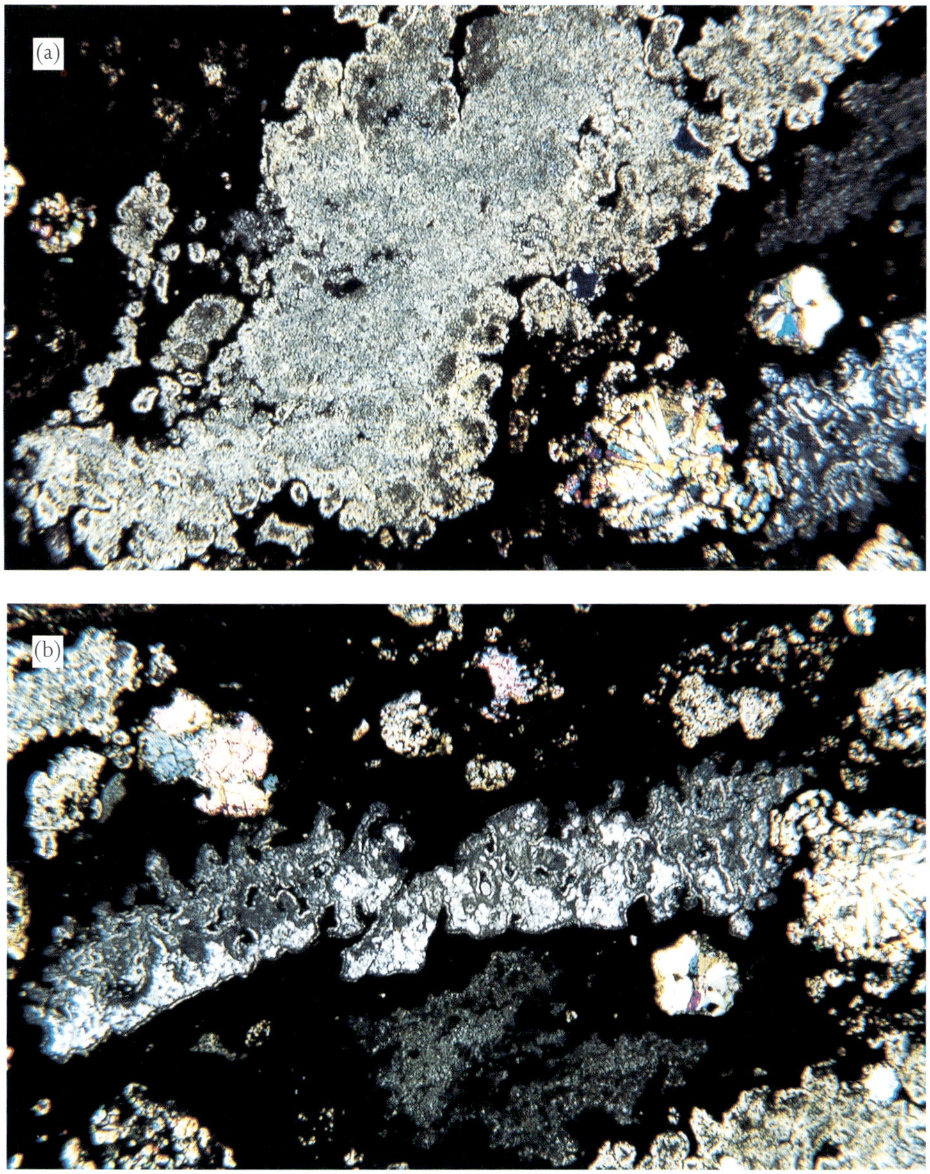

Fig. 7.17 (a) A large fine-grained calcium–aluminum inclusion 5.5 mm long from an Allende CV3 specimen. It contains melilite (a silica-poor feldspathoid), spinel, and fassaite. Melilite is the white bordering material. (b) A coarse-grained CAI from an Allende CV3. The white areas are melilite that has been partially altered to andradite (a Ca–Fe silicate) and grossular, appearing dark gray to black. The faint rims around the andradite is diopside. Inclusion is 1.7 mm long. (Photos by O. Richard Norton and Tom Toffoli.)

The mineralogy of CAIs is complex, being composed in part of highly refractory oxides such as melilite ($Ca_2Al_2SiO_7$); spinel ($MgAl_2O_4$); and perovskite ($CaTiO_3$), and the silicates clinopyroxene and anorthite ($CaAl_2Si_2O_8$). This mineralogy is comparable to that of the anorthite–forsterite–spinel chondrules. A long list of unusual minerals in CAIs, some not seen before in meteorites, has been compiled by researchers. That they are rich in highly refractory elements suggests that they formed under high temperatures at a very early stage in Solar System history. They must have been formed from the first condensates. The rims on some of the coarse-grained CAIs show that they were involved in reactions with Mg-rich olivine that condensed on them while still in a molten state. Radiometric dating shows them to have the oldest age of any Solar System material (interstellar grains excepted).

The significance of the CAIs lies in their isotopic composition.[8] To understand this significance, we need to digress for a moment to refresh our knowledge of the atom, especially its nuclear content. The signature of a chemical element is defined by the number of protons in its nucleus. For example, potassium (K) always has 19 protons in its nucleus. This defines its *atomic number* and the position it occupies in the Periodic Table of Elements. Since virtually all of an atom's mass lies in its nucleus, the *mass number* of an element is defined as the total number of heavy particles in the nucleus. Most potassium atoms in terrestrial rock have an atomic mass of 39 which means that there must be 20 additional particles in its nucleus, each with the mass of a proton. These additional particles are neutrons, subatomic particles with the same mass as the proton but with no electric charge. Most chemical elements are represented in nature by more than one form. There are actually three atomic forms, or *isotopes* of potassium, each with 19 protons but differing in atomic weight: ^{39}K with 20 neutrons; ^{40}K with 21 neutrons, and ^{41}K with 22 neutrons. These are all potassium atoms since they have the requisite number of protons. They differ only in the number of neutrons and therefore in their atomic mass. For years, scientists have studied the elemental composition of Earth as a terrestrial planet seeking to determine the naturally occurring isotopic abundances of each element. For example, we know the percent isotopic abundance of potassium: 93.26% for ^{39}K, 0.0117% for ^{40}K, and 6.73% for ^{41}K. What is usually done is to calculate ratios of one isotope with respect to the others, that is, $^{40}K/^{39}K$ and $^{41}K/^{39}K$. Generally, the elemental isotopic ratios of the elements in both Earth's crustal rocks and in meteorites are very nearly the same. This should come as no surprise since both the terrestrial planets and meteorites formed from the same material present in the solar nebula 4.6 billion years ago. However, a closer look at chondritic meteorites, especially the carbonaceous chondrites show the presence of *isotopic anomalies*, deviations in the isotopic ratios that can only be explained by introducing rare isotopes into the solar nebula from outside the system, that is, from interstellar space.

Some of the first-formed Al-rich minerals (primary minerals found in CAIs) such as anorthite ($CaAl_2Si_2O_8$), melilite ($Ca_2Al_2SiO_7$), and spinel ($MgAl_2O_4$), contain excess ^{26}Mg in their crystal lattices suggesting the radioisotope ^{26}Al originally occupied the space. Of the radioisotopes used in dating meteorites, ^{26}Al has the shortest half-life, 720 000 years; that is, half of any given quantity of ^{26}Al will decay in 720 000 years. This isotope is now extinct, leaving behind the diagnostic decay product ^{26}Mg in its place. The prevailing view is that ^{26}Al existed as a precursor isotope in the interstellar medium in and around the solar nebula immediately before the refractory condensation sequence commenced. The short half-life of this isotope places severe time constraints on the collapse of the solar nebula and the subsequent condensation of refractory minerals in the early Solar System. This isotope may have been deposited by a nearby supernova event injecting this and other isotopes into the surrounding medium within a million years or so before condensation of the highly refractory minerals occurred, allowing time for the radioisotopes to be incorporated into the minerals before they decayed and became extinct.[IX]

The importance of isotopic anomalies in meteorites cannot be overstated. They furnish the best clues to the conditions existing within the solar nebula at the time the meteorite components were forming. Isotopic anomalies among the various meteorite groups suggest that initially each formed from zones within the solar nebula that had different isotopic compositions. In other words, the solar nebula was not homogenous at the time the meteorites formed. This inhomogeneity may have been the result of a nearby supernova that produced elements with different isotopic ratios than those found ubiquitously within the solar nebula. The most important of these anomalies involves the element oxygen. Like potassium, oxygen has three stable isotopes: ^{16}O, ^{17}O, and ^{18}O. The relative abundances of the three isotopes in a sample are compared by calculating the difference between the $^{17}O/^{16}O$ ratio of the sample (referred to as $\delta^{17}O$) and a standard source, and then plotting this difference against the difference between the $^{18}O/^{16}O$ ratio ($\delta^{18}O$) and the standard. Both $\delta^{17}O$ and $\delta^{18}O$ are normalized to the standard with values in parts per thousand or per mil. The reference standard for comparing oxygen isotopic ratios is terrestrial ocean water referred to as *standard mean ocean water* or SMOW.

There is a natural isotopic sorting process which causes the proportions of the three oxygen isotopes to change as a

[IX] The vast majority of the aluminum atoms incorporated into the early minerals was the stable isotope ^{27}Al. Although the ratio $^{26}Al/^{27}Al$ is deceptively small (5×10^{-5}), ^{26}Al is not only sufficiently plentiful to be readily detectable in CAIs as its daughter isotope ^{26}Mg, it may have been plentiful enough to act as a heat source within early-formed asteroids and protoplanetary bodies leading to differentiation. This point is still controversial. ^{26}Al as a possible heat source in differentiated asteroid parent bodies is discussed further in Chapter 10.

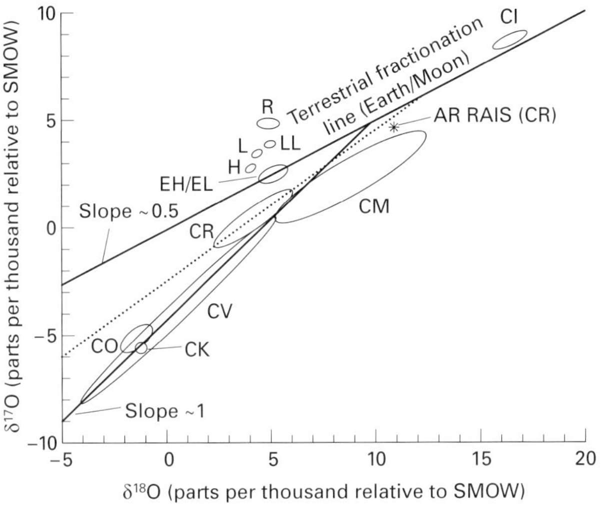

Fig. 7.18. This is a plot of the three oxygen isotopes found in all meteorites. The ratios of the three isotopes, ^{17}O, ^{17}O, ^{18}O are plotted relative to a standard value (standard mean ocean water or SMOW). All terrestrial and lunar rocks plot along the terrestrial mass fractionation line (TMFL) with a slope of 0.5. The ordinary chondrites however, plot along a line with a slope of ~1 displaced above the TMFL while the carbonaceous chondrites deviate substantially below the TMFL indicating an enrichment of ^{16}O (see text).

result of their difference in mass. This occurs in many mass-dependent physical processes such as chemical reactions, crystallization, vaporization and condensation, and diffusion to mention a few. All of these processes must have taken place in the solar nebula. This *mass fractionation* (mass separation of the three oxygen isotopes) is proportional to their mass differences. For a given physical process, the $^{18}O/^{16}O$ ratio changes twice as much as the ratio of $^{17}O/^{16}O$ reflecting their difference in mass. As a result, when the deviations of $\delta^{17}O$ is plotted against the deviations of $\delta^{18}O$ for terrestrial and lunar samples, the values reflect this fractionation and plot along a line with a slope of ~0.5 (Fig. 7.18.) Any process that separates isotopes according to their masses will produce such a distribution line. This line is the *terrestrial mass-fractionation* line from which all oxygen isotopic ratios for different meteorite groups are compared. If the meteorites and the Earth–Moon system formed in a homogeneous solar nebula where the oxygen isotopic reservoir were the same, we would expect the meteorites to plot along the terrestrial mass-frationation line, their deviations being accounted for by the mass fractionation process alone. This, however, is not the case. Figure 7.18 shows each meteorite group occupying different positions and different slopes with respect to the terrestrial fractionation line. For example, the ordinary chondrites (H, L, LL) lie in fields that deviate above the terrestrial mass-fractionation line. The *mean* of the three ordinary chondrites fall on a straight line of slope ~1.0. They must represent parent bodies that formed in different oxygen reservoirs with different proportions of oxygen isotopes. The E chondrites lie on the terrestrial mass-fractionation line suggesting that this group shared the same oxygen resevoir as the Earth and Moon. The C chondrites show a large deviation off the terrestrial line. They are widely dispersed but most striking is the position of the CV3 chondrites. They and the ureilites (a carbon-rich achondrite discussed in the next chapter) also plot along a line with a slope of ~1. This plot came primarily from chondrules and high temperature minerals found in Allende CAIs. They are enriched in ^{16}O relative to the other two O isotopes. Many researchers believe the excess ^{16}O is stellar material that originated outside the Solar System – the culprit again may have been the explosion of a nearby supernova. ^{16}O is formed at the time of the supernova outburst and at the expense of the other two isotopes, most of which are destroyed in the process. Thus, the oxygen isotopic compositions of Allende resulted from a mixture of two distinctly different materials; one component with oxygen isotopic abundances found in Solar System material, and another component enriched with nearly pure ^{16}O, the product of nucleosynthesis in a nearby stellar explosion.

CO chondrites

Although one can argue that there is sufficient chemical difference between CV and CO chondrites to make them separate groups, the most outstanding difference lies in their textures. Figure 7.19 shows a typical chondrule field of a CV chondrite compared to that of a CO chondrite. The photographs are made to the same scale. While the average chondrule diameter of a CV is about 1 mm, the average for a CO chondrite is about 0.15 mm. The sharpness of the chondrules along with the relatively low degree of thermal metamorphism show COs to be unequilibrated petrographic Type 3. The average chondrule size is closer to CM chondrules but the volume percent in COs differs markedly: 38 vol% for CO verses only 12 vol% for CM.

The chondrule compositions are close to those of CV chondrites. Most are porphyritic olivine chondrules with olivine made of nearly pure forsterite. Higher percentage combined iron is found in barred olivines and radial pyroxene chondrules. These can contain as much as Fa_{50}. Like the chondrules in CV chondrites, those in CO chondrites are closely similar to the high-temperature components in ordinary chondrites.

There is a higher percentage of olivine aggregates in CO chondrites than in CV chondrites, up to 15 vol% in the former and 8–10 vol% in CV chondrites. The reverse is true of CAIs. There is twice the volume percent in CV3 vs CO3, even though the first CAIs to be studied was in a CO chondrite. Scattered as inclusions in the matrix and sometimes in the chondrules is nickel–iron metal which averages about 6 wt% in CO while it is only about 0.5 wt% in oxidized CV chondrites. Thus, the CO group is less oxidized than the CVs. (Reduced or unaltered CVs have much more metal.) The matrix is essentially identical in both CV and CO chondrites. Among all the C groups, CO contain the least volume percent

Fig. 7.19 This figure compares the texture and chondrule size between (a) a CV3.0 chondrite (Axtell) and (b) a CO3 chondrite (Dar al Gani 067). Both photographs were made to the same scale, 9.0 mm across. (Photos by O. Richard Norton.)

matrix ranging from 50% to as little as 20%. CO comprise about 15% of the total known C chondrites. Of all the observed meteorite falls, CO chondrites amount to only 0.5%.

CK chondrites[9] (CV4–6)

In 1990, there were 13 recognized CK chondrites. All but two were found at five different sites in Antarctica. This group is the first to have been defined almost entirely through studies of meteorites from Antarctica. Most of the specimens were quite small; seven out of the 11 finds were under 50 g. By 1994, the discovery rate in Antarctica had increased enormously, reaching 59 by the end of the year, and the numbers continue to increase. (By the beginning of the year 2000, 73 have been found. Of these, only two were witnessed falls.) The first CK chondrite witnessed to fall landed in Karoonda, South Australia, in 1930. Karoonda is the type specimen for this group (Fig. 7.20). The largest CK chondrite, a 34 kg weathered mass, was found in 1974 near Maralinga in the Nullarbor Plain of South central Australia. Maralinga is the most readily available CK chondrite. One of the most noteworthy characteristics of the group is that most of its members are equilibrated showing petrographic Types from 4 to 6 with the great majority (75%) being Type 5.[10] These are the only C chondrites that show petrographic types beyond 3. All show signs of metamorphism with some containing dark shock veins. The most recognizable characteristic appears on cut faces. They appear dull and sooty, usually described as blackened, making their interior structures difficult to distinguish with a hand lens. The blackening is due to fine (0.3–10 μm) particles of pentlandite and magnetite that have permeated the interiors of the normally clear silicate minerals of the matrix, and to a lesser extent the chondrules and isolated silicate fragments. This blackening, called *silicate-darkening* is so pronounced that transmitted light is greatly reduced through thin sections (Fig. 7.21). The

Fig. 7.20. The type specimen of a CK carbonaceous chondrite from Karoonda, South Australia. The specimen is 64 mm longest dimension. (Robert A. Haag collection. Photo by O. Richard Norton.)

Fig. 7.21. A thin section of the Maralinga CK4 carbonaceous chondrite showing its general field viewed in transmitted light. Silicate-darkening over much of the field and evident in this specimen is typical of CK chondrites. The thin section horizontal field of view is 10 mm. (Photo by O. Richard Norton.)

cause of this silicate-darkening is not known but is certainly a clue to the history of these remarkable meteorites. The presence of black shock veins up to several millimeters in length may be related to the blackening of the olivine. The veins are filled with tiny rounded blebs of magnetite and pentlandite set in a glassy to microcrystalline mesostasis. These tiny blebs may have been mobilized through shock metamorphism during one or more impact events, quickly injecting into the silicate grains. All the known specimens show various shock stages from S2 to S5. Blackening is not unique to CK chondrites. Some ordinary chondrites also show blackening and shock veins, both injected with metal phases and troilite. It is possible that the original opaque phases of CK chondrites were iron–nickel metal and troilite, common in a reduced meteorite but later were chemically transformed into magnetite and pentlandite during an oxidizing event that oxidized the entire meteorite parent body.[11]

The chondrule density is only 15 vol%. Their average diameters are intermediate between CO and CV chondrules, averaging 0.80 mm. Most of the chondrules are porphyritic olivine with a few that are barred olivine chondrules. Pyroxene is relatively rare and only one radial pyroxene has been observed. Chondrule rims are not observed. The matrix

Fig. 7.22. A 28 g CR2 carbonaceous chondrite, Acfer 059, from the Sahara desert, shows sharply defined chondrules and an abundance of metal, uncommon to most C chondrite groups. The specimen's longest dimension is about 6 cm. (Robert A. Haag collection. Photo by O. Richard Norton.)

is black and opaque, containing an abundance of the above opaque minerals. FeNi metal is extremely rare or nonexistent. The lack of metal and the presence of magnetite, troilite and iron-bearing silicates attests to the high degree of oxidation of CK chondrites. There is an abundance of small isolated olivine grains within the matrix. They have a large range of sizes from 0.025 mm to 1 mm. Refractory inclusions (CAIs) are extremely rare. Two have been found in Karoonda, a few in Maralinga. CK chondrites were previously classified with the CV chondrites (CV4-6) since they are chemically similar, but the textural features described here strongly support the conclusion that these C chondrites represent a new, rare group.

CR chondrites (CV2)

CR chondrites are a relatively new group of carbonaceous chondrite. Like the CK chondrites, most were found in Antarctica. The type specimen called Renazzo fell in Ferrara, Italy, in 1824 but its unique nature was not recognized until similar meteorites were collected in Antarctica in the late 1980s and in the Sahara in the early 1990s. As of the year 2000, 78 CR chondrites were known: three witnessed falls, 12 from the Sahara Desert, 62 from Antarctica, plus one anomalous CR fall, Al Rais (Medina, Saudi Arabia, 1957). About 14% of the known C chondrites are from group CR.

In my description of the various carbonaceous chondrite groups, I have concentrated primarily on the whole-rock structures since they are readily seen with the naked eye, hand lens and petrographic microscope. What makes each group interesting are the similarities and differences between them. The differences are especially interesting since they establish the criteria for any new group. In the C chondrites studied so far, we have seen highly oxidized structures and minerals with little reduced metal. The CO chondrites were the only ones to show some metal, averaging 6 wt% in the matrix and chondrules. CR chondrites announce themselves as a new group with an amazing addition – an abundance of metal. (I remember the first time I laid eyes on a CR chondrite. I thought it was a blackened ordinary chondrite by the metal it displayed.) Metallic FeNi accounts for 5–8 vol% with sulfide (pyrrhotite and pentlandite) an additional 1–4 vol%. This amounts to between 100 and 160 mg/g of whole rock. At the time the CRs were recognized as a group, the metal content was one of their most distinguishing characteristics. For a few years they held the record for the highest iron content of any carbonaceous chondrite.

About 50% of the metal is found associated with low-iron porphyritic olivine chondrules. It appears as rounded grains within the interiors of olivine grains as well as on chondrule surfaces near the matrix. Interior metal has a relatively high and uniform Ni content throughout the grain whereas iron on the chondrule surface and near the chondrule/matrix boundary show zoning with low Ni toward the matrix and higher Ni toward the interior. This curious Ni distribution is interpreted as an indication of a reduction of combined iron at the chondrule/matrix boundary during an episode of thermal metamorphism.[12]

CR chondrites are classified as petrographic Type 2. The chondrule field is similar in appearance to CM2 chondrites in terms of population. The chondrules are intermediate in size between CM and CV chondrules with an average 0.7 mm diameter (Fig. 7.22). (For these reasons, they were at first classified as CV2.) The majority are low-FeO porphyritic olivine chondrules with the remaining being low-FeO porphyritic pyroxene and barred olivine chondrules. Present in low numbers are Ca–Al-rich chondrules similar to those found in Allende. Radial pyroxenes and cryptocrystalline chondrules are very rare. Many of the chondrules contain coarse-grained rims of chondrule minerals often with

Fig. 7.23. A 53 g CH carbonaceous chondrite Hammadah al Hamra 237 from the Libyan Sahara. Chondrules are sparse but easily distinguished and large metal fragments and metal chondrules stand out with their metallic sheen. The specimen is 75 mm long. (Robert A. Haag collection. Photo by O. Richard Norton.)

included FeNi grains. Others have rims of phyllosilicates. Most of the FeO-poor chondrules contain blebs of iron metal while the few FeO-rich chondrules contain none. The interiors of the chondrules are contradictory. About half show phyllosilicates in their interiors with their interstitial mesostasis hydrothermally altered. The other half shows no alteration of the chondrule glass, which remains clear. These chondrules have abundant inclusions of FeNi metal and sulfides. That you can have it both ways – reduced and oxidized in the same meteorite – seems contradictory. What this may be telling us is that hydrothermal conditions may have varied through the parent body or that perhaps metamorphism reduced the FeO in the refractory minerals to metallic Fe within the chondrules and matrix. These were then later hydrothermally altered to various degrees. Numerous chondrule fragments scattered throughout the matrix show disruption of the chondrules either through impact or hydrothermal alteration. There is no consensus among meteoriticists on this issue.

Refractory inclusions are present in much smaller numbers than in the CV group. Olivine aggregates and CAIs are found but are diminutive in size compared to CV chondrites. The matrix is opaque and includes magnetite and phyllosilicates as major constituents with a mixture of ferromagnesian mineral fragments, sulfides and small grains of evenly dispersed metal.

CH chondrites

This is the newest recognized group of the C chondrites. The *Catalogue of Meteorites* (2000) lists 11, all finds, making them the second rarest of the class. (CI holds that distinction with a total of five.) Seven were found in Antarctica and the remaining four in the Sahara Desert. The type specimen is ALH 85085 and all others are referred to as "ALH 85085-like".

CH chondrites hold the record as the most metal-rich of all the chondrite groups. This includes the E chondrites. The metal is easily seen in hand specimens (Fig. 7.23). Total bulk Fe content amounts to ~400 mg/g of whole rock, clearly three times or more the Fe of CR chondrites. Actual metal varies from 5 to 40 vol%. It is not unusual to see large metal fragments in the matrix. CH are pyroxene-rich amounting to 40–50 vol% with olivine 5–15 vol%. Oxidation levels indicated by iron-bearing fayalite and forsterite are $Fa_{<10}$ and $Fs_{<10}$ respectively.

Chondrules are sparce and small, averaging <0.1 mm in diameter. The largest are about 0.4 mm diameter and are porphyritic pyroxenes. Smaller chondrules are cryptocrystalline in texture. The most prevalent grains are mineral fragments, CAIs and dark inclusions embedded in a fine-grained matrix. The matrix varies from ~5 to 25 vol%.[13] Thus far, there are no petrographic types assigned to CH chondrites.

Organic carbon

We cannot end this chapter on C chondrites without saying a few words about their most intriguing quality – organic carbon. I hasten to say that by "organic carbon" I do not mean biogenic carbon, that is, carbon compounds made by living organisms. The organic chemistry of CI and CM chondrites is complex and a detailed discussion by necessity must be left to the bichemist, but the story of the search for life beyond Earth is too interesting not to summarize here.

The first to find organic compounds in meteorites was the Swedish chemist Jons Jacob Berzelius in 1834, but analytical methods then in organic chemistry were very crude. It would take a century before any progress would be made on their identification. An additional problem from working with the little meteoritic material available was the likelihood of contamination, not only by the natural environment but by simply handling the material during chemical analysis. Numerous times in the remaining part of the nineteenth century and through the mid-twentieth century, scientists fell into the contamination trap as they conducted their analyses and published their papers. The identification of long-chain hydrocarbons and fatty acids along with amino acids in the last decades of the nineteenth century fueled great excitement in the scientific community. The possibility of biological materials from space seeding Earth with life had great appeal. Meteorites could provide the first direct evidence of life beyond Earth. The stakes were high. European scientific society was shaken by controversy. Some prominent scientists supported claims that actual fossils had been found in meteorites. The "fossils" turned out to be natural inclusions taking on curious shapes (Fig. 7.24) and the amino acids and hydrocarbons were shown to be contaminants. Many years would go by before techniques would improve sufficiently to gain the confidence of meteoritical scientists that organic compounds important to life on Earth did exist in extraterrestrial material.

The notion lay dormant through the first few decades of the twentieth century. It briefly resurfaced in the 1930s when University of California biochemist Charles B. Lipman reported to have discovered bacteria in meteorites. As always, the announcement created a momentary sensation in the news media only to die down again when other scientists showed Lipman's bacteria to be common terrestrial forms. Contamination had once again reared its ugly head. The idea of life or life materials in meteorites seemed to be destined for the scientific scrap heap.

Then in 1961, armed with vastly improved equipment and better analytical techniques, Fordham University scientist Bartholomew Nagy, along with Esso Research Corporation scientists Douglas Hennessy and Warren Meinschein, hit the jackpot. Using the CI chondrite Orgueil, they found hydrocarbons closely resembling the byproducts of living organisms. The scientists were cautious in their announcement, recognizing that their scientific positions were at stake. Past history had not been kind to researchers claiming such discoveries. Although they were not willing to

Fig. 7.24. Chondrules and inclusions often take on curious shapes that, with a little imagination, can convince the observer that they are fossil life forms. This thin section in crossed-polarized light shows a remarkable object that has a convincing "moth" shape (below center). In actuality it is a radial pyroxene chondrule that has been heavily eroded or altered. Viewed at approximately 20× magnification. (Photo by Andrzej Pilski.)

openly state they had found life materials in Orgueil, their public statement was nevertheless provocative: "We believe that wherever this meteorite originated something lived!". These courageous scientists had once again opened the life in meteorites controversy. This time it would never be closed again. The time was right finally to proceed with the search for life in extraterrestrial materials.

Two events occurring almost simultaneously opened the door. America was destined to land on the Moon and bring back Moon rocks for study. NASA was well aware of the possibilities, though rather miniscule, that these rock samples could harbor bizarre forms of life. They built special laboratories at the Johnson Manned Spacecraft Center in Houston, Texas, to receive the rocks under sterile conditions, in an effort to avoid the contamination that had plagued researchers for years. Then, in 1969, the year of the first lunar landing, the fortuitous fall of Allende in February and Murchison in September provided scientists with pristine material on which to test their analytical procedures, including a search for organic matter, especially in Murchison. The results, no longer questioned by the scientific community, were nothing short of astounding. In Murchison, nonvolatile ($>C_{10}$) aliphatic hydrocarbons (saturated and unsaturated open long-chain hydrocarbons) and amino acids were found in 1970. Then followed a host of other organic compounds identified over the next five years: aromatic hydrocarbons (1971); carboxylic acids (1973); dicarboxylic and hydroxycarboxylic acids (1974); nitrogen heterocycles (1971–75); and aliphatic amines and amides (1975). Of particular interest was the discovery that specific amino acids of identical composition and structure were optically *racemic*, that is, they exhibited optical isomerism in that they were right-handed or left-handed. There were equal numbers of both right-handed and left-handed optical forms. Since life evolved on Earth with a preference for left-handed forms, finding equal numbers of left- and right-handed amino acids in a meteorite strongly suggests their extraterrestrial origins. Furthermore, an excess of left-handed amino acids would indicate terrestrial contamination. Some 74 amino acids have been identified in the Murchison meteorite (1988). Of these, eight are involved in protein synthesis in terrestrial lifeforms. An additional 11 are also biogenic though less commonly seen in biological systems. Though none of these compounds are thought to be the result of life processes in space, they showed that complex organic molecules could and indeed did form beyond Earth's environment. The majority of the 74 amino acids have no counterparts on Earth and are truly extraterrestrial.[14]

Of the total molecular carbon in Murchison, about 70%, is in the form of insoluble molecular material not completely identified. The remaining 30% is soluble organic compounds. Within this 30% are the amino acids. The concentration of amino acids in Murchison is on the order of 60 p.p.m.

Summary

Now that we have looked at all the groups of the carbonaceous chondrites, one characteristic stands out – variety. The important characteristics denoting each group is summarized in Table 7.2.

We saw with the ordinary chondrites a sameness among the petrographic types that was almost monotonous. Their textural and chemical differences were primarily the result of thermal metamorphism. Still, the ordinary chondrites must have come from at least three parent bodies. Their oxygen isotopes, chondrule sizes, bulk compositions and rarity of mixed OC breccias indicates at least three separate bodies. The carbonaceous chondrite groups on the other hand show intrinsic variations each probably representing individual parent asteroids. Although there has been some thermal metamorphism along with aqueous alteration in most of the groups, they remain the most primitive of the chondrites, retaining much of their primordial characteristics. Unlike their cousins the ordinary chondrites, the C chondrites must have formed in a variety of locations in the solar nebula. In a later chapter we will explore the known asteroids in search of the primitive parent bodies of the chondritic meteorites.

	CI	CM	CV	CO	CR	CK
chondrule size (μm)	–	270	1000	150	~700	700
aggregational chondrules	0	rare	com	rare	abnt	rare
opaque-rich PO chondrules	–	abnt	abnt	abnt	abnt	abnt
nonporphyritic chds (vol%)	0	0.5–1	0.1	0.8	0.3	rare
CAI + AOI (vol%)	–	2–8	6–13	10–18	0.1–1	~4
matrix/chd modal ratio	–	1.4–7	0.5–1.2	0.4–0.7	0.6–1.1	~6
metallic Fe–Ni (vol%)	~0	≤0.1	0–4.6	1.3–5.9	4.8–7.7	~0
metallic Fe–Ni (mg/g)	~0	≤0.2	~10	~60	~120	~0
aqueous alteration	high	high	low	low	mod	abs
magnetite framboids	pres	abs	abs	abs	pres	abs
coarse-grained rims	–	rare	abnt	rare	abnt	rare

Abbreviations: chds = chondrules; abnt = abundant; abs = absent; pres = present; com = common; mod = moderate; AOI = amoeboid olivine inclusion.
Source: Data from Kallemeyn *et al.* (1994), see reference 12 at the end of this chapter.

Table 7.2. **Petrographic characteristics of carbonaceous chondrites**

This table summarizes the important petrographic characteristics found in the carbonaceous chondrite clan. Compared to the ordinary chondrites, a much greater variety in textural and structural characteristics are seen among the groups of the carbonaceous chondrites.

References

1. Clarke, R.S., Jarosewich, E., Mason, B., Nelen, J., Gómez, M. and Hyde, J.R. (1971). The Allende, Mexico, Meteorite Shower. *Smithsonian Contributions to the Earth Sciences. No. 5*, Smithsonian Institution Press, Washington, DC.
2. Zolensky, M. and McSween, H.Y., Jr. (1988). Aqueous alteration. *Meteorites and the Early Solar System*, University of Arizona Press, Tucson, p. 137.
3. Dodd, R.T. (1981). *Meteorites: A Petrologic-Chemical Synthesis*. Cambridge University Press, New York, pp. 36–38.
4. Richardson, S.M. (1978). Vein formation in the CI carbonaceous chondrites. *Meteoritics* **13**, 141–159.
5. Yeomans, D.K. (1991). *Comets: A Chronological History of Observation, Science, Myth, and Folklore*, John Wiley & Sons, Inc., New York, p. 287.
6. Cruikshank, D.P., Hartmann, W.K. and Tholen, D.J. (1985). Color, albedo, and nucleus size of Halley's Comet. *Nature* **315**, 122–124.
7. Whipple, F. (1963). On the structure of the cometary nucleus. *The Moon, Meteorites, and Comets*, B.M. Middlehurst and G.P. Kuiper (eds.), University of Chicago Press, Chicago, pp. 639–664.
8. Podosek, F.A. and Swindle, T.D. (1988). Extinct radionuclides. *Meteorites and the Early Solar System*, University of Arizona Press, Tucson, pp. 1093–1113.
9. Wasson, J.T., Rubin, A. and Kallemeyn, G.W. (1991). The compositional classification of chondrites: V. The Karoonda (CK) group of carbonaceous chondrites. *Geochimica et Cosmochimica Acta* **55**, 881–892.
10. Norton, O.R. (1998). *Rocks From Space*, 2nd edition, Mountain Press Publishing Co., Missoula, Montana, p. 196.
11. Rubin, A.E. (1992). A shock-metamorphic model for silicate darkening and compositionally variable plagioclase in CK and ordinary chondrites. *Geochimica et Cosmochimica Acta* **56**, 1705–1714.
12. Wasson, J.T., Rubin, A.E. and Kallemeyn, G.W. (1994). The compositional classification of chondrites: VI. The CR carbonaceous chondrites. *Geochimica et Cosmochimica Acta* **58,** 13, 2873–2888.
13. Bischoff, A., Schirmeyer, S., Palme, H., Spettel, B. and Weber, D. (1993). Mineralogy and chemistry of the carbonaceous chondrite PCA 91467 (CH). *Meteoritics and Planetary Science* **29**, 444.
14. Cronin, J.R., Pizzarello, S. and Cruikshank, D.P. (1988). Organic matter in carbonaceous chondrites, planetary satellites, asteroids and comets. *Meteorites and the Early Solar System*, University of Arizona Press, Tucson, pp. 819–857.

(a) Dar al Gani 844 found in the Libyan Sahara desert in 1999. This slab cut form a 143 g main mass shows a lack of chondritic texture and is therefore an achondrite. The meteorite is permeated with dark xenoliths and other crystalline inclusions that seem to bear little relationship to the matrix. It has been classified as a polymict breccia, either a eucrite (most probable) or a howardite. The white grains are plagioclase and the rusty object in the lower left corner is an iron grain, relatively rare in HED achondrites. The speciment measures 49 mm in its longest dimension. (Specimen provided by Ronald N. Hartman. Photo by O. Richard Norton.)

(b) Northwest Africa (NWA) 900. The interior of this meteorite, found in the Moroccan desert in 2000, reveals light colored xenoliths trapped within a primitive chondrite. It is relatively rare to see such a mixture of chondrite and achondrite material. The meteorite has yet to be classified but the texture of the dark material shows distinct chondrules and grains of iron–nickel suggesting an L4 petrographic type. The lighter clasts show relatively large white crystals, probably plagioclase and/or clinopyroxene and a fine-grained matrix typical of basaltic achondrites. (Photo courtesy of Matteo Chinellato.)

CHAPTER EIGHT

Differentiated meteorites: the achondrites

In the previous two chapters we examined the most primitive of the meteorites, the chondrites. We saw that there are differences between them and that these differences are used to sort them into major groups. Truly primitive meteorites are those that exhibit the physical and chemical characteristics acquired during preaccretion and accretion phases of their parent bodies. These are their *primary* characteristics. Variations in their primary characteristics reflect differences in their formation environments and are characterized by their petrology and chemistry. The size, texture and volume percent of their chondrules, the matrix to chondrule ratios, and high-temperature refractory inclusions are all examples of primary petrologic variations acquired during their accretion phase. Variations in volatile and nonvolatile elemental abundances, precursor FeO/(FeO + MgO) ratios, isotopic ratios, matrix and chondrule compositions are primary chemical characteristics that reflect the chemistry of that part of the solar nebula in which they formed. So primary characteristics tell us much about the earliest conditions existing in the solar nebula at the time of parent body formation.

We saw that many of these primary characteristics have either not survived or have survived with considerable modification. In the chondrites we saw thermal metamorphism change primitive unequilibrated heterogeneous meteorites (Type 3) into increasingly equilibrated homogeneous meteorites (Type 4–6). We saw the textures change as thermal metamorphism recrystallized their interiors, modifying or destroying the chondrules. Low-temperature aqueous alteration of the textures and mineralogy of the carbonaceous chondrites was another form of metamorphism that changed the primary characteristics. What we see today in the chondrites are *secondary* characteristics mixed with primary. It is in part the secondary variations that have led to the current classification of chondrites. Secondary variations are no less important since they tell us about events occurring within the parent bodies after they accreted. A problem for the meteoriticists is to be able to distinguish primary and secondary variations within the groups of the chondrites.

Finally, after the primitive parent bodies accreted and secondary features began to appear, the parent bodies were subjected to yet another phase that produced *tertiary* variations. These are features produced by impact with other bodies, including shock features, brecciation and impact melting. Some tertiary variations were so major that they completely changed the chondritic nature of the original parent body. At least some of the parent bodies suffered extensive internal heating due to the decay of radioisotopes and reached a point of complete melting. These bodies differentiated and their meteoritic offspring became the differentiated meteorites. The *achondrites* form one of the groups of differentiated meteorites.

Major characteristics of differentiated meteorites

Differentiated meteorites are relatively rare. They include stony meteorites with little if any free iron, iron meteorites with or without silicate inclusions, and stony-iron meteorites with roughly equal amounts of metal and silicates. Taken totally, about 14% of all witnessed falls are differentiated meteorites.[1]

All differentiated meteorites lack chondritic textures although before melting they were originally derived from chondritic parent bodies with typical textures and compositions. Melting of the parent bodies changed their compositions so their elemental abundances are no longer solar. This is easily seen if we plot elemental abundances of siderophile, chalcophile, and lithophile elements normalized to silicon ($Si = 10^6$) for a typical achondrite against the solar abundances of the same elements much as we did for the CI chondrites in Chapter 5 (Fig. 5.2). When this is done for a eucrite, a calcium-rich basaltic achondrite, we see a wide scatter on either side of the solar abundance line (Fig. 8.1). The siderophile and chalcophile elements are considerably depleted relative to solar abundance which we would expect in a fractionated and differentiated parent body. At the same time, the lithophile elements (elements preferring the silicate phase found in crusts of differentiated bodies) are enriched by a factor of 10 over solar abundance.

The minerals in differentiated meteorites are magmatic, that is, they formed and crystallized from a melt. In Earth geology, rocks that crystallize from a melt are termed *igneous* rocks. Thus, differentiated meteorites are igneous rocks of other worlds.

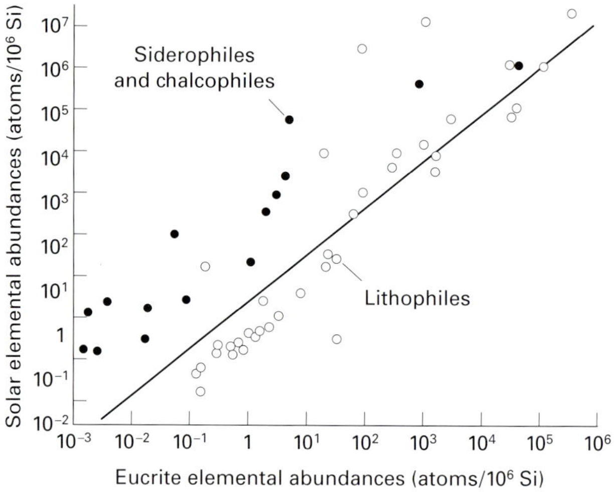

Fig. 8.1. Plot of elemental solar abundance vs elemental abundances for a eucrite (basaltic achondrite) normalized to $Si = 10^6$. The line represents solar abundance. Since achondrites come from differentiated parent bodies in which their original chondritic compositions have changed, they no longer reflect near solar abundance as in Fig. 5.2. The siderophile (metal enrichment) and chalcophile (sulfide enrichment) elements represented by solid circles are depleted by a factor of 100 in the basaltic achondrites and show a wide scatter around the solar abundance line. The lithophile (silicate enrichment) elements (open circles) show a 10-fold enrichment relative to solar abundance and plot near the solar line. (Data from Ross, J.E. and Aller, L.H. (1976). *Science* **31**, 1223.)

[1] Grady, M.M. (2000). *Catalogue of Meteorites*, 5th edition, Cambridge University Press, Cambridge, 720pp.

Differentiation

Before we see how the parent bodies of meteorites might have differentiated, we can learn something of the process by looking at the most familiar terrestrial planet – Earth. We envision Earth during its accretion phase being homogeneous throughout and chondritic in composition. Differentiation can occur only if the planet is heated internally. Early in the accretion process the first few hundred kilometers was heated through impact melting. In fact, it is probable that widespread melting of the outer few hundred kilometers occurred over and over again almost continuously as Earth accreted large planetesimals. There is a clear record of this early cratering period on the Moon. At the same time, Earth's rapid accumulation of mass must have compressed the interior, releasing heat in the process. Together, accretion and gravitational contraction raised the internal temperature in the first few million years to about 1000 °C. As overlying mass built up it furnished an insulating blanket that acted to retain internal heat. After a few tens of millions of years the accretion period gradually subsided. Earth had reached its present size and mass and had a chondritic composition.

Once sufficient mass had accumulated, an additional process was adding to Earth's internal heat. Locked within the mineral grains composing the rocks were heavy element radioisotopes (^{235}U, ^{238}U, ^{232}Th, ^{40}K) with half-lives measured in the millions to billions of years. Decay of these isotopes liberate alpha and beta particles that are readily absorbed by the surrounding rock, heating the rock in the process. Since rock is a poor conductor of heat, heat loss from the interior was less than the heat gain and the interior gradually heated to near the melting point of iron. Melting points of rock, or in this case, iron, increases with depth within the Earth because the surrounding pressure of the rock likewise increases with depth. Iron melting at approximately 1800 °C occurred first in a zone between 400 and 800 km down. Just how long it took to reach the melting point of iron at this depth depends upon which model is used. Some models predict only a hundred million years or less after Earth accreted. Other models place the iron melting phase at about a billion years. Once this zone of iron melting was created, the melted iron began to gravitate toward the center displacing the less dense, lighter rock in the process. This was the beginning of a catastrophic displacement of the iron that resulted in a liquid iron core. Enormous amounts of gravitational potential energy must have been released in the process raising the temperature an additional 2000 °C near the center, more than sufficient heat to melt or at least partially melt most of Earth.

The formation of the iron core began the differentiation process. Earth became layered. Under molten or semi-molten conditions the more dense material separated from the less dense. Heavier material settled toward the core to become the mantle while less dense material rose to near surface to become crust. This was also a chemical separation. Iron, nickel and the siderophile elements accumulated in the core; iron-bearing olivine and pyroxene in the mantle; and iron-poor feldspars and other silica-rich minerals in a thin crust.

Returning now to the parent bodies of meteorites, we need to ask how homogeneous chondritic rocky bodies in the range of fifty to several hundred kilometers diameter could melt. They were certainly subject to impacts which must have resulted in local melting but the impactors must have been small masses with small relative velocities to allow the parent bodies to accumulate material. Large impacts would have fragmented the growing parent bodies. What heat was generated near the surface by impact would have dissipated rapidly. Likewise, gravitational compression with the concomitant release of gravitational energy was probably not a factor since these bodies were not massive enough to sufficiently compress the accumulating material. Yet, we know from the achondrites, irons, and stony-iron meteorites that either partial or complete melting did indeed take place. Most astonishing is strong evidence that basaltic magma reserves existed within at least one asteroid (4 Vesta) 520 km in diameter.

Fractional crystallization and the igneous process

To understand the structure of achondrites, we need to review some geochemistry as it relates to the igneous processes responsible for these meteorites. Since achondrites come from a melt, let us begin with a melt within a deep magma chamber in Earth's crust. The elemental composition of the melt contains the eight most abundant elements making up most of the rock-forming minerals on Earth. These are listed in Table 8.1 along with their weight percent, volume percent and element percent composition in crustal rock.

It is an amazing fact that of the 2000+ minerals found in Earth's crust only about 20 are considered common and less than half of these, all made of the above eight elements, form over 90% of the rocks. Meteorites as rocks are actually simpler yet.

All igneous rock solidifies from a magma. Magma is a melt or a liquid solution of elements at high temperatures. As magma cools, the elements combine to form different minerals. With further cooling minerals grow and interlock with each other to eventually form a solid rock. This seems simple enough, but for many years geologists were puzzled by the observation that a magma of a certain composition could and usually did form a number of different rocks with different mineral compositions. How could a magma of a specific composition form rock of a different composition? This remained a puzzle until igneous petrologist Norman L. Bowen at the Carnegie Institute of Washington, D.C., wrote a landmark paper in 1922 that established fundamental principles on the origin of common igneous rocks.[1]

Element (atomic number)	Weight %	Volume %	Atom %
Oxygen (8)	46.40	94.05	62.17
Silicon (14)	28.15	0.88	21.51
Aluminum (13)	8.23	0.48	6.54
Iron (26)	5.63	0.48	2.16
Calcium (20)	4.15	1.19	2.22
Sodium (11)	2.36	1.11	2.20
Magnesium (12)	2.33	0.32	2.05
Potassium (19)	2.09	1.49	1.15

Table 8.1. **This table shows the eight most abundant elements making up rock-forming minerals on Earth**

These are arranged in order of decreasing abundance. The first two elements, oxygen and silicon are by far the most abundant, making up nearly 75 wt% of the silicate rocks. Similar abundances are seen in the achondrites.

In a series of laboratory experiments working with a silicate melt of an olivine basalt composition, Bowen noted that as the magma cooled, minerals of different compositions would crystallize out in a definite sequence which depended upon the rate of cooling of the magma and whether or not the early-formed minerals remained in the melt or settled out of the remaining magma as it continued to crystallize. He found that as a mineral forms and remains in the melt without settling out, it will continue to react with the melt and be converted into a different mineral upon further cooling. He further noted that there were two separate reaction series operating simultaneously: continuous and discontinuous (Fig. 8.2). The continuous series involves the solution series of plagioclase feldspar. The first-formed mineral in the series is Ca-rich feldspar (anorthite), which if continuing to react in a falling temperature becomes more and more Na-rich until it becomes sodic feldspar (albite). Plagioclase goes through its solid solution, continuously changing composition but sharing the same crystal structure. The discontinuous series involves ferromagnesian minerals with olivine, the first-formed mineral. Here, if allowed

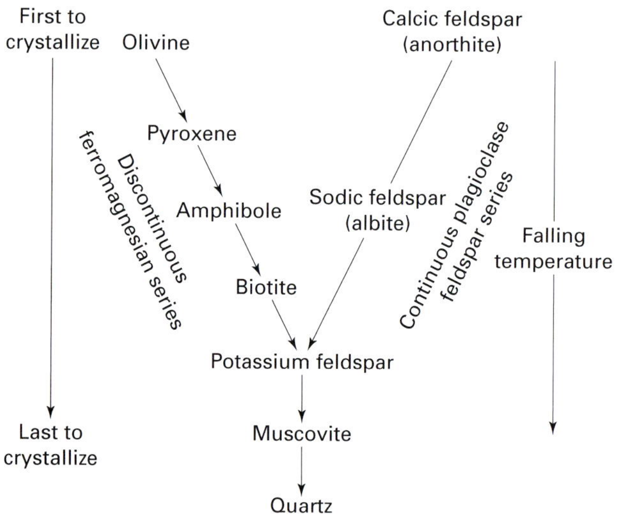

Fig. 8.2. Bowen's Reaction Series. The diagram shows the fractional crystallization of minerals in a basaltic melt at progressively lower temperatures. The reaction series consists of a continuous series with calcic plagioclase crystallizing at high temperature and continually reacting with the melt to produce more sodic plagioclase as the temperature falls; and a discontinuous reaction series where the change in mineral composition occurs in discrete steps beginning with the first-formed olivine at high temperatures and passing through other related minerals in lowering temperatures. In meteorites, amphibole is very rare and biotite is absent as they both form in the presence of water.

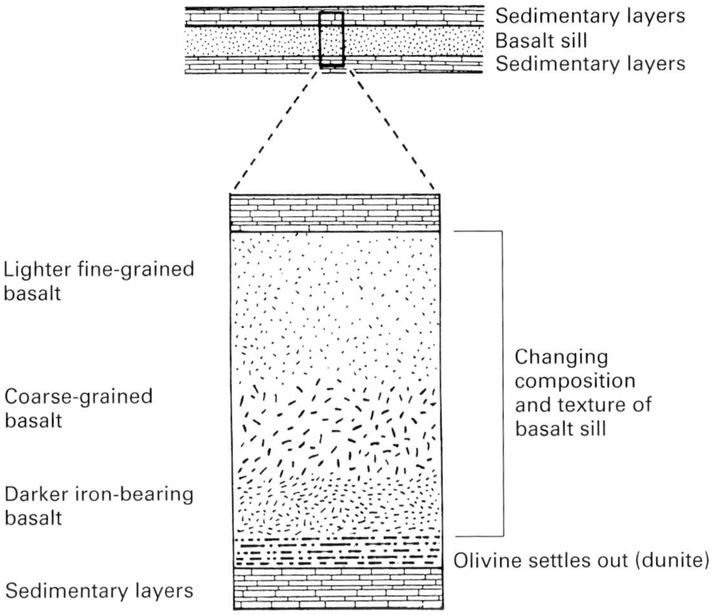

Fig. 8.3. Cross section of a basalt sill. The composition and texture of the basalt changes as the minerals crystallize out of the cooling magma in a specific sequence. Olivine, the first-formed mineral, settles to the bottom of the sill forming an olivine cumulate similar to dunite. (Adapted from *Rocks From Space*, 2nd edition (1998), Mountain Press Publishing Company, Missoula, Montana.)

to react in a falling temperature, the changes from one mineral to the next occur in discrete steps, from olivine to augite to amphibole to biotite. In this series new minerals form with different compositions and different crystal structures.

If the early-formed minerals settle out of the melt, the remaining melt, now changed in composition, will continue to react to form the remaining minerals of the series in both cases but will make a different rock. If the newly formed minerals continue to settle out, the remaining magma becomes increasingly silica-rich until all that is left is quartz. From the diagram it becomes easy to see why olivine and quartz are almost never seen together in a rock since they form under very different conditions. The settling out of minerals is called *fractional crystallization* or *fractionation*. We will see evidence of fractionation in some of the achondrites. Fractionation is a gravity driven mechanism that is very commonly seen in terrestrial rocks. Often in basalt dikes and sills one can see layers of single and multiple mineral crystals that have settled out of the melt and accumulated on the bottom of the dike, sill, or on the floor of a magma chamber (Fig. 8.3). Igneous rocks composed of such accumulated minerals are called *cumulates*. Cumulates are relatively common in terrestrial rocks and we will see them in achondrites also.

One last detail should be mentioned before we proceed with our examination of the achondrites. This has to do with the rate of cooling of a magma and the growth of crystals. We have referred many times to the texture of a meteorite that speaks of the general appearance of its interior, specifically, the fineness or coarseness of the chondrules or if examining the crystalline structures within the chondrules, the grain size. Likewise, the texture of an igneous rock usually refers to the size and shape of the interlocking mineral crystals. If a magma cools very quickly there may be too little time for crystal growth. This is often the case with basalt flows. The result is a fine or *aphanitic* texture with crystals too small to see without a microscope. If the magma is allowed to cool very slowly, there may be sufficient time for crystal growth to reach a few millimeters or more, easily visible to the unaided eye. This texture is referred to as *granular* or *phaneritic*. The crystals are usually equal in size. Chondrule textural terms differ somewhat from terrestrial igneous petrology. Chondrule textures are called granular if they are composed of grains equal to or less than 5 μm in size. Larger crystal sizes in chondrules are called porphyritic. In a slowly cooling melt of basaltic composition, often early-formed crystals will grow to exceptionally large size and then suddenly be expelled onto the surface of the Earth in a basalt flow. The rock that forms will be aphanitic but with the large crystals or *phenocrysts* embedded in the fine-grained or even glassy groundmass (Fig. 8.4) The groundmass remained glassy because it did not have time to arrange its atoms into an orderly crystalline structure. The resulting rock is called a *porphyry* and its texture *porphyritic*. We commonly see porphyritic or microporphyritic textures in the chondrules of ordinary chondrites, E and R chondrites and carbonaceous chondrites, appearing usually as large olivine or pyroxene crystals often among much smaller grains of the same composition, all set in glass showing that the reaction had not concluded. Notice that we have been speaking of both large crystals (meaning visible to the eye) and small crystals (meaning microscopic) while using the same terminology. It is the relative size of the grains that determines the texture.

It should be clear now that by noting texture we can say something about the time constraints of the crystallizing rocks and the minerals composing them. We can further say something about the conditions existing within the magma chamber and whether it was a deep-seated constrained melt

8.4. A terrestrial basalt rock with a glassy fine-grained matrix and embedded large phenocrysts of plagioclase. This rock is a basalt porphyry with a porphyritic-aphanitic texture. The rock is about 25.5 cm long. (Courtesy Science Graphics.)

or a volcanic flow near or on the surface of the host body. We will be looking for these textures as we preview the achondrites.

Achondrites are the largest class of differentiated meteorites, accounting for meteorites that come from asteroids, the Moon, and the planet Mars. Today there is some justification for subdividing the achondrites into these three subclasses – *asteroidal; martian; lunar* – since their origins and therefore the conditions under which they formed were different. An additional class, the *primitive achondrites*, are very rare meteorites that show signs of partial melting and partial differentiation. They are subdivided into three subclasses: *acapulcoites*, and *lodranites* and *winonaites*. Figure 8.5 shows the achondrite classification. Of all the meteorites the martian and lunar meteorites are most closely analogous to terrestrial igneous rock. They are much more coarse-grained than chondrites, strongly suggesting that they came from a melt and they contain little or no iron as metal. Their primary minerals are pyroxene, olivine and feldspar in various proportions. In 1920, Prior divided the achondrites into calcium-rich and calcium-poor subclasses. The calcium-rich achondrites contained between 5 and 25% CaO and the calcium-poor achondrites from 0 to 3%. He further gave each member of the subclass a chemical name much like the chondrites. Thus, eucrites and howardites were called pyroxene–plagioclase achondrites; diogenites were called hypersthene achondrites (remember the hypersthene chondrites) and so on. This scheme is no longer used but like the old chondrite nomenclature it is still a useful addition since it reminds us of the major minerals found in each.

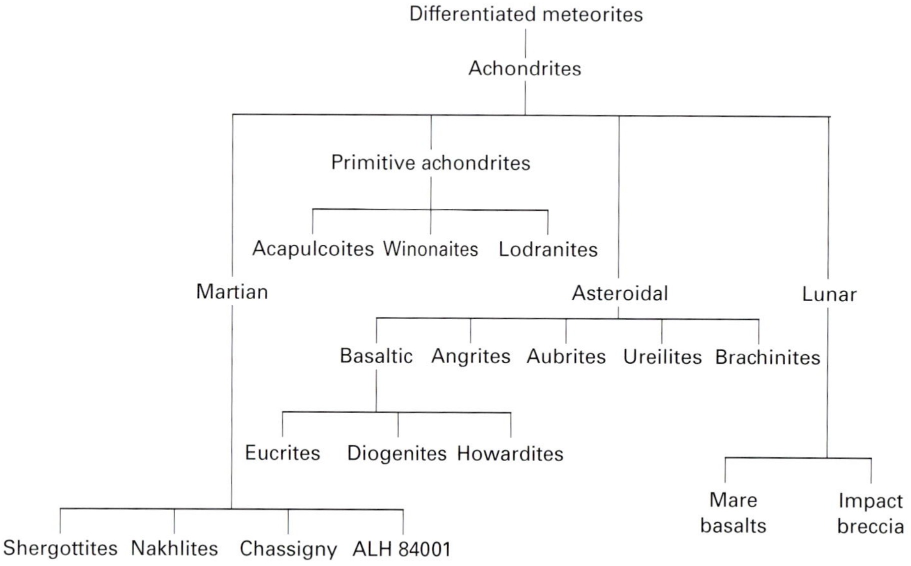

Fig. 8.5. Flow chart of the achondritic meteorites including asteroidal, martian, and lunar origins. See also Appendix A.

Asteroidal achondrites

The HED association

Today we recognize *associations* of achondritic meteorites that bear important similarities and probably similar origins, possibly coming from the same asteroid parent body. HED stands for *howardite*, *eucrite*, and *diogenite* achondrites. They are frequently called *basaltic* achondrites because of their magmatic origin and similarity to terrestrial basalts.

Eucrites

These are the most common of the achondrites. About 3% of the witnessed falls of all meteorite types are eucrites, which makes them the fourth most common meteorite to fall. Of the HED meteorites, eucrites are by far the most common, about 52%. Of the non-Antarctic meteorites, 55 were known in 1985 (*Catalogue of Meteorites*, 1985) but the Antarctic finds plus the finds in the Sahara Desert to the date of writing (2000) has increased the number to 200. Until the Antarctic meteorites became available with their large cache of eucrite finds, eucrites were defined as monomict breccias. The large number of eucrites recovered that show a wide variation of lithic fragments, unlike the fragments found in howardites, has prompted meteoriticists to accept eucrites as either monomict *or* polymict. (See Chapter 8 frontispiece.)

The most obvious external characteristic of a freshly fallen eucrite is its very black and lustrous fusion crust compared to the dull black crust of a chondrite. Eucrites are calcium-rich and this combined with the usually present small amount of iron gives these meteorites a "wet" look (Fig. 8.6). A broken face would expose a light gray interior. Meteoriticists frequently point out the similarities of eucrites to terrestrial basalts. They are speaking chemically and petrographically, but just looking at the broken face does not bring to mind terrestrial basalt. When was the last time you saw a white basalt rock? Terrestrial basalt is dark gray to black due to the presence of iron-rich pyroxenes. They contain the dark clinopyroxene, ferroaugite, that gives the entire rock a dark gray appearance. Eucrite textures are also fine-grained, but their major pyroxene is pigeonite, a medium to light gray clinopyroxene. Figure 8.7 shows a polished thin section of a eucrite from Millbillillie, Australia, with *glomeroporphyritic* texture – light gray clumps of phenocrysts visible to the unaided eye set in a dark gray groundmass. This is typical of terrestrial volcanic rocks that have cooled more slowly, producing glomerocrysts of interlocking plagioclase and pyroxene crystals. If basaltic lava contains dissolved gases that suddenly erupts onto the surface of the Earth, the sudden reduction in pressure releases the gas which quickly forms bubbles that make their way to the top of the flow (Fig. 8.8). The eucrite, Ibitira, one of the few unbrecciated eucrites known, shows a remarkable *vesicular* texture similar to that seen in a terrestrial lava flow (Fig. 8.9).

Microscopically, the resemblance of eucrites to terrestrial

Fig. 8.6. The shiny black crust on this eucrite from Camel Donga, Western Australia, is typical of calcium-rich eucrites. Note the contraction cracks through the crust. The specimen measures 5 cm in its longest dimension. (Photo by O. Richard Norton.)

Fig. 8.7. This thin section of a eucrite from Millbillillie, Western Australia, viewed in crossed-polarized light shows glomerocryst texture best seen near the upper edge as an aggregate of long white laths of plagioclase (anorthite) crystals easily visible to the unaided eye. The field of view is about 10 mm horizontal dimension. (Photo by O. Richard Norton.)

basalts is most striking. The two illustrations in Fig. 8.10 are photomicrographs of thin sections of a terrestrial olivine basalt (top) and an unbrecciated eucrite (bottom). The olivine basalt shows large colorful phenocrysts of olivine surrounded by tiny white needles of calcium-rich plagioclase and augite. The plagioclase shows simple twins. Eucrites have fine-grained textures with microbreccia fragments set in a matrix of the same mineralogy. Typical of eucrites is an *ophitic* texture usually manifested by a chaotic arrangement of plagioclase crystals set in a matrix of coarse clinopyroxene, pigeonite and/or augite. Such a texture (distinct from glomeroporphyritic) is commonly seen in dikes and sills on Earth where the cooling rate is intermediate between slow cooling as in a deep-seated magma chamber and the rapid cooling of a surfical basalt flow. This suggests that eucrites are shallow intrusives, that is, they originated in a relatively shallow chamber or in sills or dikes. On the other hand, the fine-textured matrix in which the feldspar crystals are set, suggests a basalt flow.

Mineralogically, the eucrites are quite simple. They are made up almost entirely of two minerals: plagioclase (30–50%) and clinopyroxene (40–60%). The clinopyroxene usually dominates by 10 or 20%.

Plagioclase is an important mineral in the HED series. Like olivine and orthopyroxene that forms an ionic substitution or solid solution series between the magnesium and iron end members, plagioclase forms a solid solution series between sodium and calcium feldspar. Figure 8.11 shows a ternary diagram with the plagioclase feldspar series occupying the base of the triangle. Pure sodium plagioclase, *albite* is the left end member and pure calcium plagioclase, *anorthite* is the right end member. The other members of the plagioclase series are noted at arbitrary places on the diagram between the two end members. Another series, the *alkali feldspars* are on

Fig. 8.8. This basalt from a lava flow in Central Oregon exhibits vesicular texture produced as the pressure on the escaping lava was reduced during eruption allowing dissolved gases to be released from the cooling magma, forming gas-filled cavities. The rock specimen is about 33 cm long. Compare this rock to the eucrite in Fig. 8.9. (Courtesy Science Graphics.)

Fig. 8.9. This unbrecciated eucrite fell in the village of Ibitira near Martinho Campos, Minas Gerais, Brazil, in 1957. It is the only eucrite known to have a vesicular texture. The millimeter-sized gas holes cover 5–7 vol% of the rock. (Photo by Michael Casper.)

the left side of the diagram from albite to the third end member, the potassium-rich feldspar, *orthoclase*. The alkali feldspars are very rare in meteorites. The plagioclase in eucrites is primarily anorthite with some bytownite. An average plagioclase composition would be about $An_{(80-95)}$.

The clinopyroxene is low calcium pigeonite with a composition that varies widely from specimen to specimen and within a given specimen. A typical pyroxene composition (wollastonite ($CaSiO_3$)–enstatite–ferrosilite) in mole percent might be $Wo_{(1-25)}$ $En_{(42-48)}$ $Fs_{(43-52)}$ (see Fig. 5.6). Minor minerals include chromite ($FeCr_2O_4$), iron–nickel, ilmenite ($CaTiO_3$) and troilite as opaque minerals, orthopyroxene, and polymorphs of silica: quartz, tridymite, and cristobalite.

Diogenites

These achondrites have been known by many names through the years. Thus: diogenite (named after the fifth-century BC Greek philosopher, Diogenes, who is given credit for first suggesting that meteorites came from space), chladnite (named for the late-eighteenth-century scientist Ernst Friedrich Chladni who was first to demonstrate that meteorites came from beyond Earth's atmosphere), shalkite, and rodite. It seems a shame that Diogenes was selected over Chladni for the honor of having named after him these rare meteorites. And rare they are indeed. Up to the year 2000, 94 are known and only 11 recovered outside Antarctica. This amounts to about 24% of the basaltic achondrites. In terms of observed falls of all meteorites, it occupies a position near the bottom of the list at $<1\%$, making diogenites some of the rarest of the asteroidal meteorites.

Diogenites are distinct from eucrites in both texture and mineralogy. They are monomineralic and coarse grained. Their primary mineral is the orthopyroxene, hypersthene. Figure 8.12 shows a broken face of the Johnstown, Colorado, diogenite. Nearly all of the diogenites are monomict breccias and this one is no different. The meteorite is composed of numerous large pale greenish angular fragments, its crushed, brecciated structure easily seen here without magnification. The fragments, composed of coarsely crystalline hypersthene, range in size from about 0.01 to 25 mm or more. In Johnstown, they are known to reach nearly 50 mm. With such a coarse texture, we're left with little doubt that diogenites are plutonic rocks formed deep in the iron-rich mantle

Fig. 8.10. (Top) Thin section seen with crossed Polaroids of an olivine basalt from Newberry Crater, Central Oregon. The large colorful grains are olivine; the white laths are plagioclase showing simple twinning. The horizontal field of view is 5 mm. (Bottom) Thin section seen with crossed Polaroids of a Millbillillie eucrite shows a strikingly similar texture to the terrestrial basalt above. White laths of plagioclase are scattered among grains of pigeonite and augite. Olivine is not seen at this scale and may not be present. The horizontal field of view is 5.0 mm. (Photos by O. Richard Norton and Tom Toffoli.)

of a parent body. Thin sections show the clasts embedded in a comminuted matrix of the same mineral (Fig. 8.13). Its singular mineral composition and coarsely crystalline texture strongly suggests that diogenites are cumulates formed on the floor of a magma chamber as the hypersthene crystals grew slowly and then settled out of the magma to form an almost pure rock of orthopyroxene.

One unusual diogenite should be mentioned here. It is Tatahouine, falling in 1931 near the village of Foum Tatahouine, Tunisia. It is the only known diogenite that is not brecciated. When I first saw Tatahouine, I found it hard to believe it was a meteorite. Most of the 12 kg collected were small fragments. There was no crust; the samples looked like nothing more than exceptionally large crystal fragments with the typical olive green color of hypersthene quite similar to that in Johnstown. However, in thin section they appear not as single crystals but as groups of randomly oriented hypersthene *grains*. This texture is something of a mystery. One hypothesis is that they may have once been very large crystals as they appear. They were somehow heated which converted them into clinopyroxene (clinohypersthene). When they cooled, they reverted back to the original hypersthene but failed to form crystals. Instead they remained as tiny grain assemblages. In any event, Tatahouine remains the coarsest diogenite on record.

The hypersthene in diogenites is remarkably homogenous

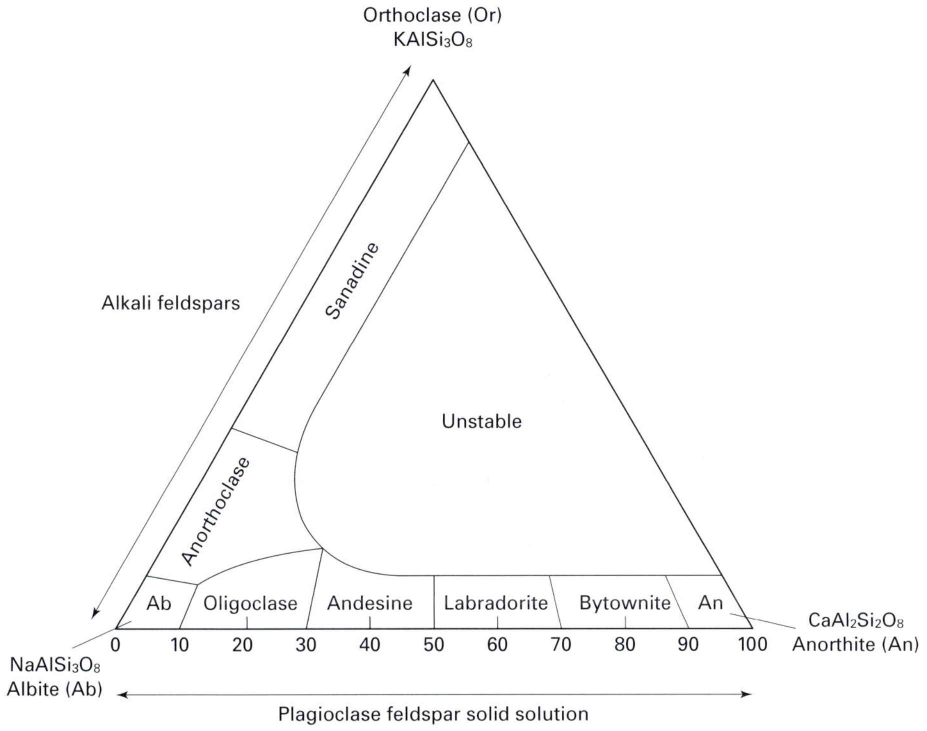

Fig. 8.11. Ternary composition diagram for the plagioclase feldspar solid solution series and the alkali feldspars.

Fig. 8.12. A diogenite from Johnstown, Colorado. The dark brown-green fragments are the orthopyroxene, hypersthene. The Johnstown diogenite texture is a coarse monomict breccia with clasts ranging from comminuted grains beyond naked-eye visibility to clasts measuring almost 3 cm across. The specimen is 9 cm wide and weighs 82.18 g. (Courtesy The Institute of Meteoritics, University of New Mexico. Photo by O. Richard Norton.)

with over 90% having an average composition of $Wo_1En_{74}Fs_{25}$ (1% wollastonite, 74% enstatite, 25% ferrosilite). The iron-rich hypersthene is about 25% ferrosillite ($Fs_{(23-27)}$), a mineral rarely if ever found on Earth. There are small amounts of anorthite, FeNi and FeS. The total iron is about 12 wt% but only 0.3 wt% is reduced metal.

Diogenites are thought to have formed in the lower crust or upper mantle of a differentiated asteroid, possibly the asteroid 4 Vesta. If the lower mantle could be sampled, a nearly pure olivine cumulate similar to terrestrial dunite is expected to be encountered (see the section on brachinites, p. 163).

Fig. 8.13. This thin section of the Johnstown diogenite viewed in crossed-polarized light shows large hypersthene fragments in different interference colors depending upon their crystallographic orientation. The large clasts rest in comminuted hypersthene with particle sizes to the limits of resolution of the picture. Some of the grains appear black, being in optical extinction. The black objects in the upper right-hand corner and upper edge center are rare metal or troilite grains. The horizontal field is 10 mm across. (Photo by O. Richard Norton and Tom Toffoli.)

Howardites

Consider the asteroid parent body that we have been describing. It was probably a body a few hundred kilometers in diameter, large enough to retain internal heat, to differentiate, and to maintain magma bodies deep in its mantle. The upper mantle cooled slowly, producing a coarse-grained crystalline rock of intrusive igneous diogenitic composition. Some of the magma erupted out onto the surface of the parent body spreading and cooling quickly into a fine-grained rock with an extrusive igneous basalt-like eucritic composition. During the first half-billion years of Solar System history, impacts by chondritic bodies pulverized the surface, sending shock waves deep into the parent body and blasting deep craters into the basalt flows, occasionally reaching the upper mantle. The mantle was overturned, exposing it to the surface conditions. Repeated impacts large and small broke and mixed the two rock types creating a regolith similar to that on the Moon. This mixing process is a form of erosion not often seen on Earth but the only substantial weathering mechanism available on the bleak surfaces of the asteroids. Micrometeorites, cosmic radiation and the solar wind continued to break down the rock into grain-sized particles that eventually welded together into a loosely consolidated soil-like rock. The net result is a new kind of rock with three different lithologies. These are the howardites, in essence, the soil of an airless world.

From the above scenario, it is not surprising that all howardites are polymict breccias. Broken faces of howardites reveal a medium gray interior and easily seen are conspicuous angular fragments of orthopyroxene (diogenitic) enclosed by much smaller rounded lithic fragments (eucritic) that make up the matrix (Fig. 8.14). In a typical thin section, microbreccia of angular clasts of plagioclase and

Fig. 8.14. A howardite from Kapoeta, Equatoria, Sudan. This interior face shows large diogenitic clasts (medium gray) and white eucritic clasts mixed with comminuted fine-grained material of the same composition making it a polymict breccia. The black grains are fragments of a carbonaceous chondrite. The specimen base dimension is 60 mm. (Courtesy Dr. Elbert King, University of Houston. Photo by O. Richard Norton.)

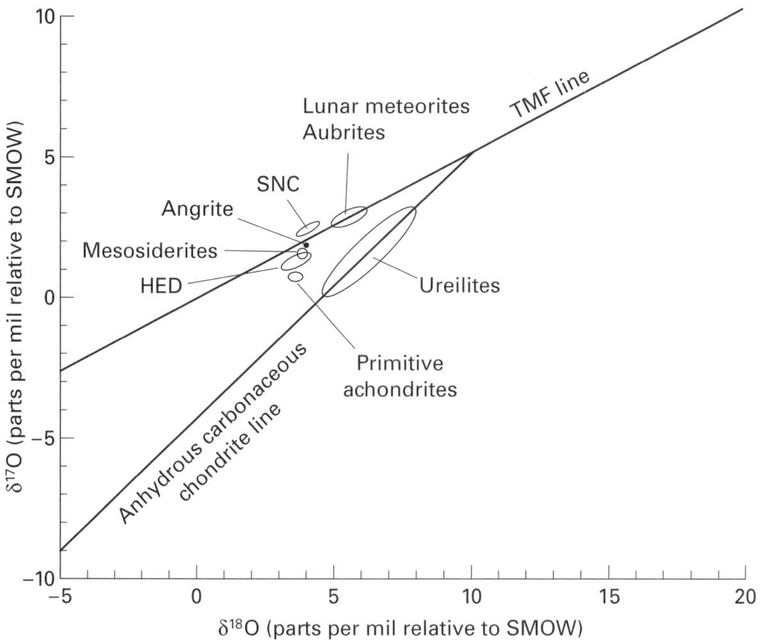

Fig. 8.15. Oxygen isotopic compositions for achondrites. The terrestrial mass fractionation line and the anhydrous carbonaceous chondrite fractionation line are included for reference. Note that the fractionation lines for HED achondrites and Mars meteorites run parallel to and on either side of the terrestrial line indicating separate parent bodies. The lunar rocks and aubrites both fall on the terrestrial line demonstrating that more than one body can have a similar oxygen isotopic composition, in this case, to Earth. This is also true of the HED and mesosiderites, both occupying the same field and therefore must be chemically related. (Data from R. Clayton and T. Mayeda of the University of Chicago.)

pigeonite are seen that sometimes retain the ophitic intergrowth found in the original eucrite. Clasts of chondritic material are often seen amounting to as much as 2–3 wt%. Interestingly, the chondritic material is carbonaceous similar in composition to CM2 chondrites. The dark inclusions in the Kapoeta specimen of Fig. 8.14 are carbonaceous. Individual crystals of orthopyroxene and plagioclase often show undulatory extinction in thin section. In addition, the shocked plagioclase grains may show the presence of maskelynite produced by the vitrification of the plagioclase. Thus, shock melting, fragmentation, brecciation and shock metamorphism all attest to a complex impact history for howardites.

There is a great deal of variability in the pyroxene composition of howardites. This can actually be seen in hand specimens by noting the color of the pyroxene, ranging from yellow-green, to brown, to black with increasing iron. The grains show chemical zoning as the composition changes across the grains. A typical pyroxene composition might be $Wo_{(1-35)}En_{(33-80)}Fs_{(18-53)}$. The plagioclase composition is eucritic with $An_{(80-95)}$.

Howardites are about as rare as diogenites; 93 are known as of the year 2000. About twice as many howardites as diogenites (20) have been observed to fall; about 2% of the total observed meteorite falls.

The mesosiderite connection

To add to the complexity of the howardites, olivine and iron–nickel metal are also found. Iron–nickel is a major component of mesosiderites, a stony-iron meteorite with stone and iron of nearly equal proportions. Accessory amounts of olivine are also present in mesosiderites. In the past, mesosiderites have been classified under the stony-irons which includes the pallasites. Superficially, they appear distinct from the basaltic achondrites because of their large metallic nickel–iron content but mineral, chemical and isotopic connections are recognized between the basaltic achondrites and the mesosiderites. Referring to the oxygen isotopic composition diagram (Fig. 8.15), we see that the HED field lies predictably very near the terrestrial mass fractionation line as we would expect of differentiated meteorites. What is interesting is that the mesosiderites share the same field, meaning that all four meteorites have similar oxygen isotopic compositions strongly suggesting that they are chemically closely related.

Mesosiderites are polymict breccias being composed of angular fragments of different mineral compositions. They are stony-iron meteorites containing nickel-rich iron metal and mafic silicates. The metal may be uniformly distributed within a silicate matrix with large silicate fragments enclosed by the metal like the Clover Springs, Arizona, mesosiderite (Fig. 8.16), or the metal may be found in clumps or aggregates surrounded by silicates like the Estherville, Iowa, mesosiderite (Fig. 8.17). In both cases the mineralogy of the silicate portion is much the same: orthopyroxene and plagioclase with minor amounts of olivine. The pyroxene, mainly hypersthene, usually dominates with smaller amounts of pigeonite. The plagioclase is anorthite. The silicate clasts show strong evidence of impact shock. A *cataclastic* texture (fragmental texture produced by mechanical stresses, in this case, impact) is evident, ranging from large fragments measuring centimeters across to granular clasts a millimeter or less. These clasts are set in a comminuted groundmass of the same minerals. In thin section (Fig. 8.18) some of the grains show undulatory extinction due to impact shock. The presence of coarse-grained plagioclase and orthopyroxene

Fig. 8.16. A slab of the Clover Springs, Arizona, mesosiderite. The meteorite is primarily iron–nickel metal uniformly distributed through the meteorite with included angular clasts of dark silicate minerals scattered through the metal. The horizontal dimension is about 11 cm. (Photo by O. Richard Norton.)

Fig. 8.17. Estherville, Iowa, mesosiderite. Here, the metal is fragmental and in aggregates scattered through the dark silicate matrix. The slab is about 100 mm longest dimension. (Robert A. Haag collection. Photo by O. Richard Norton.)

suggest thermal metamorphism and subsequent recrystallization. The iron portion is either scattered more or less uniformly throughout the meteorite or forms nodules. It has a uniform composition unlike iron meteorites (referring here to trace element siderophiles) with a nickel content of between 7 and 10%. This is the nickel content of octahedrite iron meteorites. Remarkably, some of the larger metal nodules when polished and etched show Widmanstätten structure typical of octahedrites (see Chapter 9).

A variation on the above mesosiderite theme is found in the mesosiderite, Bencubbin, a 54 kg mass found in 1930 near Bencubbin, Western Australia. This meteorite (Fig. 8.19) contains substantial amounts of enstatite and olivine appearing as clasts in a metal matrix. There are also clasts of chondritic material. The metal contains about 6.6% nickel, lower than most other mesosiderites. Bencubbin may represent a new class of enstatite–olivine mesosiderite.

So what can we say about the origin of mesosiderites? Their bulk silicate compositions are very close to howardites, i.e. a eucrite/diogenite mix. For this reason, many meteoriticists include mesosiderites with the HED association, which is why I include them here instead of in the iron meteorite class of Chapter 9. What doesn't conform to the HED composition is the large amount of metal. There seem to be no mineral relationship with the metal. Researchers refer to the iron as *exotic*, meaning that it is not related to the original eucritic parent body but was later introduced after the parent body differentiated. What is evident is that impact played a major role. A eucritic parent body was impacted by another body of similar composition. This impact may have penetrated to an

Fig. 8.18. Thin section viewed with crossed Polaroids of the mesosiderite, Bondoc, from the Bondoc Peninsula, Luzon, Philippines. Large anhedral grains of orthopyroxene (white and dark gray) and small colored grains of pigeonite dominate the field. Grains of plagioclase can be recognized with stripes (twinning planes) across their faces. One is interlocked with pyroxene in the center. The dark blebs are metal. Photographed at 39×. (Photo by O. Richard Norton.)

Fig. 8.19. Bencubbin mesosiderite from Western Australia. The silicate clasts within a metal matrix have variable composition with both enstatite and olivine components. Some stony inclusions appear to be chondritic. This anomalous meteorite may be a new class of mesosiderite. Specimen measures 114 × 93 × 11 mm. (Robert A. Haag collection. Photo by O. Richard Norton.)

iron core body or core/mantle boundary, remelting the iron and injecting iron and some olivine into impactor and recipient. The presence of Widmanstätten structure in the metal requires a slow cooling rate of a few degrees Celsius per million years. This suggests that after the impact the metal had to have experienced deep burial or at least heavy rocky insulation near the surface to retard its cooling to the above rate. A slightly different interpretation sees impact mixing of basaltic silicates from a differentiated asteroid with metallic iron from the naked core of another differentiated asteroid. The result is much the same.

Many authors have referred to mesosiderites as "waste buckets" since they appear to be composed of unassociated mineralogies that make little sense. This writer sees the mesosiderites rather differently. To me, they tell the story of a violent period in early Solar System history following differentiation of chondritic parent bodies. They are formed by impact and mixing on the surface and within the deep interior of colliding worlds. They will ultimately tell us much about the interiors of asteroids that orbit the Sun today. Unfortunately, we have yet to learn enough to read the story they are telling us.

Aubrites – the enstatite achondrites

If the mesosiderites present an enigmatic problem, the aubrites are also problematic. In Chapter 5 we looked briefly at the E chondrites. Here we found an almost monomineralic

Fig. 8.20. Norton County, Kansas, aubrite. This is a monomict breccia with coarse clasts of enstatite and a comminuted matrix of the same mineral. Iron–nickel is present as rusty brown grains on the edges of the specimen. The face measures 32 mm across. (Photo by O. Richard Norton.)

silicate fraction, the iron-poor orthopyroxene enstatite. In this respect, the aubrites are strikingly similar to their chondrite cousin. Aubrites are named after the fall of the type specimen near Aubres, France, in 1836. About 8% of all known achondrites are aubrites. They are monomict breccias with very coarse fragments of enstatite many millimeters across. These are embedded in a matrix of crushed and comminuted enstatite (Fig. 8.20). Unlike E chondrites that average more than 20 wt% metal, aubrites have only small amounts of iron averaging <2 wt% metal with the remainder as troilite. Their lack of iron results in a light brown to almost clear glassy fusion crust unique to these meteorites (see Fig. 3.11d). The interior is typically brecciated. In thin section a similar story is told: large irregular grains of enstatite dominate the field and occasional blebs of iron can be found (Fig. 8.21).

The fact that E chondrites and aubrites have nearly identical silicate mineralogies, that they are both highly reduced and that their oxygen isotopic compositions are nearly identical, with both differing substantially from the O-isotopic compositions of ordinary chondrites (Fig. 8.15), strongly suggests that they are closely related, perhaps formed from the partial melting of a single parent body. The scenario was thought to go something like this: We can imagine that the precursor was an E chondrite. Partial melting of the E chondrite body deep in its interior released the iron which formed a core body. The liquid mantle composed now of an almost pure enstatite magma (with a little metal) recrystallized under moderate pressures (>6 kbar) forming enstatite again, but now as an iron-poor aubrite. Very slow cooling allowed large enstatite crystals to form much as we

Fig. 8.21. This is a thin section viewed under crossed Polaroids of the Cumberland Falls, Kentucky, aubrite. Large grains of enstatite (gray and yellow) are emersed in a matrix of tiny grains of enstatite. The black blebs in the upper right and lower left corners are iron–nickel metal. The horizontal field is 5.1 mm across. (Photo by O. Richard Norton and Tom Toffoli.)

see in aubrites. The unmelted but partially metamorphosed E chondrite material formed a crust with substantial metal and a chondritic texture between EH3 and EH5. The EL chondrites, more highly metamorphosed and partially depleted of metal, may have formed beneath the EH layer with a dominantly EL6 texture.

But in a definitive paper[2] in 1989, Klaus Keil, then at the Institute of Meteoritics, University of New Mexico, showed that it is unlikely that the three enstatite-rich meteorites could have evolved in a single parent body. First, the EH and EL both have chondritic textures (chondrules) and their petrographic sequences, as in ordinary chondrites, represent a metamorphic sequence. In the late 1980s enstatite chondrites representing all petrographic types were not known, but by the end of the 1990s, all petrographic types from EH3 and EL3 to EH6 and EL6 had been found. Again, like the ordinary chondrites, they have not been melted or even partially melted but metamorphosed through solid-state recrystallization well below melting temperatures. Moreover, about 25% of EH and EL chondrites are breccias, suggesting that these meteorites existed at or near their parent body surfaces. If both EH and EL chondrites formed in the same parent body through the original accretional process, then we should expect continuing impacts to thoroughly mix the two types so that both should be found in a single rock as a polymict breccia. Instead, all the breccias are monomict breccias with clasts of either all EH or all EL but not both. There are a few polymict E chondrite breccias but the breccias are CI carbonaceous chondrites clasts. It seems, therefore, that EH and EL chondrites represent separate parent bodies. Keil further supported this conclusion by pointing out that many researchers believe that compositional differences between the two enstatite chondrites occurred by processes within the solar nebula before their formation by accretion. This suggests that they formed from material in different parts of the solar nebula where the compositions differed and would therefore require separate parent bodies. Forming in the same part of the solar nebula where the nebula was homogeneous would produce thorough mixing during accretion resulting in the formation of one parent body of intermediate composition. But other researchers were quick to point out that the oxygen isotopic compositions of both were very similar, meaning that they must have formed in the same oxygen reservoir in the solar nebula. But the isotopic ratios of ^{129}I to ^{127}I showed a formation interval (see Chapter 10) of approximately four million years between EH4–5 and EL6 chondrites. In other words, the EH chondrites were four million years older than the EL chondrites. They may still have formed in a similar oxygen reservoir over that relatively short time interval.

Enstatite achondrites or aubrites are clearly of igneous origin formed from a magnesium-rich melt in which sizeable crystals formed. They are also all breccias (with one exception – Shallowater). The parent body of aubrites must have melted, recrystallized and then fragmented by collision, later to be reassembled into an enstatite rubble pile. Just what heat source could have melted the original aubrite parent body is still an unanswered question. Both E-chondrites and achondrites are thought to have formed in a highly reducing environment relatively close to the Sun. At that location (under 2 AU) one could think of mechanisms that could conceivably melt the aubrites – magnetic induction heating as the Sun's magnetic field sweeps through the forming parent body; intense but temporary increase in output of the Sun as it entered the T-Tauri phase of its formation, for example. But why should the aubrites melt and the EH and EL chondrites remain unmelted in the same location and under the same conditions in the solar nebula? This question has yet to be answered.

In summary, EH and EL chondrites and the aubrites appear to have formed as separate parent bodies in a zone about 2 AU from the Sun. They formed in different parts of the solar nebula at that distance which was chemically heterogeneous but which retained a similar oxygen isotopic reservoir.

Ureilites

These rather strange meteorites are named after the fall of its prototype on a farm near the village of Novo-Urei in Central Russia in 1886. Several stones were recovered by villagers and at least one was eaten by the locals. After scientists had a look at this first of a kind meteorite they may have secretly wished all of them had been eaten. They are among the rarest achondrites after the aubrites and diogenites. Over the past 20 years, the US Antarctic meteorite program and, more recently, the finds in the hot deserts have substantially increased the number of ureilites from about a dozen in the 1970s to over 90 individuals at the present. Ureilites are totally unique and have little in common with the other achondrites. They are igneous rocks composed of olivine and the clinopyroxene pigeonite. The silicates are coarse-grained, suggesting a deep-seated origin, and many are elongated and appear to be oriented in a preferred direction. This suggests that they may have formed as cumulates on a magma chamber floor. The most noteworthy characteristic is the presence of black opaque carbon-rich material running between the grains. The material consists of graphite, the low-pressure polymorph of carbon, nickel–iron metal (more than in any other achondrite not counting mesosiderites – 15.64% total Fe), cohenite (iron–nickel carbide), troilite and the high pressure carbon polymorphs, lonsdaleite (hexagonal form of diamond) and diamond (cubic form). Meteoritic diamonds were discovered first in the Novo-Urei meteorite in 1888. The diamonds are microscopic averaging only a few microns across. Much larger diamonds have been found within graphite inclusions in Canyon Diablo iron meteorites. These can reach 0.5 mm or larger. The diamonds are usually in the form of *carbonados*, black and opaque aggregates of minute diamond particles forming a rounded noncrystalline mass. A few, however, have been found with an octahedral habit.

As much as 2 wt% carbon is found in ureilites. We are immediately reminded of the CI and CM2 carbonaceous chondrites which contain about the same amount of carbon. Might ureilites be related to carbonaceous chondrites? Looking at the O-isotopic composition of the ureilites (Fig. 8.15), we find that they lie on the C chondrite mixing line being most closely allied to the CV and CO chondrite field. Some of the silicate crystals show various stages of shock, and the presence of high-pressure carbon polymorphs strongly suggests that these meteorites have suffered impact shock sufficient to transform graphite into diamond. If a carbonaceous chondrite impacted the ureilite parent body carbonaceous material could have been injected into the parent body. That the carbonaceous material is interstitial (filling spaces between grains) to the coarse grains of olivine and pyroxene is curious since we would expect that the material would have impregnated the grains through fractures that typically occur, especially in olivine. In thin section (Fig. 8.22), the coarse grains appear to have been tightly fitted as one would expect of a typical plutonic rock. But the opaque material between the grains appears to have forced the grains apart as much as a quarter of a millimeter in some places creating a colorful mosaic tile-like pattern with black "grout" between. Many of the olivine grains show more Fe-rich composition in their cores (Fa_{13}) than in their rims (Fa_6). The carbon matrix has reacted with the olivine, reducing the iron to its metallic state.

Angrites

The first recognized and most famous angrite is the Angra dos Reis, a single stone of about 1.5 kg that fell into the Angra dos Reis bay in January, 1869, in the State of Rio de Janeiro, Brazil. It remained a unique meteorite for over a century. Through the years much of the original stone vanished, with only about 10% still existing in museums. The largest piece is at the National Museum in Rio de Janeiro, Brazil. It weighs 101 g. Over three consecutive years three more angrites were found and recognized from the Antarctic ice fields: LEW 86010, a 6.9 g piece collected in 1986; LEW 87051, a 0.6 g, piece collected in 1987; and A881371, a 11.2 g piece collected in 1988. The first two are preserved at the Johnson Space Center in Houston, Texas, and the last is housed in the National Institute of Polar Research in Tokyo. The most recent and the fifth known angrite was found in May, 1999, in the Libyan Sahara Desert (Fig. 8.23a). This remarkable specimen, Sahara 99555, is a single stone weighing 2.71 kg. It remains in private hands with a small amount available for study and trade.

The Angra dos Reis meteorite is an ultramafic igneous rock composed of 95% fassaite, a Ca–Al–Ti-rich clinopyroxene. In the old tradition of naming meteorites by their most common mineral, it is often referred to as a *fassaite achondrite*. (Since about 1988 *fassaite* was no longer an approved mineral name for the clinopyroxene, although it is still widely used in the literature. The International Mineralogical Association recommends the name *ferrain aluminian diopside!*) Accessory minerals include Ca-rich olivine and anorthite. Angrites show considerable variation in mineral composition. Angra dos Reis is the most anomalous of the known angrites with the highest fassaite composition. The newest angrite, Sahara 99555, shows a more normal texture and mineralogy (Fig. 8.23b) with modal abundances of 33% anorthite; 24% clinopyroxene (fassaite); 23% Mg-rich and 19% Fe-rich olivine. The olivine is usually zoned with Mg-rich cores and

Fig. 8.22. Thin section viewed under crossed Polaroids of the Australian ureilite Hughes 009. The large grains are olivine and pigeonite both of which show extensive fracturing. Black opaque carbonaceous matter fills the spaces between the grains. The white grains to the left of center appear to have been forced apart by introduced carbonaceous material. The horizontal field is 6.1 mm wide. (Photo by O. Richard Norton and Tom Toffoli.)

Fig. 8.23. (a) This angrite slab, Sahara 99555, was cut from a 2.71 kg stone, the largest angrite ever found. It measures about 15 cm across. It is an ultramafic rock composed of three major minerals: plagioclase (anorthite); clinopyroxene (fassaite); olivine. Note the gas holes. (Specimen provided by Labenne Meteorites. Photo by O. Richard Norton.) (b) Thin section viewed with crossed polarizers showing the internal texture and mineralogy of the angrite Sahara 99555. The long white and light blue laths are anorthite. Plagioclase is commonly seen as graphic intergrowths with colorful olivine (the blue olivine grains near the bottom center with light blue plagioclase.) The clinopyroxene, fassaite, is the large burnt orange grain and largest light gray grain above center. The horizontal dimension is 5.0 mm. (Photo by Tom Toffoli. Specimen provided by David Mouat, Desert Research Institute, University of Nevada System.)

more Fe-rich zones around the cores. The mineral composition sounds familiar – like the calcium–aluminum–titanium inclusions in CV3 chondrites. If there was a chondritic precursor for these meteorites, the closest would be the CV3 chondrites. The O-isotopic composition for the angrites places them below the terrestrial fractionation line but near the basaltic achondrites (Fig. 8.15).

Brachinites

These curious meteorites were first classified simply as anomalous achondrites since they did not fit any known group of achondrites. At last count, there are seven known brachinites, all but one found in either Australia or Antarctica. Brachinites take their name from the first recognized specimen, a 202.85 g stone found near Brachina, South Australia, in 1974. Their mineralogy is quite simple; almost entirely olivine. When first found it was understandably mistaken for a rare Mars meteorite called Chassigny, also composed of olivine.[3] Brachina's 4.5 billion year age compared to Chassigny's relative youth of 1.3 billion years argues against any relationship between the two olivine rocks. Ultramafic terrestrial rocks composed primarily of olivine are mantle rocks called *dunites*. Figure 8.24 shows a thin section of a brachinite of remarkably equigranular crystals with unmistakable bright interference colors typical of olivine. About 90% of the meteorite is olivine, with the clinopyroxene diopside next in abundance. Orthopyroxene is usually not found. Minor amounts of troilite and chromite make up the opaque phases. The meteorite has a total iron content of nearly 20 wt% but it is so highly oxidized that all of it is locked in the Fe-rich olivine (Fa_{33}). The O-isotopic composition of brachinites overlap the basaltic achondrites and angrites raising suspicions that the three may be related.

Fig. 8.24. Thin section of the brachinite Reid 013. All of the colorful interlocking grains seen here under crossed Polaroids is olivine. The dark blebs are olivine grains in optical extinction. The specimen is 15 mm in length. (Thin section provided by Marvin Killgore. Photo by O. Richard Norton and Tom Toffoli.)

Primitive achondrites

Three additional small achondrite groups are currently recognized and should be briefly mentioned to complete our inventory of the achondrites. They are called *primitive achondrites*, and as a group are distinguished from the differentiated achondrites by having been only partially melted and therefore not completely differentiated. They have achondritic textures but still retain something of their chondritic composition. Thus, these are chondritic meteorites that began their conversion to achondrites but the process was aborted before completion. They are therefore transitional forms between the two great classes of stony meteorites (chondrites and achondrites). The primitive achondrites include *acapulcoites*, *lodranites*, and *winonaites*.

Acapulcoites and lodranites[4]

Since these two primitive achondrites are closely related chemically and mineralogically, they are usually considered together. The type specimen for the acapulcoites fell near Acapulco, Mexico, in 1976. It was a confusing meteorite since it had chondritic composition but an achondritic texture. It was at first classified as an anomalous chondrite. A second meteorite that fell a few miles east of Lodran in Punjab, Pakistan, a century earlier (1868) turned out to be very similar in all characteristics to acapulcoites. These two anomalous stony meteorites remained unique until others of their kind began to be recovered in Antarctica. Today, about 20 are known, most from Antarctica with a few from the Sahara.

Both are heterogenous in texture, mineral composition, and oxygen isotopic composition, so much so that distinguishing among them is sometimes difficult. The acapulcoites are texturally finer grained than lodranites. This is intrepreted to mean that the lodranites are more recrystallized and therefore more highly metamorphosed. It is not uncommon, however, to see variable textures within any one meteorite. Both experienced metamorphic temperatures sufficiently high to melt Fe–Ni metal and FeS, but in some acapulcoites, relic chondrules have been found suggesting that acapulcoites had overall lower temperatures. The lodranites, however, experienced some silicate melting.

Both meteorites have chondritic compositions with olivine (Fa_{13}) and orthopyroxene (Fs_{16}) their main minerals in equal amounts. They are both magnesium-rich but their Fe/(Fe + Mg) is sufficiently high (oxidized iron) to place their mineralogy between the H and E chondrites. The high amount of Fe–Ni metal (~20 wt%) and its heterogeneous distribution in the meteorites convinced some early investigators that these were a kind of stony-iron meteorite but with chondritic composition, unique among this group. Oxygen isotopic compositions cluster between the terrestrial and carbonaceous chondrite fractionation line, showing their lack of homogeneity. Differentiation and igneous processes were apparently insufficient to homogenize them.

In summary, these meteorites are residues of varying degrees of partial melting, probably on the same parent body. They afford us a unique look into the differentiation process captured in progress, so to speak. The lodranites received more heating and melting and therefore probably originated at greater depth in the parent body. Some of the melted Fe–Ni as well as silicate may have made their way up into and through the acapulcoite layer enriching the layer with basalt-like magma and metal. A single collision event expelled the meteorites from different depths in the parent body, a conclusion reached by their very close cosmic-ray exposure age of about 6.5 Ma.[II]

Winonaites

This last group of primitive achondrites is of great interest because it is closely associated with the IAB/IIICD iron meteorites. For that reason winonaites will be briefly discussed in Chapter 9.

[II] The cosmic-ray exposure age or CRE is the time elapsed between the release into space of meter-sized or smaller bodies from the parent body and their eventual fall to Earth. During this time, meteorites are exposed to cosmic rays, producing primary and secondary isotopes from which their length of stay as meteorites in space can be determined. CREs are discussed in Chapter 10.

Martian meteorites – the SNC group

If I were compelled to name the single most important difference between the achondrites and chondrites I would say it in one word – diversity. Although the chondrites among themselves show differences that sufficiently justify placing them in clearly defined groups, still they are all chondrites with chondritic texture and mineralogy. Although we will discuss the ages of meteorites in Chapter 10, suffice it here to say that all the chondrites show the same ages within narrow limits – 4.5 Ga. The achondrites were either fully melted and differentiated into mineral zones within each parent body or they show partial melting in an aborted attempt to differentiate. This heating probably occurred within the first million years after their precursor chondritic parent bodies accreted. So the ages of chondrites and achondrites are quite similar. The achondrites formed by igneous activity on small parent bodies that produced distinctively different meteorites with different textures and mineralogies. We saw that some of these meteorites seem to have little in common with each other. The SNC group of achondrites differ so distinctly from the other achondrites that one is forced to concede that they must have originated on a very different parent body. Three achondritic meteorites originally made up this group. A fourth was later added. The acronym *SNC* (referred to as *Snick*) is the first letter of the three type specimens after which they are named: the "S" is for *shergottite*, named after a meteorite that fell in 1865 in Bihar, India, near the town of Shergotty; the "N" stands for *nakhlites* after the type specimen that fell in Nakhla, Egypt, in 1911; the "C"is for a single meteorite that fell in *Chassigny*, France, in 1915. These are called chassignites. Most of the SNC meteorites are cumulate igneous rocks that must have formed in a magma chamber. (Some of the shergottites are probably not cumulates but rather basalt flows.) In this respect, they are much like the HED group.

Shergottites

By far, the most common of the SNC group are the shergottites (Table 8.2). Until quite recently there was a known total of eight: Shergotty; Zagami (north-central Nigeria), five from the Libyan Sahara Desert (all paired though found as much as 30 km apart), and five from Antarctica.[III] Then, two meteorites found together somewhere in the Mohave Desert east of Los Angeles 20 years ago by Robert Verish were recognized as the fourteenth martian meteorite and the ninth shergottite. The meteorites appear very fresh, each with a dark glossy fusion crust typical of basaltic meteorites with high calcium content. One was aerodynamically oriented (Fig. 8.25a). The two meteorites were stored along with other terrestrial rocks until "rediscovered" in October, 1999. Suspecting them to be meteorites, Verish took the two specimens to UCLA where Alan Rubin and Paul Warren verified in December, 1999, that the two rocks were indeed meteorites. Within half an hour, the UCLA researchers concluded that they were paired martian meteorites, specifically, shergottites. Several indicators brought them to that conclusion. First, the meteorites matched the gas isotopic compositon unique to the martian atmosphere. Second, their deuterium (heavy hydrogen) to hydrogen ratio was high. This is because Mars, only 11% the mass of Earth, cannot gravitationally hold onto the lighter hydrogen which escapes into space leaving the heavier deuterium behind. This was all consistent with the measurements made by the Viking-Mars Landers in 1976. Third, they have a crystallization age of 1.3 billion years, which was far too young to be asteroidal; but their age matched several of the other known shergottites.[5] These are the second martian meteorites found in the continental United States. (The first was the Lafayette, Indiana, nakhlite. See below under "Nakhlites".)

Figure 8.25b shows the Zagami shergottite. Compared to the LA 001 and 002, Zagami has a fine-grained structure and demonstrates the structural variations seen among the known shergottites.

All shergottites have basaltic compositions with pigeonite, augite and maskelynite as major mineral constituents. The maskelynite phase comprises about 23 vol% and is the result of conversion of Na-rich plagioclase during an intense shock event. Minor minerals include titanium-rich magnetite, ilmenite, and small amounts of quartz, Fe-rich olivine and pyrrhotite. In thin section shergottites show a definite orientation with their grains arranged along a preferred plane, the result of accumulation in a magma chamber (Fig. 8.26). All

[III] It is risky to include numbers of known SNC specimens since the numbers usually become dated rather rapidly. The Sahara Deserts of Libya, Egypt and Oman are currently being searched by many groups, and finds of rare meteorites are becoming increasingly commonplace. The last Sahara Desert entry dates from a 1999 find. Most recently, an additional shergottite from Oman, Sayh al Uhaymir 094, was recovered which is probably paired with the other four known from the same region. Note also that two Mars meteorites from Northwest Africa were recovered in late 2000; one, Northwest Africa 817, is a rare nakhlite, only the fourth known. This brings the list of known martian meteorites to 18 (as of May, 2001).

Meteorite name	*Location found*	*Date found*	*Total mass (g)*	*Type*
Chassigny	Chassigny, France	Oct. 03, 1815	~4000	Chassignite
Shergotty	Shergotty, India	Aug. 25, 1865	~5000	Shergottite
Nakhla	Nakhla, Egypt	June 28, 1911	~10 000	Nakhlite
Lafayette	Lafayette, Indiana	1931	~800	Nakhlite
Governador Valadares	Gov. Val., Brazil	1958	158	Nakhlite
Zagami	Zagami, Nigeria	Oct. 03, 1962	~18 000	Shergottite
ALHA 77005	Allan Hills, Antarctica	Dec. 29, 1977	482	Shergottite
Yamato 793605	Yamato Mt., Antarctica	1979	16	Shergottite
EETA 79001	Elephant Moraine, Antarctica	Jan. 13, 1980	7900	Shergottite
ALH 84001	Allan Hills, Antarctica	Dec. 27, 1984	1940	Orthopyroxenite
LEW 88516	Lewis Cliff, Antarctica	Dec. 22, 1988	13.2	Shergottite
QUE 94201	Queen Ann Range, Antarctica	Dec. 16, 1994	12.0	Shergottite
Dar al Gani 476*	Sahara Desert, Libya	May 01, 1998	2015	Shergottite
Dar al Gani 489*	Sahara Desert, Libya	1997	2146	Shergottite
Dar al Gani 735*	Sahara Desert, Libya	1996–1997	588	Shergottite
Dar al Gani 670*	Sahara Desert, Libya	1998–1999	1619	Shergottite
Dar al Gani 876*	Sahara Desert, Libya	1998	6.2	Shergottite
Los Angeles 001**	Los Angeles Co., Ca.	1999	452.6	Shergottite
Los Angeles 002**	Los Angeles Co., Ca.	1999	245.4	Shergottite
Sayh al Uhaymir 005***	Oman	Nov. 26, 1999	1344	Shergottite
Sayh al Uhaymir 008***	Oman	Nov. 26, 1999	8579	Shergottite
Sayh al Uhaymir 051***	Oman	Aug. 01, 2000	436	Shergottite
Sayh al Uhaymir 094***	Oman	Feb. 08, 2001	233.3	Shergottite
Dhofar 019	Oman	Jan. 24, 2000	1056	Shergottite
Northwest Africa 480	Algeria?	Nov., 2000	28	Shergottite
Northwest Africa 817	Morocco	Dec., 2000	104	Nakhlite

Notes:

*All five of these shergottites from Dar al Gani are paired, i.e., all fragments from the same meteorite and therefore counted as a single meteorite.

**Both Los Angeles Co. shergottites are paired and counted as one meteorite.

***All Sayh al Uhaymir shergottites are probably paired.

Table 8.2. **Table of known SNC meteorites (compiled in May, 2001)**

This table of known SNC meteorites has been arranged in order of discovery. The first six discovered between 1815 and 1962 were not recognized as probable martian meteorites until meteorites with similar characteristics were discovered in Antarctica beginning in 1977 with ALHA 77005 at Allan Hills. The next five were all found in Antarctica. EETA 79001 was the first to yield direct evidence of martian origin in 1983. The last Antarctic martian meteorite was found in 1994. Discoveries of the remaining SNC meteorites shifted to the hot deserts of Libya and Oman with the single exception of the recently recognized shergottites from Los Angeles County. A total of 18 SNC meteorites are now known.

shergottites show signs of shock and shock metamorphism leading to glass formation. Their augite and feldspar grains exhibit undulatory to mosaic extinction (Fig. 8.27). As basalts, shergottites resemble Earth basalts in that they contain similar highly volatile elements not found in eucrites. Moreover, they are much more oxidized than eucrites apparently forming in a more highly oxidizing environment.

Nakhlites

Until recently, only three nakhlites were known, none of which were from Antarctica. Besides Nakhla, the type specimen from Alexandria, Egypt (Fig. 8.28), there is Lafayette (Tippecanoe County, Indiana; Fig. 8.29), and Governador Valadares (Minas Gerais, Brazil). Then in December 2000, a 104 g specimen was discovered in Morocco, labeled

Fig. 8.25 (a) Los Angeles 001 (left) and Los Angeles 002 (right), discovered in the Mohave desert are paired Mars meteorites, both shergottites. LA001 is a squarish mass weighing 452 g with a shiny black fusion crust and large augite and plagioclase grains easily visible in the exposed interior. LA002 is a beautifully oriented mass weighing 225 g. Cube scale is 1 cm^3 (Photo by Ron Baalke.) (b) Shergottites show a wide variation of textures. This photo of Zagami shows a shiny black fusion crust typical of all shergottites. The minerals are very small, best seen with a magnifier. The LA shergottites (a) have similar-appearing fusion crusts but their interiors are quite different texturally. Zagami is fine-grained, while the LA meteorites show a coarse-grained texture with their minerals easily seen with the unaided eye. Textural differences attest to different conditions of formation. The Zagami specimen measures 76 mm across. (Phote courtesy of Darryl Pitt.)

Northwest Africa 817, now bringing the total of these still very rare meteorites to four. Augite is the primary cumulate mineral in nakhlites amounting to about 80 wt% and giving the interior a greenish cast. In thin section the augite grains are long prisms and, like shergottites, often show simple twinning and preferred orientation. Fe-rich olivine (Fa_{65-68}) is a minor mineral at 5–10 wt%. Of special interest is the presence of *iddingsite*, often in veins running through the olivine. Iddingsite is an alteration product of iron-rich olivine occurring usually in the presence of water. It is seen in all three nakhlites. Between the augite and olivine grains is a mesostasis of tiny crystals of plagioclase, pigeonite, and orthoclase (K-feldspar) among other minor minerals. K-feldspar is extremely rare in meteorites. The large prisms of augite suggest a slow cooling rate but the mesostasis suggest rapid cooling. This is understandable if there were two stages of cooling. Slow cooling formed the large crystals of augite and olivine which settled out of the magma. The mesostasis is the glassy last-formed material surrounding the major grains before eruption and rapid cooling on the surface of the parent body. Unlike the shergottites and the chassignites, nakhlites show only minor signs of shock.

Chassignites

There is still only one chassignite known; the original that fell in Chassigny, France, in 1915 (Fig. 8.30). Chassignites are very similar to brachinites and at one time were included

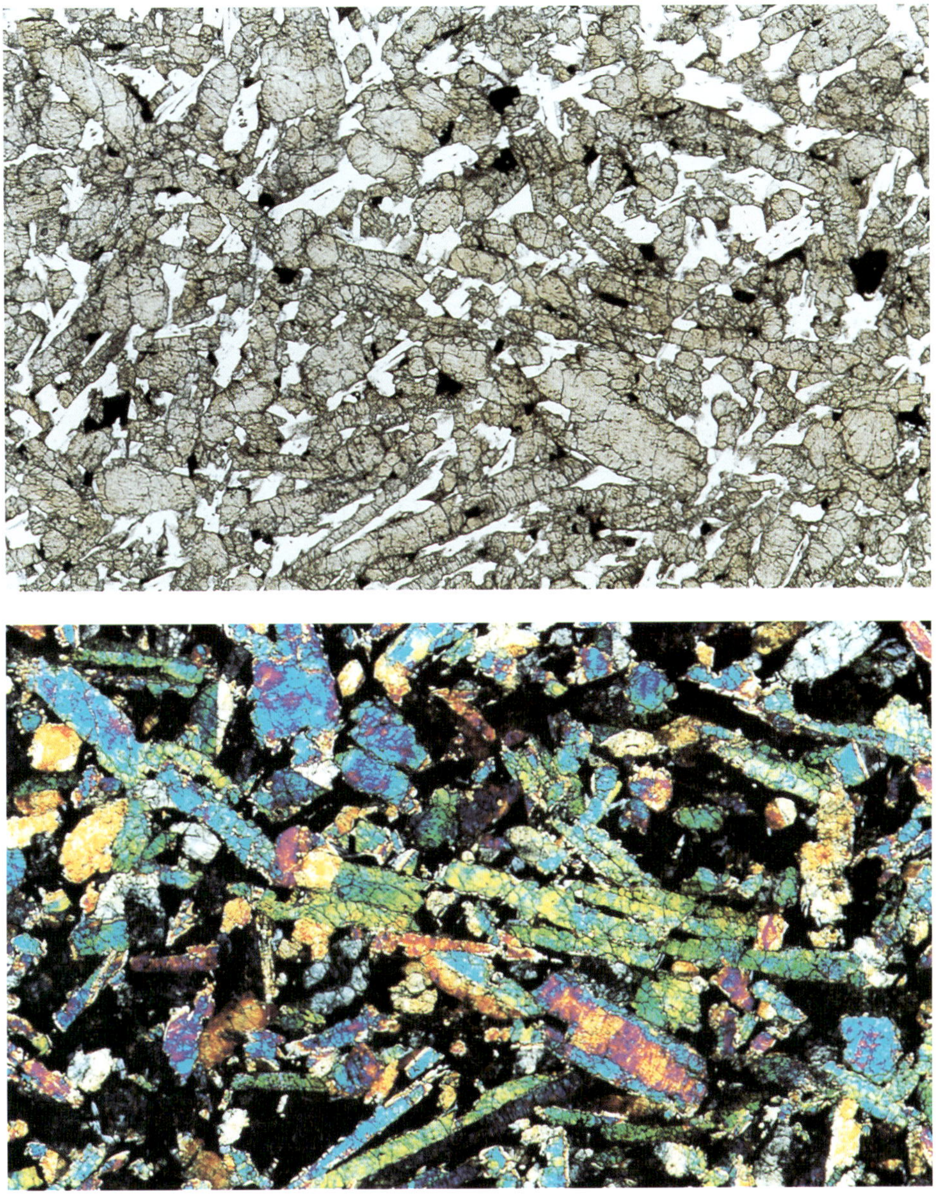

8.26 (Top) Thin section of the Mars shergottite, Zagami, from Katsina Province, Nigeria, viewed in transmitted light. Large augite prisms are oriented along a preferred plane suggesting a cumulate. White spaces between the grains is glassy maskelynite. The field is 9.15 mm across. (Bottom) The same specimen at the same scale viewed in crossed Polaroids. Note that the glass, being isotropic, has gone to extinction. (Photos by O. Richard Norton and Tom Toffoli.)

Fig. 8.27. A large grain of feldspar (gray mass in center) in the Zagami shergottite shows a mosaic extinction pattern under crossed Polaroids, a sure sign of intense shock. The horizontal field of view is 3.0 mm. (Photo by O. Richard Norton and Tom Toffoli.)

Fig. 8.28. This is a 4.02 g nakhlite sample from one of forty stones totaling 40 kg that fell in Abu Hommos, Alexandria, Egypt, in 1911. Called Nakhla, it is the prototype of the nakhlite martian meteorites. The specimen is 19 mm across. (Courtesy the Institute of Meteoritics, University of New Mexico. Photo by O. Richard Norton.)

among them. Chassignites are composed of about 90% Fe-rich olivine (Fa_{32}). In this respect they are very similar to terrestrial dunites. Dunites almost always contain a few percent chromite and likewise chassignites contain about 2 wt%. Small amounts (2%) of albite-rich plagioclase (An_{16-37}) and (5%) orthopyroxene (Fs_{12-28}) are found enclosed by olivine. Much of the feldspar occurs as diaplectic[IV] glass which indicates that chassignites are strongly shocked (Fig. 8.31). This, along with undulose extinction and planar fractures in olivine indicates an S5 shock stage in the Stöffler shock stage classification.[6] The nearly pure olivine cumulate meteorite must have formed in an ultrabasic magma chamber. Small amounts of the water-bearing amphibole mineral *kaersuitite*, a titanium-containing hornblende, have been found enclosed in olivine grains. This mineral is unique to chassignites. It bears the message that chassignites must have formed in a volatile-rich oxidizing environment and in the presence of water.

Origin of the SNC meteorites

In all, 18 SNC meteorites are known (see Table 8.2). They are all cumulates crystallized in ultramafic magma chambers. They resemble the mineralogy and chemistry of magmatic rocks formed on Earth in a volatile-rich and much more oxidizing environment than we expect to find on bodies the size of asteroids. This immediately suggested to researchers that SNC meteorites may have formed on a much larger body. There are only two larger bodies from which to choose: the Moon or Mars. An important clue that set the SNC meteorites apart from the other achondrites, and that was critical in establishing the origin of SNC meteorites, was their crystallization ages. Virtually all of the meteorites, both chondrites and achondrites with the exception of the SNC and lunar meteorites, have formation ages of around

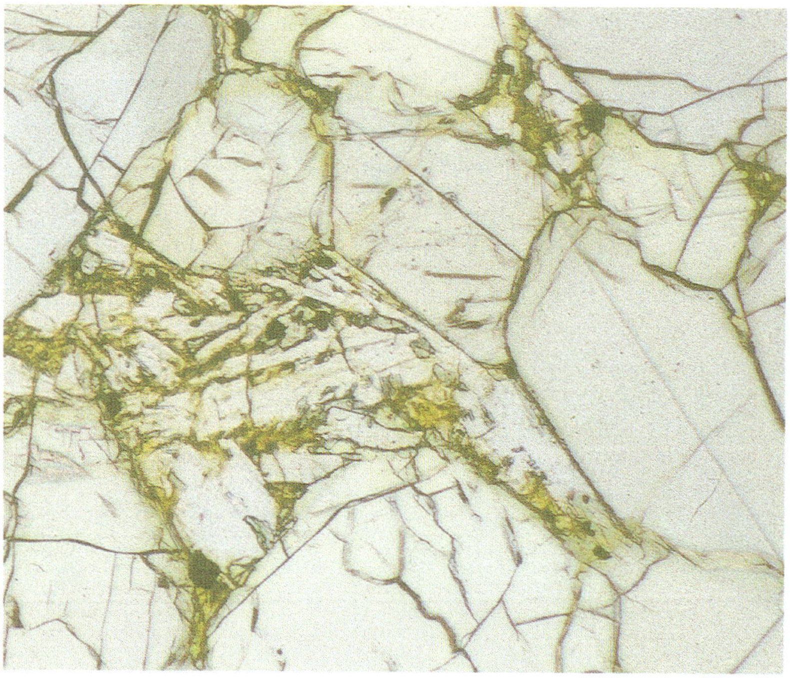

Fig. 8.29. This is a thin section viewed in plane-polarized light showing a true-color image of the Lafayette nakhlite. The large, slightly greenish-gray cracked material is the high calcium clinopyroxene, augite, showing simple twinning and a preferred orientation. The whiter small grains and laths (above center) are plagioclase. The tiny black grains in the plagioclase field are titanomagnetite and the iron sulfide, pyrrhotite. The orange material is iddingsite, an intergrowth of iron-rich clays and iron oxides (goethite and hematite). Iddingsite formed when olivine and glass (bluish grains among the feldspar laths) was altered by groundwater. Interestingly, the iddingsite is melted at the meteorite's fusion crust, showing that it must have formed before it reached Earth. This is strong evidence that water existed when the Lafayette meteorite was still emplaced on Mars. The field of view is 0.7 mm across. (Photo courtesy of Dr Allan H. Treiman, Lunar and Planetary Institute, Houston, Texas.)

[IV] The term diaplectic comes from the Greek, diaplesso, meaning to destroy by striking. Thus, a diaplectic glass is created with melting but without flowing, by shock waves that disorder and distort the original crystal structure of a silicate. The glass produced are pseudomorphs of the original crystals. Thus, plagioclase is converted to a diaplectic glass, in this case, maskelynite, through strong impact shock.

Fig. 8.30. This is a thin section view of the Chassigny meteorite taken under crossed polarizers. The colorful grains are all olivine, by far the most abundant mineral in Chassigny. The black grain on the right extending toward the center is a subhedral crystal of olivine near optical extinction. Surrounding this extinguished grain are grains of orthopyroxene with typical multiple twinning and low interference colors. The small black grains scattered in the olivine are chromite. The field is 2.8 mm across. This image was made from thin section No. 624-1 from the collection of the US National Museum. (Photo courtesy of Dr Allan H. Treiman, Lunar and Planetary Institute, Houston, Texas.)

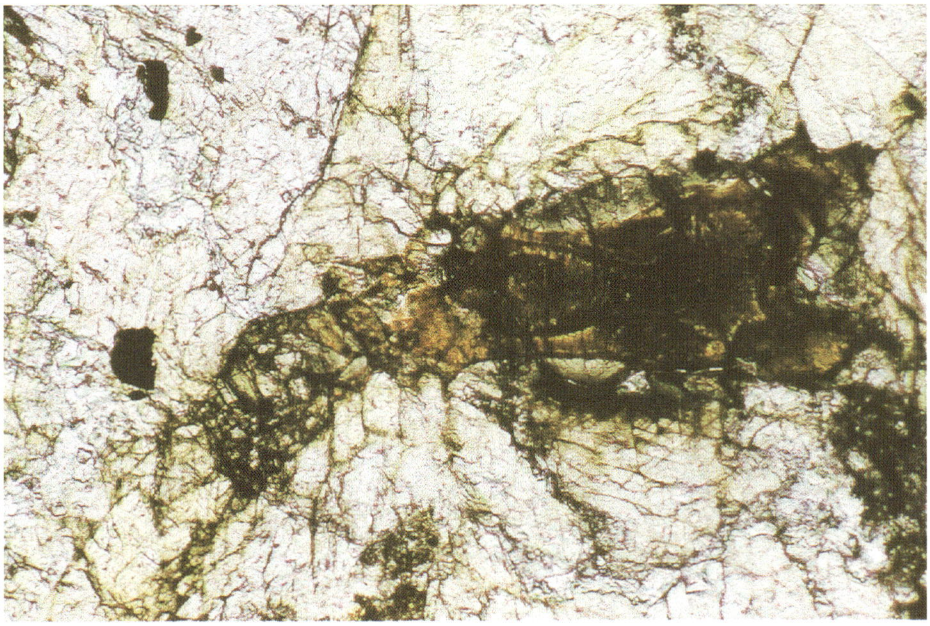

Fig. 8.31. Chassigny shows evidence of shock. This thin section viewed under plane-polarized light shows a large brown area of diaplectic glass of olivine composition, quenched from shock-melted olivine. Some fine, elongated crystals of olivine have grown from the shock-melt before it quenched to glass. The clear, white areas surrounding the glass is crystalline olivine; the black grains are chromite. The horizontal field of view is 0.7 mm across. (Courtesy Dr Allan H. Treiman, Lunar and Planetary Institute, Houston, Texas.)

4.5 Ga.[V] Even the eucrites that probably originated as basalt flows on 4 Vesta show ages of 4.4 Ga, only one hundred million years after differentiation and after which igneous activity probably ceased. Lacking the mass necessary to retain and sustain high temperatures required by igneous processes within achondritic parent bodies led to the conclusion that all formative activity requiring a heat source ceased in these asteroids shortly after their accretion.

When radioisotopic studies were made of the nakhlites, a red flag suddenly appeared in the otherwise monotonous list of meteorite ages that always fell between 4.4 and 4.6 Ga. (See Appendix B.) All three showed crystallization ages of between 1.22 and 1.34 Ga! It was immediately obvious that here were samples of igneous rocks that formed as a crystalline mass and erupted out onto the surface of a large body only a little over a billion years ago. Since nakhlites show only minor amounts of shock metamorphism, most researchers believe that these are true crystallization ages and not due to post crystallization shock metamorphism. This igneous activity had to have taken place within a large plan-

[V] Meteoriticists define several chronological ages for meteorites. The formation age can be defined as that age in which the asteroid/meteorite ceased crystallization, either as separate components such as the minerals in chondrules or as differentiated and fractionated asteroids. This is determined through studies using a number of different isotopic methods (Rb/Sr, Sm/Nd, U/Pb and others). The dating of meteorites is discussed in Chapter 10.

etary body that could maintain magmatic activity through radioactive heating long after the asteroids cooled. The oldest lunar rock samples collected during the Apollo program came from the lunar highlands where bedrock was found. These rocks are complex anorthositic breccias made up of clasts of both earlier and later formed rock. The oldest clast measured about 4.51 Ga, significantly close to the crystallization ages of the asteroid parent bodies. The youngest rock came from the basalt-filled impact basins. These basalts, 3.1 Ga, established the final phase of igneous activity on the Moon, ending nearly 2 billion years before the nakhlite rocks formed. This eliminates the Moon and points to the Earth-like planet, Mars.

The shergottites have a less conclusive story to tell, however. Remember that shergottites have been modified by major shock metamorphism. The maskelynite in the matrix was probably the result of impact shock and vitrification of plagioclase. Recall also that the augite grains show undulatory to mosaic extinction and planar fractures. Shock metamorphism resets radioisotopic clocks to the most current recrystallization date. Shergottites show a recrystallization date of only about 180 Ma, far short of the 1.3 Ga age of their SNC cousins. This must record the time when the shergottite block was blasted off Mars, shocking the rock in the process and erasing its earlier isotopic signature.

Although SNC meteorites were suspected of being rocks from Mars as early as the mid-1970s, 1983 marks the first direct evidence for Mars origin. Martian noble gases were found locked within glassy nodules in the matrix of a shergottite from Antarctica, EETA 79001.[7] These noble gases compared closely to the noble gases found in the martian atmosphere at ground level by the Viking 1 and 2 landers. Further work on the abundances and isotopic compositions of these gases as well as the larger nitrogen and CO_2 components, essentially matched the earlier Viking data (1976).

The saga of martian meteorite ALH 84001

It seems only fitting at this point to recount a story that has not yet ended; a story of the world's most famous rock. There is yet another SNC meteorite that is substantially different from the three discussed above. It is an Antarctic meteorite bearing the label ALH 84001, which is the specimen number (001) found in 1984 (84) at Allan Hills (ALH), Antarctica. Today this 1.94 kg meteorite is classified as an ultramafic intrusive igneous rock called an orthopyroxenite, but when it was first looked at, its large slightly olive-green crystals reminded researchers of the orthopyroxene hypersthene, so they casually classified it as just another diogenite. This was in 1984 not long after it was found. It remained in storage at the Johnson Space Center in Houston for nearly 10 years, mislabeled and nearly forgotten. Then, in 1993, David Mittlefehldt looked more closely and found trace minerals that do not normally belong in diogenites. Further analysis showed the meteorite to be related to the SNC group of

Fig. 8.32. Mars meteorite ALH 84001. A sliced section from the 1.94 kg total mass. The cube scale is 1 cm^3. (Courtesy the Johnson Space Center, Houston, and NASA.)

achondrites but with a sufficiently different composition to represent a new SNC type (Fig. 8.32).

The new Mars meteorite turned out to have an interesting history. Its coarse crystalline texture suggests that it crystallized from a melt within the initial martian crust 4.4 Ga ago. Some time between 3.8 and 4.0 Ga ago, the shock of nearby impacts produced *in situ* fractures throughout the rock. Then there was a period in which the rock was subjected to an aqueous environment, possibly submerged in water several times. The water would have been charged with CO_2, a major component of Mars' atmosphere and seeping along the fractures, depositing tiny beads of white carbonate minerals. An asteroid impact ejected ALH 84001 into space along a trajectory that eventually carried it to the inner Solar System and across Earth's orbit. For 16 million years it crossed paths with Earth and finally entered Earth's atmosphere, landing in Antarctica where it remained undisturbed and frozen in the ice for another 13 000 years. Roberta Score of the US Antarctic Meteorite Recovery team recovered it on December 27, 1984.[8]

Within a year of Mittlefehldt's recognition of the true nature of ALH 84001, a team of nine scientists, led by Johnson Space Center's David S. McKay and Everett K. Gibson, Jr., quietly studied strange structures hiding within its green crystals. They found carbonate deposits along the fractures in unlikely association with magnetite and iron sulfide, organic compounds typical of decaying organic matter on Earth, and strange "fossil-like" structures that morphologically closely resembled fossil bacteria found in 3 billion year old terrestrial rock. After an intense two year study, the team announced to the world on August 7, 1996, that they had found evidence for microbial life in ALH 84001. A week later their research was published[9] in the American journal *Science*.

Fig. 8.33. Compounds rich in magnesium and iron (black) encircle carbonate globules (yellow) of diameter 50 μm in ALH 84001. These mineral assemblages strongly resemble those produced by primitive Earth bacteria. The carbonates may have harbored early martian life. (Courtesy the Johnson Space Center, Houston, and NASA.)

The carbonate deposits (Fig. 8.33) became the source of great interest and later, vigorous debate, among the researchers. They are yellow-orange globules averaging about 50 μm across and rimmed with alternating black and white magnesium and iron-rich layers. Within the layers were found tiny magnetite and iron sulfide grains in close proximity. Normally, oxidized iron (magnetite) and reduced iron (FeS) should not chemically coexist, especially in the presence of carbonates. The tiny magnetite crystals only 0.05 μm long appeared as nearly perfect hexagonal forms strikingly similar to magnetite crystals manufactured by some Earth bacteria. Their similarity and the fact that they are incompatible with the FeS strongly suggested they were deposited in the carbonate by martian bacteria. It was pointed out by opposing scientists that close proximity of iron oxide and sulfide does in fact exist on Earth in places where temperatures of formation reached several hundred degrees Celsius, much too hot for life forms to exist. But interestingly that's not the end of it. Measurements of the oxygen-isotope ratios within the carbonates can be interpreted *either* as resulting from fluid temperatures hotter than 300 °C *or* ground water temperatures as low as 25 °C. McKay and his coworkers continue to believe that these magnetite forms cannot be produced by inorganic processes. Most researchers, however, disagree. John P. Bradley of McCrone Associates Inc., point out that screw dislocations[VI] found in the magnetite negated an organic origin. This impasse will probably not be resolved until we can know something of the context in which these minerals formed. As meteoriticist and Mars geologist Allan Treiman so casually puts it, "If the rock were from Earth we would go back and poke around where it was found for more clues." This is not likely to happen for another decade or so.

The carbonate deposits also contain organic compounds collectively called polycyclic aromatic hydrocarbons, or PAHs. PAHs have been the most controversial of all. Dead organisms can be converted to PAHs by gentle heating. The scientists are fond of pointing out that there are plenty of PAHs on steak seared on a barbeque. The NASA group interpreted the meteorite's PAHs as originating from gently cooked martian organisms. Contamination was suspect and much of the research concentrated on this hypothesis. McKay showed that the PAHs increased from the surface inward, which is just opposite of what one would suppose if the meteorite was contaminated while on and in the Antarctic ice. There was one researcher who showed that PAHs in the Antarctic ice was similar to that found in the meteorite and could therefore find their way into the meteorite during brief thawing periods. Simon Clemett of McKay's group, using very sensitive analytical instruments, could find no PAHs in the ice. Even more interesting, Clemett found that no other Antarctic meteorite contained the amount of PAHs found in ALH 84001. PAHs are commonly found in carbonaceous chondrites. It is therefore possible that carbonaceous chondrites could have entered Mars's atmosphere and contaminated the rocks with a residue of PAHs. Definitely, ALH 84001 has been contaminated with terrestrial organisms. It has been shown to contain organisms whose amino acid complement is identical to terrestrial life, but the PAHs origins remain unanswered.

An even more provocative statement in McKay's paper was the announcement that bacteria-like structures had been found in the carbonate deposits. These appeared identical to many terrestrial bacilli, only they were a tenth the size of the smallest bacteria known on Earth (Fig. 8.34). Many researchers pointed out that these "fossil bacteria" are too small to contain the large and complex organic components necessary to propagate life as we know it. The NASA group finally

[VI] A *screw axis* is a combination of a rotation axis with a translation parallel to the axis. A *screw dislocation* is the axis around which a crystal plane seems to spiral as it grows. The growth of the magnetite crystals is along this axis and the pathway along which the crystals grow is helical.

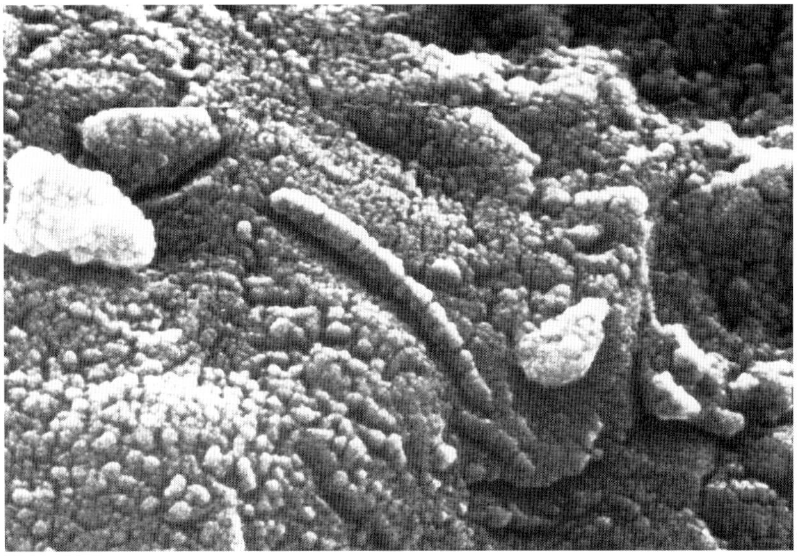

8.34. High-resolution SEM image within a carbonate globule in ALH 84001 reveals a tiny bacteria-like structure, 100 nm in length, that may be the remnants of early martian bacteria. These structures are equivalent in size to the smallest bacteria found in 3.5 billion-year-old rocks on Earth. (Courtesy the Johnson Space Center, Houston, and NASA.)

conceded that this was probably the case but pointed out that they may be fragments of originally larger bacteria.

The war of PAHs, fossil bacteria and perfect magnetite crystals goes on. At stake here is a challenge no one recognized earlier. Do we really know enough about terrestrial life to recognize alien life forms when we see them or their remains? Could we distinguish martian microbes from Earth microbes? Daunting questions, indeed. If the saga of ALH 84001 is any indicator, then perhaps we need to reevaluate our understanding of primitive life and the traces it leaves behind. A striking example of the need for this reevaluation appeared in a paper in early 1999 in the journal *Geology* by principal researcher, Kathie Thomas-Keprta.[10] Bacteria were collected 0.5 km beneath the surface of the Columbia River Basalts in Washington State, considered a possible terrestrial analog for a martian subsurface ecosystem. These live bacteria were grown in chambers matching the original natural growth conditions. They readily grew, reproduced and subsequently died and fossilized within eight weeks. These bacteria and their parts became fossilized by mineral deposits that completely encircled them, creating hollow mineral structures matching closely in size and shape similar structures in ALH 84001. About 30% of the bacteria had appendages that also mineralized, forming structures too small to be bacteria but mimicked the small structures found in ALH 84001. Thomas-Keprta and her coworkers concluded that their experiments clearly showed that structures similar in size and shape to those in the martian meteorite do form on Earth and do exist in the terrestrial fossil record. Thus, the structures in ALH 84001 may be simply the fossil remnants of very small cells or structures from larger cells, a supposition McKay had offered during debates several years earlier.

Following closely on the heals of the Thomas-Keprta paper, Lars Borg, a geochemist at the Institute of Meteoritics, University of New Mexico, announced in a paper in *Science*[11] (September, 1999) that he had established a reliable age for the carbonates deposited after the martian meteorite formed. Recall that the rock formed about 4.4 Ga ago. The carbonates with their cache of bacteria were deposited 500 million years later at a time when Mars' atmosphere was much denser and the conditions were much milder. At that time Mars was relatively warm, perhaps similar to Earth conditions today, and water flowed on the surface. This would place the first life forms on Mars at about 3.9 billion years ago, at the same time life was just getting a foothold on another planet of the Solar System – Earth. This simultaneity is probably no accident. If bacteria formed on Mars during this early period it would have been after this time that Mars lost most of its atmosphere and water to space or to the frozen ground while Earth, a more massive planet, retained its evolving atmospheric gases and continued its relatively mild condition conducive to the first life forms.

Finally, even if these structures were fossil bacteria, to prove that they came from Mars remains a formidable task probably not solvable until a search can be made by robots or humans. It is not likely that martian microbes will be alive in any samples returned to Earth from Mars. Nor do we expect to fine live bacteria hitching a ride on a Mars meteorite destined for Earth. ALH 84001 has been a wonderful exercise for researchers much like the Allende and Murchison meteorites were for geochemists just months before lunar samples were returned. When uncontaminated Mars rocks are returned to Earth in the next half dozen years, we can only hope that these exercises will have paid off.

Lunar meteorites – lunaites

Less than a half million kilometers away lies a source of extraterrestrial material, the Moon. If sufficiently energetic impacts of asteroid fragments occur on a low but more or less steady rate on the Moon, then we would expect fragments dislodged from the Moon to reach Earth as lunar meteorites on a low but regular basis, much as pieces of asteroids have continued to pelt Earth since it formed 4.6 billion years ago. Today, we look upon the lunar craters as striking evidence of a violent impact history. The sizes of the craters alone tell of enormous energies released by these impacts a half billion years after the Moon formed and consolidated. The remnants of these craters represent the terminal cratering event of the most violent period in the Moon's early history, ending about 2.7 billion years ago.[12] Today they make up the lunar *highlands* or *terrae*, a vast field of overlapping craters occupying the southern part of the Earth-facing side and nearly all of the Moon's back side (Fig. 8.35). After this initial period of 0.8 billion years, cratering events gradually subsided but with occasional periods of resurgence. Impacting did not cease altogether. Some time toward the end of this period, the Moon suffered huge impacts that created large circular basins on the Earth-facing side. These basins, mostly filled with basalt, all but erased the earlier cratering record in these areas but hold a record of impacts after basalt inundated them. Studies of the cratering record within these younger basalt-filled areas, the Moon's *lowlands* or *maria*, attest to a low but more or less continuous impact history to the present.

What does it take to dislodge a piece of the Moon and send it into space on an escape trajectory to Earth? Typically, meteoroid initial velocities before entering Earth's atmosphere are between 10 and 15 km/s. Impact velocities on the Moon for average meteoroids of, say, 1 to 10 tons would be higher than on Earth due to the Moon's lack of a shielding atmosphere. Initial cosmic velocities and masses would be retained to impact. These impact velocities are much higher than the 2.38 km/s needed to escape the Moon. Although most of the kinetic energy would be expended in vaporizing the impactor, some of the energy would be utilized in accelerating fragments into a symmetrical ejecta blanket surrounding the newly formed crater. The newest large craters appear quite prominent due to their white ejecta blankets and ray systems that extend for hundreds of kilometers across the Moon's face. If a large impactor happened to strike the Moon at a low angle to the surface, the resulting glancing impact could send some of the ejecta into space with the requisite escape velocity. Some of the rock might be captured by Earth immediately after escaping the Moon's gravitational field, but most of the rock would establish elliptical orbits

8.35. This full Moon image shows the two terrains from which all of the Moon rocks and lunaites came. The white areas are the lunar highlands; the dark circular areas, the lunar basins or maria. The highlands are older cratered terrain that probably existed over the entire lunar surface before the large circular basins formed. The basins are the largest impact features on the Moon. When they formed they obliterated the earlier cratered terrain, leaving only remnants of the older terrain. After they formed, sometime between 3.9 and 2.7 billion years ago, fractures resulting from the impacts acted as conduits that carried magma to the surface filling them with dark basalt. The side facing away from Earth lacks the maria and is almost entirely covered with highland terrain. The large C-shaped structure near the top is the Imbrium Basin where Apollo 15 astronauts collected Moon rocks. The Imbrium Basin opens into an equally large basalt-filled terrain, Oceanus Procellarum, where Apollo 12 astronauts landed and collected specimens.

Fig. 8.36. This is the Moon at exactly first quarter. Along the terminator line where the Sun is just rising above the horizon, long shadows are cast by the crater walls of the highland terrain bringing them out in high relief. The highland terrain is represented in the Moon rocks and lunaites by anorthositic regolith breccia. It is here where impacts over millions of years have fragmented the surface mixing it with both highland and basin terrain fragments forming a regolith several meters thick. The maria are much younger basalt-filled terrain from which four samples of lunar basalts have been found on Earth, three in Antarctica and one in Northwest Africa.

around the Sun with aphelia very near Earth's orbit which they would pass across about once a year. It may take millions of years intersecting Earth's orbit before Earth was favorably situated to receive them. Cosmic-ray exposure ages, the length of time in space after leaving the Moon (see Chapter 10), are as long as 9 Ma for some lunar meteorites found in Antarctica. Judging from the paucity of lunar meteorites, most may remain space-bound for many more millions of years.

History's lesson

What seems so obvious to us today was not so obvious just a few decades ago. For years, astronomers battled among themselves over competing hypotheses of lunar crater origins. Through a telescope, over 30 000 craters are visible on the Earth-facing side (Fig. 8.36), ranging from less than a kilometer wide to huge circular basins hundreds of kilometers across. These are landforms that seemed to be lacking on Earth. The only structures that geologists were aware of with circular form similar to these craters were volcanoes.[VII] During the eighteenth and nineteenth centuries, and even into the early twentieth century, the word "crater" was synonymous with volcanoes.

The hypothesis that the lunar craters were volcanic extends back more than three centuries to the time of Robert Hooke (1665). Toward the end of the eighteenth century, a rash of lunar observations were made and published by the most prominent astronomers of the day. William Herschel and Neville Maskelyne in England, Johann Schröter and Johann Bode in Germany, and Joseph de Lalande and Jean-Dominique Cassini in France all reported observing volcanic activity on the Moon. In all of these cases, their "volcanic" bias tainted their visual observations. In his book, *Treatise on Astronomy* (1851), Sir John Herschel continued to propagate this fallacious belief by casting attention onto the central mountain peaks prominent in the larger lunar craters referring to them as possessing , "the true volcanic character, as it may be seen in the crater of Vesuvius." With this impressive list of supporters backing the volcanic hypothesis, there was little disagreement among the scientists. Craters on the Moon were volcanic.[VIII]

This conclusion complemented nicely an eighteenth-century hypothesis on the origin of meteorites: meteorites come from lunar volcanic eruptions. On numerous eruptive occasions scientists had observed Vesuvius throwing rocks many kilometers. By analogy, the much larger lunar "volcanic" craters should be able to throw the stones clear of the

[VII] Terrestrial meteorite craters were unknown in the eighteenth and nineteenth centuries. The pioneering work of Eugene Shoemaker in the mid-twentieth century (1957) finally established the Barringer crater as the first proven terrestrial meteorite crater. Following in his footsteps, geologists had accounted for over 150 terrestrial impact craters by the close of the twentieth century (see Appendix G.)

[VIII] We know today that the central mountains found in the larger craters are rebound features produced by momentary compression of the elastic crust during impact, and a rebounding of the crust immediately thereafter. The crater floor moves inward and upward at the same time causing a rebound peak to form in the center and the surrounding rim to slump and fracture into terraces.

lunar gravitational field. The mineralogy of meteorites seemed to support this notion. Compared to terrestrial rocks, meteorites contained elemental iron showing they had to come from an airless or at least an oxygen-free environment. Moreover, there was a sameness among the meteorites in texture and chemistry that suggested they came from one specific location in the Solar System. The density of the stony meteorites was very close to the density of terrestrial volcanic rocks which also matched the calculated density of the Moon. All of these observations supported a lunar origin.

But there was a serious problem. It seemed reasonable that violent Vesuvian-type eruptions from the much larger lunar craters could hurtle rocks to the lunar escape velocity, but meteorites were observed to enter Earth's atmosphere many times faster. How could they be accelerated to such high speeds from the Moon? Indeed, these velocities resembled orbital velocities of planetary bodies. This observation alone was sufficiently serious to eventually cast doubts on the volcanic origin of meteorites; and by the mid-nineteenth century, lunar volcanoes had been all but rejected as a source.

Baldwin's Moon

The volcanic crater hypothesis, however, stubbornly persisted well into the twentieth century. In the second quarter of the century, a lone voice insisted that the Moon's craters were formed by impact. Ralph Baldwin, a young astronomer fresh out of a doctorate program in astrophysics from the University of Michigan (1937), had a part-time job lecturing at the Adler Planetarium in Chicago. Lining the walls outside the planetarium were large photographs of the Moon which he found irresistible every time he walked by them. Baldwin was especially intrigued by line-like markings cutting across the surface that, when extended along great circles, crossed into Mare Imbrium, a large C-shaped feature in the northern hemisphere of the Moon's visible face (see Fig. 8.35). This, he concluded, had to be ejected material from Imbrium and this ejecta could only have been the result of a massive explosion. Baldwin was the first to suggest that the large circular areas on the Moon were huge impact basins made by impacts of large asteroids. These were the Moon's largest impact craters. Moreover, he further reasoned, by analogy, that since Earth's largest volcanoes were only a few kilometers across, the vast majority of craters on the Moon, being much larger, were not volcanic constructs.[13] Though Baldwin found few supporters of his impact hypothesis, he continued to gather volumes of data which only strengthened his convictions. At the time, shortly after the Second World War, he was one of the few scientists studying the Moon. His ideas were finally published in his book, *The Face of the Moon* (1949).[14] This book was a wake up call, a key element that finally provided the stimulus for astronomers to turn their telescopes once again to the Moon, an object of study that, with few exceptions, had lain dormant for a century. Twenty years later, humans would travel to the Moon and return with the first Moon rocks. In those twenty years, there was a complete reversal of attitudes and beliefs. The volcanic hypothesis finally passed into history and impact cratering became the front runner. But Baldwin's impact hypothesis now extended far beyond the Moon. Space probes visiting all but one planet (Pluto) showed that impact cratering as a natural mechanism sculpturing planets and/or their moons was prevalent throughout the Solar System. Virtually all of the solid bodies of the Solar System showed impact features.

Once impact cratering was understood as an important geological mechanism modifying planetary surfaces, it seemed a reasonable hypotheses that at least a few of the rocks dislodged from the Moon by impact could escape the Moon's gravity and through circuitous routes eventually make it to Earth. But how does one recognize a lunar meteorite on Earth? Not easily. It is no surprise that the discovery of lunar meteorites or *lunaites* on Earth had to await the cache of Moon rocks to be brought back from the Moon. This was a time for geochemists and meteoriticists to educate themselves to the new rocks from a nearby world.

A history of discovery

Japanese scientists from the National Institute of Polar Research found the first lunar meteorites on November 20, 1979, on the ice fields near the Yamato Mountains in Antarctica (Yamato 791197). A second lunaite was found the same year also near the Yamato Mountains (Yamato 793169); and one additional specimen was recovered in 1980 (Yamato 793274). But none of these was recognized as lunar meteorites until after 1982 when an American team found the first *recognized* lunar meteorite near Allan Hills in Victoria Land on January 18, 1982. Designated ALHA 81005, this was the fifth meteorite found in the Allan Hills area during the 1981–2 collecting season (Fig. 8.37).

Over the next eight years, four additional lunaites were recovered in Antarctica. Then, in 1990, an American collector, Robert A. Haag, joined the exclusive lunar meteorite finder's club. He had received a cache of meteorites collected by an Australian meteorite hunter searching for Millbillillie eucrites near Wiluna, Western Australia. Unknown to the collector and Haag, there was a very different meteorite among the Millbillillies. Haag soon recognized the meteorite masquerading as a eucrite. It appeared very similar to photographs he had seen of ALHA 81005, both texturally and mineralogically. In the months to follow, the specimen was officially recognized by William Boynton at the University of Arizona's Lunar and Planetary Laboratory in Tucson as a lunar meteorite. The first non-Antarctic lunaite, called Calcalong Creek, had now been found (Fig. 8.38). Calcalong Creek was also the first lunar meteorite to fall into private hands. That distinction was to change soon.

The last lunaite to be found in Antarctica was in 1998. A year earlier a French dealer in meteorites, a member of the

Fig. 8.37. This is ALHA 81005, the first lunar meteorite to be discovered and recognized on Earth. It was found by an American field party on January 18, 1982, near Allan Hills, Victoria Land, Antarctica. It is a 31.5 g anorthositic regolith breccia from the lunar highlands. The fusion crust is still in tact but has been altered to a light brown by terrestrial chemical weathering. The large white clasts are anorthositic rock fragments composed of the calcium-rich plagioclase, anorthite embedded in a dark matrix. The matrix is composed of an assortment of broken and partly melted mineral grains, some dark clasts of basaltic composition and partially melted and recrystallized clasts of basalt and anorthositic rock. Glassy aggregates are also very common. The specimen is 38 mm high. The cube is 1 cm on each side. (Courtesy NASA.)

Fig. 8.38. This is the first lunar meteorite found outside Antarctica, the 19 g Calcalong Creek anorthositic regolith breccia discovered near Wiluna, in Western Australia. The similarities in texture between ALHA 81005 and Calcalong Creek are striking. White anorthositic fragments are embedded in dark fine-grained matrix. Black shock veins run through the specimen like all the regolith breccias from the lunar highlands. The specimen is 34 mm wide. (Robert A. Haag collection. Photo by O. Richard Norton.)

Labennes family, had located the first find from the Libyan Sahara Desert, a 513 g specimen (Dar al Gani 262) similar to Calcalong Creek and several of the Antarctic specimens. From that point on the remaining lunaite finds were from the hot deserts of Libya, Oman and Morocco. Much of this material has been divided into gram-sized chips and sold to collectors, museums, and research facilities worldwide. Table 8.3 summarizes the lunar meteorite finds.

Anorthositic regolith breccias

Like Earth and Mars, the Moon is a differentiated planetary body. Thus, we would expect to find rocks from the Moon that demonstrated igneous crustal activity. We saw earlier that the Moon possesses basically two major landforms: the lunar highlands and the lowlands or basins. The lunar rocks collected by astronauts during the Apollo missions sampled both of these landforms. This provided a reference collections that proved invaluable in recognizing the lunaites as specimens from the Moon. The lunar meteorites are also samples of these two landforms. From Table 8.3 we can see that most of the lunaites are anorthositic regolith breccias. Many of the Moon rocks collected by the Apollo astronauts were also anorthositic regolith breccias nearly identical to the lunaites. *Anorthosites* are a group of igneous rocks. On Earth, they are nearly monomineralic plutonic igneous rocks composed almost entirely of plagioclase feldspar, usually labradorite or bytownite. For that reason, anorthositic lunaites are often referred to as feldspathic lunaites. On the Moon, the lunar highlands rocks are regolith breccias rich in the calcium plagioclase mineral, anorthite. As much as 75–80% of the rocks are composed of this mineral, giving the rock its anorthositic (or sometimes, feldspathic) classification. Figure 8.39 shows a slab of Dar al Gani 400, a typical lunar anorthositic regolith breccia and the largest lunaite found to date. The anorthositic clasts are easily recognized as white, fragmented clasts embedded in a darker fine-grained matrix. The matrix is composed of minor amounts of ilmenite ($FeTiO_3$), pyroxene, and olivine mixed together with partially recrystallized clasts of basalt composition

Meteorite name	Location found	Date found	Mass (g)	Lunar rock type
Yamato 791197	Antarctica	1979	52.4	Anorthositic regolith breccia
Yamato 793169	Antarctica	1979	6.1	Mare basalt
Yamato 793274	Antarctica	1980	8.7	Anorthositic mare regolith breccia
ALHA 81005	Antarctica	1982	31.4	Anorthositic regolith breccia
Yamato 82192/82193/86032*	Antarctica	1982/86	37/27/648	Anorthositic fragmental/regolith breccia
EET 87521/96008*	Antarctica	1987/96	31/53	Mare polymict breccia
Asuka 881757	Antarctica	1988	442	Mare basalt
MAC 88104/88105*	Antarctica	1989	61/663	Anorthositic regolith breccia
Calcalong Creek	Australia	1990	19	Anorthositic/mare regolith breccia
QUE 93069/94269*	Antarctica	1993/94	21.4/3.1	Anorthositic regolith breccia
QUE94281	Antarctica	1994	23	Anorthositic/regolith breccia
Dar al Gani 262	Libya	1997	513	Anorthositic regolith breccia
Dar al Gani 400	Libya	1998	1425	Anorthositic impact-melt breccia
Yamato 981031	Antarctica	1998	186	Anorthositic/mare (?) regolith breccia
NW Africa 032/479*	Morocco	1999/2001	300/156	Mare basalt
Dhofar 025	Oman	2000	751	Anorthositic regolith breccia
Dhofar 026	Oman	2000	148	Anorthositic impact-melt breccia
NW Africa 482	Algeria(?)	2000	1015	Anorthositic impact-melt breccia
NW Africa 773	West Sahara	2000	633	Olivine norite with regolith breccia
Dhofar 287	Oman	2001	154	Mare basalt breccia
Dhofar 081/280*	Oman	1999/2001	174/251	Anorthositic fragmental breccia

Note: *These are paired and represent fragments of a single meteorite.

Table 8.3. **Table of known lunar meteorites (compiled May, 2001)**

This table of known lunar meteorites is arranged in order of discovery. The first discovered, though not recognized as a lunaite, was Yamato 791197, on November 20, 1979. The first meteorite recognized as a lunaite was Allan Hills 81005, found on January 18, 1982. The first discoveries were all made in Antarctica with the single exception of a 19 g specimen found among a box of Millbillille eucrite specimens picked up by a meteorite hunter searching near Calcalong Creek, Australia, in 1990. The Calcalong Creek meteorite was recognized as a lunaite by collector Robert A. Haag of Tucson, Arizona, and became the world's first privately owned lunaite. Beginning in 1977 lunaites began to be recovered by private collectors from the hot deserts of Libya, Oman, and Morocco. As of May, 2001, privately owned lunaites by weight outnumber the total weight of all Antarctica finds by a factor of 2.4 to 1 or 5539 g vs 2294 g respectively. (Table based upon a compilation by Randy L. Korotev, Department of Earth and Planetary Sciences, Washington University.)

(large black clasts) and agglutinates (mineral grains bonded by impact-melted glass), typically found in lunar regolith breccias. Both ALHA 81005 and Calcalong Creek show this texture. It comes as no surprise considering their origin that all of the anorthosites from the highlands show signs of intense shock and brecciation.

Mare basalts

The huge circular impact basins on the Moon were filled with runny flood basalts beginning about 3.9 billion years ago and ending as late as 1.3 billion years ago, probably as a result of impact fracturing extending to the mantle providing conduits to the surface. Mare basalts are by far the rarer of the two lunar meteorite types. This is to be expected since the maria cover only about 17% of the lunar surface. Four lunar basalt meteorites are known; three from Antarctica and the fourth and newest, the non-Antarctic lunar meteorite, Northwest Africa 032 (NWA 032) found in October, 1999, near the Algerian border in Morocco. It is classified as an unbrecciated olivine–pyroxene basalt as it contains up to 12 vol% olivine and 5 vol% pyroxene (primarily pigeonite and augite). Figure 8.40 shows a small slab cut from the original mass. The largest and most conspicuous clasts up to 0.2 mm across are yellow-white olivine standing out against the dark matrix in exposed interior fragments. The dark ground mass shows elongated feathery-textured bundles of plagioclase (An_{86}) crystals (lighter purple lamellae) radiating from nucleation sites and elongated pyroxene laths similar to but darker than the plagioclase (Fig. 8.41). Accessory

Fig. 8.39. This slab is a sample of the non-Antarctic lunaite Dar al Gani 400 found in the Libyan Sahara desert in 1998. It is the largest yet discovered with a total weight of 1425 g. Like the other highland meteorites, this is an anorthositic regolith breccia with light anorthositic clasts in a comminuted dark matrix. Black clasts of basaltic composition are scattered in the matrix along with dark glass and veins filled with terrestrial carbonates. The specimen measures 64 mm across. (Photo by Walt Radomsky, specimen supplied by R. A. Langheinrich, Meteorites, NY.)

minerals include ilmenite, troilite and trace amounts of elemental iron. The meteorite is heavily shocked with whole-rock melt veins crossing the entire meteorite. These veins are filled with silica glass and appear black and irregular cutting across the matrix. White fractures have been permeated with terrestrial carbonates. In thin section the olivine shows undulatory and mosaic extinction and some of the plagioclase has been converted to diaplectic glass (maskelynite). This lunar basalt is similar to those collected during the Apollo 15 mission to Hadley Rille in Mare Imbrium, but NWA 032 contains a higher concentration of olivine phenocrysts. Lunar basalts collected from Oceanus Procellarum during the Apollo 12 mission have a higher whole-rock MgO than NWA 032. These differences suggests that NWA 032 is a sample from another unknown area of the Moon.[15]

The chemistry of lunaites distinguishes them from other achondrites such as the asteroidal HED suite and the martian SNC suite, and terrestrial basalts. Ratios of iron to manganese in both olivine and pyroxene are diagnostic for lunar rocks and matches Apollo moon rocks from both the highlands and the basins (Mare Imbrium and Oceanus Procellarum). Equally distinctive is the whole-rock oxygen isotopic ratios which places all lunaites directly on the terrestrial fractionation line.

Fig. 8.40. This lunar meteorite specimen was cut from the 300 g Northwest Africa 032 mare basalt found October, 1999, near the Algerian border in Morocco. It is classified as an olivine–pyroxene basalt after the two most prominent mineral components. The large yellow-white fragments are fractured olivine comprising about 12% by volume. The matrix consists of small phenocrysts of pyroxene and chromite (black) among laths of feldspar. Shock veins permeate the specimen and contain shock induced glass. Some veins contain terrestrially deposited carbonates (upper right). A curving shock vein with black glass is prominent on the right side. Note the large phenocryst of olivine near the intersection of the horizontal and vertical scale bars that appears bisected and dislocated. A close-up photo (Fig. 8.41) shows this phenocryst in greater details. The field of view is about 37 mm. (Photo by Walt Radomsky.)

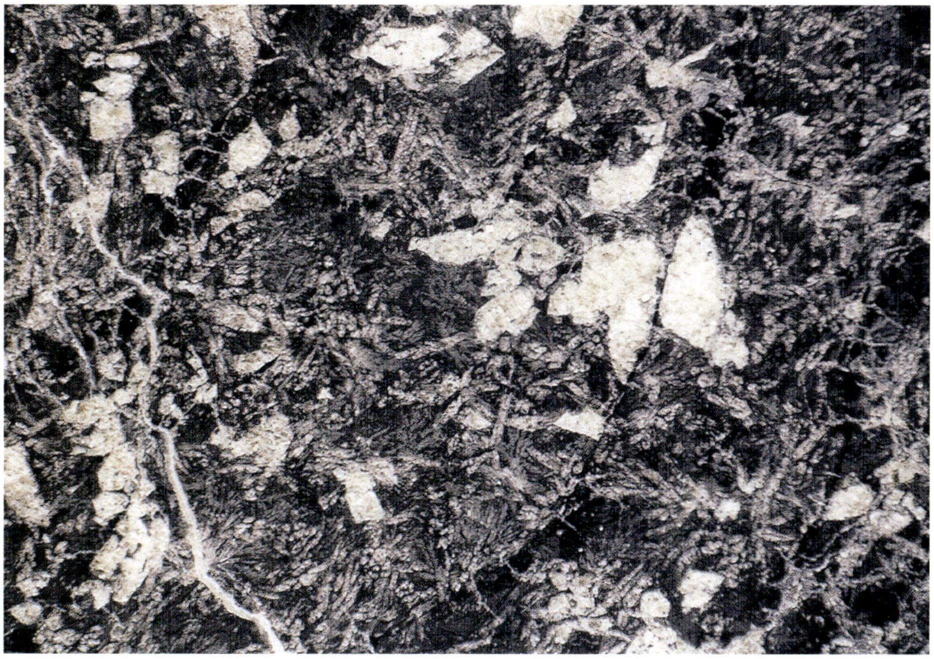

Fig. 8.41. This is a close-up photo of the central area of the slab of Northwest Africa 032 (Fig. 8.40). The feathery plagioclase laths appear prominently and black chromite phenocrysts are scattered throughout the matrix. The large olivine grain right of center has been bisected and dislocated by a fine black shock vein running through the crystal. Pale pink deposits of carbonates within a white vein are clearly seen at the top of the image and running up the left side. The field of view is 8 mm. (Photo by Walt Radomsky.)

References

1. Bowen, N.L. (1922). The reaction principle in petrogenesis. *Journal of Geology* **30**, 177–198.
2. Keil, K. (1989). Enstatite meteorites and their parent bodies. *Meteoritics* **24**, 195–208.
3. Nehru, C.E., Prinz, M. *et al.* (1983). Brachina: a new type of meteorite, not a chassignite. *Proc. 14th Lunar Planetary Science Conference, Journal of Geophysics Research* **88**, B237–B244.
4. McCoy, T.J. *et al.* (1996). A petrochemical and isotopic study of Monument Draw and comparison with other acapulcoites: evidence for formation by incipient partial melting. *Geochimica Cosmochimica Acta* **60**, 2681–2708.
5. Schiff, J. (2000). Mars Invades Los Angeles. *Meteorite!* **6**, 5–6.
6. Langenhorst, F. and Greshake, A. (1999). A transmission electron microscope study of chassigny: evidence for strong shock metamorphism. *Meteoritics and Planetary Science* **34**, 43–48.
7. Bogard, D.D. and Johnson, P. (1983). Martian gas in an Antarctic meteorite? *Science* **221**, 651–654.
8. Treiman, A. (1999). Microbes in a martian meteorite. *Sky and Telescope* **97**, No.4, 52–58.
9. McKay, D.S. *et al.* (1996). Search for past life on Mars: possible relict biogenic activity in martian meteorite ALH 84001. *Science* **273**, 924–930.
10. Thomas-Keprta, K. *et al.* (1999). Bacterial mineralization patterns in basaltic aquifers: implications for possible life in martian meteorite ALH 84001. *Geology* **26**, No. 11, 961–1056.
11. Borg, L., Connelly, J.N., Nyquist, L.E., Shih, C.Y., Wiesmann, H. and Reese, Y. (1999). The age of the carbonates in martian meteorite ALH 84001. *Science* **286**, No. 5437, 90–94.
12. Cohen, B.A., Swindle, T.D. and Kring, D.A. (2000). Support for the lunar cataclysm hypothesis from lunar meteorite impact melt ages. *Science* **290**, No. 5497, 1754–1756.
13. Baldwin, R.B. (2000). Pulling back the curtain. *Meteoritics and Planetary Science* **35**, A13–A18.
14. Baldwin, R.B. (1949). *Face of the Moon*, University of Chicago Press, Chicago, Illinois, USA, 239pp.
15. Fagan, T.J., Bunch, T.E., Wittke, J.H., Jarosevich, E., Clayton, R.N., Mayeda, T., Eugster, O., Larenzetti, S., Keil, K. and Taylor, G.J. (2000). Northwest Africa 032, a new lunar mare basalt. *Meteoritics and Planetary Science, supplement* **35**, A51.

Chapter Nine

Differentiated meteorites: irons and stony-irons

The first meteorite most people experience is an iron and their first impression is that it is heavy, as they had always imagined a "real" meteorite should be. It looks like the meteorite their mind's eye had always visualized – black with deep cavities. They are surprised to learn that they are hefting a relatively rare-type fragment from space. Of the meteorites seen to fall, only about 4% are irons. Many more than that, however, are represented in collections because they are far easier to recognize among the terrestrial stones than the stony meteorites. Some 40% of all finds are irons. Generally they tend to be larger than stony meteorites since they are stronger and more easily pass through the atmospheric gauntlet with less fragmenting than their stony cousins. Even though they are subject to rusting, being composed of iron–nickel alloys, they are still much more resistant to weathering than the stones. The iron meteorites found around Meteor Crater, Arizona, have lain on the surface or been buried for 50 000 years and yet a bit of cleaning to remove the rust and caliche rejuvenates them. Even the irons around the Odessa Meteorite Crater in west Texas, with a terrestrial age of around 200 000 years, can still be found in good condition after rust removal. In the same locations most stony meteorites would have completely weathered. It is of little wonder then that irons are more often found than stones.

Irons are differentiated meteorites that represent the cores of asteroid parent bodies. Think of the history iron meteorites must have experienced. First, the precursor refractory minerals had to condense out of the solar nebula. Passing through a brief period of heating, chondrules formed and crystallized. Then, the original chondritic parent body formed through accretion and agglomeration of these chondrules and other solid condensates including Fe–Ni grains. After building up to perhaps 100–200 km in diameter, the chondritic body melted and differentiated to produce a layered differentiated parent body. Then, the overburden of crust and mantle rocks was stripped away to expose the metallic core. Multiple impacts over millions of years were required to denude the core. Finally, the iron core which could measure several kilometers across had to be fragmented, an even more daunting task considering that the density of iron meteorites is very nearly 8 g/cm^3, more than twice the average density of chondritic meteorites. It has been estimated that about 65% of the recovered iron meteorites have been subjected to shock pressures greater than 130 kbar.[1] With such a formidable history, it is surprising that iron meteorites are as plentiful as they are.

Iron meteorites play a vital role in our efforts to understand the differentiation of the Solar System's first solid bodies. Their textures and chemistry tell us much about the thermal history of their parent bodies; and from the cooling rates of their central cores, estimates of the sizes of their original parent bodies can be made. In some parent bodies a central iron core may not have formed. Instead, there may have been partial melting and differentiation on the floors of impact craters or within smaller pools dispersed throughout the parent body. In all cases, their cooling rates were a function of their depth of burial.

Mineralogy of meteoritic iron–nickel

The mineralogy of iron meteorites is much simpler than chondrites or achondrites. When the cut face of an iron meteorite is finely ground and polished it appears featureless, its apparent uniformity broken only by occasional inclusions. But a hidden texture lies beneath the polished surface. It was a serendipitous moment that led to its discovery by William Thompson (Lord Kelvin) in 1804 and independently rediscovered by Count Alois von Widmanstätten in 1808.[1] The texture, called the *Widmanstätten* structure is unique to meteorites, the result of the intergrowth of two iron–nickel minerals: low-nickel *kamacite* and high-nickel *taenite*. Figure 9.1 shows the Widmanstätten structure in a typical etched iron meteorite (octahedrite). The wide bands are kamacite, the dark thin lamellae bordering the kamacite bands is taenite. The pattern appears upon etching with dilute nitric acid or ferric chloride (see Appendix D) because the kamacite dissolves more readily in the acid than taenite. A highly magnified cross-sectional view would show the bordering taenite standing above the kamacite. The bands are actually lamellae or three-dimensional narrow plates seen in cross section. In other words, the meteorite is made of narrow plates extending into the specimen until it is interupted by another plate. These plates are readily seen if the specimen is dissolved in nitric acid for several days (Fig. 9.2). A third iron–nickel component, *plessite*, occurs in areas bounded by the kamacite and taenite lamellae. Typically, they have triangular or polygonal shapes with their concave sides facing outward, terminating against the kamacite plates. Plessite is not a mineral but a mixture or fine-grained intergrowth of kamacite and taenite. In the example of Fig. 9.1, many plessite fields appear coarse-grained and triangular in shape. Iron meteorites that display the Widmanstätten structure are called *octahedrites* since the kamacite/taenite plates are arranged parallel to the eight equilateral triangular faces of a pyramidal structure called an *octahedral dipyramid* or simply an *octahedron* (see Fig. 9.6). There are four sets of parallel plates (two each) defined by the eight faces of the octahedron. This orientation of plates is uniform throughout the iron meteorite, which shows that it is built upon a single octahedron crystal. Occasionally, weathered octahedrite iron meteorites will shear across these crystal boundaries so that octahedral-shaped crystals can be extracted, showing the dipyramidal

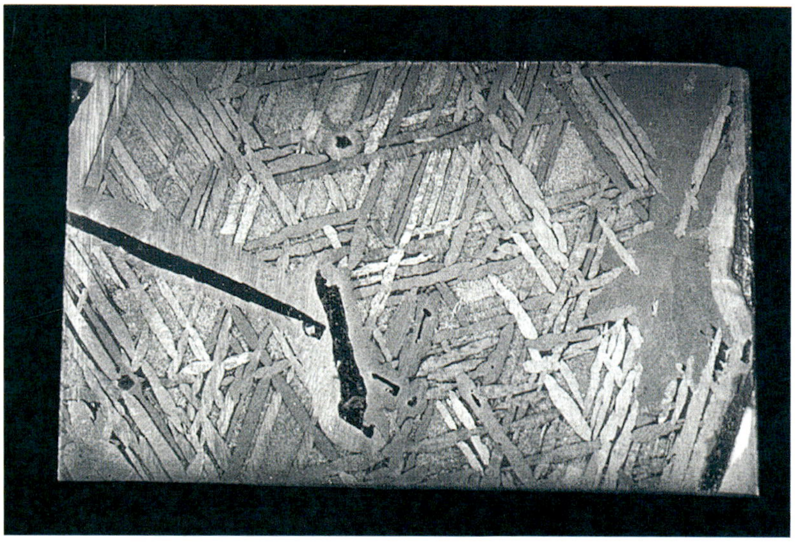

Fig. 9.1. Widmanstätten structure in the IIIB medium octahedrite from Bear Creek, Jefferson County, Colorado. Kamacite bands are the elongated fine-grained structures making ~60° angle triangles with each other. They are enclosed by a thin layer (black) of high-nickel taenite. Swathing kamacite encloses needles of schreibersite (black). The enclosed triangles contain coarse-grained mixtures of kamacite and taenite called plessite. Plessite is also found between parallel-running kamacite plates. The slab measures 34 × 20 mm. (Photo by O. Richard Norton.)

[1] Thompson was treating the metal in a piece of the Krasnojarsk pallasite with dilute nitric acid in an attempt to prevent rusting. The nitric acid etched the metal into a distinctive pattern that revealed three different kinds of iron–nickel. Each alloy was dissolved by the nitric acid at different rates. The pattern was composed of a broad plate or lamella (kamacite) bordered by a bright thin lamella (taenite) and a mix of the two in triangular sections (plessite). Thompson published his discovery in French in *Bibliothèque Britannique* (1804) and again in Italian in the *Atti dell' Academia delle Scienze di Siena* (1808). Yet he was never given due credit for the discovery. Four years after Thompson's discovery, Count Alois von Widmanstätten of Vienna, the director of the Imperial Porcelain Works, rediscovered the pattern by gently heating several iron meteorites (Toluca, Elbogen, Lenarto, and Krasnojarsk) which brought out the pattern. The three different iron alloys oxidized at different rates causing a difference in luster. Interestingly, Widmanstätten did not publish his discovery, only claiming it through oral communication. Nevertheless, Widmanstätten was given full credit and Carl von Schreibers, the director of the Vienna Mineral and Zoology Cabinet named the structure after Widmanstätten. In all fairness, this unique texture should have been called the *Thompson* or *Kelvin* structure (Burke, 1986).[2]

Fig. 9.2. This 171 g iron meteorite, a IA coarse octahedrite from Odessa, Texas, has been treated for several days in strong nitric acid. The thin taenite plates have dissolved away leaving the kamacite plates in high relief showing that they are individual plates that extend through the meteorite until interrupted by another plate. (Photo by O. Richard Norton.)

shape. This is most commonly seen in Gibeon irons (Fig. 9.3).

Crystal structure of meteoritic iron–nickel minerals

To understand how the Widmanstätten structure forms, we need to look more closely at the two iron–nickel *minerals:* kamacite and taenite. Metallurgists refer to these as alpha (α) and gamma (γ) iron respectively. Both alloys are members of the isometric crystal system with a *cubic* or *hexahedron* form. If we could see the space lattices of a single unit cell of the alloys, they would appear as in Fig. 9.4. Notice that the arrangement of iron and nickel atoms in the two crystal lattices is quite different. The kamacite lattice is *body-centered* in that an atom (Fe or Ni) occupies the center of the lattice as well as all corners. It can contain up to 7.5% nickel. The taenite lattice is more tightly packed with atoms because it is a *face-centered* cube, that is, atoms of Fe and Ni occupy the centers of the six faces as well as the corners of the lattice. Its Ni content can vary between about 20 and 45%. The two arrangements result in different numbers of atoms – kamacite (9), taenite (14). In both cases, Ni can substitute for Fe in the lattices, but because there are more attachment sites in the face-centered arrangement, the taenite can have and always does have more Ni atoms in its lattice than kamacite. The migration of Fe and Ni atoms in a cooling solid state environment determines the final texture of the iron meteorite.

Iron–nickel stability phase diagram

Attempts to understand the behavior of Fe–Ni alloys in a homogeneous crystallizing mass below the melting point (solidus) began in the early decades of the twentieth century. Numerous phase equilibrium diagrams were constructed in an attempt to explain the Widmanstätten structure but not until mid-century was real progress made, with refinements continuing to be applied to the present. In 1979, the behavior of Fe–Ni below solidus was redetermined experimentally by A.D. Romig and J.I. Goldstein.[3] The Fe–Ni phase diagram used today in meteoritics is the result of their refinements

Fig. 9.3. An individual octahedron crystal weathered out of a IVA fine octahedrite from Gibeon, Namibia. The crystal shows equilateral triangular faces defined by an octahedron. (Photo by O. Richard Norton.)

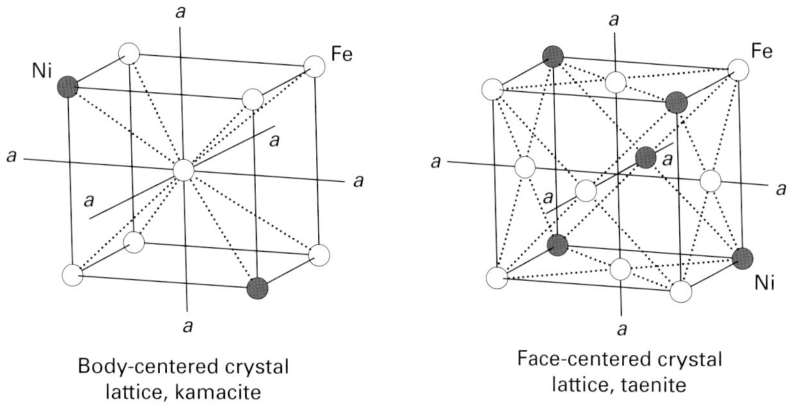

Fig. 9.4. (Left) The body-centered crystal lattice of kamacite. In a close-packed arrangement, each atom is surrounded by eight other atoms. (Right) The face-centered crystal lattice of taenite. In a close-packed arrangement, each atom is surrounded by 12 neighboring atoms.

and led to an understanding of how the Widmanstätten structure forms in meteorites with varying nickel compositions. This is summarized in Fig. 9.5 which shows the Fe–Ni stability phase diagram. The diagram predicts the stability (or equilibrium) of taenite and kamacite at specific temperatures and Ni compositions (wt% Ni). When the metal mass cools below about 1400 °C it crystallizes into the face-centered structure of taenite. Between 1400 and 900 °C only taenite is stable. As the temperature drops below 900 °C, the situation changes. Three stability fields appear into which the cooling metal may enter. What field is chosen depends upon the bulk nickel content of the metal and the cooling rate. The large field extending from the upper left downward toward the lower right is constrained by a sloping line. Only taenite is stable above that line. A much smaller field on the far left side defined by a nearly vertical curving line is the field where only kamacite is stable. Between the two is a third field in which both kamacite and taenite coexist as an intergrowth.

Fig. 9.5. Iron–nickel stability phase diagram which predicts three stability fields of kamacite, taenite, and kamacite + taenite for various temperatures and nickel compositions. Above 900 °C only the taenite phase of the iron-nickel alloy is stable. As the temperature drops stability shifts to the taenite + kamacite field or the kamacite field depending upon the weight percent nickel of the alloy. See text.

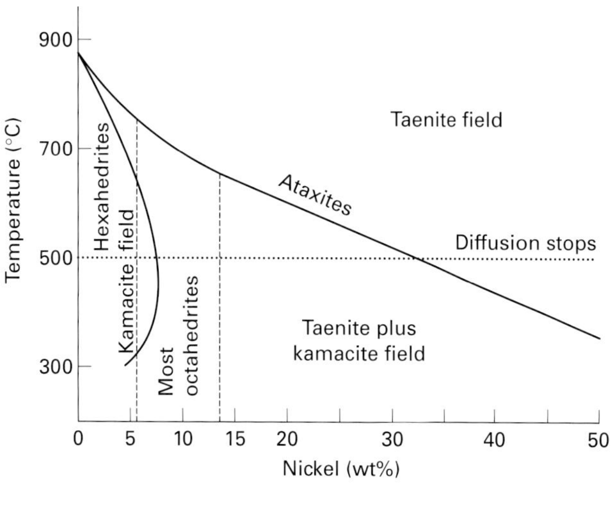

Here, as the temperature drops, the taenite begins to change from a face-centered cubic structure to a body-centered structure separating into two coexisting phases, taenite and kamacite. This separation occurs as the metal crosses the sloping boundary line between the taenite and taenite + kamacite fields. The temperature at which this transformation takes place depends upon the bulk nickel content of the metal, illustrated by the sloping line. Note that as the temperature decreases both taenite and kamacite phases gain in nickel composition but as the temperature continues to decrease, the relative proportions of kamacite and taenite change; the taenite phase decreases while the kamacite increases, that is, kamacite increases at the expense of the taenite.

The transformation of taenite into kamacite occurs through a process of solid-state diffusion in which nickel and iron can migrate into the crystal structures of both alloys. The rate of diffusion is very temperature dependent. Diffusion is more rapid at high temperatures and becomes less and less efficient as the metal cools. Moreover, the rate of diffusion is not the same for both alloys. Kamacite can more easily receive atoms into its crystal structure than taenite and the difference in efficiency becomes greater as the temperature continues to decrease. An important constraint is the point at which diffusion ceases. This occurs at different temperatures for each alloy but for this discussion, we set that point at below 500 °C.

We are now ready to use the Fe–Ni stability phase diagram to look at a few important temperatures and nickel compositions that result in the three groups of iron meteorites: *hexahedrites*; *octahedrites*; and *ataxites*. We begin with a bulk nickel content of around 5% and a temperature of above 900 °C. As we saw, only pure taenite can exist at that temperature. It remains stable until the temperature drops to about 750 °C whereupon the taenite enters the kamacite–taenite field. At this point the taenite releases some of its nickel and low-nickel kamacite begins to form, containing only a few percent nickel. As the temperature drops further to about 650 °C, the metal enters the kamacite field where, at a 5% nickel content, only the kamacite structure can exist. Both kamacite and taenite have gained nickel during the cooling process. Meteorites with 4.5–6.5% nickel content are made of nearly pure kamacite and without an intergrowth of taenite cannot form the Widmanstätten structure. These irons are called *hexahedrites*. Notice that kamacite begins with very low nickel, only 2 or 3%, and increases up to nearly 7% Ni as the temperature decreases. The kamacite stability curve then becomes negative at about 450 °C and kamacite actually begins to lose nickel.

If the original nickel content lies somewhere between 7 and 13%, the cooling taenite metal releases some of its iron (and nickel) and enters the field where both taenite and kamacite coexist. It is in this field where most of the iron meteorites crystallize, and it is here where the kamacite begins to grow on the taenite cube leading, as we will see, to the Widmanstätten structure. These meteorites are the *octahedrites* and account for more than 80% of all iron meteorites. Note that as the bulk nickel content gets higher and higher, the taenite field gets closer to the 500 °C line where diffusion becomes very sluggish or stops altogether. This means that less kamacite will be able to form. For that reason, the Widmanstätten structures gradually narrows until it finally grades into the last group of iron meteorite, the *ataxites*. These meteorites appear to be without visible structure, but they actually grade into a microscopic Widmanstätten structure. Beyond about 15% Ni, the meteorites are composed of a fine-grained intergrowth of taenite and kamacite, *plessite*. Beyond 30% Ni, the metal of the meteorite remains pure taenite since diffusion stops before kamacite begins to form. Pure taenite meteorites are extremely rare. The highest nickel content known in an iron meteorite (Oktibbeha County, Mississippi) is 62 wt%.

Formation of the Widmanstätten structure

The taenite + kamacite field is the most interesting. Here, kamacite begins to grow at specific sites on the taenite cubic lattice (Fig. 9.6a–c), truncating the lattice at 45° angles on its eight corners. The kamacite slowly thickens and lengthens into plates, as much of the taenite is replaced. The final structure depends not only on the nickel percent but equally important on the cooling rate. Solid-state diffusion is more efficient at high temperatures but a slow cooling rate at lower temperatures can be equally as effective at promoting diffusion. The result is an increase in the width of the kamacite plates and a corresponding decrease in the taenite margins.

The growing kamacite plates are arranged with their cross sectional area paralleling the eight faces of an octahedron. Notice that both the original taenite hexahedron and the kamacite octahedron share the same crystallographic axes; therefore the octahedron is still based upon the original cubic taenite space lattice. As kamacite continues to grow, the original six-sided hexahedron structure is gradually converted to an eight-sided octahedron structure as the faces meet along the axes. Octahedrites acquire their name from the crystal structure paralleling an octahedron.

Octahedrite meteorites are found as fragments that in outward appearance hardly resemble either an octahedron or hexahedron. Moreover, when an octahedrite is cut and etched, the kamacite plates are revealed in what usually appears to be a random orientation. However, they are far from random. Since their orientation is parallel to octahedral faces, the Widmanstätten pattern should change in a predictable way as the meteorite is cut along different planes. The kamacite plates appear to cross each other in two, three, or four directions depending upon how the meteorite is cut. Figure 9.7a–c shows three octahedrites cut along different planes with respect to the axes of the octahedron. If the meteorite is cut exactly perpendicular to any of the three axes, i.e. along any face of the cube from which the octahedron is derived, two parallel systems of kamacite plates will appear oriented at right angles to each other like the Cape York, Greenland, IIIAB iron meteorite shown in (a). When the meteorite is cut parallel to any of the eight octahedral faces, three parallel systems of kamacite plates will appear making angles of 60° (or 120°) to each other. The St. Genevieve Co., Missouri, IIIF iron meteorite (b) has been cut parallel to one of the faces. Cutting an octahedrite randomly along any plane between an axis and a face of the crystal will produce four parallel systems of plates with differing angles of intersection as in the Smithville, Tennessee, IAB iron specimen (c). Note that although there are eight faces to a regular octahedron, opposite faces are parallel; therefore there cannot be more than four sets of parallel kamacite plates perpendicular to the faces of an octahedron.

Octahedrite texture is somewhat delicate. It is formed below the subsolidus temperature over a history of very slow cooling sufficient to allow taenite to convert to kamacite through the diffusion process. If the meteorite is reheated to the 900 °C point where only taenite can exist, the texture will change. In fact, this change takes place when an iron meteorite heats in the atmosphere. Within their brief atmospheric passage, iron meteorites reach and exceed this temperature. Iron metal conducts heat more efficiently than

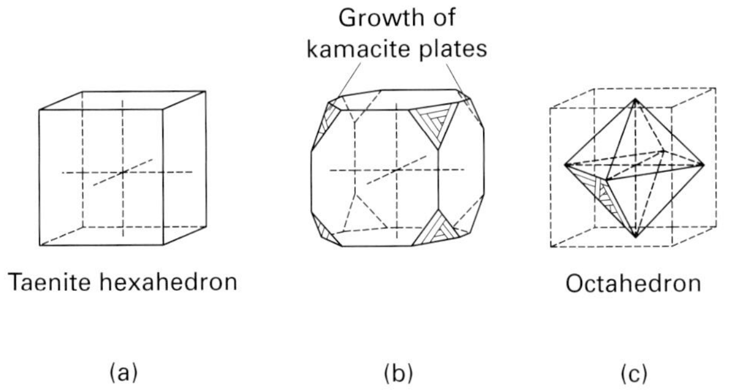

Fig. 9.6. Kamacite growth takes place on the eight corners of the taenite hexahedron. Growing on faces at 45° to the six taenite faces, the hexahedron structure is gradually converted to an eight-sided octahedron, composed of both kamacite and taenite.

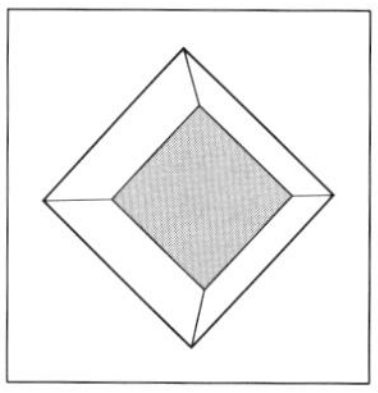

(a) Cape York

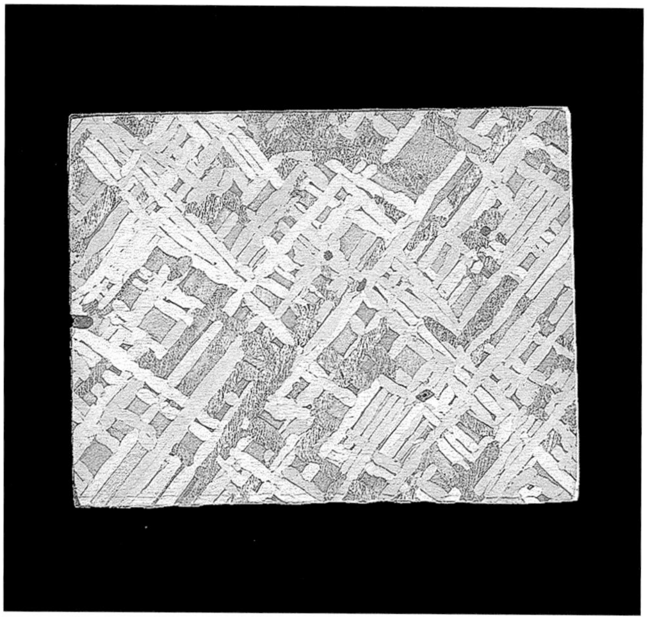

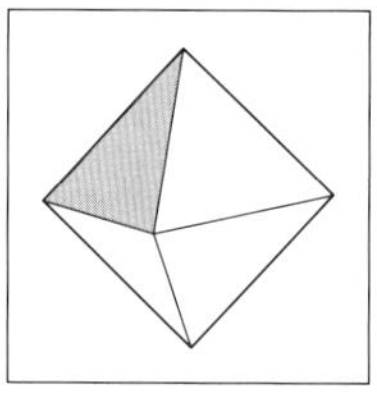

(b) St. Genevieve

Fig. 9.7. Three octahedrite meteorites have been cut along different planes with respect to the faces of the octahedron. (a) The Cape York, Greenland, IIIAB medium octahedrite cut perpendicular to one of the three axes result in two parallel systems of kamacite plates running in two directions at right angles to each other; (b) The St. Genevieve County, Missouri, IIIF medium octahedrite cut parallel to one of the octahedral faces showing three sets of kamacite plates running in three directions at 60 or 120° angles to each other; (c) The Smithville, Tennessee, IAB coarse octahedrite cut at an angle between an octahedral face and an axis producing four parallel systems with different angles of intersection. (Photos by O. Richard Norton.)

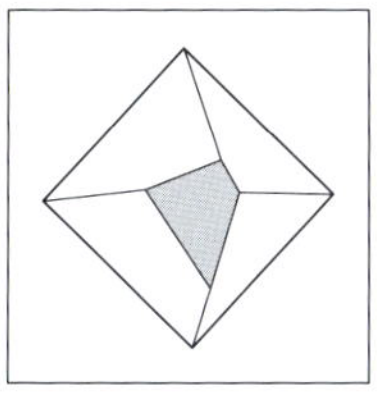

(c) Smithville

silicate minerals and is therefore more affected by atmospheric heating. We saw in Chapter 3 that if a newly fallen iron is cut and etched, a zone around the meteorite reaching depths of as much as several millimeters shows that heating was sufficient at this depth to destroy the Widmanstätten structure (see Fig. 3.14). Iron meteorites such as Canyon Diablo or Odessa that have lain on or within the ground for centuries show no zone of heating because rusting has effectively removed the zone. Artificial heating with a propane torch or even a hot oven is more than sufficient to damage the structure. Unfortunately, many of the large irons discovered centuries ago were used by their discoverers as anvils or in other ways subjected to artificial heating which effectively destroyed their structures to a depth of several millimeters.

Classification of iron meteorites

There are two classifications of iron meteorites based upon either structural or chemical criteria. Earlier classifications were based upon the structural appearance of the cut and etched iron. This was before the days of trace element analysis. Today, trace element (and bulk) compositions of irons are determined by neutron activation analysis. The chemical classification is much more analytical and therefore more productive to the science, but both schemes are necessary because it becomes immediately apparent that the chemical composition and structure bear important relationships to each other as we will see.

Structural classification of iron meteorites

The structural classification is based upon the texture of the Widmanstätten structure, that is, how wide or narrow the kamacite plates are. The width depends upon the bulk nickel content. Table 9.1 gives the structural classification based upon these two criteria: band width and nickel content. This is primarily an octahedrite classification with two other meteorite groups defining the low and high end of the nickel composition. The textural groups in this classification gives us the impression that they are distinct from each other. In actuality, they are not well-defined groups but grade smoothly from one texture into another. The nickel content boundaries listed for each textural type are somewhat arbitrary.

Group	Symbol	Band width (mm)	Nickel (%)
Hexahedrite	H	>50	4.5–6.5
Octahedrites	O		
Coarsest	Ogg	3.3–50	6.5–7.2
Coarse	Og	1.3–3.3	6.5–7.2
Medium	Om	0.5–1.3	7.4–10.3
fine	Of	0.2–0.5	7.8–12.7
finest	Off	<0.2	7.8–12.7
Plessitic	Opl	<0.2	Kamacite spindles
Ataxite	D	no structure	>16.0

Table 9.1. **Structural classification of iron meteorites**

Iron meteorites were originally classified according to their internal structure. This table shows a structural division of the irons into nine groups based upon their kamacite band width and their percentage of nickel. Most are octahedrite subgroups, ranging from coarsest to finest kamacite plate widths as their nickel content increases. Meteorites with the lowest nickel percent are hexahedrites with bandwidths that often equal or exceed the width of the meteorite itself. The entire meteorite is composed of one or more large cubic crystals of low-nickel kamacite. Widmanstätten structure seen in the octahedrites is absent. At the opposite end of the structural classification are the ataxites. Here the finest Widmanstätten structure grades smoothly into a structureless mass with a nickel content of over 16%. They are essentially high-nickel taenite.

Hexahedrites (H)

Hexahedrites are those iron meteorites that contain the least amount of nickel, usually less than 6%. They are nearly pure kamacite. Their name is derived from the six-sided hexahedral shape of its crystals. In the purest form, these meteorites are essentially large cubic crystals of kamacite. When cut, they appear featureless internally and when etched do not show a Widmanstätten pattern. Instead, the cut face commonly exhibits several sets of fine parallel lines 1–10 μm wide that represent very thin twinning lamellae on the planes of the kamacite lattice (Fig. 9.8).[II] This structure was first described by Franz Ernst Neumann in 1848 and bears his name, *Neumann lines*. Normally these planes are not seen with the naked eye but if the meteorite has suffered intense shock, weak points such as twinning planes in the crystal lattice show a slippage along planes parallel to the faces of the hexahedron. This mechanical twinning becomes visible as multiple sets of fine parallel lines.

Octahedrites (O)

Octahedrites are the most plentiful of the iron meteorites and also show the greatest textural variation due to their variation in nickel and intergrowth of kamacite and taenite phases. They are divided into six different subgroups based upon their nickel content and kamacite band width. The coarsest bands (widest) have the least bulk nickel but there is substantial overlap of nickel content for each subgroup. Table 9.1 and a millimeter scale are all that are necessary to classify octahedrites as coarsest (Ogg), coarse (Og), medium (Om), fine (Of) or finest (Off). Figures 9.9a–d show representative examples of textural grades all made to the same scale. Some

[II] Twelve sets of twin lamellae are possible on the kamacite body-centered space lattice but usually only two–four sets of parallel lamellae are observed in a given hexahedrite.

Fig. 9.8. Hexahedrite (IIAB) iron meteorite from Calico Rock, Arkansas, showing sets of parallel shock planes (Neumann lines) running in several directions through the specimen. The specimen contains 5.4 wt% nickel. The horizontal dimension is 43 mm. (Photo by O. Richard Norton.)

of the kamacite bands show Neumann lines, especially clear in coarse octahedrites (Fig. 9.10). Sometimes thick kamacite plates in the coarsest octahedrites will separate from the main body of the meteorite either as they pass through Earth's atmosphere or by weathering on the ground. These plates appear cubic with 90° cleavage planes and are crossed by Neumann lines superficially mimicking a hexahedrite.[III] Neumann lines on kamacite plates suggest a shock history for the octahedrites. As the nickel content increases to about 13%, the Widmanstätten texture becomes microscopic. At higher percentages the texture becomes *plessitic* (Fig. 9.11), showing microscopic spindle-shaped laths of kamacite surrounded by taenite. These are designated Opl in Table 9.1.

Ataxites (D)[IV]

This last group of iron meteorites is perhaps the least interesting based upon their texture – there isn't any. The finest-textured octahedrites grade smoothly into the ataxites as the nickel content increases beyond 13%. At this point, microscopic Widmanstätten figures may be seen but with increasing nickel content (>16%), the meteorite becomes basically all taenite; therefore there is no Widmanstätten texture. Microscopically, ataxites show tiny crystals of taenite cloaked in a thin layer of kamacite. These are in turn embedded in a matrix of fine-grained plessite. Although ataxites are the least common of all the iron meteorites, the group can claim the distinction as the world's largest single iron meteorite – the 60 ton Hoba iron (see Fig. 2.21).

Anomalous irons

This simple structural classification works fine for most of the irons but there are a few that cannot be pigeon-holed into any of the groups. They often show unusual structures and/or chemistry that do not conform to the classification and therefore must be considered unique. These are referred to as *anomalous or ungrouped irons*. The Tucson Ring meteorite and the Australian iron, Mundrabilla, are well known examples of ungrouped anomalous irons (Fig. 9.12). Often these meteorites display a random-appearing polycrystalline texture (crystals of taenite and kamacite) and some contain large populations of troilite nodules and/or silicate inclusions. Typically, anomalous irons can show any of the structural types described above.

Chemical classification of iron meteorites

Classification of irons by structure goes back to the Rose–Tschermak–Brezina system. Between 1885 and 1904 Brezina

[III] In my youth many years ago, I received just such a meteorite in trade from the Soviet Academy of Sciences' Committee on Meteorites. It was a specimen from the famous Sikhote-Alin fall which had an (Ogg) classification. The specimen had fragmented on impact along natural cleavage planes giving it a blocky cubic appearance and etching revealed distinct Neumann lines. In my confusion I wrote the Committee on Meteorites suggesting that Sikhote-Alin was a hexahedrite, not an octahedrite. Their reply was patient if somewhat blunt: "Look closer." I did, and with some embarrassment noticed a very thin lamellae of taenite still clinging to the edges of the kamacite plate, revealing its true octahedrite nature.

[IV] The symbol (D) for the ataxites is somewhat obscure. Tschermak called ataxites *dicht eisen* meaning compact or densely structured irons, referring to the lack of Widmanstätten structure or, in fact, any structure visible to the unaided eye. The "D" in *dicht* became the symbol for ataxites in the Tschermak classification.

Fig. 9.9 (a) Coarsest octahedrite (Ogg – 10 mm) from Sikhote-Alin, Class IIAB, 6.3 wt% Ni; (b) Canyon Diablo, Arizona, coarse octahedrite (Og – 2 mm), Class IAB, 6.98 wt% Ni; (c) Henbury, Northern Territory, Australia, medium octahedrite (Om – 0.9 mm), Class IIIAB, 7.47 wt% Ni; (d) Gibeon, Namibia (Of – 0.30 mm), Class IVA, 7.68 wt% Ni. ((a) Photo by Andrzej Pilski; (b, c, d) by O. Richard Norton.)

expanded upon Rose's earlier classification and was first to distinguish the octahedrites according to the width of the kamacite bands. This system remained with little substantial change up to the mid-twentieth century, much as it is displayed in Table 9.1. The irons are divided into eight groups of which six are octahedrite subgroups but the structural classification based upon cooling rate[V] and bulk nickel content alone does not produce distinct subgroups because, as we saw, they grade into each other. Moreover, their compositions have considerable variation within a given bandwidth, showing that bandwidth alone does not necessarily group genetically related irons. These structural parameters, however, *combined* with several other parameters (chemical) allow a definite classification into meaningful genetic groups. One of the strong points of the newer chemical classification is that it divides the irons into distinct groups that probably represent different parent bodies. If there is a disadvantage to the chemical classification it is that the meteorites cannot be classified visually but must be chemically analyzed using neutron activation techniques, but, as we will see, since the chemical classification is also based upon nickel content, there is at least a general correlation between the structural and compositional groups (Table 9.2).

The chemical classification uses nickel and trace element compositions of the meteorites. Trace elements are usually found in quantities of a few atoms per million atoms of iron. Certain elements have an affinity for iron alloys, that is, they easily fit into the crystal structure of the alloys. These are the

[V] The role of cooling rate and diffusion in subsolidus Fe–Ni metal was not recognized until the latter half of the twentieth century when a correct phase equilibrium diagram for meteoritic Fe–Ni finally appeared.

Fig. 9.10. Neumann lines across plates of kamacite in the Lake Murray IIAB coarsest octahedrite (10 mm). The horizontal dimension is 60 mm. (Photo by O. Richard Norton.)

siderophile or iron-loving elements and nickel is certainly the outstanding major siderophile element in irons. The trace elements gallium (Ga), germanium (Ge), and iridium (Ir) are used in iron meteorite classification and these in turn are plotted against Ni abundance usually in mg/g (milligrams of nickel per gram of metal) or Ni wt%.

Let us see how trace elements are used to subdivide iron meteorites into distinct chemical groups. In a slowly cooling Fe–Ni melt, the first iron to crystallize out of the liquid contains less nickel than an equal volume of the liquid melt itself. This is because the nickel tends to accumulate and concentrate in the liquid phase rather than in the solid phase. This means that the first solid Fe–Ni accumulating at the developing core of a differentiated asteroid is relatively low in nickel. As the crystallization process continues, both the melt and crystallizing solid metal become richer in nickel although the metal is still less rich in nickel than the melt. This *fractional crystallization* results in metal with various nickel compositions. The key to the trace element classification is that trace siderophile elements have different affinities for iron as a solid versus iron as a liquid. Compared to the major elements, fractional crystallization affects trace elements much more strongly, making them more desirable to use as parameters for analysis. Iridium provides a good example of this. It prefers to combine with the first crystallizing metal which, as we saw, is low in nickel. With increasing nickel concentration in the melt, the amount of iridium in the crystallizing metal decreases since most of it was tied to the first metal to crystallize out. If iridium abundance in the crystallized metal is plotted against nickel content (for a given melt with an initial bulk of nickel), we see a distinct field appear in which the data points are confined to a narrow area with a negative slope (Fig. 9.13). As the percentage of nickel in the metal goes down, the iridium content goes up dramatically. Melts

Fig. 9.11. A 7.5 g slab from the Cowra Opl iron meteorite found 1888 in New South Wales, Australia. This anomalous meteorite has a fine-grained plessitic texture with a nickel content of 12.94%. The narrow border around the specimen is the result of differential etching. (Photo by Dr Richard Herd, Curator, the National Collections, Geological Survey of Canada.)

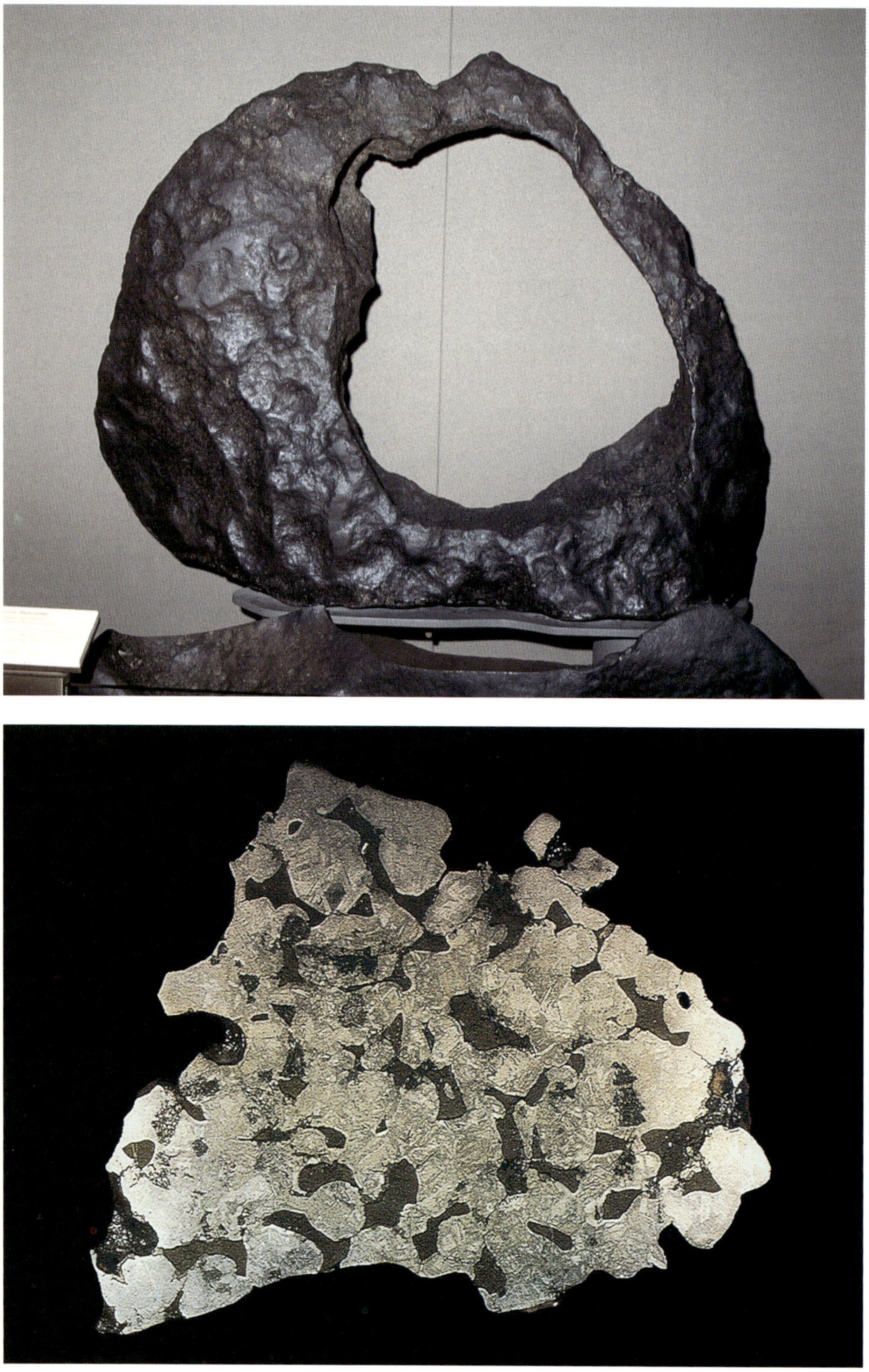

Fig. 9.12. Anomalous irons. (Top) The 688 kg ataxite, the Tucson Ring meteorite. (Bottom) Cut slab of the Mundrabilla iron from the Nullarbor Plain, Western Australia, showing numerous bronze-colored troilite nodules in its interior. Specimen is 22.6 cm longest dimension. (Tucson Ring from the Smithsonian Institution collection; Mundrabilla from the Robert A. Haag collection. Photos by O. Richard Norton.)

with higher initial bulk nickel content produce different confining fields, displaced to the right of the low nickel field. Each field is a distinct chemical iron group that represent a different melt and probably a different parent body.

Historically, the first indications that trace elements could be used to classify meteorites into distinct genetic groups occurred with the work of Edward Goldberg and associates in 1951.[4] These researchers were among the first to use neutron activation analysis[VI] to detect minute quantities of trace elements in iron meteorites. They measured the amount of the trace element gallium along with nickel in some 40 iron meteorites. They noted that the amount of gallium varied with

[VI] Neutron activation analysis is a technique used to detect the presence of trace elements in rock samples. The sample is placed in a reactor and bombarded with neutrons which produces unstable radioisotopes. These isotopes decay giving off emissions with characteristic energy of the sought-after trace elements. Typically, this analysis yields sensitivities in parts per billion.

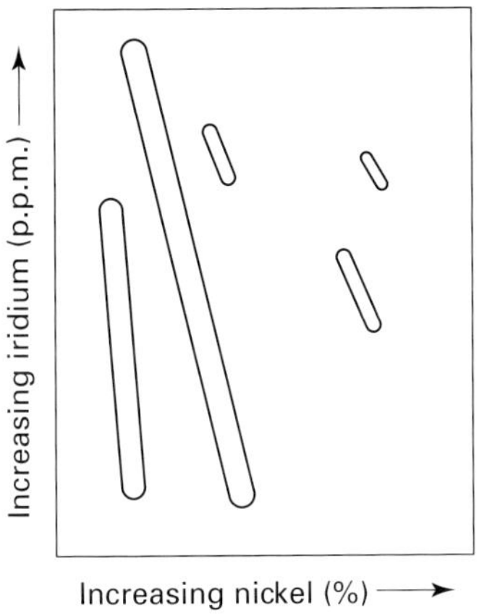

Fig. 9.13. A hypothetical plot of the iridium abundance vs nickel content in crystallizing melts of various nickel concentrations. Since most of the iridium will crystallize with the first formed low-nickel metal, the resulting low-nickel field will show an abundance of iridium which increases with decreasing nickel. For other melts with higher initial nickel abundance, different fields will appear, each representing a different parent body with a different core composition.

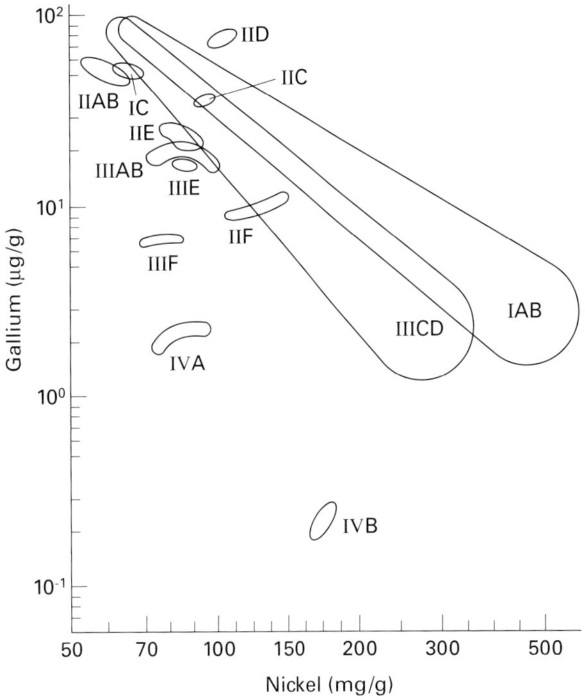

Fig. 9.14. A plot of the abundance of gallium vs nickel in iron meteorites. Thirteen fields are distinguished, each representing a different chemical group and, presumably, a different parent body. These small but distinct fields allow researchers to classify most iron meteorites into one of the 13 chemical groups. About 14% do not fall into these groups and remain anomalous. (Used with permission of Dr John T. Wasson, University of California, Los Angeles.)

the bulk nickel content and found three distinct fields which represented three genetically separate iron groups. In 1957, another group lead by John Lovering[5] measured both gallium and germanium concentrations in 88 irons and found a fourth distinct group. These four groups were labeled I, II, III, and IV. Only one meteorite type seemed to correlate with structural class; a low-nickel hexahedrite that appeared to fit Group II. Not all the meteorites fitted neatly into any of the four groups. These were set aside and labeled "anomalous".

Beginning in the mid-1960s and continuing to the present day, John Wasson and his colleagues at UCLA studied over 500 iron meteorites using more advanced techniques so that lower concentrations of both gallium (Fig. 9.14) and germanium could be measured. They were able to resolve Lovering's Group IV into two groups which they labeled IVA and IVB. Structurally, they were quite different. Group IVA was a fine octahedrite (kamacite band width of 0.3 mm) with a nickel content varying from 7.4 to 9.4 wt%. The IVB group consisted of ataxites with 16–26 wt% nickel. Likewise, Lovering's Group III was resolved into two genetically related groups they labeled IIIA and IIIB. In addition to the gallium and germanium plots, Wasson studied iridium concentrations (Fig. 9.15) that generally showed much more extensive fields but which still confirmed the new groups in the germanium and gallium plots. The small distinct fields in the

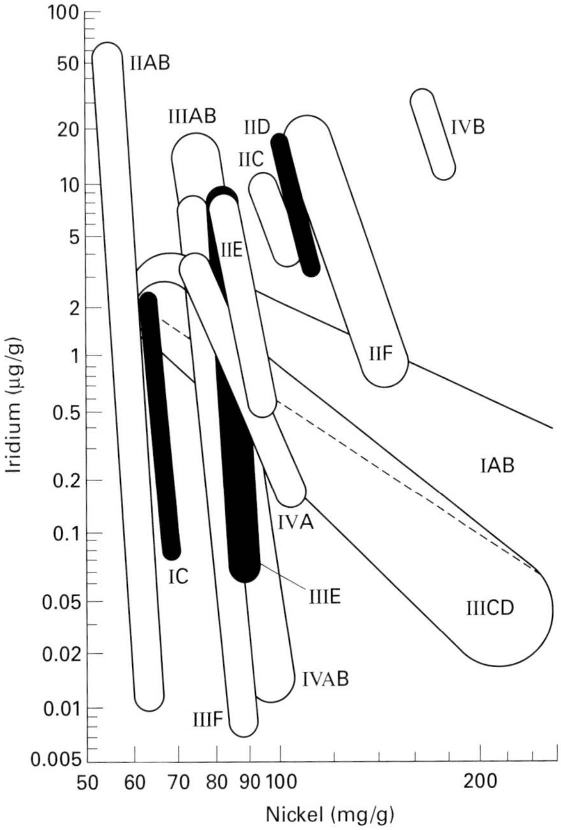

9.15. A plot of iridium abundance vs nickel abundance in iron meteorites. Note that the range of the fields are much more extensive than in the gallium vs nickel plot. This allows for a higher resolution within the fields so that meteorites of the same group can be distinguished from each other based upon their iridium concentrations. Thus, two meteorites found in close proximity and suspected of being paired can be tested for iridium and an accurate determination made based upon their iridium concentration. (Used with permission of Dr John. T. Wasson, University of California, Los Angeles.)

germanium and gallium plots are most useful when classifying new meteorites. The more extensive fields of the iridium plot has the requisite resolution to allow researchers to distinguish meteorites within the same group. In all, Wasson's investigation found 16 groups. (Another group, IIF, was added later.) Wasson retained Lovering's original four Roman numerals and added alphabetical letters A to F to distinguish the groups. Subsequently, three sets of groups have been combined since transitional members were found, connecting them linearly on the plots. Thus, IA and IB are now IAB; IIA and IIB are IIAB; IIIA and IIIB are IIIAB; IIIC and IIID are IIICD, reducing the number of groups to 13. This is interpreted to mean that these groups represent the cores of thirteen separate parent bodies.

Of the over 500 meteorites studied, about 430 (86%) could be assigned to one of the 13 (or 17 if not combined) groups. The remaining 60+ could not be assigned to any of the groups. Like Lovering's leftovers, Wasson labeled these "anomalous". These have since been renamed, *ungrouped*. These show distinct chemical and structural differences sufficient to represent separate parent bodies. Altogether the irons may represent more than 60 additional parent bodies.

Chemical vs structural classification

We have already seen that when Lovering plotted gallium and germanium against nickel percent, he found a single related iron, a hexahedrite that seemed to fall into Group II. Wasson's work involving more than five times the number of irons and with superior equipment and techniques allowed him to distinguish 17 groups. These chemical groups do indeed appear correlated with their structure (kamacite band width) and nickel content. Table 9.2 shows the structural groups and how they relate to the chemical groups.

We see first that Group IIIAB is by far the most populous with 32.3% of the irons. The next most common irons are the IAB irons with 18.7%. Both groups are coarse to medium octahedrites. Notice that some of the chemical groups are defined by relatively narrow structural variations. For example, IVA irons are fine octahedrites; IVB irons are ataxites; IIIE irons are coarse octahedrites; and IIIAB are medium octahedrites. All occupy narrow nickel percent ranges. The paired groups if split show a continuous structural variation like IIICD with IIIC as fine octahedrites and IIID as finest octahedrites; IIAB with IIA as hexahedrites and IIB as coarsest octahedrites. These occupy opposite ends of their group's field. Other chemical groups have remarkably broad structural variations like IIIF irons that vary from coarsest to fine octahedrites. It seems curious that of the 13 chemical groups two groups, IAB and IIICD, stand out on the Ga, Ge, and Ir plots because their fields are very extensive and they both show large negative slopes suggesting a relationship between them. In 1995 it was argued that they were genetically linked and should occupy the same field which has been called IAB/IIICD.[6] This group now varies structurally from coarse

Chemical groups	Frequency (%)	Bandwidth (mm)	Ni (wt%)	Structural groups
IA	17.0	1.0–3	6.4–8.7	Ogg, Og, Om
IB	1.7	0.01–1.0	8.7–25	Om → D
IC	2.1	<3	6.1–6.8	Anom, Og
IIA	8.1	>50	5.3–5.7	H
IIB	2.7	5–15	5.7–6.4	Ogg
IIC	1.4	0.06–0.07	9.3–11.5	Opl
IID	2.7	0.4–0.8	9.6–11.3	Om, Of
IIE	2.5	0.7–2	7.5–9.7	Ogg → Om (Anom)
IIF	1.0	0.05–0.21	8.4–10	Of → D
IIIA	24.8	0.9–1.3	7.1–9.3	Om
IIIB	7.5	0.6–1.3	8.4–10.5	Om
IIIC	1.4	0.2–3	10–13	Ogg → Off
IIID	1.0	0.01–0.05	16–23	Off → D
IIIE	1.7	1.3–1.6	8.2–9.0	Og
IIIF	1.0	0.5–1.5	6.8–7.8	Ogg → Of
IVA	8.3	0.25–0.45	7.4–9.4	Of
IVB	2.3	0.006–0.03	16–26	D

Table 9.2. **Structural and chemical relationships in iron meteorites**[VII]

Using trace element analysis, iron meteorites have been divided into 17 groups. (This can be reduced to 12 groups. See text.) There seems to be at least a general relationship between the earlier structural classification and the newer chemical classification. This table lists the 17 known iron chemical groups and relates them to kamacite band width and percentage of nickel. A few of the groups seem related to each other. For example, IIIA and IIIB irons both have very narrow structural variations, both related to the structural group Om. Thus, they are usually combined as IIIAB irons. Other chemical groups are paired in similar fashion.

[VII] Data from Wasson, J.T., 1985. *Meteorites: Their Record of Early Solar System History*, W.H. Freeman and Co., New York.

to finest octahedrites and reduces the number of genetic groups to 12. Furthermore, recently the 12 groups have been divided into those that show magmatic characteristics, *magmatic* groups, and those that do not, *nonmagmatic* groups. *Magmatic* means that the slopes of their bulk element-to-Ni trends are consistent with having formed by fractional crystallization. Anyone struggling with this complex classification would undoubtedly breathe a sigh of relief to have any simplification, but there seems to be no room for optimism. There are approximately 90 ungrouped irons. It is unclear which of these are magmatic, but most of them probably are. This means that most of them formed by magmatic processes within separate differentiated parent bodies just waiting to be upgraded to a fully fledged group status.

Silicate inclusions in iron meteorites

Iron meteorites are differentiated meteorites. Since iron and silicates are immiscible and tend to separate in a melt, we would not expect to find irons associated with silicates. In fact, the relative rarity of silicate inclusions in iron meteorites is a strong argument that iron meteorites were formed by magmatic processes within differentiated asteroids resulting in core formation. Yet, as unlikely as it seems, we do find silicate inclusions in at least three groups: IAB/IIICD; IVA; and IIE. These meteorites are called *silicated irons* and they do not seem to be associated with magmatic processes (melting leading to differentiation). Unlike the other iron groups that show large variations in their trace element plots (especially Ir), silicated irons show only weak trends or slopes inconsistent with fractional crystallization and are therefore considered *nonmagmatic*.

The two nonmagmatic groups, IAB and IIE, show significant differences in their mineralogy and probably their origin. Group IAB irons are coarse octahedrites with irregular and often angular inclusions of dark silicates. They can occur as tiny single crystals buried in the kamacite plates up to large aggregates several centimeters wide. Well-known IAB irons such as Campo del Cielo, Argentina, Odessa, Texas, and Landes, West Virginia (Fig. 9.16), frequently exhibit them. Their compositions are generally chondritic with granular low-FeO pyroxene (Fs_{4-9}) and olivine (Fa_{1-4}). The minerals are homogenized similar to recrystallized chondritic material with substantial amounts of plagioclase (An_{9-22}). Their oxygen isotopic compositions lie below the terrestrial mass fractionation line far removed from the ordinary chondrites but occupying a position near the primitive achondrites, the *winonaites* (Fig. 9.22). Winonaites are primitive achondrites thought by some to be samples of a parent body that only partially differentiated (see Chapter 8). It is likely that both the IAB irons and the winonaites came from the same parent body. Their textures and compositions suggest that they are a mix of chondrite and metal breccia. The silicates were heated sufficiently to recrystallize them or in some cases melt them but the metal was not completely melted. The metal shows Widmanstätten structure which suggests that the parent body was sufficiently insulated to allow slow cooling.

Group IIE irons are altogether different. The silicate inclusions are usually rounded or amoeboid in shape as though they had been melted and then trapped while still liquid in cooling metal such as the Miles IIE silicated iron in Fig. 9.17.

Fig. 9.16. Landes, West Virginia, IAB iron meteorite, a coarse octahedrite with dark silicate inclusions. The slab is 70 mm longest dimension. (Photo by Michael Casper.)

Fig. 9.17. A IIE silicated iron meteorite from Miles, Australia. The silicate inclusions are rounded and amoeboid in shape. Evidence of shock melting of augite suggest that IIE irons are the product of impact melting. The horizontal dimension is 52 mm. (Photo by O. Richard Norton.)

The major silicate minerals are Fe-rich low-calcium orthopyroxene implying a greater oxidation state; augite, often in large euhedral crystals that show shock deformation lamellae; and fine-grained to cryptocrystalline plagioclase matrix that also shows evidence of shock melting. Olivine is usually not found. All of this suggests that these inclusions are melted chondritic material and that the parent body was subjected to intense impact shock. The oxygen isotopic compositions (Fig. 9.22) lie above the terrestrial fractionation line and within the ordinary chondrite field nearer the LL chondrites.[7]

IIE irons do not appear to have been derived from a differentiated core body. All evidence seems to point to IIE irons as an impact product. A possible formation scenario may go something like this. Initially, two chondritic precursor bodies, one considerably larger than the other impacted each other. This impact was a crater-forming event on the larger body. The smaller body was totally disrupted and partially melted upon impact and mixed with partially melted chondritic mineral phases from the larger body. This melted material differentiated *within* the crater; metal and sulfide segregated from the silicate material and settled to the crater floor. Melted silicate material became trapped within the metal melt pools. Since IIE metal shows well developed Widmanstätten structure (Fig. 9.17), the impact must have reached considerable depth so that the melt was insulated allowing the metal to cool slowly. Thus, IIE meteorites are probably impact melts from chondritic parent bodies.

UCLA meteoriticist Alan Rubin and his coworkers have developed a somewhat different scenario that also involve two colliding chondritic bodies. Shock compression preferentially melted the plagioclase forming alkali-rich melts. These became the silicates of the IIE irons. The chondritic metal also melted and pooled on the floor of the crater on the larger body trapping the silicates in the process. Interestingly, melt pocket glasses in ordinary chondrites have similar bulk compositions to IIE silicates.

Cooling the core

Hidden within the Widmanstätten structure is a natural recording thermometer that has permanently recorded the rate at which the core body cooled. This thermometer became available when in 1951 a new analytical tool, the electron microprobe, was developed by Raymond Castaing in France. This nondestructive tool allows the user to analyze selected spots across a meteorite with a sensitivity of a few parts per billion. To do this, a polished section of a meteorite (usually a thin section) is placed in a vacuum and in the path of an electron beam. The electrons are absorbed by the elements present which give off X-rays whose energies are characteristic of each element. The intensity of the X-rays for each element is measured, giving the concentration of the element within a spot as small as 1 μm, when compared to a standard of known chemical composition

We learned earlier by looking at the Fe–Ni stability phase diagram (Fig. 9.5) how the Ni-rich taenite is slowly converted to low-nickel kamacite as the metal cools. We further saw through the diffusion process that both kamacite and taenite gain nickel atoms in their structures with decreasing temperatures. Finally, we learned that diffusion is more efficient at higher temperatures and becomes sluggish and inefficient as the temperature decreases. This simply means that the slower the cooling rate, the more efficient is the diffusion of nickel and therefore the more homogenous kamacite and taenite become. For a given temperature, nickel atoms more easily enter the kamacite structure than the taenite structure. The difference in diffusion rates of the two alloys becomes greater with ever lowering temperatures up to the point where all diffusion ceases. In this case, time is the great homogenizer. If the cooling rate is sufficiently slow, then time is available for the less efficient taenite diffusion to homogenize the taenite plates. A microprobe scan for nickel in completely homogenized taenite would therefore look something like Fig. 9.18 (above and middle) with nickel distributed uniformly throughout the taenite plate. A completely flat nickel distribution profile like this is never seen, however, because the cooling rate of differentiated parent body cores is never sufficiently retarded to homogenize the taenite.

In the early 1960s using the electron microprobe, meteoriticists scanned across the kamacite and taenite plates to determine the nickel distribution. Instead of a flat peak, the trace across the taenite plates took on the appearance of an "M" (Fig. 9.18 (below)) showing more nickel on the edges of the plate and progressively less toward the center. This is simply because nickel atoms had piled up on the edges due to the difference in rate of diffusion between the taenite and kamacite.

The evolution of the Widmanstätten structure as a function of bulk nickel concentration and cooling rate is shown schematically in Fig. 9.1a–d; each sketch is a stage in the forming Widmanstätten structure at specific but slowly dropping temperatures. Above 700 °C (a), only the taenite plate is seen with the nickel content equal to the bulk nickel content. As the temperature drops, nickel diffusion begins,

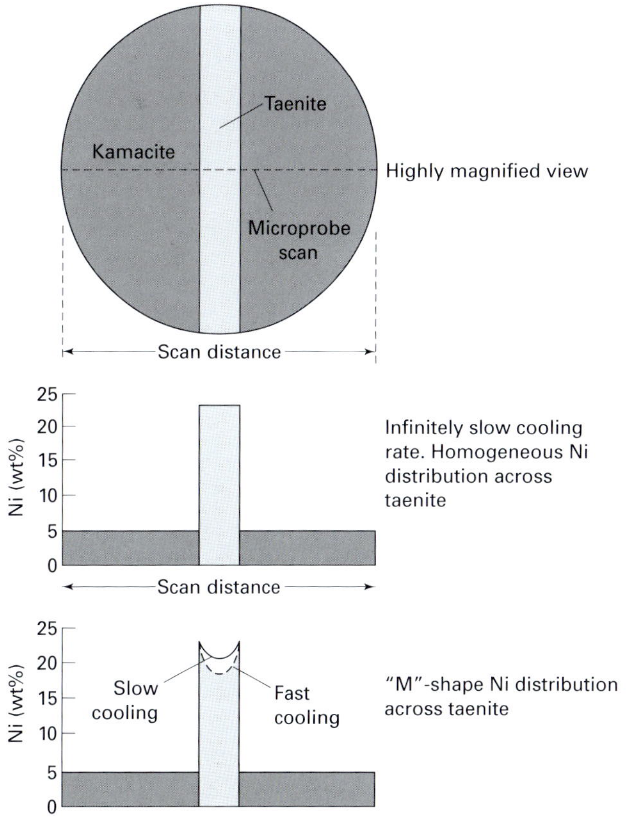

Fig. 9.18. A hypothetical microprobe nickel trace across kamacite and taenite plates (above). If the rate of cooling of the core is sufficiently slow to allow diffusion to distribute nickel uniformly into the taenite plate, the taenite plate would become homogenized with respect to nickel and the trace would appear flat (middle). This situation is never found. Instead, the trace would appear M-shaped (below). The trace shows a peak distribution of nickel on the edges of the taenite plate adjacent to the kamacite and then a progressive decrease in nickel toward the center making the characteristic "M" shape distribution curve. The depth of the "M" is a statement of the rate of diffusion of nickel into the taenite which is determined by the cooling rate. A slow cooling rate produces a flatter "M", a fast cooling rate deepens the "M".

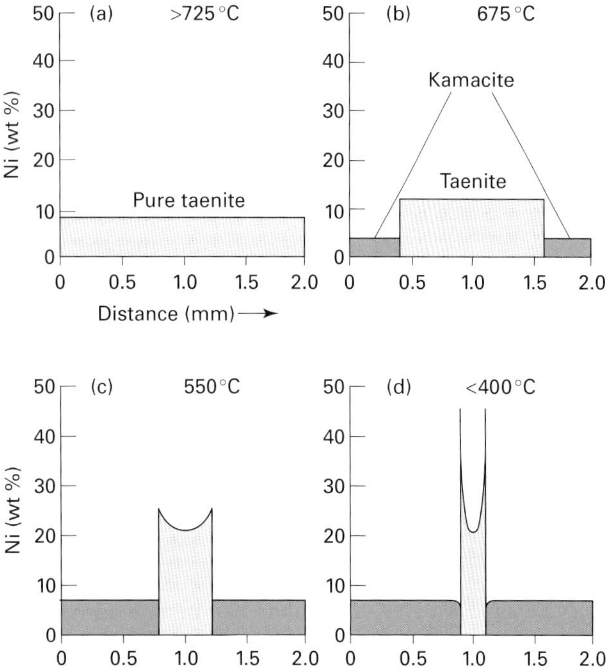

Fig. 9.19. This illustration shows stages in the evolution of the Widmanstätten structure. Each stage shows an electron microprobe trace across the taenite boundary measuring the changing nickel composition in wt% across the taenite as the temperature drops. The rate of temperature drop is assumed to be 2 °C per million years. (a) Initially, above 700 °C, the iron meteorite is composed of pure taenite. Nickel is homogeneous in the taenite illustrated by the flat trace. (b) As the temperature drops below 700 °C, the metal enters the kamacite/taenite stability field and thin kamacite plates begin to form on either side of the taenite. The taenite shrinks in size, reducing its volume and increasing its nickel content at the same time. (c) At this stage, the temperature has reached ~550 °C. The kamacite has continued to grow at the expense of the taenite. Diffusion begins to get sluggish at this temperature and taenite fails to accept the nickel into the interior. Instead, the nickel builds up on the kamacite/taenite margins creating a nickel gradient across the taenite plate. (d) When the temperature drops to ~400 °C, the characteristic M-shape trace appears with high nickel concentration at the margins of the taenite but dropping steeply as the probe cuts across the interior of the taenite. At the same time, kamacite, better able to accept nickel, becomes homogenized. (Based upon the work of Kaare Rasmussen, University of Odense, Denmark.)

slowly transforming taenite into kamacite. At (b), the temperature has dropped below 700 °C and kamacite begins to grow on the taenite structure as the metal enters the kamacite + taenite stability field and the nickel content of both increases. Since the nickel content of the original metal is constant, the nickel content of the taenite can increase only if its volume decreases (it loses both iron and nickel). The taenite plates therefore become thinner with increasing nickel content while, at the same time the low-nickel kamacite plates increase in width (c). Finally, as the temperature drops below 500 °C, diffusion becomes so sluggish that the nickel builds up at the margins of the taenite plates not able to penetrate to the interior (d). A microprobe scan across the kamacite plate shows the low-nickel metal to be homogenized with nickel evenly distributed. The scan, however shows the taenite nickel distribution to be zoned. The plate margins are nickel-rich but further into the plate the nickel content drops steeply to a minimum near the center.

The distribution of nickel in the taenite and kamacite plates is an indicator of the cooling rate of the parent body's core. A nearly flat nickel profile is indicative of very slow cooling somewhere between 1 and 10 °C per million years while steep profiles indicates a more rapid cooling rate of 100 °C or more per million years. The cooling rate is also an indication of the size of the parent body. The rocky silicate mantle around the cooling core provides an effective insulating blanket. Providing there is no other heat source, slower cooling rates imply larger parent bodies. Parent bodies with diameters between 200 and 400 km would have cooling rates that fall into the 1–10 °C per million year range. Large planetary bodies such as Earth have such slow cooling rates that their cores are still, in part, molten.

Stony-irons: the pallasites

It wasn't so long ago that meteoriticists divided meteorites into three primary classes: stones, irons, and stony-irons. This simple division quickly proved to be inadequate as important relationships among them began to appear. We have now seen how these classes have been subdivided into groups and subgroups. The stony-iron class, with meteorites composed of both silicate minerals and iron–nickel mix, were divided into four groups: *silicated irons*, which despite their silicate inclusions were never considered stony-irons and now fall into the iron class under the catagory, nonmagmatic irons; *lodranites* and *acapulcoites*, which contain ~20 wt% iron and are now classified as primitive achondrites; and *mesosiderites*, which contain iron–nickel and silicate minerals in roughly equal proportions that are now grouped with the achondrites since they show a close relationship to the basaltic achondrites. So the stony-irons as a major class of meteorites have been reduced to a single meteorite type, the *pallasites*. Pallasites are differentiated meteorites with an igneous origin and because they contain a large iron–nickel component, should rightfully be classified under the irons since their metal compositions and textures place them with the irons.

Texture

Pallasites are among the most beautiful of the meteorites. They are the result of an improbable mixture of two immiscible mineral components: almost pure olivine and a matrix of Fe–Ni alloys. Figure 9.20 shows cut examples in which the ratio of olivine to iron is roughly 2:1 by volume. But this ratio can vary enormously even within a single specimen. The Brenham, Kansas, pallasite is known to have sections of pure iron–nickel and no olivine at all, while other sections may show large areas of almost pure olivine. Generally, metallic Fe–Ni can vary from 28 to 88 wt%. Pallasitic texture is unique. If you could dissolve away all of the olivine from a hand specimen, what would be left would be a complete connected network of iron metal. The metal is primarily low-nickel kamacite with thin lamellae of taenite and plessitic mixtures of both alloys. Where the metal area is sufficiently large, a medium octahedrite Widmanstätten structure appears when etched.

The network encloses transparent yellow to yellow-green crystals of olivine. They may be single crystals ranging in size from a few millimeters to a centimeter or two or up to several centimeters for polycrystalline masses. Olive-green olivine so common in terrestrial basalts (Fig. 9.21) is relatively rare in pallasites. The crystals take on the shape of the metal enclosures, sometimes having rounded edges and other times angular borders. Often thin metal veins separate adjacent crystal faces sharing the same orientation. In almost all cases the olivine is extensively fractured, but in rare instances nearly perfect crystals may be found, from which gem-quality jewelry can (and has) been made.

Unfortunately, many pallasites have been subjected to severe weathering. The Brenham pallasite was buried for an undetermined period of time in a shallow crater, exposed to constant moist conditions. Typically, the outer few centimeters are severely altered, rusting the metal and converting the olivine to opaque iron oxides. Fragments of the Imilac pallasite found on the surface of the Chilean Atacama Desert have been reduced to small networks of iron with only traces of olivine. The largest intact pallasite in the world, the 1411 kg Huckitta pallasite from Northern Territory, Australia, was found in 1924 with 900 kg of iron shale surrounding a relatively small core. This ancient pallasite is mostly completely oxidized to hematite and the yellow olivine grains have been converted to opaque black masses.[VIII]

Composition and classification

The two main mineral phases, olivine and Fe–Ni, constitute about 95 vol% of pallasites. The olivine is magnesium-rich and the Fe–Ni has between 7 and 16 vol% Ni. Low-Ni kamacite and high-Ni taenite are intergrown although the taenite is usually too thin to be apparent without magnification. The remaining 5 vol% include three accessory minerals usually

[VIII] For more than a century, the Port Orford, Oregon, pallasite was considered the world's largest intact pallasite. A sample of the pallasite was taken from the main mass, reported to be lying half buried on "Bald Mountain" some 40 miles east of the coastal town of Port Orford. The find was reported by Dr John Evans who happened across it in 1858 during his government sponsored geological survey of the Oregon territory. From Evans's measurements of its size, a mass of over 10 tons was estimated. Evans died unexpectedly in 1861 and his report of the meteorite mysteriously disappeared from the US Government Printing Office. For 120 years numerous expeditions were made to the approximate location of the meteorite but all failed to relocate the great pallasite. Then, in 1986, Howard Plotkin, a scientific historian from the University of Western Ontario, Canada, in a brilliant piece of detective work, showed that the story was fabricated by Evans for personal financial reasons. Roy Clarke, Jr., of the Smithsonian Institution's Division of Meteorites and Vagn F. Buchwald of the Department of Metallurgy, Technical University, Lyngby, Denmark, analyzed Evans's fragment and found it to be identical to the Imilac pallasite. The infamous Port Orford meteorite is now considered a fraud (Plotkin and Clarke, 1993).[8]

Fig. 9.20. (Top) Pallasite from Esquel, Chubut, Argentina. It was found as a single mass of 1500 kg in 1951 and remained in the possession of the finder until 1992 when it was purchased by the American meteorite collector, Robert A. Haag. The specimen measures about 405 mm base dimension. (Photo by Michael Casper.) (Bottom) Marjalahti pallasite. This is one of only three witnessed pallasite falls. It fell in Viipuri, Karelia, USSR, in June, 1902. A single stone weighing 45 kg struck a granite outcrop and broke into many pieces. (Courtesy Darryl Pitt, Curator, The Macovich Collection.)

found in the metal bordering the olivine: troilite (2.3 vol%); schreibersite (1.2 vol%); and chromite (0.4 vol%). Two groups of pallasites have been recognized based upon their nickel vs. trace element content and their oxygen isotopic compositions. The largest is the *main-group* (MG) pallasites and includes almost all of the well known specimens. A small group called the *Eagle-Station Trio* (ES), is named after an 1880 find in Carroll County, Kentucky. The other two members come from Cold Bay, Alaska (found 1921), and Itzawisis from Namibia (found 1946). In terms of their mineralogy, the two groups are distinguished by the iron content of their olivine and the nickel content of their metal. The olivine in the MG is more magnesium rich (Fa_{11-19}) while the ES group is more iron rich (Fa_{20-21}), both with a very narrow range. Generally,

Fig. 9.21. Olive-green olivine on a terrestrial basalt. While common on Earth, especially in basalts, green olivine is rare in pallasites. The specimen is 9.5 cm widest dimension. (Photo by O. Richard Norton.)

the MG metal contains between 14 and 16 vol% Ni compared to 7.8–11.7 vol% for the ES group. Unfortunately, none of these characteristics can be used to visually distinguish them. They are best distinguished by the oxygen isotopic composition of their olivine. Figure 9.22 shows substantial differences in their oxygen isotopic ratios, with the MG pallasites tightly grouped along the terrestrial mass fractionation line in the IIIAB iron field. This suggests that they are closely related to the IIIAB core and their nickel-trace element ratios actually extends the IIIAB field (McSween, 1999).[9] The ES group plots along a much steeper slope in a field far removed from the MG group. These two subgroups must have formed in separate parent bodies.

A new pallasite "grouplet" has been recently recognized. The main mineralogical feature of these pallasites is their content of about 5 vol% orthopyroxene, compared to the MG pallasites with only a trace. These have been designated *pyroxene pallasites*. The pyroxene is usually found rimming the olivine crystals or as millimeter-sized inclusions within the larger olivine grains. Recognition of this new grouplet came with the discovery of a 27 kg find from central Kansas called Vermillion. This pallasite has 86 vol% Fe–Ni with only 14 vol% silicates. The metal shows a fine octahedrite Widmanstätten structure instead of the medium octahedrite texture of the MG pallasites. A second pyroxene pallasite find from Antarctica, labeled Yamato 8451, contains two low-Ca pyroxenes with different Ca content. O-isotopic data clearly resolve the pyroxene pallasites from the MG and ES groups plotting well removed from their fields (Fig. 9.22), demonstrating that these pallasites represent yet a third parent body.[10]

Formation of pallasites

Pallasites can be thought of as an immiscible emulsion, like oil and water. During differentiation, fractional crystallization should separate the two major minerals so that they crystallize separately and in different parts of an asteroid parent body. But yet, here they are mixed together to form the most beautiful of meteorites. How can this be? Olivine is an ultramafic mineral that forms and accumulates deep within an asteroid parent body. We can imagine it gathering as a cumulate in a layer at the core/mantle boundary, perhaps completely rimming the core (Fig. 9.23). To mix iron core material with pure olivine cumulate requires one of two processes. Either solid crystalline olivine must settle

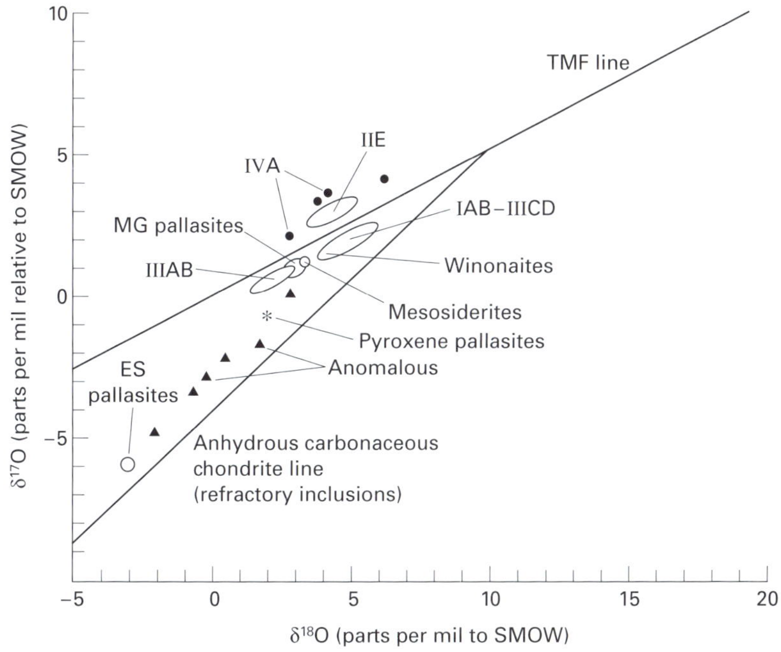

Fig. 9.22. Oxygen isotopic composition for silicated irons, pallasites, and mesosiderites. Comparing this graph with the oxygen isotopic compositions of the chondrites (Fig. 7.18) and achondrites (Fig. 8.15) shows apparent relationships with the silicate fractions of the irons and stony-iron meteorites. For example, the main group pallasites and the IIIAB irons , the mesosiderites, and the main group pallasites lie close to the HED field. The IIE and IVA irons occupy fields very close to the ordinary chondrite mass-fractionation line even though their mineralogy does not include olivine.

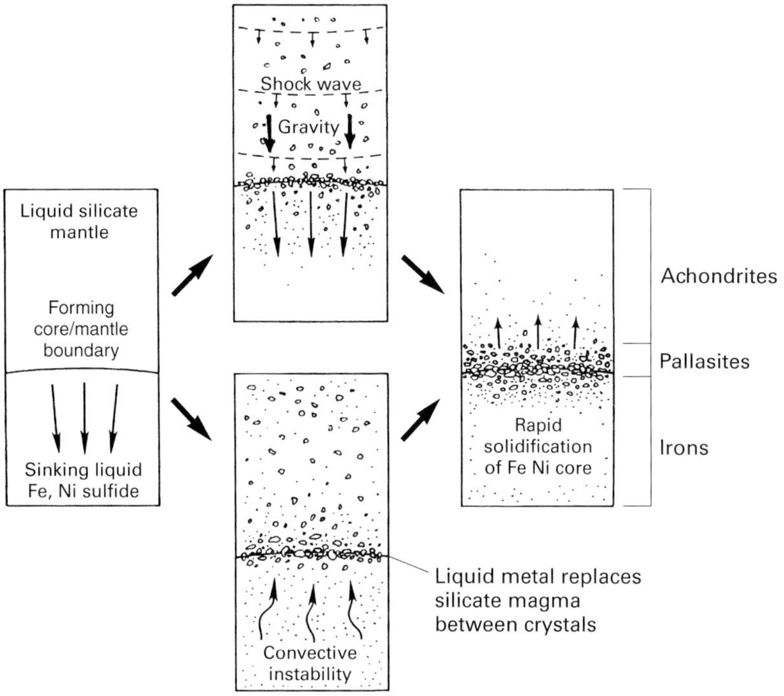

Fig. 9.23. Formation of pallasites. Olivine crystallizes out of the magma and settles onto the forming iron-nickel core. Gravitational compression along with occasional shock waves from impacts could temporarily force the olivine into the liquid core material. Convective instability within the core could force liquid iron into the olivine layer temporarily mixing the two minerals. In both cases, rapid solidification of the iron is necessary to maintain the mix.

into the still molten Fe–Ni core or liquid core metal must be forcibly injected into the olivine cumulate layer. In both cases, both silicate and metal have a strong tendency to quickly separate. It does not seem likely that gravitational compression alone would provide sufficient force to insert the olivine into the core. It could have been assisted, but briefly, by shock waves generated from surface impacts. Another possibility is that there may have been convective instability in the partially molten core that could drive the metal into the interstices of the olivine cumulate. Both processes would be short-lived and only temporary since separation would begin anew, which means that the core metal must have solidified quickly, trapping the olivine before separation commenced. This had to occur within the time constraints of a nearly solidified core. So pallasites formed after differentiation but before complete solidification of the core. If olivine formed a nearly pure cumulate around the core body, we would expect to have meteorite samples of nearly pure olivine from that part of the layer above the olivine/core mixing line. The brachinites may be such a candidate but one would expect them to be much more common than they are.

Metallic inclusions in iron meteorites

Whenever I cut an iron meteorite I always do so with a sense of anticipation. Unlike most chondrites that appear strikingly uniform in texture, with few if any significant inclusions, irons often have metallic minerals that sometimes display stunning textures, often forming beautiful patterns with other metallic minerals. They are always different; they make every iron meteorite unique. Fortunately, most metallic inclusions have easily identifiable characteristics that separate them from each other. Characteristics such as shape, luster, color, and relationships with other inclusions are easily observed and remembered.

Swathing kamacite

We have encountered kamacite as the primary Fe–Ni alloy in all iron and stony-iron meteorites. We have seen how it forms by iron exchange with taenite resulting in the growth of plates of varying widths on the cubic lattice of the taenite (Widmanstätten figures). Kamacite appears in another textural form in virtually all iron meteorites and is especially prevalent in IIAB irons. It is called *swathing kamacite* because it appears to wrap around inclusions of troilite, graphite, schreibersite, and silicates, totally enclosing them (see Fig. 9.9a). It can vary in width from microscopic (a few microns) to several millimeters and like Widmanstätten structure, depends upon the nickel content and cooling rate. This is the first kamacite to form from taenite solid state transformation. Inclusions act as nuclei for kamacite growth. Not all inclusions are surrounded by kamacite, however. This first-formed kamacite is randomly formed without taenite lamellae and with no apparent relationship structurally with the Widmanstätten plates that form thereafter; the Widmanstätten plates terminate against the swathing kamacite.

Troilite (FeS)

Troilite is the most common accessory mineral and is ubiquitous in iron meteorites, pallasites and in chondrites. In irons it is found as randomly distributed subspherical nodules often surrounded by other accessory minerals. Troilite has a characteristic bronze color (Fig. 9.24), it has a hardness of 4 on the Mohs Hardness Scale being easily scratched with a knife, and it has a specific gravity of 4.6. Its terrestrial counterpart is *pyrrhotite*. Pyrrhotite differs from troilite in that it has a deficiency of iron at its octahedral crystal lattice sites with respect to its sulfur. This is implied in its chemical formula, $Fe_{1-x}S$, with (x) between 0 and 0.2. Troilite occurs in meteorites with an abundance of iron (Fe^{+2}) so that its structural sites are completely filled with iron and its chemical formula, FeS, describes its true chemical composition. Pyrrhotite is magnetic but with varying intensity. Oddly, it increases in magnetism as the deficiency in iron increases. Under natural conditions within meteorites, troilite is nonmagnetic; but if it is melted and cooled, it becomes magnetic. Troilite is easily dissolved in a 10% aqueous solution of nitric

Fig. 9.24. Bronze-colored troilite nodules in a IIICD medium octahedrite from Nantan, Guangxi, China. The troilite nodule to the far right is lined with silvery cohenite. Silicate material is mixed with troilite in the irregular-shaped dark nodule below center. The inclusion to the lower left is surrounded by black-appearing swathing kamacite. The horizontal dimension is ~10 cm. (Specimen prepared by Bill Mason, Uncommon Conglomerates of Wisconson. Photo by O. Richard Norton.)

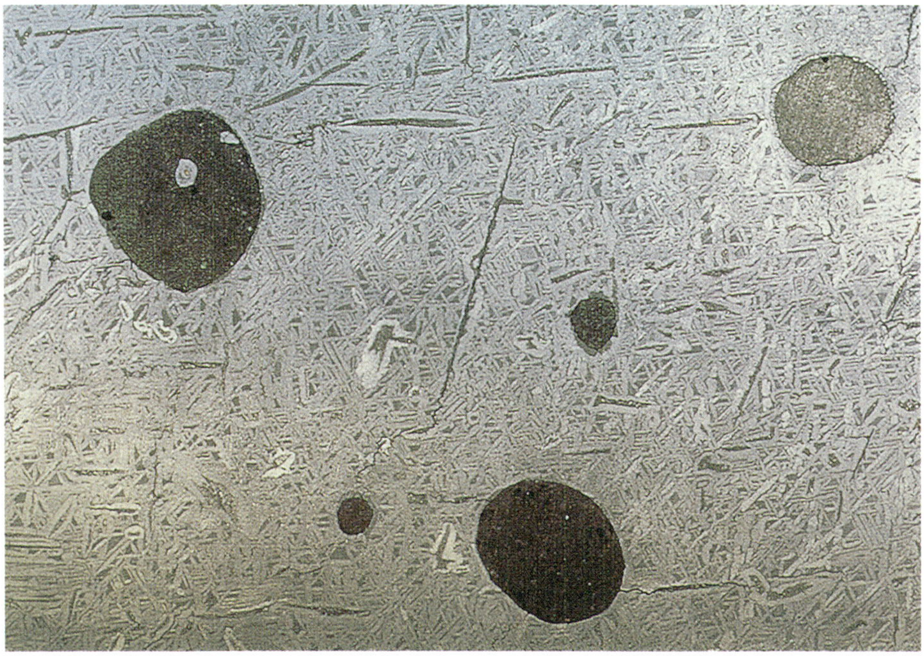

Fig. 9.25. Long straight needles of schreibersite encircled by kamacite permeate this IIIAB medium octahedrite (0.7 mm) from Tambo Quemado, Ayacucho, Peru. This meteorite has exceptionally large troilite nodules surrounded by thin shells of schreibersite. The largest (upper left) measures 40 mm across. The horizontal dimension is 207 mm. (Robert A. Haag collection. Photo by O. Richard Norton.)

acid, a typical etching solution, leaving an undesirable brown stain. Meteorites with troilite inclusions should be etched with nitric acid mixed in 100% ethanol (see Appendix D.)

Troilite often is seen in intimate association with small inclusions of other minerals. It can be immersed in a graphite nodule or sometimes shares a cavity with a graphite nodule. In octahedrites it may form thin plates that lie parallel to the Neumann lines within the kamacite plates. These were first noted by the German mineralogist Karl Ludwig von Reichenbach in 1861 and confirmed in 1880 by Brezina, who named them *Reichenbach lamellae*. The best examples of troilite in the nodular habit may be seen in the Cape York, Greenland, IIIAB iron and the Mundrabilla, Western Australia, anomalous iron (Fig. 9.12).

Schreibersite ($(Fe,Ni,Co)_3P$)

Schreibersite is a phosphide of iron–nickel. Like troilite it is a "cosmic" mineral since it occurs very commonly in iron meteorites and pallasites but not in terrestrial native iron (except in very very rare occurrences like Disko Island, Greenland), which in itself is rare. It is a magnetic mineral, brittle and granular-appearing in texture, shiny in luster, silvery-white in color when freshly cut but tarnishing to a bright yellow-bronze tone as it oxidizes. It is much harder than troilite, with a hardness of 6.5 on the Mohs Hardness Scale. The Canyon Diablo IAB iron has a considerable admixture of schreibersite. A typical elemental composition of schreibersite for Canyon Diablo is: Fe, 58.54 wt%; Ni, 26.08 wt%; P, 15.37 wt%; Co, 0.05 wt%.

Berzelius was the first to describe schreibersite in 1832. Charles Upham Shepard (1804–1886) also described a mineral he had isolated in 1846 as a "sesquisulphuret of chromium" which was a brown and black striated mineral. This he named *schreibersite* in honor of Carl Franz Anton von Schreibers (1775–1852). Schreibersite, however, does not conform to this description, raising suspicions that Shepard was not studying what we call schreibersite today. He did isolate another mineral that had a similar appearance which he labeled *dyslytite*. Other workers (Adolphe Patera and William Karl Haidinger) isolated yet another residue in 1847 that they also named schreibersite. Confusion reigned supreme while the chemists continued to add to the confusion. Reichenbach found a silvery-white substance he called *lamprite*. Then, Gustav Rose in 1863 described schreibersite and showed it to be Reichenbach's lamprite and further described another similar mineral he termed *rhabdite*. Rhabdite typically appears as nearly microscopic silvery thin needles. Finally, Tschermak, studying the Braunau iron, concluded that schreibersite and rhabdite were the same mineral with the rhabdite showing slender rhombohedral prisms.[IX]

Schreibersite's growth habit varies considerably within a given meteorite. It often surrounds troilite and graphite nodules as silvery shells that appear as a ragged border in cross section (Fig. 9.25). It can also appear as subround inclusions in kamacite plates or as elongated thin plates. Long, impressively straight needles are also fairly common. The needles and inclusions always appear granular but a microscope shows tiny flat crystal faces. Growth habits of schreibersite as very fine tetragonal prisms or plates oriented parallel to the Neumann lines (parallel to the hexahedral faces) within the kamacite is

[IX] A detailed description of the confusion early workers had concerning the discovery and naming of schreibersite is present in John G. Burke's masterpiece, *Cosmic Debris* (1986). University of California Press, Chapter 4, pp. 135–138.

Fig. 9.26 A graphite nodule 20 mm in diameter with a wedge of included troilite in a IAB coarse octahedrite (2.0 mm) Canyon Diablo, Arizona, iron meteorite. The graphite is rimmed by coarse-grained schreibersite with finer-grained cohenite over the schreibersite. (Photo by O. Richard Norton.)

the form called rhabdite. They seem to be chemically associated with the kamacite and may be exsolution lamellae.

Cohenite ($(Fe,Ni,Co)_3C$)

Cohenite is an iron–nickel carbide and is often associated with schreibersite. Unfortunately, cohenite and schreibersite appear so similar that they are sometimes difficult to distinguish. Under the microscope cohenite and schreibersite are more easily distinguished. Schreibersite is isotropic and under reflected light with the analyzer (upper polarizing filter) in place it does not vary in shade but appears equally bright from all directions. Cohenite is anisotropic and reflects light preferentially so that in reflected light with the analyzer in place cohenite can be seen to vary significantly in brightness. Cohenite is brittle with a hardness of 6, silvery-white when fresh, and like schreibersite is magnetic. Cohenite was first described in 1889 by Ernst Weinschenk who named it after Emil William Cohen (1842–1905). The mineral seems to be constrained to iron meteorites with 7 wt% nickel or less. Thus, it is found most commonly in coarse octahedrites and hexahedrites. Cohenite is found as plates oriented to the octahedral plates of coarse octahedrites. In Canyon Diablo irons cohenite has an elemental composition of: Fe, 91.31 wt%; Ni, 1.77 wt%; Co, 0.25 wt%; C, 6.67 wt%;.

Cohenite has similar growth habits to schreibersite and often appears in physical contact with it. H.H. Nininger describes a nondestructive chemical test first attributed to O.C. Farrington that can be performed to distinguish the two minerals.[11] A slightly acidic solution of copper sulfate is brushed across a freshly prepared surface. The kamacite and cohenite will react with the solution, reducing the copper to its elemental form and coating both minerals. Scheibersite does not reduce the copper and remains uncoated. Using this chemical test on Canyon Diablo irons, Nininger found that cohenite predominates over schreibersite by a ratio of five to one. It is good to keep in mind that, unlike schreibersite, cohenite usually does not occur in iron meteorites having over 7 wt% nickel.

Graphite (C)

Four polymorphs of carbon are found in meteorites: graphite; diamond; lonsdaleite; and carbon. In iron meteorites graphite is the most common form. It is usually found as subspherical inclusions called *graphite nodules*. These nodules are often intergrown with troilite, and sometimes cohenite and schreibersite surround the nodule (Fig. 9.26) making a very attractive combination. Graphite nodules can be impressively large, reaching several centimeters across. They are easily weathered from iron meteorites. Gibeon and Canyon Diablo irons often show large spherical cavities where graphite nodules once resided (Fig. 9.27). Several graphite nodules the size of baseballs have been found buried around Meteor Crater. Running through these nodules is a reticulum of shocked metal veins (Fig. 9.28). Graphite is soft, with a hardness of only 1–1.5. It is a member of the hexagonal crystal system and usually crystallizes in tabular crystals in hexagonal outline. It has a metallic luster and is greasy to the touch. It has perfect cleavage in one direction and is sometimes found as plates or foliated masses but it is rarely found as hexagonal crystals in meteorites. It is more often found in an irregular form.

Graphite is the low-pressure form of carbon in meteorites. Its high-pressure state is diamond. Diamond has an isometric crystal form, usually octahedral but sometimes cubic or dodecahedral. Although some of these crystal forms have been found in graphite nodules, many are hexagonal

Fig. 9.27. A 14 kg Canyon Diablo iron meteorite with two large cavities, once housing graphite nodules. They probably weathered out of their cavities in the terrestrial environment. One has weathered through, producing a hole. (Courtesy Darryl Pitt, Macovich Collection.)

Fig. 9.28. A large graphite nodule found buried 20 cm deep within two miles of the rim of Meteor Crater. The interior shows veins of shocked metal, probably the result of an impact while in space. The specimen is ~12.5 cm across. (Photo by O. Richard Norton.)

polymorphs of diamond. Diamonds were first found in Canyon Diablo irons by the mineral dealer, A.E. Foote in 1891, long before their associated crater was recognized as a meteorite crater. They are usually found in irregular carbon or in graphite nodules. The diamonds are very small, usually not obvious to the eye. More often they are found during cutting operations in which they provide strong resistance to the saw. In all cases they are found in carbonaceous inclusions furnishing a soft matrix that allows them to be excised from the meteorite. Usually the diamonds appear black, but clear, white, and yellow diamonds have also been found. Most require a microscope to see their form, which is usually irregular. They range in size from about 0.1 mm to, in rare instances, over 1 mm diameter. H.H. Nininger, who probably cut more Canyon Diablo irons than anyone else, found statistically one diamond for every 10 cm^3 of metal, making Canyon Diablo irons one of the richest sources of extraterrestrial diamonds. Other diamond meteorite sources are the two irons: the IAB iron ALH 77283; and in the Chuckwalla iron. Diamonds are also found in ureilites and as SiC interstellar dust grains in stony meteorites. The origin of meteoritic diamonds is generally considered to be impact shock, either by impact in space or impact on Earth. The conversion of graphite to the hexagonal form of diamond requires shock pressures of at least 700 kbar and temperatures well above 1000 °C. In 1967, scientists at the General Electric Company produced the first hexagonal polymorph of carbon that had characteristics very close to natural diamond. Almost simultaneously, the first hexagonal diamond was recognized in a Canyon Diablo iron. Both are given the name *lonsdaleite*. The majority of the diamonds in Canyon Diablo are black, opaque and have no cleavage. These are termed *carbonado*.

References

1. Wasson, J. T. (1975). *Meteorites: Classification and Properties*, Springer-Verlag, New York, Chapter 6, p. 73.
2. Burke, J.G. (1986), *Cosmic Debris: Meteorites in History*, University of California Press, Los Angeles.
3. Romig, A.D., Jr., and Goldstein, J.I. (1979). Determination of the Fe–Ni and Fe–Ni–P phase diagrams at low temperatures (700–300 °C). *Metallurgical Transactions Acta* **IIA**, 1151–1159.
4. Goldberg, E., Uchiyama, A. and Brown, H. (1951). The distribution of nickel, cobalt, gallium, palladium, and gold in iron meteorites. *Geochimica Cosmochimica Acta* **2**, 1–25.
5. Lovering, J.F., Nichiporuk, W., Chodos, A. and Brown, H. (1957). The distribution of gallium, germanium, cobalt, and copper in iron and stony-iron meteorites in relation to nickel content and structure. *Geochimica Cosmochimica Acta* **11**, 263–278.
6. Choi, B., Ouyang, X. and Wasson, J.T. (1995). Classification and origin of IAB and IIICD iron meteorites. *Geochimica Cosmochimica Acta* **59**, 593–612.
7. Rubin, A.E. *et al.* (1985). Properties of the Guin ungrouped iron meteorite: the origin of Guin and of group IIE irons. *Lunar and Planetary Science Letters* **76**, 209–226.
8. Plotkin, H. and Clarke, R.S. (1993). The Port Orford, Oregon, meteorite mystery. *Smithsonian Contributions to the Earth Sciences*, No. 31.
9. McSween, H.Y., Jr. (1999). *Meteorites and Their Parent Planets*, Cambridge University Press, Cambridge.
10. Boesenberg, J.S. *et al.* (1995). Pyroxene pallasites: a new pallasite grouplet. *Meteoritics* **30**, No. 5, 488.
11. Nininger, H.H. (1952). *Out of the Sky: An Introduction To Meteoritics*, Dover Publications, Inc., New York, p. 94.

Chapter Ten

Meteorites and the early Solar System

Meteoritical science, with its modern analytical techniques borrowed from many other scientific disciplines, holds the promise of revealing the origin and early history of our Solar System. Unlike their cousins, the astronomers, who can look back in time to the early universe imaged from great distances and see for themselves what the universe was actually like billions of years ago, meteoritical scientists must surmise the conditions that prevailed in the solar nebula through more indirect but perhaps equally effective methods. Astronomers have, however, crossed paths with meteoritical science. Using great astronomical instruments like the Hubble Space Telescope, they have imaged dust disks surrounding other stars in the Orion Nebula 1600 light years away that are undoubtedly close analogs to our own Solar System as it must have been billions of years ago. As enticing as these images are (Fig. 10.1), they are no substitute for pieces of the "real" thing.

Meteorite researchers are very fortunate. They have in their laboratories ancient rocks from the early Solar System that can be studied with every kind of analytical tool at their command. Over 50 years ago when researchers first began to apply methods of isotopic dating to meteorites, it became clear that the chondritic meteorites were very primitive; their constituent mineral assemblages dated among the first formed out of the solar nebula. To date these mineral assemblages is, in essence, to write the history of the early Solar System. The hope of the meteoriticist is that dating methods will extend our samples back to the "beginning" with sufficient resolution, so that the primary characteristics in our meteorite samples can be placed in chronological order.

Throughout this book we have alluded to meteorites as primitive, meaning that they formed very early in the evolution of Solar System bodies. From what point do we measure the age of meteorites or the age of the parent bodies that spawned them? It's like asking from what point to date the

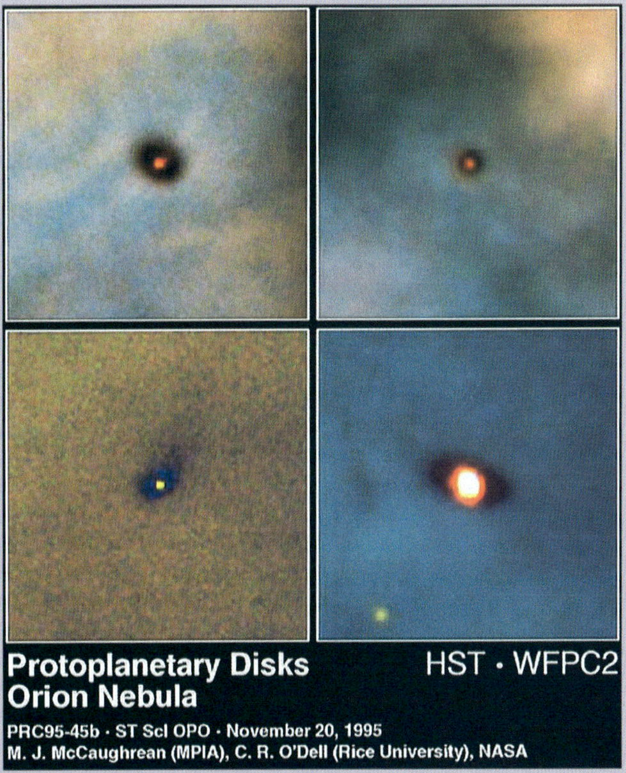

Fig. 10.1. Four dusty protoplanetary disks discovered in 1995 around young stars in the Orion Nebula located 1600 light years away, photographed by the Hubble Space Telescope. The red glow in the center of each is a young, newly formed star. The dark disks range in size from two to eight times the diameter of our Solar System. Though the disks are circular, they appear elliptical since each is tilted toward Earth at different angles. These disks represent an early stage of planetary formation. (Courtesy Mark McCaughrean, Max Planck Institute for Astronomy; C. Robert O'Dell, Rice University; and NASA.)

age of a human being. From physical birth? From conception? From the formation of the first zygote cells? From the maturation of precursor haploid or gamete cells? We could continue almost indefinitely but the fact is, our choice of a starting point is arbitrary and not the same in all societies. Likewise, where do we begin counting the age of meteorites? From the metamorphism and differentiation of chondrite parent bodies? From the accretion of their precursor chondritic parent bodies? From the formation of the constituents that make up the primitive parent bodies, the chondrules? The chondrules have had a varied and complex history in themselves so there is a history of chondrules further back in time. From the precursor relict grains that formed before the chondrules? From the CAIs, some of the first condensates of the solar nebula? We could go back further yet to the formation of the cocoon-like stellar incubator that was destined to give birth to the proto-Sun long before solid bodies of any kind existed. Or, for that matter, we could go back to the formation of the refractory elements deep within the thermonuclear furnaces in hot giant stars or from the explosion of a nearby supernova that some believe polluted the pre-solar nebula with short-lived radioisotopes. As it turns out, researchers are interested in all of these "starting" points for together they write the history of the Solar System.

Radioisotopes – some basics

In Chapter 7 we briefly discussed the concept of isotopes of elements and saw that nearly every element is represented by more than one isotope, that is, each element has several members with the same atomic number defined by the number of protons in the nucleus but each differs by the number of neutrons in the nucleus.

Some of these isotopes are not stable but spontaneously emit particles from the nucleus that transforms them into a different element. One way this is done is by a neutron disintegrating into a proton, a beta (β^-) particle (a particle identical in mass and charge to an electron) and an antineutrino (ν^-). This process, called *beta decay*, is the most common mechanism of radioisotope decay and usually involves nuclides with neutron excesses. For example, the radioisotope rubidium-87, the *parent* isotope, spontaneously emits a beta particle changing a neutron into a proton, thus converting the rubidium isotope (atomic number 37) into the next higher atomic number element, strontium (38), its *daughter* isotope. Thus:

$$^{87}\mathrm{Rb} \Rightarrow {}^{87}\mathrm{Sr} + \beta^- + \nu^-$$

Note the mass number remains the same but there is a net gain of one proton with a concomitant loss of a neutron.

Another example of elemental transmutation has been encountered briefly several times in this book: $^{26}\mathrm{Al} \Rightarrow {}^{26}\mathrm{Mg}$. Aluminum has 13 protons, magnesium has 12. Here, instead of gaining a proton, a proton has been lost and a neutron gained. This occurs when an electron near the aluminum-26 nucleus is captured by the nucleus converting a proton into a neutron. This process of *electron capture* is essentially the reverse of beta emission but it is rarely seen and is not involved in radioisotopes used in the absolute dating of meteorites. It may have been very important in the first few million years of the Solar System's formation but, because the half-life of aluminum-26 is only 720 000 years, all vestiges of it have long ago become extinct. We will see later that the short half-life of aluminum-26 presents an opportunity to study the earliest formed minerals (CAIs) and relate them to chondrules.

One additional decay process, called *alpha* (α) *decay*, involves the emission of an alpha particle composed of two protons and two neutrons. An α-particle is essentially the nucleus of a helium atom. Here, the mass number decreases by four and the atomic number decreases by two. This type of decay is especially important in isotopic decay of heavy elements such as uranium-238.

Of the naturally occurring elements there are 339 isotopes of which 70 are unstable or *radioactive*. All isotopes of elements beyond atomic number 82 (lead) are naturally radioactive while only a few of the naturally occurring elements that have atomic numbers less than 83 are radioactive. Interestingly, two elements within this list (technetium-43 and promethium-61) have no known stable isotopes.

When we look at the forces involved within the atom, we wonder how nuclides stay together at all. There is a strong repulsive force among the protons of the nucleus because they have like charge. Just why some isotopes are radioactive is no longer a mystery. Since most stable nuclides tend to stay together, it seems clear that there must be another force, an attractive force stronger than the electric force at the nuclear level, that acts to hold the nucleons (protons and neutrons) together. This is the *nuclear force*. It operates only over very short distances. It is very strong between nucleons if they are closer than about 10^{-15} m but it quickly drops to zero if the particles are separated by greater than this distance. One curious characteristic of the nuclear force is that it is sensitive to the ratio of protons to neutrons in the nucleus, that is, only certain ratios of protons and neutrons will yield stable nuclides. If a nuclide contains too many or too few neutrons relative to the number of protons, the nuclear force is weakened. As the electric repulsive force increases with increasing atomic number (more protons) a greater number of neutrons are required to maintain stability. For example, the first 20 elements with stable nuclides have proton to neutron ratios of 1:1. Above element 20 (calcium), the ratio gradually increases to about 1.5:1 for element 82, lead. For the largest atomic numbers (bismuth to uranium) no increase in neutrons can overcome the increased electric repulsion produced by greater numbers of protons. Thus, beyond element 82, the nuclides spontaneously adjust themselves by emitting from or absorbing particles to the nucleus.

Decay constant and half-life

Radionuclides can be used as geologic clocks because they disintegrate or decay at specific and constant rates. Because radioactive decay is related only to the nucleus of the atom, this rate is fixed for any given isotope and is not affected by any changes in chemical or physical conditions (pressure, temperature, chemical reactions) in the radionuclide's environment. Each radionuclide of a given element has its own decay rate. This rate is expressed in terms of the number of atoms, (n), that decay over a period of time relative to the total number of atoms (N) in a given sample. This ratio, n/N, is the *decay constant* (λ); it is the fraction of the radionuclide that disintegrates per unit time. Since the total number of radionuclides (N) is constantly decreasing in a given sample, the number of atoms that decay over a given successive time interval must also decrease by the same proportion.

Another way of expressing the decay of a radionuclide is in terms of its *half-life* ($t_{1/2}$). The half-life is the time required for exactly one half of a specific amount of a radionuclide to decay. It is related to the decay constant (λ) by an inverse proportion; that is, the longer the half-life of a radionuclide, the slower the decay rate. The mathematical relationship between half-life and decay constant is:

$$t_{1/2} = 0.693/\lambda.$$

To illustrate the concept of half-life let us assume a hypothetical radionuclide has a half-life of eight years. We begin with 32 g of this radionuclide. In Table 10.1 we see six columns showing the decay of the parent radionuclide as well as the accumulation of the daughter isotope over five half-lives.

Note that the amount of parent radioisotope in grams is reduced by half for each half-life. At the same time, the daughter isotope produced by the decaying parent accumulates at the same rate. Thus, the sum of the parent + daughter isotopes at any time is equal to the original amount of the parent isotope. After five half-lives, the remaining parent isotope has been reduced to 1 g. If we plot the time in years against the parent mass remaining and the daughter mass accumulating after each half-life, we obtain the two curves in Fig. 10.2, showing the decrease in parent isotope and a corresponding increase in the daughter isotope. Both curves follow an exponential path: the parent follows one of exponential decay and is asymptotic to the time axis and the daughter isotope follows a curve of exponential growth and is asymptotic to the line $m = 32$. What this means is that for very large numbers of atoms encountered in nature, the decay is a never-ending process and that the parent isotope never completely disappears. There are limits of detectability, however. Half-lives of radioisotopes range from fractions of a second to 49 billion years. Radioisotopes with short half-lives measured in months or a few years decay below detectable limits after a dozen or so half-lives have passed. Isotopes such as carbon-14 with a half-life of 5700 years is useful for dating events back to about 50 000 years, beyond which the isotope becomes essentially extinct. Aluminum-26, with its half-life of 720 000 years, became extinct before the Solar

Half-lives	0	1	2	3	4	5
Lapsed time (years)	0	8	16	24	32	40
Remaining parent (g)	32	16	8	4	2	1
Accumulating daughter (g)	0	16	24	28	30	31

Table 10.1. **This table illustrates the concept of half-life**

A hypothetical isotope has a half-life of eight years and goes through five half-lives. The starting mass of the parent isotope is 32 g. As the parent decays, the daughter isotope accumulates at the same rate. Thus, the sum of the parent plus the daughter isotopes is always equal to the original mass.

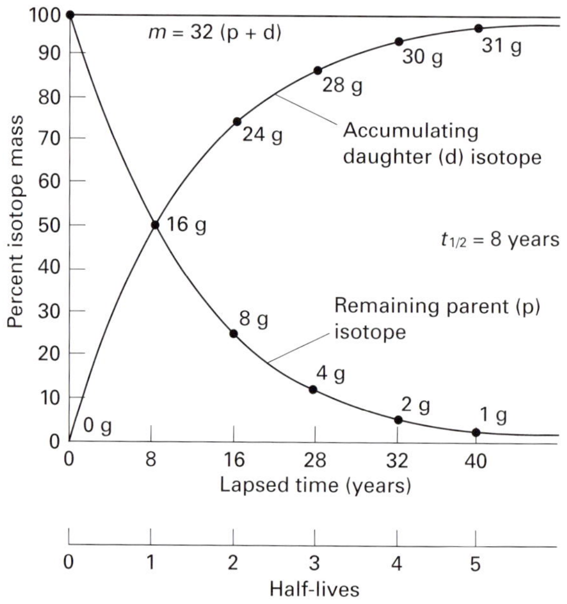

Fig. 10.2. Graph shows the decay of the parent isotope and the corresponding accumulation of the daughter isotope beginning with 32 g of parent isotope with a half-life of eight years. The decay and accumulation curves are exponential curves and are asymptotic to the horizonal axes of the graph. The mass of the parent + daughter isotope is always equal to the original mass (32 g) of the parent isotope.

Parent	Stable daughter	Decay constant ($year^{-1}$)	Half-life (years)	Primary use
*^{26}Al	^{26}Mg	9.62×10^{-7}	7.20×10^{5}	Earliest minerals (CAIs)
*^{129}I	^{129}Xe	4.08×10^{-8}	1.70×10^{7}	Formation intervals
^{40}K	^{40}Ar	5.81×10^{-11}	1.25×10^{9}	Gas retention age
^{87}Rb	^{87}Sr	1.42×10^{-11}	4.88×10^{10}	Formation age
^{232}Th	$^{208}Pb + He$	4.95×10^{-11}	1.40×10^{10}	Formation age
^{235}U	$^{207}Pb + He$	9.84×10^{-10}	7.04×10^{8}	Formation age
^{238}U	$^{206}Pb + He$	1.55×10^{-10}	4.47×10^{9}	Formation age
^{147}Sm	^{143}Nd	6.54×10^{-12}	1.06×10^{11}	Formation age

Note: *These two radionuclides are extinct today because of their relatively short half-lives. They can, however, indicate the relatively short interval of time since the nucleosynthesis of the elements and their incorporation into the first highly refractory minerals condensing from the solar nebula.

Table 10.2. **Radioisotopes used in dating meteorites**

A number of radioisotopes are used to determine the formation age of meteorites. Such radionuclides must have half-lives that are suitable for measuring ages in billions of years; that is, their half-lives must be a large fraction of the formation age of the meteorites so that their remaining parent isotope does not fall below the limits of detectability. This table shows six radioisotopes used as clocks to date meteorites. Of these, the rubidium/strontium clock is perhaps the most used since it is less subject to resetting due to impact heating. Also, rubidium/strontium has the longest half-life of the six, making it most suitable for dating objects billions of years old. The radioisotopes aluminum-26 and iodine-129 have very short half-lives but are suitable for measuring short time intervals (the first 10 million years) in the early Solar System.

System formed completely. We know of its existence because of its daughter isotope magnesium-26 locked into the crystal structure of the early-formed minerals. (Aluminum-26 is produced during supernovae explosions and, more gently, by cosmic rays interacting with meteoroids in interplanetary space, i.e. it is not entirely extinct.) Clearly, then, only the radioisotopes with the longest half-lives are suitable for determining the ages of meteorites. Moreover, there must be measurable amounts of both parent and daughter isotopes. Fortunately, virtually every meteorite contains small amounts of long-lived radioisotopes ideal for age determination.

There are six long-lived radionuclides used as radioisotopic clocks. Table 10.2 lists these with their decay constants, half-lives, and parent and daughter isotopes.

Rubidium/strontium method

Just how radioisotopes are used to determine the age of a meteorite can be seen using the decay of rubidium-87 to strontium-87. The principles examined here are applicable to other systems in use today. The rubidium/strontium method is a favorite of meteoriticists. Notice that several of the radioisotopes listed in Table 10.2 have gases as daughter products. The accumulation of these gases can be used to date the meteorite but if the system did not remain closed, that is, if the parent body was heated by impact shock after crystallization, it may lose some of its accumulated gases, effectively reducing the accuracy of the method. Rubidium and strontium tend to remain as trace elements in the meteorite in that they do not easily form compounds. They are firmly locked within the crystal structures of the minerals so there is little chance that either rubidium or strontium will be lost during metamorphism or impact heating.

We saw earlier that rubidium-87 decays to strontium-87 through simple beta decay with a half-life of about 49 billion years. As rubidium decreases, its daughter product strontium-87 increases by the same amount. One would think that knowing the half-life or decay constant of rubidium-87 and measuring the amount of the daughter product would lead directly to the time it took to accumulate the daughter product. The problem with this simple accumulation clock is that there may have been an initial amount of strontium-87 which, if not taken into account would lead to a calculated age that is too great. The problem of determining initial strontium-87 is a formidable one but, happily, there is a method that does not require knowledge of initial strontium-87. (However, quantitative information about the initial amount of daughter isotope can be determined by this method.)

If $^{87}Rb_0$ represents the initial amount of rubidium-87, then the amount of ^{87}Rb present at time t is given by the exponential decay formula

$$^{87}Rb = {}^{87}Rb_0\, e^{-\lambda t}.$$

Also, if $^{87}Sr_0$ represents the initial amount of strontium-87, then

$$^{87}Sr = {}^{87}Sr_0 + ({}^{87}Rb_0 - {}^{87}Rb).$$

Substituting into this formula the preceding expression for $^{87}Rb_0$ gives

$$\begin{aligned} ^{87}Sr &= {}^{87}Sr_0 + ({}^{87}Rb e^{\lambda t} - {}^{87}Rb) \\ &= {}^{87}Sr_0 + {}^{87}Rb\,(e^{\lambda t} - 1). \end{aligned}$$

Dividing both sides by the quantity ^{86}Sr which remains constant with time, yields

$$^{87}Sr/^{86}Sr = (e^{\lambda t} - 1)\ (^{87}Rb/^{86}Sr) + (^{87}Sr_0/^{86}Sr).$$

This equation is of the form:

$$Y = MX + B,$$

a straight line, which in this case is known as an *isochron*, since any values lying on the line represent the same age (time) t.

In practice, at least two whole-rock samples of the meteorite are analyzed in a mass spectrometer to determine the ratios $Y = {}^{87}Sr/^{86}Sr$ and $X = {}^{87}Rb/^{86}Sr$. The points (X,Y) are then plotted on a graph so that the straight line connecting them can be drawn (Fig. 10.3). The slope of the line, $M = \Delta Y/\Delta X$ can then be obtained, and since $M = e^{\lambda t} - 1$, the time t (formation age) is readily solved for since λ is a constant.

The Y-intercept where the isochron intersects the Y-axis is the value of the unmeasurable initial quantity $^{87}Sr_0/^{86}Sr$.

We can confirm the value of the half-life of rubidium-87 by the following sample calculation. Suppose the initial amount of rubidium-87 is $^{87}Rb_0 = 1\ \mu g$ and let us take an elapsed time of $t = 49 \times 10^9$ years. Then, given the decay formula $^{87}Rb = {}^{87}Rb_0 e^{-\lambda t}$, where $\lambda = 1.42 \times 10^{-11}$ (Table 10.2),

$$\begin{aligned} ^{87}Rb &= (1\ \mu g)\ e^{-(1.42 \times 10^{-11})(49 \times 10^9)} \\ &= (1\ \mu g)\ e^{-0.6958} \\ &= 0.4987\ \mu g. \end{aligned}$$

This is one-half of the original quantity of rubidium-87, showing that $t_{1/2} = 49 \times 10^9$ years (Table 10.2).

Figure 10.3 gives an example of an actual isochron diagram constructed using data from $^{87}Rb/^{86}Sr - {}^{87}Sr/^{86}Sr$ measurements for 38 whole rock chondrites. All of the chondrites were undisturbed and they all fell on the isochron to within the accuracy of the data. The slope of the isochron gives an age of 4.498 ± 0.015 Ga. The Y-axis intercept gives the initial $^{87}Sr_0/^{86}Sr$ value of 0.698 85 ± 0.000 10. This study was conducted by J.F. Minster and coworkers at the University of Paris (1982).[1]

Fig. 10.3. $^{87}Rb/^{86}Sr - {}^{87}Sr/^{86}Sr$ whole-rock isochron for 38 H, LL, and E chondrites from a study by J.F. Minster, J.L. Birck, and C.J. Allègre of the University of Paris (1982). All 38 chondrites give nearly identical ages (they all fall on the isochron) so that they are treated as a single isotopic system with a common age of 4.498 ± 0.015 Gy. Adapted from Minster *et al.* (1982).[1]

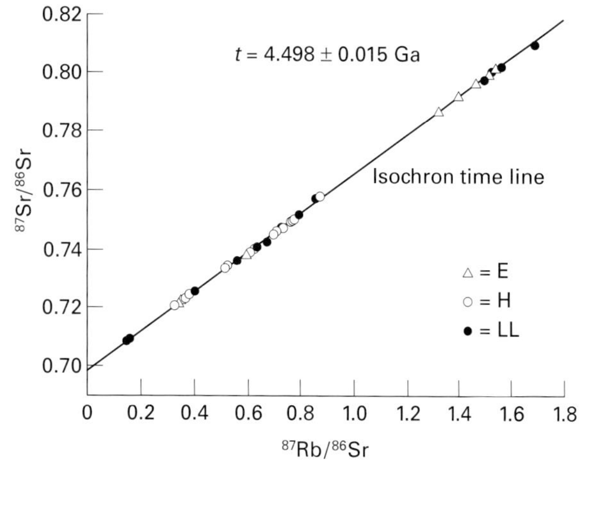

Frequently, *whole rock* samples rather than individual mineral grains are measured using the Rb–Sr method. Whole rock samples can be defined as samples sharing the same genetic characteristics, i.e. the same chondrite group. Sometimes the entire rock specimen has remained a closed undisturbed system whereas individual minerals have not. In such a case, the whole rock measurements will yield formation ages, and the data points will lie on an isochron. The disturbed minerals will give discordant ages that do not lie on the isochron. Usually both sample types are taken which will reveal both concordant and discordant elements. Some care must be taken as whole rock isochrons tend to obscure small differences in the samples' history that may prove to be important.

Types of meteorite ages

At the beginning of this chapter, I implied that there are many meteorite ages used in meteoritics. All of them date specific events in the history of the meteorite. The foregoing chapters bear witness to a complex history of the meteorites. Most meteorites show that their parent bodies have undergone considerable change since their origin. A few, such as the petrographic Type 3 chondrites, are pristine, showing very little change since their formation. These are the meteorites that are used to determine the earliest accretionary period when these bodies were forming in the primitive solar nebula. Others show heating, melting and differentiation that must postdate the formation of chondritic parent bodies. Still others show one or more episodic histories of impact revealed in their brecciated and impact melt textures. The breakup of parent bodies release smaller fragments into space that are irradiated by cosmic rays that produce certain isotopic signatures giving us an estimate of the period between breakup and arrival at Earth.

Below is a list of possible episodes in a meteorite's history. There may be others. These are not all conclusive in that some meteorites may have undergone some of these episodes but not all of them. Others have undergone only two or three. Moreover, one can easily invent other episodes that may have occurred during brief unusual or extraordinary events that may have happened in certain regions of the solar nebula affecting some parent bodies but not others.

1. Formation of earliest precursor materials.
2. Formation of chondrules and matrix material.
3. Accretion of chondritic parent bodies.
4. Thermal metamorphism and differentiation.
5. Impact melting, brecciation.
6. Lapsed time since parent body fragmentation and liberation of meteorites.
7. Terrestrial aging or time since arrival on Earth.

As we peruse the list, it is evident that we are considering great lapses of time in some cases and relatively short lapses of time in others. Episodes 1–3 go back billions of years with episode 4 following close behind episode 3, probably occurring within a few million years after accretion. Episode 5 can occur several times over the life of the parent body. Episode 6 on the average lasts a few million years, and episode 7 typically a few thousands of years. These wide variations of lapsed time require the use of specific isotopes that best fits the period of time for each event. Thus, for the oldest age, the formation of chondrules and accretion of chondritic parent bodies, requires the use of very long half-life isotopes. Furthermore, the primary isotopes must be of sufficient abundance to be detectable after billions of years and the half-life must be such that a sufficient amount of daughter isotopes accumulate over that period to be detectable with current methods. With this list in mind, we now look at the long history of the meteorites specifically as clues to the early history of the Solar System. Without the early record within these relicts of the past it would be difficult if not impossible to write this history.

In the beginning . . .

It is a great complex journey from a molecular cloud to a one solar-mass star with orbiting planets. Yet we know such a journey has been made at least once. Our own Solar System attests to that. Until quite recently the existence of other solar systems was a mere statistical and philosophical speculation. But researchers now concede that this journey has been made countless times throughout the Milky Way Galaxy – and it still goes on today. The first indications of newly born stars came through infrared and 1 mm radio wavelength observations of molecular clouds in the Orion region and elsewhere within the Milky Way. Infrared light penetrating the dense dust-laden clouds revealed hidden stars with characteristics of extreme youth, stars not over a few million years old at most. In the mid-1970s the first visible light image of a dust disk encircling a star was made using an optical telescope at the Las Campañas Observatory in the Chilean Andes. The star, Beta Pictoris, located in the southern constellation of Pictor, is some 15 parsecs (1 parsec = 3.26 light years) from the Sun. The image showed a near edge-on dust disk measuring roughly 1000 AU across. This is ~10 times the diameter of Pluto's orbit (Fig. 10.4). Then in 1983, the first infrared astronomy satellite, IRAS, was launched into Earth orbit and almost immediately detected infrared excesses around many nearby stars. Planck spectral energy distribution curves of stars of one to a few solar masses show that the infrared energy coming from these stars is far greater than expected for their masses and temperatures. The only plausible answer was that these stars were enveloped in a shroud of dust-like material that absorbed visible and ultraviolet energy coming from the stars and reradiated this energy in the infrared.

The decade of the 1990s brought important astronomical discoveries applying directly to problems of the origin of the Solar System. The discovery of the first extra-solar planetary bodies orbiting nearby stars was made in 1996, and removed any remaining doubt that planetary systems populated the Galaxy. Our Solar System was not alone nor was it unique.

Fig. 10.4. Beta Pictoris was the first star discovered with an encircling dust disk. The upper image in visible light shows the dust disk nearly edge-on to our view. The particles are ices and silicate grains shining by reflected light from the central star, blocked out here. The disk diameter is ~10 times the diameter of the Solar System. The lower image in false colors show details of the disk. The white and pink zones of the inner edge of the disk is tilted slightly with respect to the red–yellow–green outer disk suggesting gravitational perturbations from one or more planets in the clear zone (obstructed). Both images were made with the Hubble Space Telescope. (Photo courtesy of Chris Burrows, Space Telescope Science Institute (STScI), the European Space Agency, G. Krist (STScI), the Wide Field Planetary Camera team, and NASA.)

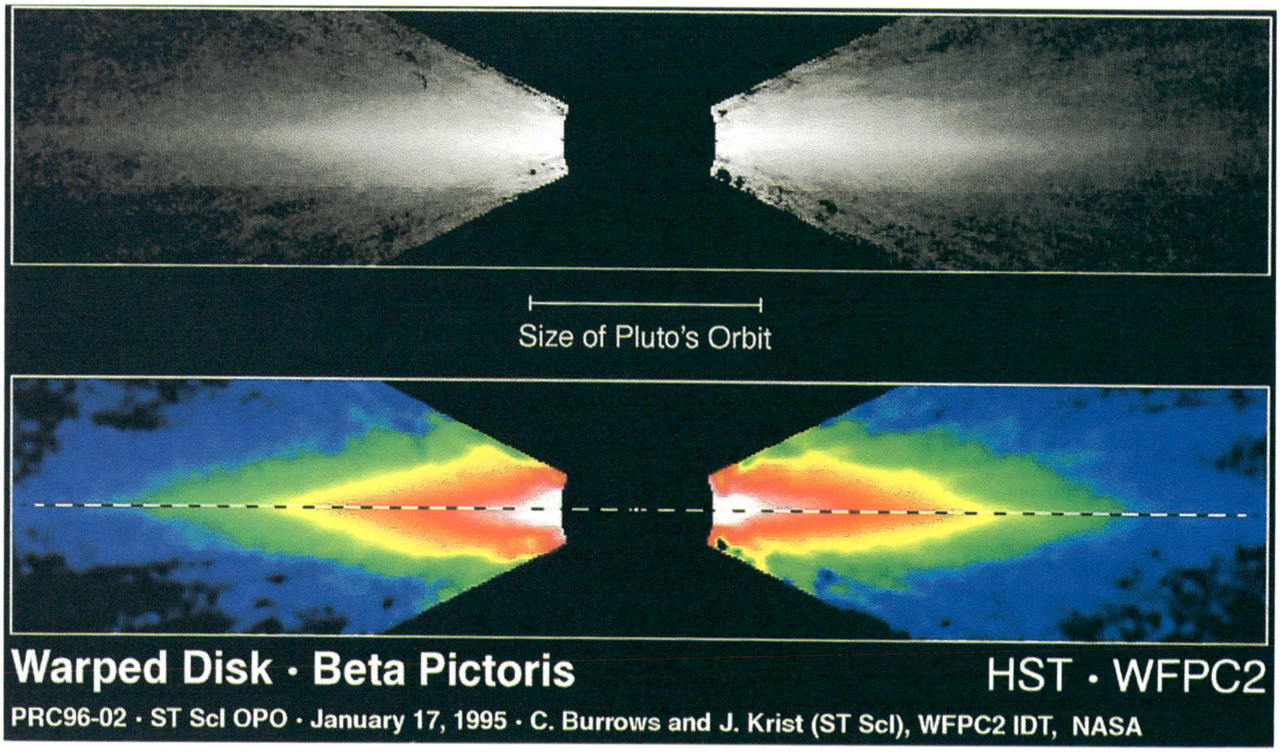

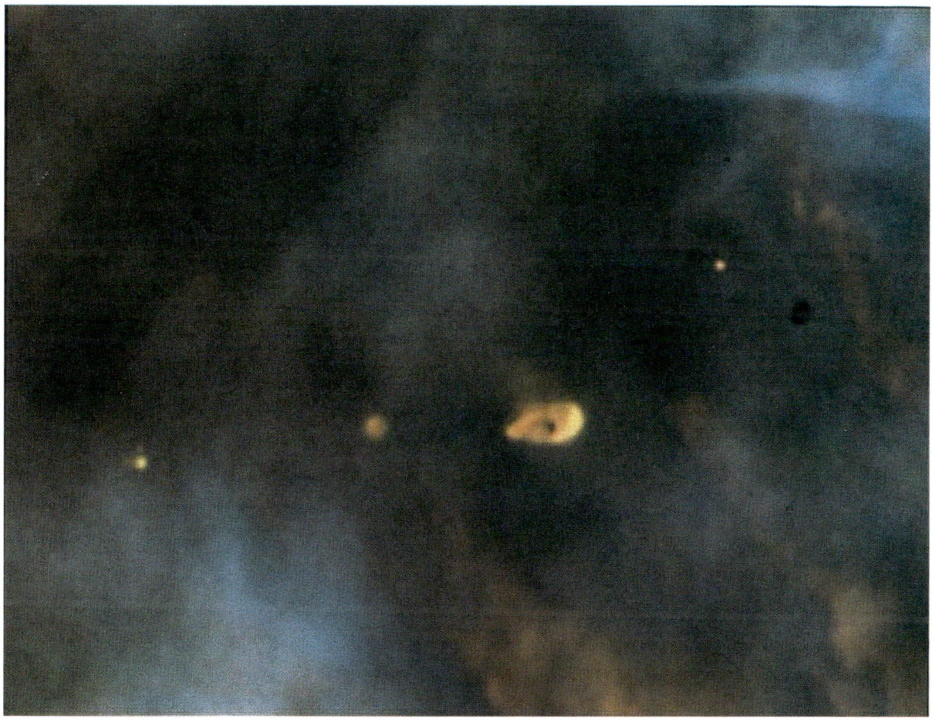

Fig. 10.5. Protoplanetary disks (proplyds) in the Orion Nebula. This Hubble Space Telescope image shows a small area within the Orion Nebula revealing five young stars all with surrounding gas and dust disks: four bright objects, one black object. The black areas are the accretion disks encircling the stars. The brightest object shows its disk nearly edge-on. This disk has a diameter of about 100 AU. The field of view is only 0.14 light years across. These are a few of the more than 150 proplyds known to exist in the Orion Nebula complex. (Photo courtesy C. R. O'Dell, Rice University and NASA.)

These newly-discovered planets were gas giants the size of Jupiter or larger. Later on in the decade, other extra-solar planets were found including planets approaching Earth in size. These planets were not directly imaged but were detected by small but measurable wobbling motions in the stars' proper motions (motion across our line of sight) caused by gravitational perturbations by the accompanying planets as both star and planets moved around their common center of mass. This first discovery was followed quickly by others so that the number of known extra-solar planets has now reached 33. In the same decade, the Hubble Space Telescope imaged in visible and infrared light dust disks around stars in the Orion Nebula. More than 150 protoplanetary disks were imaged in the central part of the Orion Nebula alone (Fig. 10.5). The Hubble Telescope gave us our first high-resolution images of a protoplanetary disk with its central star hidden by dense dust but revealed in infrared images (Fig. 10.6). Now meteoriticists and cosmologists had the best of both worlds. They had images of developing planetary systems in various states of evolution and they had bits and pieces of the first solid material to develop in our own Solar System that could be analyzed and dated. It seemed that we were finally ready to begin writing the story of our origin.

The solar nebula

Scattered along the plane of the Milky Way Galaxy are massive clouds of gas and dust. They cover huge expanses of interstellar space, typically 10–50 parsecs across. They are dark – there is no light coming from them – and cold, typically only about 10 K. Their presence becomes known visually as black silhouettes when they lie in front of bright diffuse hydrogen clouds set aglow by ultraviolet radiation from nearby hot stars (Fig. 10.7). These are *dark molecular clouds* composed of hydrogen molecules, helium, organic molecules and dust. The dust component is composed of heavier elements such as iron, carbon, silicon carbide and various silicate minerals forming cores or nuclei around which water-ice and other ices form. These heavier components are not native to the molecular cloud but are contaminants from nearby massive stars that manufacture heavier elements and scatter them as tiny grains (10^{-5} cm) through strong stellar winds. The mean density of the clouds is about 10^4 molecules/cm^3 and a typical mass often exceeds 10^5 solar masses. It is here within these great dark masses that the Solar System had its beginnings.

Within a molecular cloud are inhomogeneities in the form of clumps or cores with much higher mean densities. A typical core contains between 1 and 10 solar masses of gas with densities of as much as 10^6 molecules/cm^3. Gravitational instabilities within the primary molecular cloud result in a collapse of the entire cloud along with collapse of the cloud cores. The cause of the initial collapse is unknown. Perhaps shock waves from a nearby exploding supernova compressed the cloud, tearing it into fragments and compressing the fragments into cores where the densities reach high enough values to trigger star formation. Confining magnetic fields within these cloud cores tend to resist collapse, but much of the gas and dust is uncharged so they are able to breach the magnetic field as sufficiently massive cores exert gravitational pull, dragging the gas and dust across the field lines and into the cores' centers.

The mass of a typical molecular cloud is so great that it could potentially spawn hundreds or thousands of solar-mass

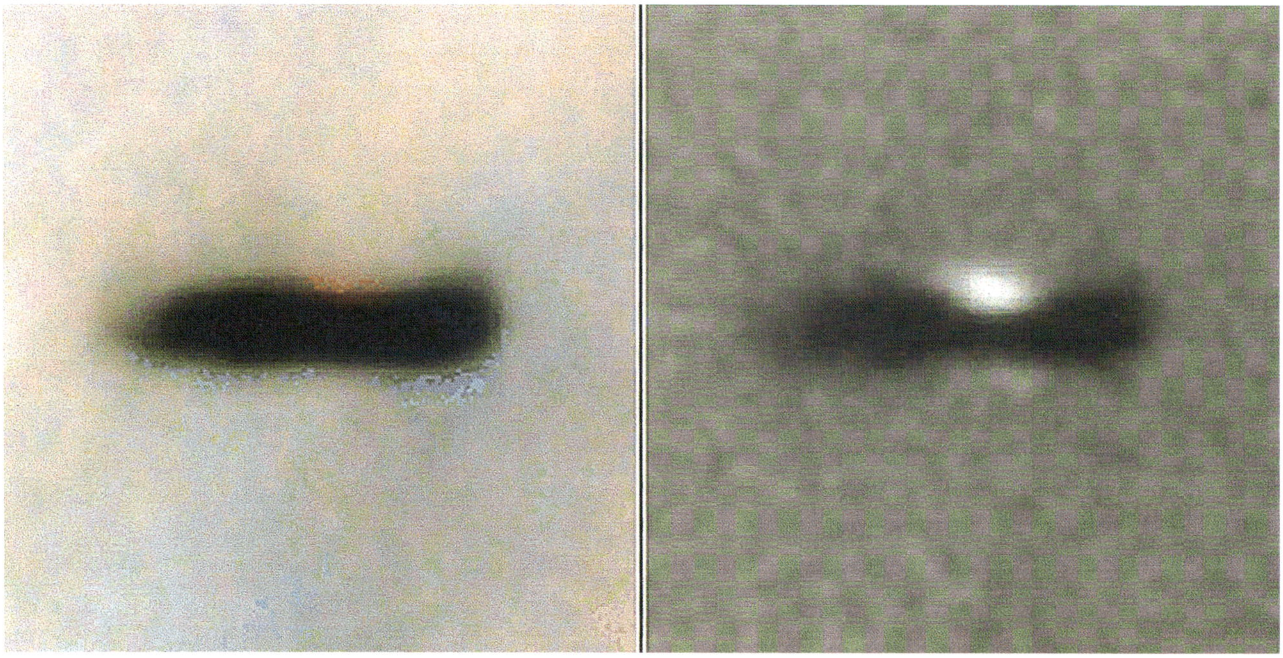

Fig. 10.6. High-resolution image made by the Hubble Space Telescope of an edge-on protoplanetary disk located in the Orion Nebula. The true-color image on the left shows an opaque dusty disk 17 times the diameter of our Solar System. The newly formed central star is hidden by the dust. The image on the right, made near the infrared, has partially penetrated the dust revealing a bright glow of light from the central hot star scattered on the surrounding nebulosity. (Courtesy Mark McCaughrean, Max Planck Institute for Astronomy; C. Robert O'Dell, Rice University; and NASA.)

stars. The Sun must have been among many other stars that formed from the same molecular cloud. Individual cloud cores, however, could form individual stars of solar mass. Yet we know that most stars of the Galaxy are not single stars but binaries or multiple star systems or members of stellar associations such as galactic star clusters. Just why the Sun is in the minority as a single star today is not known. If the Sun was born a member of a galactic (open) cluster with a population of a few hundred stars and a diameter of several tens of

Fig. 10.7. The Horsehead Nebula in Orion. This is part of a large, dark molecular cloud made visible because of red hydrogen emission from hydrogen gas behind the nebula. Most of the material in the Horsehead is composed of silicate grains coated by ices. A young, hot star shines through the dust in the lower left. Molecular clouds such as this are the birthplaces of stars. (Copyright Association of Universities for Research in Astronomy Inc. (AURA). All rights reserved.)

Fig. 10.8. Open star cluster, Messier 52 (NGC 7654), in Cassiopeia. This rich cluster is composed of hundreds of young, blue-hot stars at a distance of about 3000 light years.

parsecs, it is possible that differential galactic rotation may have created tidal forces that tore them apart after a few million years. Another possibility is that the cluster was gravitationally disrupted by a passing molecular cloud. The effects would be the same, to scatter the cluster members. The cluster stars including the Sun, now released from the gravitational bonds that held them to the cluster, went their separate ways as binaries or multiple systems or as single stars. Open star clusters are young, most only tens of millions of years old (Fig. 10.8). The youngest clusters, yet to be disrupted by galactic tidal forces, contain considerable gas and dust and an abundance of heavy elements (any element beyond helium is considered a heavy element and called a metal in astronomers' jargon.) The heavy elements in particular were manufactured by earlier generations of massive stars that ejected them into the molecular cloud from which the Sun formed.

Returning again to the solar nebula, a central condensation, the protoSun, begins to form as collapse of the core continues, building up mass and density and heating the protoSun as gravitational energy is converted to thermal energy. Once this *free-fall contraction* begins, it proceeds rapidly and it isn't long before the protoSun begins to radiate weakly in the infrared. A spherical gas cloud with an initial uniform density of $\sim 10^3$ particles/cm^3 collapsing radially toward the core center has a free-fall time that depends upon the rate of change of the density. Assuming no other influences, the free-fall time lasts about 10^6 years.[2] The core, as it existed in the original molecular cloud, must have possessed a slow rotation. As the collapse continues and the radius decreases, both the protoSun and the surrounding core material must rotate faster to conserve angular momentum within the closed system. But the rotating core material soon possesses too much angular momentum to collapse directly onto the protoSun but instead begins to flow around it forming a thick disk (Fig. 10.9). Now the disk carries most of the angular momentum of the system. The disk for a one solar mass star forms within the first 100 000 years. The Sun, still a forming protostar but now surrounded by a developing disk, is buried deep within the original cloud core. At this point the core is referred to as a *cocoon* because it effectively hides the forming star from direct view. After this period of 100 000 years, the protoSun begins to show signs of instability. It becomes a *T-Tauri* star. These stars are named after the prototype star, T-Tauri, found in the constellation of Taurus. T-Tauri stars are chaotic variable stars still in the protostar

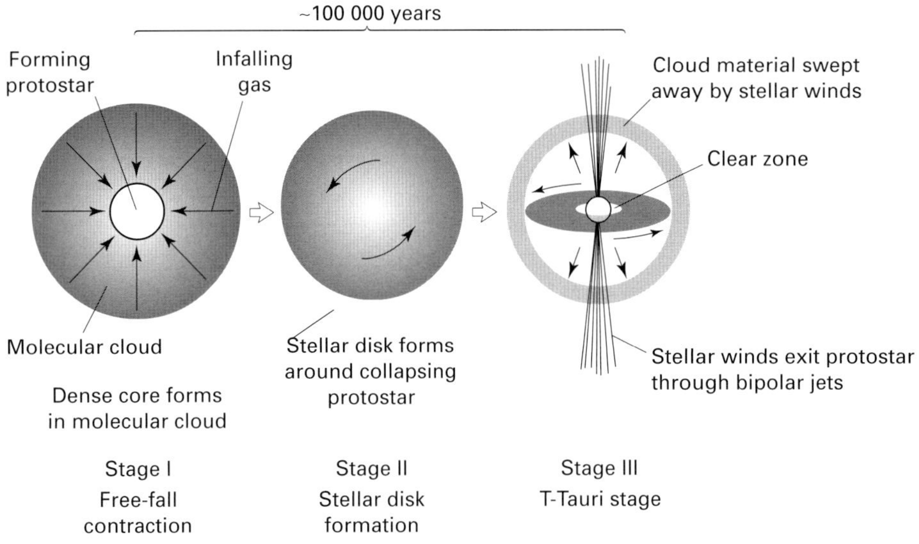

Fig. 10.9. Stages in the collapse of a cloud core within a molecular cloud and the formation of a circumstellar disk and protoSun. See text for details.

stage that show unpredictable outbursts of energy. They are very young stars that are just beginning to emerge from their cocoons. Apparently all stars with masses between 0.75 and 3 solar masses pass through the T-Tauri stage during their pre-main sequence period.[1] This period of instability occurs before thermonuclear reactions commence in their centers. As the accretion disk grows denser and the protostar hotter, the protostar begins to produce a strong stellar wind in which mass loss can reach as much as 10^{-8} solar mass per year. At the same time, its luminosity becomes quite variable and unpredictable. The winds blow from all over the star but the thick disk prevents the winds from escaping along the disk plane. Instead, most of the wind particles are directed through the rotational axis of the protostar, forming bipolar jets or outflows. The jets involve matter falling onto the innermost region of the disk about 0.1 AU from the protostar, which eventually clears the area of matter. Although there is little concensus about the cause of this bipolar outflow of gas and dust there is no question that it happens. Using the Hubble Space Telescope, astronomers have found numerous outflow jets involving accretion disks and T-Tauri stars. In particular, in 1997, the Hubble Telescope photographed an extraordinary protoplanetary disk located in a molecular cloud on the Taurus–Auriga border. Here, a protostar in its T-Tauri stage is still obscured by the thick accretion disk but bipolar jets are easily seen extending through both rotational poles at right angles to the disk (Fig. 10.10). Besides losing a significant amount of mass during the bipolar outflows, the material may be coupled to the magnetic field of the protostar which, when lost in the outflows, acts to reduce the angular momentum of the protostar and at the same time increases the angular momentum of the outer parts of the disk. T-Tauri stars have much slower rotational velocities than one would expect of a collapsing one solar mass body. The Sun's slow rotation today must be a result of the loss of angular momentum in its formation years. Today the planets contain 98% of the Solar System's total angular momentum.

The first-condensed refractory minerals

As soon as an accretion disk settles around the evolving protoSun, a temperature gradient forms. The infall of gas and dust from the surrounding cocoon not only heated the protoSun during the first 100 000 years but also the inner parts of the accretion disk. The temperature in the inner part of the disk (the terrestrial zone where the four inner planets currently reside), reached or possibly exceeded 2000 K, so that all elements were essentially in the gaseous state. The vast majority of the gaseous material in the disk was hydrogen and helium. The remaining material, only 0.001 as abundant as hydrogen, was composed of heavy elements. As small as this sounds, heavy elements were relatively abundant in the solar nebula. These elements were manufactured through thermonuclear processes within the cores of giant stars and released as strong winds or through supernovae. The abundance of heavy elements in the Sun suggests that the Sun is a third-generation star, that is, it was formed in a solar nebula contaminated with heavy elements ejected into space by two other generations of stars.

As the disk began to cool in the terrestrial zone, the first highly refractory elements and minerals began to condense,

[1] The *main sequence* period begins when a star's core temperature has reached a sufficiently high value to begin and to sustain thermonuclear reactions. This is a period of great stability and the longest lasting stage in the evolution of a star. This period begins when hydrostatic equilibrium is achieved; when the weight of the overlying layers of a star is balanced by the opposing gas pressure that supports the layers. This state begins when thermonuclear reactions commence in the star's core with core temperatures of about 10^7 K. Stars more massive than the Sun reach the main sequence within a million years or so. One solar mass stars take as long as 30 million years to evolve to the main sequence. The Sun will remain on the main sequence as long as core hydrogen burning continues, a period of nearly 10 billion years.

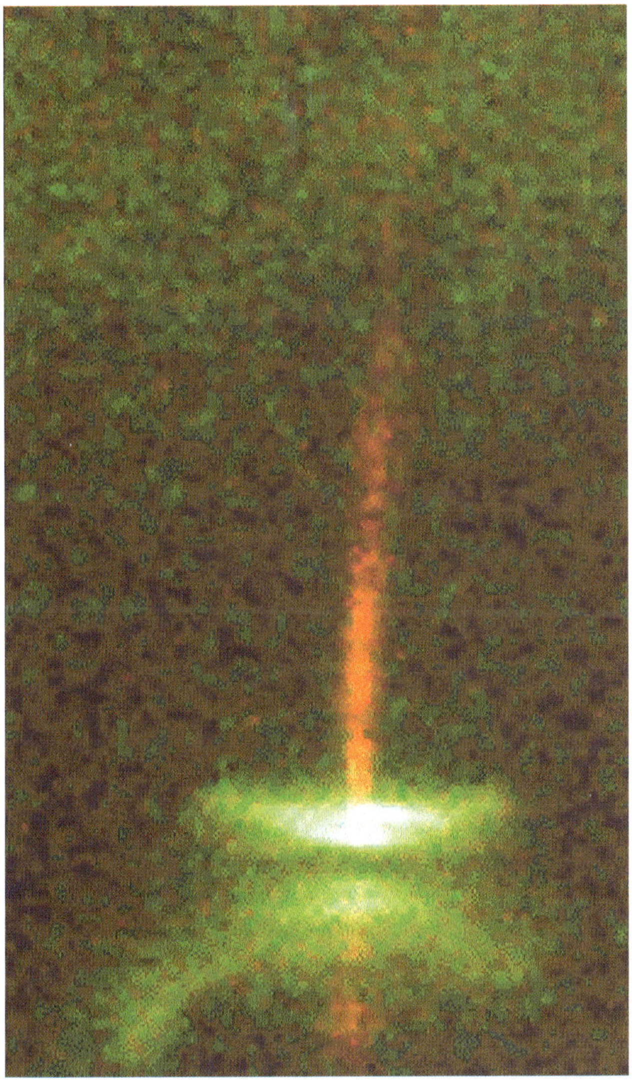

Fig. 10.10. This Hubble Space Telescope image in unprecedented detail reveals an edge-on disk of gas and dust encircling a newly born star in its T-Tauri stage. The star remains hidden in the disk but it lights the top and bottom of the disk. A bipolar jet (red) is seen escaping along the rotational axis of the star at right angles to the disk. (Courtesy Space Telescope Institute, Johns Hopkins University, and NASA.)

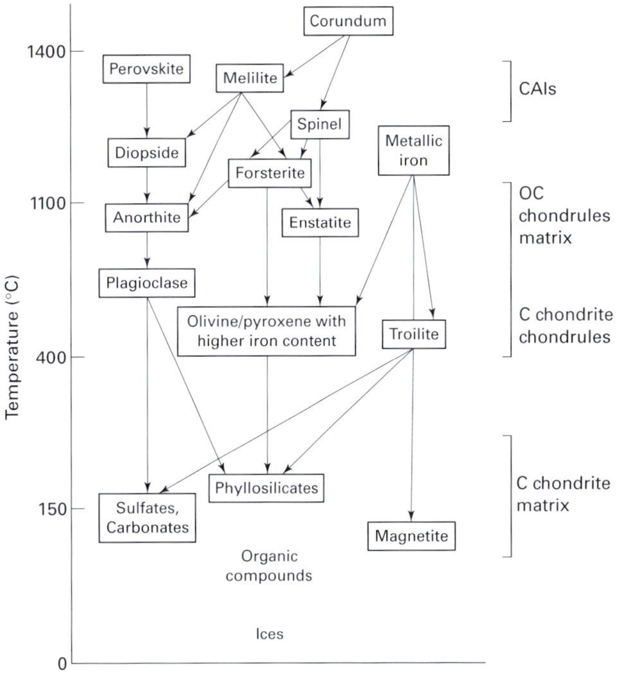

Fig. 10.11. Condensation sequence in the early solar nebula. This chart is a theoretical construction showing the order in which minerals would condense out of a nebula of solar composition in gradually cooling conditions. The highest-temperature minerals, the oxides, probably condensed out of the original nebular gas. These first minerals continued to react with residual gases to produced further minerals at lower temperatures. The reaction pathways are noted by arrows. The sequence can be divided into three major segments, all involved in chondritic meteorites: first generation minerals as oxides condensing at the highest temperatures appear as CAIs eventually to be incorporated into the most primitive chondrites; the second generation minerals produced at lower temperatures make up the chondrules; the third generation minerals, formed at lower temperatures still, make up the matrices of C chondrite meteorites. (Adapted from L. Grossman, University of Chicago.)

changing directly from a gas to a solid. Thermodynamic studies are used to predict the condensation sequence one would expect to find in the cooling accretion disk. This is a natural fractionation process that involves various elements from the solar nebula in a specific sequence. Figure 10.11 gives a general picture of the sequence and relates it to the first formed minerals found in meteorites. These are the precursor minerals that were processed in the solar nebula to form chondrules and CAIs. The first minerals were formed in the presence of oxygen producing oxides of Al, Ca, Ti, and Mg. We find melilite, spinel, perovskite and corundum in the most primitive chondrites. In the hot disk, these first minerals continued to react with the gas to produce a second generation of minerals such as diopside, anorthite, forsterite, and enstatite, all iron-poor minerals and all found in chondrites. Iron begins to oxidize and enters the sequence at about 1200 K, to produce FeO-rich minerals such as olivine and pyroxene, both major components of chondritic meteorites. Along with these precursor minerals were tiny grains of interstellar dust less than 1 μm in size containing carbide nitrides and diamonds, both older than the Solar System. They could only come from the interiors of giant stars. These mixed with the first-formed minerals that eventually found their way into chondritic bodies.

The earliest-formed meteoritic inclusions are thought to be the calcium–aluminum inclusions or CAIs. These are found primarily in carbonaceous chondrites with small amounts found in ordinary and enstatite chondrites. Several of the early-formed minerals are aluminum-rich and magnesium-poor but anorthite or calcium feldspar is the mineral most often cited to have contained the now extinct radionuclide, ^{26}Al. If you recall, ^{26}Al has the shortest half-life of any of the radionuclides used for dating meteorites, a mere 720 000 years. Its daughter product, stable ^{26}Mg, is found replacing the aluminum in some of the anorthite grains in

CAIs. As early as 1955, Harold Urey predicted the presence of ^{26}Al in the early solar nebula and a corresponding ^{26}Mg isotopic anomaly in chondrites, but the quantities were so minuscle that analytic techniques at that time were not sensitive enough to detect the ^{26}Mg excess.

The abundance of the aluminum radionuclide in the solar nebula is stated by the ratio of ^{26}Al to its stable nuclide, ^{27}Al or $^{26}Al/^{27}Al = 5 \times 10^{-5}$. This means that for every ^{26}Al radionuclide incorporated in a feldspar crystal lattice there are a half million ^{27}Al stable nuclides. This appears to be a miniscule amount, but because of its widespread distribution throughout the region where CAIs formed and because of its rapid decay, it is thought by most researchers that this amount is sufficient to actually melt and differentiate chondritic parent bodies over the lifetime of the radionuclide. The fact that chondritic parent bodies exist today constrains the original abundance of ^{26}Al. It could not have been too high; otherwise there would not be any chondrites left!

The importance of ^{26}Al is that its short half-life allows us to look back to a time when the first solid bodies were forming, a period of time that could not have lasted over a few million years. Thus, ^{26}Al could not have been created too long before being trapped within meteorites. This time between the creation of the radionuclide and its incorporation into the parent body of the meteorite is called the *formation interval* and establishes the earliest interval of Solar System history. Most researchers believe that the lifetime of the solar nebula is only about 10 million years. Aluminum-26 would become extinct long before the end of this period.

Formation of the first chondrules

Of all the issues being studied in meteoritics today, the origin of chondrules, their heating mechanisms and their period of formation are by far the most vigorously debated but also the most enigmatic. Chondrules have been studied for two centuries using every means available at the time to understand their mineralogy, their structures, and their history. Today, there is still no concensus among researchers as to how they formed or when and where they formed. There have been two important conferences dealing with the chondrule question: the first at the Lunar and Planetary Institute in Houston in 1982 with the title, *Chondrules and Their Origins*;[3] the second in 1994 at the Institute of Meteoritics, University of New Mexico, with a more encompassing title, *Chondrules and the Protoplanetary Disk*.[4] The participants' list reads like a Who's Who of meteorite researchers, with people representing universities and laboratories worldwide. This was the "chondrule club" extraordinaire! Those fortunate enough to attend the conferences witnessed an extraordinary display of the way science works. These were the scientists whose task it was to tell us about our origins, a formidable responsibility indeed. The task that lay ahead for these researchers was to try to find consensus for the origin of the solar nebula, the protoplanetary disk, CAIs and chondrules. In particular, the origin of CAIs is about nothing less than the origin of the Solar System's first solid bodies, which carries with it strong implications for the origin of everything else in the Solar System.

We have spent considerable time looking at the structure of meteorites and for good reason. It is from these ancient bodies that the processes operating in the early Solar System will come to light. How were chondrules created in the solar nebula? What came before chondrules? Let's look at the second question first. We saw that in a cooling solar nebula, certain precursor minerals condensed out of the solar nebula while others were created by reactions with these first-formed materials. Virtually all of the mineral components of meteorites existed as tiny dust grains settling toward the midplane of the accretion disk. Mixed with these were interstellar grains we mentioned earlier. Imagine these grains forming loose aggregates of tiny solid crystals no larger than a micron or two in diameter, perhaps originally held together by ices or other volatiles collecting in the interstices. The consistency of these fluffy "dust balls" (as chondrule people like to call them) must have been something like a snowflake, which makes a good analogy. Snowflakes are composed of a tangle of six-sided ice crystals. The next time you are in your car in a snow storm, stop at the side of the road (as long as it is safe to do so!), turn off the windshield wipers and watch the snowflakes as they strike the window. They sit there for a second or two and then suddenly flash out of existence, turning into tiny spherical water droplets as the heat passing through the windshield is absorbed by the snowflakes. Likewise, as the dustballs were settling toward the midplane they must have suddenly experienced a "flash" heating event of short duration. Temperatures experienced by these aggregates must have approached their liquidus point, somewhere between 1500 K and 2000 K. Many of these dustballs must have experienced complete melting. Others along the disk experienced only partial melting, especially the more coarse-grained materials. The heating event was very rapid followed by an equally rapid cooling rate, something on the order of ~1000 K/h. This means that they would have cooled to near the solidus temperature within minutes. (The length of time needed to melt a chondrule is a very controversial area. Some believe seconds, some minutes to hours.) How do we know this? Recall the microscopic structure of the chondrules described in Chapter 6. The barred olivine chondrules in particular showed a texture that confirms this melting scenario. In many, parallel plate structure of one or more olivine crystals show that they had an igneous origin. Some chondrules exhibit a skeletal, incomplete appearance in the forming olivine laths. They are immersed in glass showing that the chondrule formed from a melted spherical drop of liquid and cooled rapidly before the chondrule structure was completely crystallized. Other chondrules such as those with radial pyroxene, cryptocrystalline and micro-porphyritic structure show similar textures and tell much the same story.

What was the heat source that mysteriously flashed into existence and then as rapidly disappeared? Lightning discharges in an ionized gas cloud? A nearby recurrent nova? Rapidly moving shock waves? This is a tough question. It may be that it was not a single mechanism but perhaps several operating together. Their individual importance probably changed in time as conditions in the solar nebula changed. Chondrules were being formed during the more violent period of solar nebula and protoSun accretion covering the first half million years of Solar System evolution. At this stage the protoSun was prone to violent outbursts called by astronomers the "FU Orionis stage" after the prototype-star in Orion. FU Orionis stars are classified as eruptive variables apparently triggered by material in-falling from the accretion disk to the star's surface. This causes a brief flare-up of the star in which the star can become as much as six magnitudes brighter (250 times brighter). It appears that FU Orionis stars are simply T-Tauri stars during an especially energetic flare-up. Many astronomers believe that during these eruptive stages, temperatures, especially inside 1 AU, are high enough to melt silicate minerals. The frequency of these FU Orionis stages in the life of the T-Tauri protostars is estimated to be roughly every 10 000 years. Activity, however, tapers off after the first-half million years. Thus, a one solar mass star will probably not see more than about 10 major flare ups in the 10 million years of a T-Tauri's lifetime.

Other mechanisms have been proposed for melting the precursor grains. A favorite mechanism among meteoriticists involves shock waves that heated the precursor solids. Large-scale shock fronts are set up as gas and dust from the collapsing presolar core cloud free-fall onto the forming accretion disk, producing shock waves that travel into the disk. As material flows through the region behind the shock fronts with velocities of 10 km/s or greater, it encounters much higher gas densities, producing high gas drag and subsequent heating sufficient to melt the precursor grains in dustballs that are 1 mm in size. Calculations show that *compact* precursor dustballs can reach temperatures greater than 1600 K for 15 min or more. Compaction of the dustballs could take place as melting begins, or, if the density of the fluffy dustball population increases by collisions.[5]

A novel idea involves the protoSun's highly directed solar winds, the bipolar jets. While astronomers are currently unable to explain how bipolar jets are generated, these jets do, in fact, exist and probably existed throughout the first million years or more of accretion disk formation. As much as 0.001 solar mass of solid material could have been ejected from the solar nebula. A large portion of this may reach escape velocity but if only 10% of this material were to fall back onto the accretion disk, this would amount to 10^{-4} solar mass or approximately the total mass of the planets. Early-formed jets are narrow with semi-angles of only a few degrees but as the accretion disk ages a steadily increasing clear zone is formed around the protoSun allowing the jets to widen until the returning solid material is moving lineally across the disk. This would tend to slow the material as it encounters gas drag through the halo of gas above the disk, dropping the velocity of the material below escape velocity. This mechanism provides a means of distributing solid material across the disk, effectively homogenizing the chemistry of the disk. According to this theory, chondrules are ablation droplets produced by bipolar jets.[6]

The number of models for chondrule origins seems to increase every time a conference convenes. Every researcher has a favored theory. In the Albuquerque meetings there was no end to the range of opinions among participating scientists. This healthy dialog forces critical thinking. If nothing else, it demonstrates the complexities of early Solar System formation mechanisms. It is indeed exciting to see how the science of meteoritics has come to involve so many other scientific disciplines. The chondrule problem will be solved (if it's solvable at all) by the combined efforts of all these participants.

Multiple melting and recycling of chondrules?

Substantial evidence exists that there has been more than one generation of chondrules. This suggests that the heating mechanism that melted the first precursor dustballs creating the first generation of chondrules repeated itself perhaps numerous times in the first million years or two of the Solar System's history. This story is told by relict grains, multiple or layered rims, igneous and dusty chondrule rims and compound chondrules.

Relict grains found in meteorites are relics of an earlier time, some formed before the first chondrules. Although they are found within chondrules, it is clear that they did not crystallize from the host chondrules. They are chondrule precursor grains crystallized from an earlier cycle of chondrule production. They did not melt during the most recent cycle of chondrule formation in which they are found. Relict grains are so characteristically different from the host chondrule grains that they tend to stand out. Most are olivine grains although some relict orthopyroxene grains are known. They are usually significantly larger than the host grains (probably because large grains are more resistant to melting). The largest relicts can exhibit almost perfect euhedral forms with some occupying as much as 90% of the chondrule volume. They are usually different in composition from the host olivine, showing that they originated from a different melt (Fig. 10.12). They are most frequently found in Type 3 ordinary chondrites and less commonly in CV3 chondrites, both within porphyritic chondrules. It was relict olivine found in Type 3 ordinary chondrites that furnished the first direct evidence of chondrule recycling. Two types of relicts are recognized: FeO-rich, metal-bearing olivine and pyroxene grains, and magnesium-rich olivine grains. The metal-bearing olivine grains are identified petrographically by a dusty appearance viewed in transmitted light. This dusty appearance is due to the presence of micrometer-sized rounded inclusions of low-nickel, iron metal dispersed uniformly through the crystal. The relict was originally more FeO-rich, but solid-state reduction of its FeO during a chondrule

Fig. 10.12. A relict olivine grain (blue) occupies the interior of a porphyritic olivine chondrule in a CO3 chondrite. The anomalous large size and shape makes them stand out among the host olivine grains. The view is in crossed-polarized light. The horizontal field of view is 2.25 mm. (Photo by O. Richard Norton and Tom Toffoli.)

melting event created iron metal within the grain. In low-FeO chondrules, relict olivine often has an enclosing jacket of FeO-poor, metal-free olivine overlying the FeO-rich interiors. This jacket probably formed during chondrule formation with the jacket growing on the relict grain. About 10% of the porphyritic chondrules have small dusty olivine grains enclosed poikilitically within pyroxene grains. They are high in FeO which suggests that they are relicts also. Magnesium-rich (almost pure forsterite) relict grains are much less common. Most are found as small grains in the matrix and less commonly in chondrules of CO, CM, and CV chondrites. In the most primitive Type 3 ordinary chondrites, ~10% contain dusty olivine relicts and ~5% bear relict forsterite.[7]

Relict grains, then, are relics of past generations of chondrules. They are principally derived from FeO-poor and FeO-rich porphyritic olivine and pyroxene chondrules as isolated phenocrysts probably released from the earlier chondrules through collisions that disrupted the chondrules. The relicts have evidently survived multiple chondrule melting events. This may be because most porphyritic chondrules were either not melted at all or only partially melted.

Along with relict olivine grains in chondrules there are many olivine crystals that are suspected of being prechondrule relicts. These are easy to spot under the microscope. They are almost always larger than chondrule olivine grains

Fig. 10.13. This yellow crystal, viewed under crossed-polarized light, is a nearly perfect euhedral olivine grain found in the matrix of an L4 Gold Basin, Arizona, chondrite. The grain measures almost 1 mm at its widest point. This may be a prechondrule relict grain that formed before the chondrules in the Gold Basin meteorites and was subsequently incorporated as an individual grain along with the chondrules into the meteorite matrix. (Photo by O. Richard Norton and Tom Toffoli.)

Fig. 10.14. A porphyritic chondrule encircled by at least three complete or partial fine-grained rims. The rim material was deposited onto the exterior of the chondrule that was subsequently heated (but not melted) and recrystallized to form the rims. Multiple rims are evidence of multiple heating events in the solar nebula. The orange-yellow area is staining due to weathering and alteration. This thin section viewed in crossed-polarized light is from a Gold Basin, Arizona, L4 chondrite. The chondrule is about 0.6 mm diameter. (Photo by O. Richard Norton and Tom Toffoli.)

and often show near perfect crystal form (Fig. 10.13). These defining characteristics are unlike some of the large relict grains in chondrules that often show partial reaction within the chondrule medium destroying the original well-defined textures. These individual grains have apparently escaped the melting process suffered by chondrule material and were never incorporated into chondrules, only trapped within a forming chondrite during the accretion process.

Chondrule rims jacketing most chondrules provide the second piece of evidence of multiple heating events in the solar nebula. There are two basic types: fine-grained rims and coarse-grained rims. Both have been investigated in Type 3 ordinary chondrites, CM2 and CV3 chondrites, since these are considered the most primitive and least equilibrated. Fine-grained rims have widths of 10–30 μm with individual grains in the submicrometer range. They have compositions and textures very close to that of the matrix material in which they are embedded, which includes FeO-rich olivine and MgO-rich pyroxene. Variable amounts of Fe–Ni and FeS are also found in submicrometer-sized grains. Like the matrix, the rims are dark and best observed in thin section under plane-polarized, or better yet, reflected, light. Rims of individual chondrules are very homogeneous but can vary widely in composition from chondrule to chondrule with rims of adjacent chondrules in Type 3 ordinary chondrites varying by as much as a factor of two in their FeO and MgO abundances and even more so in their lithophile elements. Fine-grained rims probably formed from the accretion of fine nebular dust that gathered around already formed chondrules. These are referred to in the literature as *accretionary dust mantles* (Fig. 10.14). Almost all of the chondrules in Type 3 chondrites have fine-grained rims. In fact, fine-grained dust rims extend around chondrule fragments, CAIs, relict grains and other coarse-grain inclusions. Most of the fine-grained rims show no signs of melting. Instead they appear to have "sedimentary" textures often with multiple layers of varying compositions.

By contrast, coarse-grained rims occur in only about 10% of the OC chondrules and 50% in CV3 chondrules. They are much easier to observe microscopically, averaging about 150 μm in thickness in OCs and ~400 μm in CV3, with a mean grain size of about 4 μm and 10 μm, respectively (Fig. 10.15). Like the fine-grained rims, the compositions of coarse-grained rims in both OC and CV3 chondrites are similar to the matrix and much the same as fine-grained rims (FeO-rich olivine, Ca-poor pyroxene, plagioclase, and Fe–Ni.) About 80% of the low-FeO chondrule rims show partial or complete melting and high-FeO chondrule rims are not far behind at >50%.[8] For this reason, meteoriticists often refer to coarse-grained rims as "igneous rims" since they must have crystallized from a chondrule heating event. A close look at the outer edge of the chondrule shows remelting of previously formed mafic silicate minerals and concomitant fusion of rim silicates with chondrule silicates clearly preserving evidence of the reheating event that produced the rim. This strongly suggests that at least some of the rim material came from the host chondrule. The fact that fine-grained rim material is frequently found mantling coarse-grained rims in both OC and CV3 chondrites shows that rim material, in part, came from dusty regions of the solar nebula.

To summarize, the coarse-grained rims probably formed after the chondrules' precursor material melted and cooled to a spherical shape but before accreting onto a parent body. The rim material was then deposited uniformly around individual chondrules as a fine dust that we know pervaded the solar nebula. A second heating event melted the dust and part of the chondrule which then recrystallized into the rim material. Some chondrules show multiple rims concentric around the

Fig. 10.15. Multiple coarse-grained "igneous" rims compose most of the chondrule seen in this thin section of an L6 chondrite from Songyuan, China. The black areas are metal and FeS. The chondrule measures 1 mm across. The section shows extensive recrystallization typical of petrographic Type 6 but the chondrule appears relatively unaltered. (Photo by O. Richard Norton and Tom Toffoli.)

chondrules. This implies multiple heating and dusting events. Figure 10.14 shows three concentric rims around a porphyritic olivine chondrule from the L4 Gold Basin, Arizona, chondrite. One interesting variation is the presence of FeS forming discontinuous rims both inside and outside the coarse-grained rims. This may have come from the interiors of iron-bearing chondrules during a heating event that must have raised the chondrule temperature to around 1200 K. Also, complete metal-jacketed chondrules are common in ordinary chondrites and are probably the product of impact shock melting of metal grains in the matrix (Fig. 10.16).

Compound chondrules are among the most intriguing of the chondrules in that they give us some understanding of the chondrule density and the random collisions that must have occurred among the newly formed chondrules in the solar nebula. Surveying a Type 3 or 4 ordinary chondrite thin section under low magnification often reveals chondrules with spherical indentations or "*chondrule craters*" (Fig.10.17). They have been a subject of debate for many years. There are two main viewpoints to explain them. One idea is that a smaller solid chondrule gently impacts a larger, still plastic chondrule. The impact is not sufficient to cause them to stick together as a compound chondrule; rather, the smaller solid body escapes, leaving behind a dent in the larger. The less-solid chondrule, considered to have formed at a later time and still plastic at the time of the collision, is the *secondary*

Fig. 10.16. This 1.75 mm diameter porphyritic olivine chondrule has a thin metal jacket completely encircling it. The view is seen in reflected and crossed-polarized light, allowing both the metal as well as the texture of the chondrule to be visible. The blue material embedding the olivine is glass. Thin section from a Richfield, Kansas, LL3.8 chondrite. (Photo by O. Richard Norton and Tom Toffoli.)

Fig. 10.17. A radial pyroxene chondrule seen under crossed-polarized light with a conspicuous spherical indentation or "chondrule crater". The solid *primary* chondrule may have impacted this large *secondary* chondrule while the secondary was still in a plastic but viscous state. The chondrule is 1 mm in diameter. From an Allende CV3.2 chondrite. (Photo by O. Richard Norton and Tom Toffoli.)

chondrule. The more-solid chondrule that retains its original shape is considered to have formed first and is called the *primary chondrule* (regardless of size). In the case of chondrule craters, only the shape of the primary chondrule is evident. Rarely do we find the impactor nudged against the impactee. One of the problems with this *elastic impact* idea is that the chondrules stay in a molten or plastic state for only seconds or at most a few minutes, requiring that the distance over which a chondrule travels before impacting another chondrule, the *mean free path*, be quite low. This further requires a high chondrule density to insure a sufficient number of impacts to account for the rather common occurrence of chondrule craters in chondrites.

Most researchers believe the mean free path of forming chondrules tended to be high so that few impacts occurred while they were in a plastic state. Instead, they accept an alternate viewpoint that involves only one fully molten chondrule containing immiscible metal plus sulfide and silicates. The chondrule is spinning, causing the denser metallic liquid to migrate to the surface of the spherical chondrule. The metal bleb is ejected from the chondrule while the chondrule is in a viscous state, allowing the chondrule to retain the spheroidal shape of the metal bleb at the exit point. Supporting this idea, one occasionally finds traces of metal and iron sulfide within the indentation, as in Fig. 10.18.

Fig. 10.18. Thin section of a barred olivine chondrule viewed with crossed-polarized and reflected light from a Brownfield, Texas, H3.8 chondrite. The "crater" in the upper left is lined with metal. Some of the metal has deposited along fractures within the chondrule. Diametrically opposite the "crater" is a small adhering chondrule appearing as a bump on the larger chondrule. The large chondrule measures about 1 mm diameter. (Photo by O. Richard Norton and Tom Toffoli.)

Fig. 10.19. Two barred olivine compound chondrules in an Allende CV3 chondrite. Both have similar textures with thick rims and barred olivine structure. They are therefore adhering siblings. The field of view across the horizontal is 1.8 mm. (Photo by O. Richard Norton and Tom Toffoli.)

If the primary and secondary chondrules are still attached we call them *compound chondrules*. If the primary and secondary have similar textures and mineralogy, the pair is a *sibling compound chondrule*; those pairs with different textures and mineralogies are *independent compound chondrules*. Some independent compound chondrules appear to be stuck together with the smaller forming a "bump" on the side of the larger chondrule. These are amply described as *adhering chondrules* since they are doing just that (Figs. 10.18 and 10.19). Often, a chondrule will be enveloped by dust much as the relict grains were and then quickly reheated. The enveloped chondrule remains in tact but the dust mantle melts and then recrystallizes around the original chondrule.These are *enveloping chondrules*. In very rare instances, a smaller more solid primary chondrule may become completely enveloped by a larger secondary as a result of impact. Figure 10.20 is an example of an enveloping chondrule, a barred olivine chondrule completely enclosed by a porphyritic olivine chondrule. In OC chondrites, siblings and independents are more or less evenly divided with perhaps a slight preference toward siblings. Of the independent types, most are BO chondrules within porphyritic chondrules such as the one shown in Fig. 10.20.[9]

What we are witnessing with compound chondrules is the accretion process in action, caught in a moment of time. Chondrules impact chondrules, some sticking, others

Fig. 10.20. A typical enveloping chondrule in an Allende CV3 chondrite. The small barred olivine chondrule has been completely enveloped by a much larger porphyritic olivine chondrule. The horizontal field of view is 3.0 mm. (Photo by O. Richard Norton and Tom Toffoli.)

bouncing off while still others are enveloped completely. These must be samples of the more energetic chondrules. Most were accreted so gently that they remain individual, if a bit distorted during accretion.

Compound chondrules pose an interesting problem. Chondrules are thought to form through a nebular flash-heating process that heats precursor mineral aggregates to the melting point. The melted phase, however, can last only a minute at most, too short for random collisions to create independent forms. Higher velocities of the chondrules would result in more collisions but would probably destroy the chondrules. Rather, the primary (solid chondrule) may have been a precursor "relict chondrule" formed and solidified earlier. A dust clump could form around the precursor chondrule. Then a second flash-heating event could have melted the dust clump around the primary but not the primary itself, producing the secondary chondrule of different composition and completely enveloping the primary. Siblings are more easily explained. Both primary and secondary chondrules formed from the same precursor mineral aggregate during a single heating event giving them the same composition. Their close proximity to each other guaranteed a collision before they both solidified.[10]

The age of the Solar System

We began this chapter by reviewing the physics of radioisotopes and briefly looked at one of the more common and useful radionuclides (rubidium/strontium) used to determine the age of meteorites, which translates to the age of the Solar System since we believe meteorites are among the oldest surviving materials (cometary material may be older). We reviewed current notions about the formation of the solar nebula out of a molecular cloud core and the development of an accretion disk and protoSun. The cooling of the accretion disk condensed the first-formed refractory minerals (CAIs) and other materials that make up the most primitive meteorites. We saw how precursor minerals rained down on the midplane of the accretion disk, formed dustballs and, with an as yet mysterious flash-heating mechanism, melted the dustballs to form the first chondrules. This is a crucial moment in the history of the Solar System. After the first minerals condensed and chondrules formed, it was only a matter of a few million years before large bodies formed through accretion and compacting of chondrules and matrix. It remains now to try to pin down when it all happened. We return to radioisotopes to give us the answers.

Aluminum-26 and the age of chondrules and CAIs

We learned earlier that ^{26}Al is the radioisotope with the shortest half-life (720 000 years) and because of this short half-life it became extinct long before the first 10 million years ended, leaving behind radiogenic ^{26}Mg in the aluminum-bearing minerals of CAIs. Unlike long-lived radioisotopes, short-lived radioisotopes with short formation intervals have the resolution necessary to establish time intervals between important very early events in the solar nebula differing by only a few million years. The time of CAI formation vs the time of chondrule formation is an important example. In studies such as these, the initial distribution of ^{26}Al is critical. Most studies assume that ^{26}Al was homogenous in the solar nebula in those areas where CAIs and chondrules formed. One study in 1995 involving over 1500 analyses of CAIs made over 25 years showed that ^{26}Al was uniformly distributed over wide areas of the nebula. The initial abundance level of ^{26}Al/^{27}Al was $\sim 5 \times 10^{-5}$ as recorded in the CAIs.[11]

When one looks at aluminum-bearing minerals (plagioclase) in chondrules of ordinary chondrites and carbonaceous chondrites, the excess ^{26}Mg miraculously all but disappears. Excess ^{26}Mg expressed by the ratio ^{26}Al/^{27}Al drops precipitously from the CAI value of $\sim 5 \times 10^{-5}$ to an average of $< 1.2 \times 10^{-5}$ for chondrules. Values can differ markedly from chondrule to chondrule and can drop to as low as $2\text{–}6 \times 10^{-6}$, these values being estimated upper limits.[12] The traditional explanation is that the difference between CAI and chondrule ^{26}Mg abundances is simply a matter of time, that is, CAIs formed with initial ^{26}Al which decayed into the ^{26}Mg signature we observe today. Two to three million years later the chondrules formed after most of the initial ^{26}Al had decayed away. Thus, no excess ^{26}Mg is seen in the chondrules. This tells us that CAIs and chondrules did not form simultaneously. CAIs are at least two to three million years older than chondrules. This also suggests that the components of chondritic meteorites formed in the first two to three million years while the solar nebula was still in its youth.

Although most researchers accept the ^{26}Mg excess in CAIs relative to chondrules as a real age difference, this idea is not without its opponents. Perhaps there was no "live" ^{26}Al at all in the solar nebula at the time, only "fossil" ^{26}Mg. But ^{26}Mg in aluminum-bearing minerals appears always to have been associated with the original parent Al isotope within the crystal structure of the aluminum-bearing minerals. Magnesium is not normally found in plagioclase minerals such as anorthite and does not easily substitute within the crystal structure. Furthermore, if ^{26}Mg were truly a fossil, it should have become homogenized in the solar nebula so that the Mg isotopic composition would appear the same in all chondritic components, especially in ferromagnesian minerals of ordinary chondrites. This is not seen.

Some still insist that ^{26}Al was not distributed uniformly in the solar nebula. Could a nearby supernova explosion scatter ^{26}Al homogeneously into the forming solar nebula? Not only does this require that the supernova remnant be spherically and uniformly ejected (they're not), it also assumes that the density of the growing solar nebula be uniform. Given sufficient time, turbulence within the solar nebula or scattering of material through the action of the bipolar jets could help to homogenize the aluminum isotopes but the short half-life of the radioisotope leaves little time for homogenizing. A heterogenous solar nebula would mean that the ratio of ^{26}Al/^{27}Al was not consistent throughout the nebula. Thus, the differences we see in ^{26}Al abundance between CAIs and chondrules would be an effect of heterogeneity and not attributable to the decay rate. If so, then ^{26}Al could not be used as a short-term chronometer but could only provide information about the distribution of ^{26}Al within the nebula. An interesting observation in a 1987 study demonstrated that an inclusion in the Efremovka CV3.2 showed a difference in distribution between the core and rim. The rim showed more ^{26}Mg than the core. The core must have formed first and would be expected to show more ^{26}Mg than the later formed rim. Since just the opposite was seen, the rim must have formed in another part of the nebula where the ^{26}Al was higher.[13]

Perhaps the most difficult problem that arises with this two to three million year gap is the problem of preserving the CAIs in the solar nebula until the chondrules form so that they can accrete together into parent bodies. We saw in Chapter 1 that a particle 1 mm in diameter at a mean distance of 1 AU from the Sun and an orbital eccentricity of 0.7 would spiral into the Sun in about 1 million years due to the Poynting–Robertson Effect. The situation is analogous except particles in and above the disk are subjected to gas drag all along the disk which would slow them appreciably. They would spiral into the Sun well within the 10 million years of the nebula's lifetime. What protects the CAI grains while waiting at the chondrule formation sites for chondrules to create themselves so they can accrete together? This provocative question has yet to be answered.

Aluminum-26 is one of a triad of extinct short-lived radionuclides that have been successfully applied to date the formation intervals of CAIs and chondrules. The other two are manganese-53 and iodine-129. Manganese-53 decays to chromium-53 through beta decay. It has a half-life of 3.7 million years, conveniently between aluminum-26 and iodine-129. Manganese-53 is the newest of the triad. It was first successfully found in Allende CAIs in 1985.[14] Of the two, iodine-129 is perhaps the most familiar or at least the most often used, so it is the radionuclide we will briefly summarize. Iodine-129 decays to its daughter radiogenic nuclide, xenon-129, by beta decay. It has the longest half-life of the three: 16 million years. It was the first extinct radionuclide to be discovered through the work of J.H. Reynolds at UC Berkeley in 1960–1 and has been successfully used to measure the formation interval or the time from nucleosynthesis (production of elements heavier than helium within massive stars) to the formation of the first solid bodies (CAIs and chondrules) in the early Solar System. Unlike the rubidium/strontium system discussed earlier in which initial ^{87}Sr was present in the first formed minerals, the xenon daughter product that must have existed in the solar nebula along with iodine-129 probably was not taken up by the first-formed solid grains. Xenon gas is inert and does not react chemically with the minerals. Iodine is much more reactive and most certainly entered the crystal structures of the CAIs and chondrules. Thus, all of the xenon-129 is thought to have come from the *in situ* decay of iodine-129. Initial abundance of iodine-129 as measured by the ratio $^{129}I/^{127}I$ is not well known but is estimated to be quite low, approximately 10^{-4}. The I/Xe method is used to determine formation intervals or apparent ages rather than absolute ages (since initial ^{129}I is not well known).

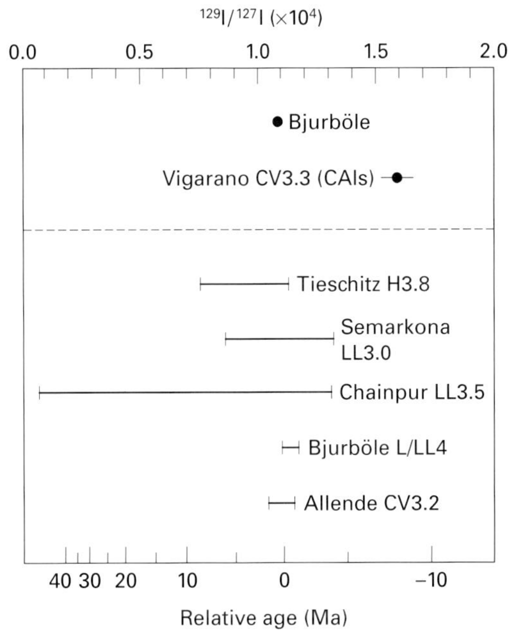

Fig. 10.21. Apparent ages determined for chondrules from five different unequilibrated chondrites using the isotopic ratio, $^{129}I/^{127}I$. The apparent ages are calculated with respect to a standard chondrite, the L/LL4 Bjurböle meteorite, which was given a relative age of zero. The horizontal lines represent the range of apparent ages determined using a number of chondrules from each chondrite. Most of the chondrites span an interval of about 10 million years. The latest formation age is from chondrules in the LL3.5 Chainpur which spans a 50 million year formation interval (Swindle *et al.* 1996).[15]

Researchers wanted to determine the timing of chondrule formation so they used unequilibrated ordinary chondrites in their studies and these were calculated relative to a standard, the L/LL4 chondrite Bjurböle. The difference in formation intervals between five meteorites was used to determine "apparent ages" of each compared to Bjurböle which was given a relative age of zero. Figure 10.21 shows the results of I/Xe studies on five unequilibrated (Type 3) ordinary chondrites using the $^{129}I/^{127}I$ estimated abundance of 10^{-4}. The apparent ages on the average cover a range of 10 million years or less with Bjurböle covering about 2 million years and Chainpur LL3.5 with the largest variation spanning 50 million years. In all five meteorites the chondrules were sampled. Using the CV3 chondrite Vigarano as the least altered carbonaceous chondrite, I/Xe ages were determined for its CAIs.[II] These are located on the diagram. CAI ages

[II] Many CAIs and chondrules were measured using the I/Xe system for direct comparison with the Al/Mg system. Unfortunately, Allende showed a large variation in $^{129}I/^{127}I$ abundance in the CAIs and chondrules which extended their apparent ages over a range of 50 million years. Some researchers suggested that this might to be the result of post-formation alteration which would affect the amount of ^{129}Xe in the samples. Vigarano was selected as a less altered CV3 chondrite for this work.

showed a much smaller variation with an average of >8 million years older than Bjurböle. This is the oldest I/Xe age so far determined.[15]

Researchers are still struggling to interpret the I/Xe measurements. Clearly, the gap between the formation age of CAIs and chondrules in the I/Xe system is much greater (>8 million years) compared to two to three million years for the Al/Mg system. Since only CAIs from Vigarano were measured, there is a clear need to measure CAIs and chondrules from other unaltered CV3 chondrites. There remains some doubt even among its users that the I/Xe system is a good chronometer. Perhaps the best that can be said of the I/Xe measurements so far is that they do confirm that chondrules and CAIs were formed over a 10 million year period of Solar System history and that the gap between CAI and chondrule formation occurs within that 10 million year period. The other short-lived radionuclides support this conclusion.

Absolute age of the Solar System

How do we define the age of the Solar System? From what point do we begin? We asked these questions earlier in this chapter and we discovered that there are several "ages" from which we can choose. The age of the Solar System is specifically called the *formation age*. But the formation of what? To reckon time from the "birth" of the Sun is ambiguous. The birth process is too lengthy. From the formation of the protoSun to the main sequence covers almost 30 million years of time for a one solar mass star. During this critical period other important events were taking place that must be accounted for in Solar System history. We've demonstrated that meteoriticists have been able to resolve the first 10 million years of Solar System history from extinct short-lived radionuclides in some of the first-formed minerals. This is only about 1/500 of Solar System history, a good point from which to reckon Solar System time. Thus, the age of the Solar System should commence at a time when the solar nebula begins to seed the area with refractory minerals; when the precursor materials of chondrules appear and chondrules begin to form.

Determining the absolute age of the Solar System requires long-lived radioisotopes with half-lives that extend over a large fraction of the age of the Solar System. They must decay sufficiently quickly to leave measurable daughter products over a few billion years but at the same time still retain detectable amounts of primary isotopes. Table 10.2 shows several long-lived radionuclides that fit these criteria. The Rb/Sr system has the longest half-life, 49 billion years, more than 10 times the estimated age of the Solar System. The $^{238}U/^{206}Pb$ system has a half-life that matches the age of the Solar System. Isotopes with such long half-lives tend to reduce the accuracy of the age determination since decay over a few million years would be undetectable.

We saw earlier using the Rb/Sr system how an isochron diagram is plotted, resulting in a straight line, the isochron time line, whose slope is a measure of the age of the rock sample. Well over a hundred meteorites have been measured using Rb/Sr and other long-lived radionuclides. Appendix B lists formation ages of selected chondrites and differentiated meteorites determined by Rb/Sr measurements. The majority of formation ages fall between 4.45 and 4.56 Ga with the older age preferred. Variations on the low side may reflect resetting of the clock during thermal metamorphism. This is certainly the case with the low values of the IIE iron Kodaikanal, which may be a severe case of impact heating. Likewise, the Stannern eucrite's young age must be the result of a shock event that reset the clock a billion years after it formed. Kapoeta also shows a low value but a second measurement falls within accepted values. Kapoeta is a brecciated howardite with the fragments showing substantial differences in formation ages. The lowest value (i.e. the youngest age) is Nakhla, an SNC achondrite. This was a mystery for several years until it was finally realized that this meteorite came from Mars and must have been part of an active eruption 1.34 Ga ago, three billion years after all igneous activity ceased among the asteroidal bodies of the Solar System. An important fact emerges from this list. We look at the chondrites as the oldest meteorites in the Solar System. The differentiated meteorites generally have lower formation ages, but they run a close second. Apparently, differentiation occurred very soon after chondritic parent bodies formed. This compilation suggests that only ~100 million years separated the formation of chondritic parent bodies from their differentiated cousins.

Impact metamorphism

Heating of parent bodies resets the isotopic clocks. Radionuclides vary widely in their ability to resist thermal metamorphism. Any radionuclide whose daughter isotope is a gas is sensitive to heating events. During such times, the gas, not being held strongly within the crystal structure of the ferromagnesian minerals, can be released and lost to space. This immediately resets the clocks and dates the time of metamorphism. Thermal metamorphism deep within a parent body, shock-induced heating due to major impacts, and igneous activity are all examples of events that can change the distribution of gaseous daughter products within parent bodies and their meteorite fragments. Dating such events depends upon the accumulation of helium-4 released from the decay of uranium and thorium and argon-40 from the decay of potassium-40. A variant of the K/Ar technique, designated $^{40}Ar/^{39}Ar$, is commonly used for dating secondary heating events.

Approximately 75% of the meteorites analyzed by the $^{40}Ar/^{39}Ar$ method show loss of argon gas. Many of these meteorites, both chondritic and achondritic, show brecciated textures, the result of parent body impacts in space one or more times in their history. These meteorites show signs of shock and subsequent heating as enormous pressures are exerted and then released. Shock metamorphism causes the release of accumulated argon gas which stops the accumulation clock. As cooling reaches the *closure temperature* (temperature at which argon begins again to accumulate) the clock is reset. The time of impact (or the age of the meteorite at the time of impact) can be determined by measuring the amount of accumulated argon gas since the impact event. This is called the *gas-retention age*. Gas-retention ages represent the most recent secondary heating events suffered by the meteorites. These vary enormously from about 4 billion years to 100 million years or less. It seems that the Solar System has had a traffic problem throughout its 4.56 Ga history![16]

As gas-retention age measurements accumulated, some interesting facts began to emerge. The gas-retention ages of LL chondrites seemed to cluster around 4.1 Ga. This is near the end of the intense bombardment of the Moon's surface resulting in the crater distribution seen there today. Might the LL chondrites have suffered impacts along with the Moon, Earth and other terrestrial planets during this period? The L chondrites showed a different picture. The first L chondrite to be measured, the L6 Bruderheim, Alberta, chondrite (which fell in 1960), showed a phenomenal loss of more than 95% of its argon about 550 Ma ago.[17] Through the years, measurements of other L chondrites show that about two-thirds had major argon losses between 400 and 500 Ma ago. This was rechecked using U/He and Th/He, both of which showed a loss of helium around 400 Ma. Thus, it seems likely that there was a major collision and a possible breakup of an (or the) L chondrite parent body about 450 Ma ago.

Cosmic-ray exposure ages

We know now that the first solid bodies formed within the first 10 million years of Solar System history. These solid bodies, the chondrules and inclusions, were probably incorporated into the first parent bodies within that period. The meteorites, fragments of other worlds, share the age of the parent bodies, 4.56 Ga. But meteorites that make their way to Earth possess yet another, relatively youthful age. All bodies in space are subject to bombardment by subatomic particles. The solar wind particles are the least energetic with values of about 1 KeV. These have penetration depths of only a few ångströms. The occasional solar flares emit higher-energy particles ranging from ~1 MeV to 100 MeV (MeV = million electronvolts) or more. These can penetrate solid rock to a depth of a few centimeters. These particles have a weathering effect on unprotected surfaces of asteroidal bodies and meteoroid fragments (see Chapter 11), but they lack the energy to produce radionuclides. This is the domain of *galactic cosmic rays*.

Cosmic rays are not "rays" as the name implies, but highly energetic subatomic particles. Nearly 90% are protons, with alpha particles and other heavy nuclei about 9% and the remaining 1% electrons. Galactic cosmic rays come from outside the Solar System. They are by far the most energetic particles to pass through the interplanetary medium. They have energies typically in the hundreds of million to billions of electron volts with an average of around 10 GeV (GeV = billion electronvolts). At these energies the particles can penetrate about 1 m of stony meteorite material and less for irons. Most astronomers believe that cosmic rays come from violent supernova explosions where subatomic particles are accelerated to near the velocity of light. Others suggest that violent events at the galactic center where strong electromagnetic forces exist accelerate particles to high energies. The galactic cosmic-ray flux seems to have been more-or-less constant over the lifetime of the Solar System.

Cosmic-ray protons interact with target nuclides in the minerals composing the asteroid or meteoroid fragment resulting in the partial disintegration of atomic nuclei, a process called *spallation*. This process can change the target elements to new elements that are both stable and radioactive nuclides. Among the new nuclides are ^{3}He, ^{21}Ne, ^{38}Ar, and ^{41}K, all stable isotopes. These build up slowly as the meteoroid is exposed to galactic cosmic rays. This amounts to a cosmic accumulation clock. The clock is set when cosmic-ray exposure begins and the cosmogenic isotopes begin to accumulate. The total amount of cosmogenic nuclides divided by the rate of production yields the time the meteoroid has been exposed in space. This is the *cosmic-ray exposure age* (CRE age).

Since cosmic rays can penetrate only a meter into an asteroid, only the outer meter of the asteroid is exposed. The remainder is shielded from cosmic rays. If the body remains intact, the surface exposure could extend back to the origin of the asteroid parent body giving a CRE of >4.4 Ga. This is not very likely since all asteroids investigated so far show a history of impacts that must have produced fragments from beneath the 1 m surface. Thus, during a collision, fresh unexposed fragments are released into space where, if they are a meter or less across, they are exposed to the full flux of galactic cosmic rays. If, by chance, such a fragment finds its way to Earth, cosmic-ray exposure ceases and the cosmogenic isotopic abundances are determined, giving the total time the fragment spent in space after fragmentation. The CRE in this case gives the time of fragmentation as well as the time spent in space. Of course, it is possible that a surface rock can fragment into pieces larger than a meter or two. The interior may not have been exposed at all while the exterior would have substantial CRE time. This anomalous meteorite would therefore have two CRE values. This nonuniformity of CRE is usually recognized as a near-surface rock exposed for millions of years on its parent body before being exhumed intact with fresh unexposed rock that gives the time of fragmentation.

Hundreds of measurements have been made on all meteorite types giving significant CRE ages. The results show a wide variation of CRE that seem to be type dependent. For example, the ordinary chondrites show an average CRE of roughly 50 Ma. This is the period of time that fragments are

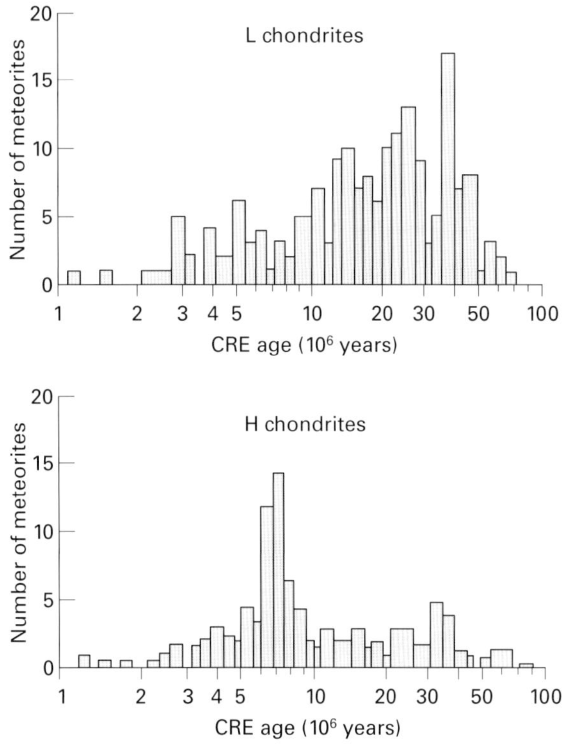

Fig. 10.22. Cosmic-ray exposure ages for L and H chondrites. The CRE is relatively uniform through the last 50 million years of L chondrite history. This is probably the result of a continuous "chipping away" of small fragments from the L chondrite parent body. The large peak near the 6 million year CRE for the H chondrites strongly suggests a major impact that catastrophically disrupted an H chondrite parent body resulting in a large number of fragments with that age. (Crabb, J. and Schultz, L. (1981). Cosmic ray exposure ages of ordinary chondrites and their significance for parent body stratigraphy. *Geochimica Cosmochimica Acta* **45**, 2151–2160.)

expected to last in Mars-crossing and Earth-crossing orbits. Of the three ordinary chondrites, the L and LL chondrites show a roughly even distribution between a few million years to about 50 Ma, where there is a slight peak. This has been interpreted to mean that the L chondrites were produced by chipping away of one or more L chondrite parent bodies throughout that period. The H chondrites, however, show an obvious non random peak at between 6 and 8 Ma. This may be showing us that there was a major collision in that interval that disrupted a large asteroidal parent body and that many of our H chondrite specimens are representatives of that collision (Fig. 10.22). The irons show the oldest CRE ages. This is not surprising since irons are much more resistant to fragmentation. They range from less than 200 Ma to more than 1 Ga with prominent clusters at about 650 and 400 Ma. These are the IIIAB and IVA irons respectively, and may represent breakups of IIIAB and IVA core bodies. The average age for irons is around 500 Ma, ten times that of the more fragile chondrites, and like the chondrites may represent the lifetime of iron bodies in Mars-crossing (and Earth-crossing) orbits.

Terrestrial age

The *terrestrial age* of a meteorite is a measure of how long the meteorite has resided on Earth. While in space, we saw that the meteorite is subject to spallation induced by galactic cosmic rays creating radionuclides that slowly decay. Assuming that cosmic radiation remained constant through the meteorite's stay in space, it will have reached an equilibrium in which radioisotopes are forming and decaying at a constant rate (*saturation level*). Once the meteorite reaches Earth it is protected from cosmic rays by Earth's atmosphere so that production of primary radionuclides will cease, but the primary radionuclides present when it landed will continue to decay. Thus the ratio of primary to daughter isotopes will decrease in time depending upon the half-life of the radionuclide. Since all meteorites reach a radionuclide saturation level in space, then any meteorites if picked up within days after falling will demonstrate this saturation level. This can be used as a standard from which to compare the abundance of remaining radionuclides in relatively old meteorites that have lain on Earth's surface for centuries. This comparison, along with the known half-life of the radionuclide, will yield the approximate terrestrial age of the meteorite. Useful isotopes for determining terrestrial ages have relatively short half-lives since most meteorite lifetimes without complete terrestrial alteration is less than 100 000 years. These include ^{39}Ar, ^{14}C, and ^{36}Cl with half-lives of 270 years, 5700 years and 300 000 years respectively. The oldest meteorite terrestrial age (outside Antarctica) is an iron from Tamarugal, Chile, found in 1903 in the Atacama Desert. Its terrestrial age is greater than 1.5 million years. Terrestrial ages of stony meteorites usually do not go beyond around 50 000 years unless special circumstances of preservation are present. Such is the case with Antarctic meteorites. Here, the meteorites have been preserved in ice for hundreds of thousands of years. The oldest ages for stony meteorites are among the Antarctic finds. Many are over 500 000 years with a few over a million years. They are the oldest meteorites as a group in the world but yet they are essentially identical to non-Antarctic meteorites. This tells us that meteorite compositions have generally not changed in the last 500 000+ years and the source of Antarctic and non-Antarctic meteorites probably is the same.

References

1. Minster, J.F., Birck, J.L. and Allègre, C.J. (1982). Absolute age of formation of chondrites by the ^{87}Rb/^{87}Sr method. *Nature* **300**, 414–419.
2. Cameron, A.G.W. (1995). The first ten million years in the solar nebula. *Meteoritics* **30**, No. 2, 138.
3. King, E.A. (ed.) (1983). *Chondrules and their Origins*, Lunar and Planetary Institute, Houston, Texas.
4. Hewins, R.H., Jones, R.H. and Scott, E.R.D. (eds.) (1996). *Chondrules and the Protoplanetary Disk*, Cambridge University Press, Cambridge.
5. Ruzmaikina, T.V. (1994). Abstract: chondrule formation in the radiative accretional shock, Conference on *Chondrules and the Protoplanetary Disk*. University of New Mexico, Institute of Meteoritics, Albuquerque.
6. Liffman, K. (1994). Abstract: The jet model of chondrule formation, Conference on *Chondrules and the Protoplanetary Disk*. University of New Mexico, Institute of Meteoritics, Albuquerque.
7. Jones, R.H. (1994). Relict grains in chondrules: evidence for chondrule recycling, Conference on *Chondrules and the Protoplanetary Disk*. University of New Mexico, Institute of Meteoritics, Albuquerque.
8. Krot, A.N. and Wasson, J.T. (1995). Igneous rims on low-FeO and high-FeO chondrules in ordinary chondrites. *Geochimica Cosmochimica Acta* **59**, 4951–4966.
9. Wasson, J. *et al.* (1995). Compound chondrules. *Geochimica Cosmochimica Acta* **59**, 1847–1869.
10. Rubin, A.E. and Krot, A.N. (1996). Multiple heating of chondrules, in *Chondrules and the Protoplanetary Disk*, Cambridge Univesity Press, Cambridge.
11. MacPherson, G.J. *et al.* (1995). The distribution of aluminum-26 in the early Solar System – a reappraisal. *Meteoritics and Planetary Science* **30**, 365–386.
12. Krot, A.N. *et al.* (1999). Abstract: mineralogy, aluminum–magnesium-isotopic, and oxygen-isotopic studies of the relic calcium–aluminum-rich inclusions in chondrules. *62nd Annual Meeting of the Meteoritical Society*, Supplement, **34**, A68.
13. Fahey, A.J., Zinner, E. and Crozaz, G. (1987). Microdistribution of Mg isotopes and REE abundances in a Type A CAI from Efremovka: constraints on rim formation. *Geochimica Cosmochimica Acta* **51**, 3215–3229.
14. Birck, J.L. and Allègre, C.J. (1985). Evidence for the presence of ^{53}Mn in the early Solar System. *Geophysical Research Letters* **12**, 745–748.
15. Swindle, T.D. *et al.* (1996). Formation times of chondrules and Ca–Al-rich inclusions: constraints from short-lived radionuclides, in *Chondrules and the Protoplanetary Disk*, Cambridge University Press, Cambridge.
16. Turner, G., Miller, J.A. and Grasty, R.L. (1966). The thermal history of the Bruderheim meteorite. *Earth and Planetary Science Letters* **1**, 155–157.
17. Bogard, D.D. (1995). Impact ages of meteorites: a synthesis. *Meteoritics and Planetary Science* **30**, 244–268.

The Near Earth Asteroid Rendezvous (NEAR) spacecraft was placed in orbit around 433 Eros on February 14, 2000, beginning a year-long study of the asteroid. Gradually the orbit was lowered, increasing the spacial resolution of the instruments at the same time. On February 12, 2001, after nearly one year in orbit, controllers at the Applied Physics Laboratory of Johns Hopkins University commanded the spacecraft to fire its engines to slow the spacecraft, allowing a slow descent to the surface. Photographs were taken to within 394 feet of the surface. This rendering shows the NEAR spacecraft moments before touchdown on Eros. The spacecraft landed safely and continued to operate for an additional 10 days from the surface. (Courtesy of APL, JHU and NASA.)

Chapter Eleven

Asteroid parent bodies

They used to be called "vermin of the sky", referring to those pesky rocks that created light streaks on astronomers' photographic plates during long-time exposures near the ecliptic plane. They were first called *asteroids* by William Herschel because of their star-like appearance. Later on they were called *planetoids,* since it was recognized that they were solid bodies of subplanetary mass. After the initial excitement of discovery was over and astronomers realized that there were many such bodies between the orbits of Mars and Jupiter, they were classified as only "minor planets", a term which carried with it implications that their importance to the scheme of the Solar System was minimal at best. Though asteroids continued to be discovered by chance through the nineteenth and early twentieth centuries, their importance in the evolution of the Solar System went unrecognized. Even planetary astronomers tended to ignore them since, unlike the major planets, they could not be resolved into extended objects with any telescope in existence. In this case, bigger was better. How ironic it is that meteorites, the "children" of asteroid parent bodies, tell us more about the early Solar System than all the telescopic studies of the major planets put together.

This attitude clearly lasted into the 1980s. I remember an incident at Kitt Peak National Observatory in Southern Arizona in the early 1970s. On the grounds of the observatory the University of Arizona had their Steward Observatory. It happened during the evening meal in the observatory cafeteria. A colleague of mine and I were scheduled to use one of the university's telescopes to search for asteroids. Another astronomer took a seat at our table and began a conversation. He was scheduled on one of the larger telescopes that night and hastened to tell us that he was studying mass distribution in galaxies. The conversation then turned to us and he inquired first what we were studying and what telescope we were using. When we told him we were interested in asteroids an awkward silence ensued. Apparently having nothing in common with us, the astronomer picked up his tray, excused himself and promptly moved to another table where there were "others of his kind". So much for asteroids at Kitt Peak in the early 1970s. Little did these astronomers suspect that asteroid studies would become not only respectable science but would make it "big time" on the mountain within a decade with the Spacewatch project under University of Arizona planetary astronomer, Tom Gehrels.

The asteroid belt

Anyone even casually studying the geometry of the Solar System will first notice that the planets are arranged with remarkable symmetry in terms of distance from the Sun. The planets Mercury, Venus and Earth occupy orbits that vary in mean distance, increasing by roughly 48 million kilometers (0.321 AU) for each planet: Mercury at 57.9 million kilometers (0.387 AU); Venus, 108.2 million kilometers (0.732 AU) and Earth, 149.6 million kilometers (1.000 AU). Mars, 0.5 AU beyond Earth, still maintains this geometry but the next planet out, Jupiter, is 5.2 AU from the Sun, twice the distance it should be if the geometry were to be maintained. Johannes Kepler, a contemporary of Galileo, was the first to notice the curious gap in the otherwise orderly arrangement of the planets, at a position 2.8 AU from the Sun. Kepler was convinced that this geometrical progression had to be maintained and insisted that an "unseen" planet had to exist in this position. Kepler was the first to correctly describe the motions of the planets and he established an empirical relationship between a planet's period of revolution and its mean distance from the Sun, the so-called Harmonic Law: $P^2 = D^3$ where P is the planet's orbital period in terms of Earth's period and D is the planet's mean distance from the Sun in terms of the Earth's mean distance in astronomical units (AUs). The planet's mean distance could therefore be calculated from observations of the planet's orbital period. Kepler could not explain why this relationship worked. He could only intuitively conclude that it was by "divine plan".

Titius–Bode Rule

There the matter rested until 1766 when the German astronomer Titius von Wittenburg brought it to the attention of the astronomical world. Titius had developed a mathematical tool that showed the distances of the planets as a mathematical progression. His rule (not a physical law) begins with a sequence of numbers starting with zero (for the first planet, Mercury), then 3 (for Venus), and then doubling each number thereafter: 0, 3, 6, 12. . . . To each number he added the number 4 and then divided the result by 10. The progression then became 0.4, 0.7, 1.0, 1.6, 2.8, 5.2, 10.0. . . . These numbers are very nearly the mean distances of the planets in astronomical units. The first four numbers are the distances of the terrestrial planets out to Mars at 1.6 AU. Jupiter is 5.2 AU from the Sun. Between Mars and Jupiter Kepler's gap appeared at 2.8 AU. Johann Bode, director of the Berlin Observatory, was impressed by Titius's rule and used it on numerous occasions to convince his colleagues that a planet was missing. Then Herschel discovered Uranus in 1781. Its mean distance, 19.6 AU, was precisely that predicted by the Titius–Bode Rule. That the rule could predict the distance of an unseen planet was proof enough of its validity. (The rule completely fails for Neptune and Pluto but these planets were not known at the time.) The rule predicted a planet at 2.8 AU, so it seemed to these astronomers that a planet must exist there.

Discovery of the first asteroids

Over the next several years there were fruitless attempts to find the missing planet. Finally, a young Hungarian astronomer, Baron Franz von Zach, decided to organize a society of astronomers with the sole purpose of locating the planet he was convinced occupied Kepler's gap.[1] Six astronomers met on September 11, 1800, to form the society which would be composed of 24 astronomers scattered throughout Europe. Each member would be assigned a region of the sky along the ecliptic in which they would search for the hypothetical planet while at the same time mapping each section with the added purpose of making a new updated star chart.

The scene shifts to Palermo on the island of Sicily where Europe's southernmost observatory had been constructed a decade before the century's end. Its director, Giuseppe Piazzi, was already hard at work revising and correcting existing star charts when he received word of the new society. He turned his attention to a region along the ecliptic in the constellation Taurus and began checking star positions, star by star, making position corrections along the way. On the night of December 31, 1800, he noticed an 8th magnitude star that was not on the chart. Noting its position he returned the next night to find that it had moved. Over the next month or so Piazzi observed the motion of the body with a final observation on February 11, 1801, when an illness brought his observations to a close. He remained secretive about the discovery, calling it a comet even though he observed no gaseous coma or tail. Moreover, the motion of the object was nearly circular, unlike most comets. He wrote a letter to Bode at the Berlin Observatory which Bode received on March 20 telling him of his observations until February 11. This convinced Bode that Piazzi had indeed located the missing planet. In the letter Piazzi had named the new planet, Ceres, after the Roman goddess of agriculture.[1]

[1] Piazzi actually named the first asteroid Ceres Ferdinandia, in honor of the king of Naples and Sicily, Ferdinand IV. This was a political move not uncommon in those days. Galileo called the four moons of Jupiter the "Medician Stars," in honor of his patron and protector, Cosimo dé Medici. Likewise, William Herschel called his newly discovered planet (Uranus) Georgium Sidus after King George III of Great Britain. Only in those countries in which the discoveries were made did these political names persist for a time, but eventually (and fortunately), they were dropped in favor of the more traditional classical names.

Fig. 11.1. The motion of asteroid 1 Ceres is seen in this photograph during an hour-long exposure made with the 13-inch astrograph at the Lowell Observatory in Flagstaff, Arizona, in 1931. Several hazy objects in the field of view are distant galaxies. (Courtesy the Lowell Observatory.)

Curiously, Piazzi had not included positional data needed to confirm the discovery. Understandably, Piazzi had wanted to calculate and publish the orbit of Ceres on his own, but he needed further position observations beyond February 11. Piazzi sent his observations to J.J. Lalande at the Paris observatory on May 31 and to Bode in Berlin on June 11 hoping that further observations could be made, but by that time Ceres had changed its position so that it could not be located at either of these observatories. It appeared that Ceres, the first asteroid to be found, was also the first to be lost. The problem of calculating the precise orbit of Ceres was a formidable one that was tackled by an equally formidable mathematician, Karl Gauss. Gauss calculated an orbit for Ceres from Piazzi's earlier positional data and furnished positions that enabled Ceres to be relocated. It was a year to the day of the initial discovery that Ceres was recovered – January 1, 1802 (Fig. 11.1).

But more discoveries followed shortly. On March 28, 1802, Heinrich Olbers, a member of von Zach's search group, found the second asteroid, Pallas. With only a year's respite, the third asteroid, Juno, was found on September 1, 1804. The final asteroid located in this initial seven year discovery period was Vesta, the brightest of the four asteroids and Olber's second find, on March 29, 1807. There the count remained for the next 38 years. Then, in 1845 a German amateur found the fifth asteroid, Astraea, and the race was on again. Toward the end of the century the count had reached over 300. All of these were found visually with the eye at the telescope. In the last decade of the nineteenth century, film sensitivities had increased to a point where moving asteroid images could be recorded. Moreover, film could record far fainter images than the eye could see. Exposures of several hours along the ecliptic would often record an asteroid in motion, leaving a characteristic light streak on the photographic plate (Fig. 11.2). With photography, the number of discoveries increased at a rapid pace. By the mid-twentieth century, well over 4000 asteroids had been discovered, catalogued and orbits calculated.

Modern discoveries

Today the search has entered the electronic arena. Electronic components called charge-coupled devices (CCDs) with sensitivities to light hundreds of times the sensitivities of the fastest films have taken over the search. Automated systems comb the skies every night searching for the elusive fragments of other worlds. As a result, the number of newly discovered asteroids has risen dramatically. Many amateur astronomers today have at their disposal relatively large aperture commercially made telescopes equipped with sensitive CCD electronics that only professional observatories had a decade earlier. With this equipment they can detect small asteroids in near-Earth orbits down to 18th magnitude or fainter (Fig. 11.3). Their dedication has added significantly to the discovery rate.

Among the score of sky-survey projects currently underway world-wide, three stand out prominently: the California Institute of Technology Near-Earth Asteroid project (Jet Propulsion Laboratory); the Massachusetts Institute of Technology's Lincoln Near-Earth Asteroid project (LINEAR); the University of Arizona's Spacewatch program (Steward Observatory.) All three of these programs are searching for small asteroids that pass near Earth's orbit. More will be said about this important group of asteroids later.

Numbering and naming asteroids

Today it is an enormous task to keep track of the asteroids being discovered. The count at the end of the twentieth century is approaching 15 000[II] and all indications are that the rate of discovery will continue to rise as progressively smaller and fainter fragments become detectable. Early attempts to keep track of the increasing population of asteroids met with limited success. After World War II, the International Astronomical Union established the Minor Planet Center, headed by Paul Herget at the Cincinnati Observatory. In 1968, Brian Marsden took over the directorship and moved its

[II] In May, 2000, in a private correspondence I asked Brian Marsden for an updated figure. His response was truly remarkable.

> There are currently 14,788 numbered asteroids, all of them . . . with orbits that are very well defined. Later this month, the total will clearly rise above 15,000; the number is in fact nowadays doubling in only three to four years. In addition, there are currently more than 26,000 unnumbered asteroids with "good" orbits, and 31,000 more with "fair" to "poor" orbits. Business is booming!

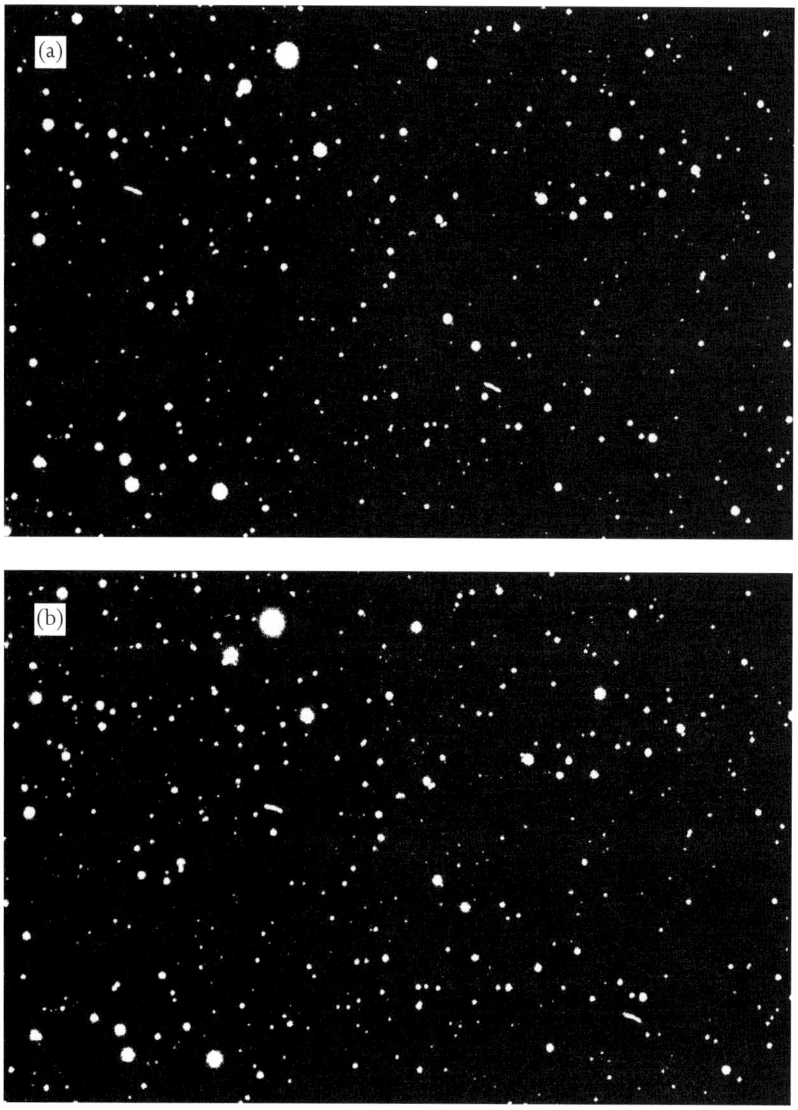

Fig. 11.2. These images show the trails of two unnamed asteroids photographed at Lowell Observatory. Both asteroids are the same brightness and their motions during an exposure lasting 2½ hours appear identical showing them to be at the same distance from Earth. The total angular motion of both asteroids among the stars during a 24 hour period can be estimated by comparing the star fields and asteroid trails of the upper and lower photographs. (Courtesy the Lowell Observatory.)

location to the Smithsonian Astrophysical Observatory adjacent to Harvard University in Cambridge, Massachusetts. Huge volumes of observational data from professional and amateur astronomers world-wide pour into the Center every month. As Brian Marsden is fond of saying, "If it moves or it's fuzzy, it goes through us." (The fuzzy refers to outgassing asteroids and comets.) Here the data on each discovery are run through computers to see if there is a match with known or suspected asteroids or comets. If the object appears to be a new asteroid, a temporary or provisional designation is given. To be eligible for a provisional designation the object must be observed over at least two nights. The provisional number is simply a combination of the year and month of discovery. For this convention, the year is divided into 24 intervals, each covering half of a given month and given an alphabetical designation. Every month has two consecutive letter designations, January (AB); February (CD), March (EF) and so on. Thus, March would be designated E or F, the E corresponding to the interval between March 1 to 15 and the F corresponding to March 16 to 31. This single letter is followed by another letter beginning with A, the first asteroid found in that interval. Thus, if an asteroid is found on March 5, 1999 and is the third asteroid to be found in the interval March 1 to 15, its designation becomes 1999EC. The number is provisional because its position must now be compared to known and catalogued asteroids or provisional objects that have been observed through at least one opposition. (Opposition occurs when the asteroid, Earth and Sun make a straight line. At that time Earth is between the Sun and the asteroid and the asteroid is at its closest point to Earth. At opposition, the asteroid rises as the Sun sets since they are "opposite" each other in the sky, separated by the Earth.) If the newly discovered asteroid cannot be linked with any known provisional asteroids, then further observations are made and from these an orbit is computed. Ideally, the object should be followed for several months as it moves along its orbit. This allows the orbit to be refined and further searches are made along the way to again attempt to link it to other provisional asteroids. If it passes these tests and no links are found, there is a very good chance that the asteroid is indeed a new find. Now a lengthy wait for

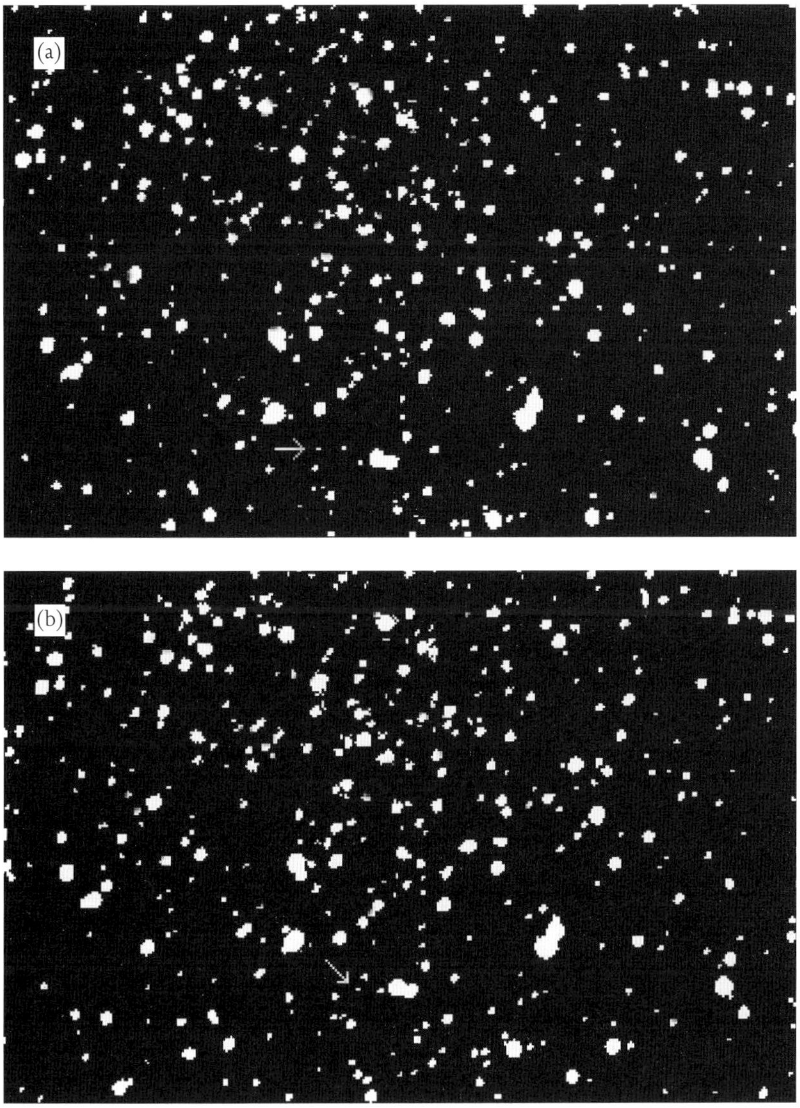

Fig. 11.3. These two photos made with a 12-inch telescope equipped with CCD electronics show the motion of a 16+ magnitude asteroid discovered by J.L. Schiff and C.J. Schiff at Takapuna, New Zealand. The time between exposures was 75 min. The star images in the constellation Sagittarius appear as clusters of pixels on the picture. The asteroid was discovered on September 27, 1999, and given the temporary designation 1999 SO9. Confirmation of the discovery soon followed and now carries the designation, 12926 Brianmason, named in honor of the New Zealand–American meteoriticist, Brian Mason. (Photos courtesy C.J. Schiff and J.L. Schiff.)

permanent status begins. Further observations must be made through several oppositions. This usually takes several years. For example, the time between consecutive oppositions for the middle main belt asteroid Ceres at 2.77 AU from the Sun is 1.28 years (synodic period). Normally, four oppositions of the new asteroid suffice to award it permanent status. Only then is the provisional designation dropped and the asteroid given a permanent number that denotes the numerical order of discovery. Thus, Ceres is designated 1 Ceres, the first asteroid to be discovered, 4 Vesta, the fourth discovered, 433 Eros, the 433rd asteroid to be given permanent status.

Naming asteroids is another matter entirely. The discoverer has the option of naming his discovery. At first female names from classical Greek and Roman mythology were chosen, fitting names for companions of the planets. This was short-lived since the growing number of asteroids soon used up all the well-known mythological characters and the names became increasingly obscure. The twentieth century saw the classical names replaced by whatever the discoverer chose and the practice quickly became subject to abuse. There were no rules. In the heavens are asteroids named in honor of scientists, artists, composers, rock stars, film stars, politicians, and historical characters, some of dubious distinction, such as 1489 Attila. Eyebrows were raised when Karl Marx was honored with asteroid 2807. Finally, in 1982 a committee was formed to examine proposed names and judge their suitability. Guidelines were drawn by the Small Bodies Names Committee of the International Astronomical Union. The guidelines state that the name should be pronounceable and expressed in a single word of no more than 16 characters. For political and/or military figures, the committee suggested they would be unsuitable until at least 100 years after their deaths. Finally (I regret to say), names of pets are discouraged.

The first asteroid discovered (1898) to cross the orbit of Mars was given a masculine name, 433 Eros, the Greek god of love, which set a convention for all Mars-crossing asteroid discoveries. Another group of asteroids was found to share Jupiter's orbit at Jupiter's L_4 and L_5 Lagrangian points (Fig. 11.5). These are the *Trojan* asteroids and are assigned names

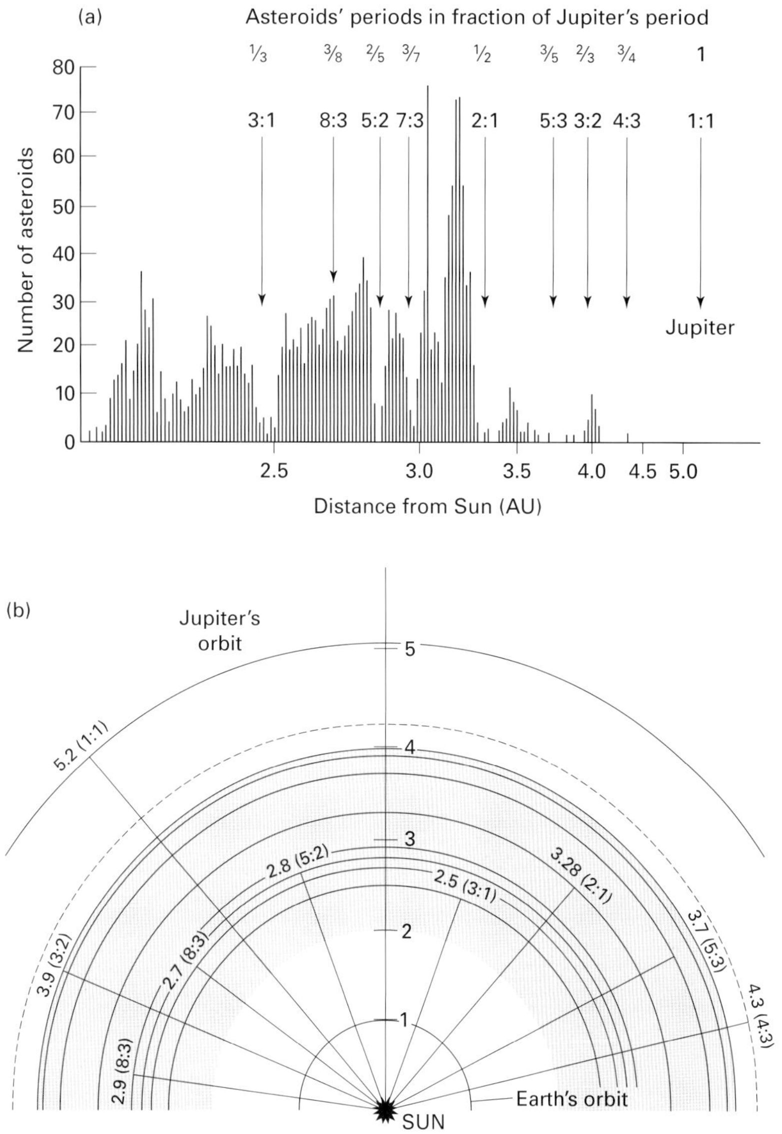

Fig. 11.4 (a) The asteroid belt showing the distribution of asteroids between 2 and 4 AU from the Sun. The distribution is not uniform but shows a series of gaps where the asteroid population is considerably diminished. These are the Kirkwood gaps. A plot of asteroid distances verses their orbital periods with respect to Jupiter's period shows the gaps to conform to simple fractions of asteroid vs Jupiter orbital periods. (b) View of the asteroid belt showing the position of the major gaps as plotted in (a).The scale is in astronomical units.

taken from the Homeric legend of the Greco-Trojan War. Those at the L_4 point are heroic Greek names and those at the L_5 point are heroic Trojan names. Still other asteroid-like objects were found far from the asteroid belt. One group exists beyond the Jovian planets but approach or cross their orbits at their perihelia. These are called *Centaurs* and are given like names.

Main belt asteroids

Most asteroids with relatively stable orbits are found within a belt or torus extending from about 2 AU to 4 AU. These are the *main belt asteroids*. This is where the majority of asteroids discovered over the past two centuries, including the largest, reside and these are the asteroids that astronomers have observed in their struggle to match meteorites with their asteroid parent bodies. By 1866 a sufficient number of asteroids had been discovered to establish the belt as a real entity, but it was apparent to American astronomer Daniel Kirkwood and others that the asteroids were not distributed uniformly within the belt. A plot of their distances from the Sun showed that there were gaps in the belt where the asteroid population dropped to near zero. Kirkwood realized Jupiter's role in herding the asteroids into a main belt but he also recognized that under Jupiter's gravitational influence some of the asteroids had unstable orbits. Especially prominent were gaps at 2.50, 2.83, and 3.28 AU respectively (Fig. 11.4a,b). Kirkwood calculated the orbital periods of the gap positions and found that they were related to Jupiter's orbital period (11.86 years). Asteroids in gap positions would have orbital periods that were simple fractions of Jupiter's period. For example, an asteroid at 3.28 AU has a period of 5.94 years, exactly one half of Jupiter's period. This means that every two years, Jupiter and the asteroid will experience a close encounter. At 2.83 AU an asteroid has a period of 4.76 years. Thus, for every two revolutions of Jupiter, the asteroid makes five revolutions, so that every 23 years the asteroid and Jupiter have a close encounter. Astronomers call this

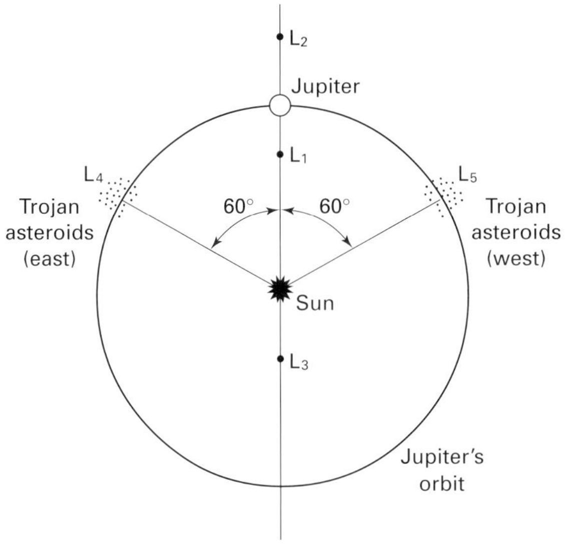

Fig. 11.5. Location of the Lagrangian points on the orbit of Jupiter. The two stable positions where the Trojan asteroids reside are at L_4 and L_5, 60° on either side of Jupiter. In a two mass system (Jupiter and the Sun) other Lagrangian positions exist (L_{1-3}) but they are metastable, subject to perturbations by Saturn. Since Jupiter's orbit is not circular, solar perturbations tend to scatter the Trojan asteroids on either side of the 60° line on Jupiter's orbit.

gravitational link *resonance*. Asteroids in *commensurate* orbits (orbits whose periods are simple fractions of Jupiter's period) experience periodic gravitational perturbations from Jupiter with much greater frequency than that experienced by main belt asteroids not in resonance with Jupiter. The *Kirkwood* gaps, as they are now called, are resonance zones controlled by Jupiter that result in unstable orbits within the belt. Any asteroid entering these zones will only be temporary residents. Over a mere few million years, their orbital eccentricities will gradually increase, eventually accelerating them along an elliptical path that will carry them out of the resonance position and across the zone of stable orbits, increasing the possibility of collisions with other main belt asteroids. If their progress is not hindered by collision, they stand a good chance of leaving the main belt altogether eventually to enter the realm of the *near-Earth asteroids*.

Trojan asteroids

Not all commensurate orbits in resonance with Jupiter produce gaps. Three are known that can concentrate asteroids. In 1906, the German astronomer, Maximilian Wolf discovered an asteroid that shared Jupiter's orbit. It was found nearly 60° ahead of Jupiter. This asteroid named Achilles occupied a position first studied by the French mathematician Joseph Lagrange in 1772. Lagrange postulated that there are five positions in an orbiting system involving two massive bodies where a third body (or a group of small bodies) could exist, sharing the motion of the gravitating bodies around a common center of mass. In this case, Jupiter and the Sun are the massive bodies sharing the center of mass (Fig. 11.5) and points L_4 and L_5 are the *Lagrangian points* in which small bodies could coexist sharing a mutual orbit with Jupiter.[III] The points are stable 60° east and west (ahead and behind) of Jupiter. The locations of the Trojan asteroids are at the 1:1 resonance point, a unique position in which small bodies can gather and remain stable for an indefinite period of time. Dozens of Trojans have been catalogued and numbered, the largest being 624 Hector, measuring ~300 × 150 km. Many more are known to exist with diameters of 15 km or less and recent surveys suggest that they number in the thousands, comparable to the main belt population of smaller objects.

There are two additional resonance concentrations. The Hilde asteroids are concentrated at 4.0 AU and have orbital periods of 8 years or 2/3 of Jupiter's period. A single asteroid, 279 Thule, at 4.26 AU, has a period 3/4 that of Jupiter.

Near-Earth and Earth-crossing asteroids

I mentioned earlier that the first asteroid discovered that had left the main belt and crossed inside the orbit of Mars was 433 Eros. Such asteroids were subsequently labeled *Mars-crossing* asteroids. These asteroids do not cross the orbits of any of the other terrestrial planets and usually their perihelia are well beyond Earth's orbit. Asteroid 433 Eros comes within 21 million kilometers of Earth's orbit. A subclass of Mars-crossers was designated after 1221 Amor was discovered in March, 1932. This asteroid had a perihelion of 1.08 AU, only 14 million kilometers from Earth's orbit. *Amor* asteroids as they are now called are defined as having perihelia between 1.0 and 1.3 AU (Fig. 11.6). Eros easily falls into this subgroup. Within a month of 1221 Amor's discovery another asteroid left its mark on a photographic plate at the Heidelberg Observatory on April 24, 1932. The long, thin trace of the fast moving object could only mean that this asteroid, today designated 1862 Apollo, was very close to Earth. At the time of its discovery, it was only 11 million kilometers from Earth's orbit – and approaching. Apollo became the prototype for asteroids that have perihelia inside Earth's orbit. In fact, Apollo crosses the orbit of Venus. Both Amor- and Apollo-type

[III] Of the five Lagrangian points, only two points, L_4 and L_5, are stable. The other three points are metastable. If a body were to occupy L_1, L_2 or L_3 it would remain indefinitely as long as there were no outside gravitational influences. If a fourth body entered the system, it would quickly be perturbed out of its position and would be lost to the system. If the L_4 and L_5 points are perturbed by an external source (Saturn or Mars) they could oscillate around the Lagrangian points drifting to either side by as much as 20°. Also, stability at the Lagrangian points requires that Jupiter have a circular orbit so that it does not vary in distance from the Sun. Since the orbit of Jupiter has an eccentricity of 0.05, this requirement is not met, which further perturbs the Trojans so that, in reality, they are rarely found exactly at the 60° Lagrangian points.

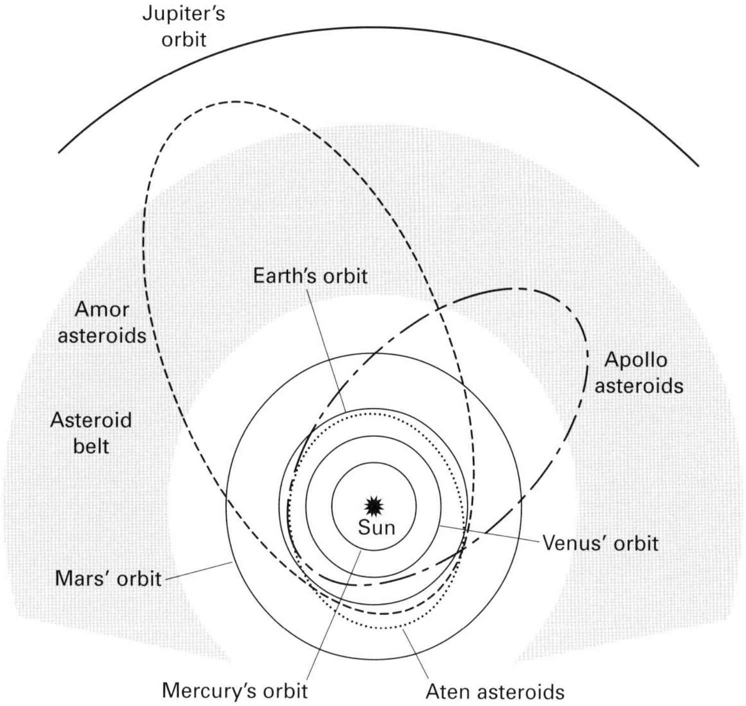

Fig. 11.6. Typical orbits of the three near-Earth asteroids: Apollo, Amor, and Aten. Both Apollos and Amors have their aphelia well inside the asteroid belt but their perihelia is either inside Earth's orbit (Apollos) or at or near Earth's orbit (Amors). Aten asteroids no longer have aphelia in the main belt. All known Atens have aphelia outside and perihelia inside Earth's orbit.

asteroids are collectively known as *Earth-approaching* or *Earth-crossing* asteroids. More than 260 Apollos and 209 Amors had been discovered and catalogued by the close of the twentieth century. Neither Amors nor Apollos have broken their ties with the asteroid belt, however. Their aphelia lie well within the belt so they return to their place of origin during every orbit. The prototype of one final group of asteroids was discovered by Eleanor Helin of the California Institute of Technology during a near-Earth asteroid survey at Palomar Mountain in 1976. This new asteroid had completely broken its ties with the asteroid belt. Its mean distance from the Sun lay within Earth's orbit. This asteroid, 2062 Aten, became the prototype *Aten* asteroids. These spend almost all of their time inside Earth's orbit but they can cross Earth's orbit for a few days as they pass through their aphelia. More than 30 are known. Altogether, the number of known near-Earth asteroids has reached over 500, estimated by some to be about 45% of the total population.

Asteroids as parent bodies of meteorites

Almost everyone in meteorite circles agrees that the vast majority of meteorites are fragments splintered off asteroid parent bodies. Meteorite collections worldwide must sample at least 135 separate asteroids and the number goes up yearly. A few very rare meteorites may be comet fragments although, as we have seen in recent years, the distinction between asteroids and comets is becoming more obscure. Other rare meteorites, the SNC suite of martian meteorites and the nearly two dozen known lunar meteorites make up the remainder. That there is a general consensus that meteorites come from the asteroid population does not prove it so, however. How do we know? A variety of approaches to this number-one problem in meteoritics has been suggested and tried in a continuing effort to establish a connection between meteorites and asteroids. These approaches run the gamut from orbital dynamics of meteoroids to comparative spectrophotometry of both meteorites and asteroids. Close encounters with a few main belt asteroids by spacecraft on their way to other planets have yielded remarkable high-resolution photographs of heavily cratered bodies. None were parent bodies; all were irregular fragments of other worlds.

Meteoroid orbits and the asteroid belt

In Chapter 2, I described the first photographic networks set up to capture on film the passage of bright, potential meteorite-dropping fireballs with the goal of determining their orbits. In three cases, ordinary chondritic stony meteorites were recovered. Figure 2.8 shows the three meteoroid orbits with aphelia at different points within the asteroid belt. To these three have been added three additional meteorite-dropping events, all whose orbits are consistent with the others and which resulted in recovered chondritic meteorites (and one iron – Sikhote-Alin).

During the lifetime of the networks, hundreds of fireballs were photographed providing sufficient data to calculate their preterrestrial orbits also. Figure 2.9 showed the preterrestrial orbits of twelve of these recorded by the European Fireball Network and randomly selected from a group of 46 that probably dropped meteorites.[IV] Most of the 46 orbits show perihelia just inside Earth's orbit and aphelia somewhere within the asteroid belt. A surprising 24% have perihelia inside Venus' orbit, 8% have aphelia inside Mars' orbit (Amor fragments), two have aphelia just outside Earth's orbit and perihelia near Venus' orbit (Aten fragments) having broken their ties to the asteroid belt altogether, and one actually crosses Mercury's orbit. It seems clear that through this pioneering work, we can be reasonably certain that Earth is being visited by asteroid fragments that have escaped the main belt. It remains now to unscramble the asteroid population and match meteorite with asteroid parent body.

Classifying asteroids through infrared reflectance spectrophotometry

If meteorites are fragments of asteroids then we can only conclude that there are at least as many asteroid types as there are meteorite types. In reality, our meteorite collections most certainly do not include all of the asteroid types that must exist. Judging from the abundance of ordinary chondrite falls, it appears that our collections are heavily biased.

If asteroids are the parents of meteorites then we should expect their mutual mineralogies to reflect that association. All the asteroids we have seen close-up have shown the effects of an impact history. Their surfaces must be a jumbled mix of fragmented rock ranging in size from boulders to tiny grains. Together they form a regolith much like the regolith on the Moon. Incident light from the Sun strikes the surface, absorbing much of the light and reflecting the remainder. The ratio of incident to reflected light, called the *albedo*, is a statement of the darkness of the asteroid, or more accurately, the reflective efficiency of the surface material. Since the surface is composed of rock fragments made of mineral grains with varying compositions, the reflectivity depends upon how each mineral responds to the visible, infrared and ultraviolet spectrum. Atoms of a mineral are arranged in a geometric crystal lattice with each corner, center and face occupied by elements composing the mineral. Each atom is surrounded by a cloud of electrons in various "orbits" or energy levels. These electrons are free to move to different levels further or closer to the nucleus given the proper energy conditions. Electronic transitions that carry electrons outward away from the nucleus requires that the atom receive and absorb an input of energy of a specific wavelength or frequency. Specificity is the word here. Each element is unique in that it requires a specific amount of energy before the atom will absorb the energy and the electron will move to a higher energy level. A continuous

[IV] Data from Table 4 of Halliday *et al.* (1996). Detailed data for 259 fireballs from the Canadian Camera Network, *Meteoritics and Planetary Science* **31**, 193.

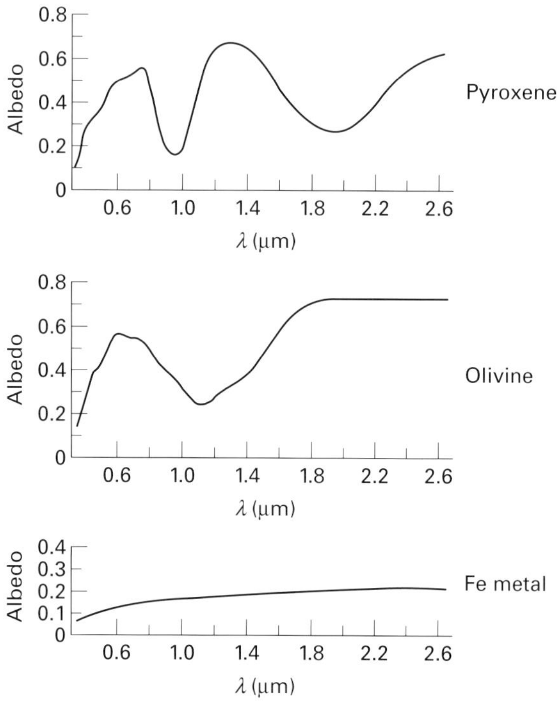

Fig. 11.7. Reflectance spectra of the three most common silicate minerals found in ordinary chondritic meteorites: olivine, pyroxene, and metallic iron. Note that pyroxene has a double valley in its absorption spectrum while olivine has only a single absorption feature. Both spectra are in the near infrared. Iron is characterized by a relatively flat curve with no absorption features.

spectrum of energy from ultraviolet through the visible to the infrared are available in sunlight and the atoms "choose" the energy best suited to the electron transitions. If energy is absorbed by the atom, energy at that wavelength is taken from the sunlight. The rejected light is the reflected light returned into space. It is the rejected light that we measure as the albedo of the asteroid surface. Visually, this selective absorption and reflection of light through the transparent surface crystals tends to darken the asteroid as well as give it a color.[V]

Specific minerals found on the surface of the asteroid can be inferred by a technique first successfully used on 4 Vesta by T.B. McCord in 1970 called *reflectance spectrophotometry*.[2] As sunlight passes through the mineral crystals, the crystals absorb specific wavelengths and reflect back a solar spectrum minus the absorbed wavelengths. Only a very thin layer (<1 mm) of surface minerals are sampled when measuring reflectance spectra. The surface mineral crystals do not produce sharply defined, high-resolution absorption lines like we are used to seeing in the solar spectrum. Instead, the absorption features are broad dark bands that are actually a composite of several minerals that compose the rock. These can be sorted out by comparing the spectra with laboratory reflectance spectra of pure minerals (Fig. 11.7). Although there may be many minerals composing the rock, only a few produce prominent absorption features in asteroid spectra. Typically a spectrum of an asteroid with a high silicate composition show one or more steep dips indicative of olivine and/or pyroxene in the near infrared while asteroids with high metal content show a relatively flat curve in the near infrared. A number of variables such as packing and grain size can affect the steepness of the dips. Generally, the shape of the total curve is more important for asteroid/meteorite comparison than the albedo of the absorption dips.[3]

From reflectance spectra, a classification of asteroids was developed. This began modestly with a few types which have since grown to a long list including numerous subtypes. Table 11.1 lists the important types giving their albedos, possible relationships to known meteorites, and approximate locations in the asteroid belt.

Distribution of asteroid types

One of the most remarkable discoveries resulting from the classification of asteroid compositional types is that there is a distinct relationship between an asteroid's composition and its position in the asteroid belt. A look at Table 11.1 again shows that there is a further distinction of position and albedo. The table is arranged in order of decreasing albedo but notice that it is also arranged in order of increasing distance from the Sun. The innermost part of the belt houses the E-type asteroids with the highest albedo. Next come the S-type asteroids extending from the inner to the middle belt. The M-type fall between the S- and C-types in albedo and are found in the middle belt. The most abundant C-type have the lowest albedo and occupy the middle belt to the outer edge. Among the darkest bodies of all, the P-type exist primarily at the outer edge and beyond; the D-type are among the Trojan asteroids at the Lagrangian points of Jupiter's orbit. Figure 11.8 tells a fascinating story. It plots the distribution of the various asteroid types within the asteroid belt with their distance from the Sun. Notice that the curves for each asteroid type are given as a percentage of the total population of that type. For example, the S-type asteroids are found from the inner belt to beyond the middle belt but nearly 80% of their total population occupies the inner belt at 2.4 AU with the remainder scattered to the middle belt and beyond. This simply says that there is a substantial overlap of the types but the greatest number of their population does, in fact, occupy a specific location within or beyond the belt. The E-asteroids have a very narrow range of distance, virtually all of them found between 2.0 and 2.2 AU from the Sun with the great majority lining the innermost belt at 2.0 AU. The flatter the curves, the more uniformly distributed they are within the

[V] In asteroid terminology, the albedo of the asteroid in the near infrared is a measure of the "redness" of the asteroid. This does not mean that the asteroid is red in color in the visual sense but only bright in the infrared part of the spectrum.

Type	Albedo (%)	Meteorite association	Location
E	25–60	Aubrites	Inner belt
A	13–40	Pallasites, olivine-rich	Main belt (?)
V	40	Eucrites, basaltic	Middle main belt, 4 Vesta and fragments
S	10–23	OC (?), mesosiderites	Middle to inner belt
Q/R	Like S	Possibly unweathered OC, with variable olivine/pyroxene	Middle to inner belt
M	7–20	E chondrites, irons	Central belt
P	2–7	Like M, lower albedo	Outer belt
D	2–5	Trojans	Extreme outer belt, Jupiter L_4, L_5 points
C	3–7	CM carbonaceous chondrites	Middle belt, 3.0 AU
B/F/G	4–9	C subtypes	Inner to outer belt

Note: *Arranged in decreasing albedo.

Table 11.1. **Asteroid classification***

Asteroids are classified by the shapes and slopes of their reflection spectra. Their surface compositions are deduced by these curves which defines the various classes. These, in turn, are compared to meteorite reflection spectra. This table shows the important classes of asteroids and their comparison to meteorites. Although the albedo or surface reflectivity does not by itself define the classes, it is an important observational quantity that relates directly to the asteroid/meteorite comparisons. For example, the V-class asteroids have a 40% albedo, similar to eucritic meteorites. Likewise, the C-class has one of the lowest albedos, 3–7%, very similar to CM carbonaceous chondrites. There also seems to be a relationship between asteroid compositions and their location in the asteroid belt, included in the far right column. This concept is further developed in the text.

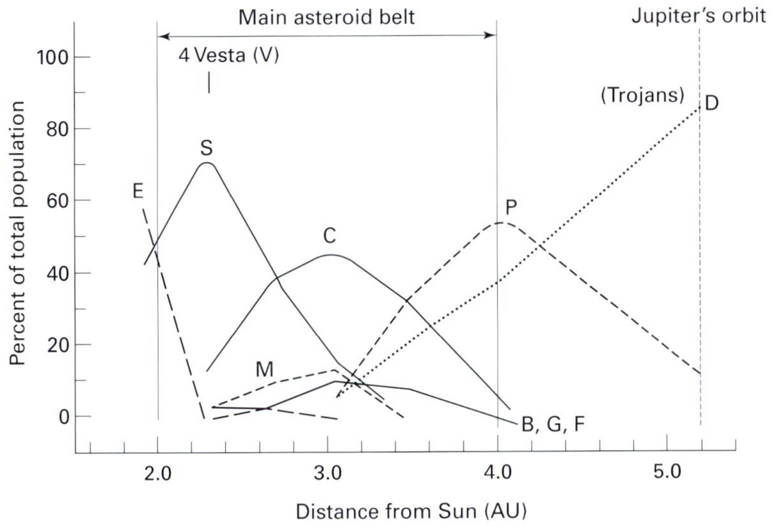

Fig. 11.8. Plot of the distribution of the primary asteroid types with their distances in astronomical units (AUs) from the Sun. Each curve shows a specific asteroid type and the distribution of their population along the belt. The plot demonstrates that the asteroid belt is chemically heterogeneous with asteroids of different mineralogies occupying relatively segregated parts of the belt. (Peter Thomas, Cornell University and NASA.)

belt. The B, F and G subtypes of the C-asteroids show a more or less even distribution from the inner belt to the outer belt.

This distribution of asteroid types is telling us something important about the conditions in the solar nebula early in its history. The inner belt contains E- and S-type asteroids made up of refractory minerals with few volatiles as we might expect being closer to the Sun. The middle and outer belts contain the C-asteroids and their relatives with more volatiles, water-bearing minerals and organic compounds typical of the icy environment at that distance from the Sun. There has been mixing due to perturbations and collisions throughout the belt's history which would tend to scatter the asteroids of any given type, but the concentrations of asteroids at specific distances where they probably originated in the solar nebula still remains. That the asteroid belt is today heterogenous and relatively unmixed may be telling us that conditions in the belt were not so violent as to homogenize the belt. It has been pointed out by J. Wasson and others that the vast majority of breccias in H, L, or LL chondrites are either monomict or genomict breccias with very few polymict breccias of different asteroid or meteorite classes. This suggests that there were collisions and mixing among each asteroid class but few interactions with asteroids of different classes, being separated by great distances within the belt.

Matching meteorites with parent asteroids

Once reflectance spectra are obtained the next step is to search for a match with a particular meteorite type. The best comparisons are made when meteorites are ground to a powder to simulate an asteroid regolith. Laboratory reflection spectra of the meteorites are then made and compared to asteroids. A few of these comparisons are seen in Fig. 11.9a, in which reflectance is plotted against the wavelength in micrometers for four main belt asteroids and the near-Earth asteroid 433 Eros. An additional near-Earth asteroid, 1685 Toro, is plotted in Fig. 11.9b. The plots are matched with the most common meteorite groups. The comparison is striking. In all cases, major meteorite groups do tend to match the spectra of both main belt and near-Earth asteroids. Of particular interest is the Apollo asteroid 1685 Toro which seems to match the spectrum of L-type chondrites (shaded area in Fig. 11.9b). In addition, asteroid 433 Eros' spectrum also lies close to the L-type meteorites, in particular, type L4. These near-Earth asteroids could be the progenitors of the L-type ordinary chondrites in Earth collections.

C-type carbonaceous asteroids

The most abundant asteroids are the C-type, comprising more than three-quarters of the main belt population. They are all dark bodies with albedos averaging about 3.5, only half the albedo of the Moon. In Fig. 11.9, their spectra appear flat and nearly featureless. But when we cross the 3.0 μm line, about two-thirds of the C-asteroids show evidence of combined water, displaying a broad absorption band between 3.0 and 3.4 μm in the infrared. This is due to hydrated minerals. The amount of water in the hydrated minerals can be judged from the depth of the absorption band. The remaining C-asteroids show far less water (CK4 or CR2) or almost none at all (CV3). The water-bearing C-asteroids may be CI and CM parent bodies while the "dry" C-asteroids may have been derived from CV parent bodies. Some workers speculate that water-bearing and dry asteroids may have originated as a single large parent body with a heated, metamorphosed core surrounded by "wet" layers representing CI and CM asteroids. These wet layers could have been stripped off the parent body by repeated impacts and lost. This idea, though interesting, does not withstand close scrutiny. There is little mixing among the carbonaceous chondrite groups which is inconsistent with what one would expect if they were derived from a single parent body. Textural differences among the CM, CO and CV chondrites such as chondrule size and proportion of textural types suggest that they formed in different solar nebular environments. One would not expect to find sorting of chondrule size and textural type on a local level from material accreting onto a single parent body. Finally, plots of the oxygen isotopic compositions of the mafic minerals in CI, CM, and CV show wide differences in position on the carbonaceous chondrite mass fractionation line with CI near the intersection with the terrestrial line and CV much further down the line (see Fig. 7.18). This strongly suggest that each formed as a separate body in chemically different locations in the evolving solar nebula.

Asteroid 2 Pallas is an interesting C-asteroid (subtype B) that shows an excellent spectroscopic match with the rare less hydrated CR chondrites. Figure 11.10 shows the reflectance spectrum of 2 Pallas superimposed over that of the CR prototype, Renazzo. The superposition of the two is striking.[4]

S-type asteroids and ordinary chondrites

S-type asteroids provide a dilemma. They are the second most abundant asteroid type and most are found near the inner edge of the belt. Most meteoriticists believe the S-type are the parent bodies of ordinary chondrites but there are differences between the two that have to be reconciled. Hundreds of S-type asteroids have been studied but only about 16% have spectra that are close to but do not match exactly ordinary chondrite compositions. Rather, some researchers interpret the S-type spectra as a best fit for the metal-rich stony-iron meteorites (mesosiderites, pallasites, 1AB irons and perhaps primitive achondrites). An absorption feature at 1 μm in the spectrum clearly was produced by a combination of olivine and pyroxene and this was superimposed over a relatively flat linear infrared spectral signature of metal. But we saw in Chapter 5 that ordinary chondrite meteorites are primarily composed of olivine and pyroxene and as much as 16% iron–nickel metal. The ratio of olivine to pyroxene is relatively uniform in each of the groups (H, L, LL) varying small amounts due to thermal metamorphism and their state of oxidation. Thus, we have both stony-irons and ordinary chondrites as contenders of the S-type asteroids.

In 1984, an important paper by Michael Gaffey, an asteroid researcher at the Rensselaer Polytechnic Institute in Troy, New York, was published that seemed to tie the S-type asteroids to the stony-irons.[5] He noted that as the asteroid 8 Flora rotated, the mineralogy of the asteroid changed showing a mineral diversity, varying iron–nickel and olivine content with both higher than in ordinary chondrite meteorites but acceptable for a differentiated asteroid. Moreover, the metal spectral signature of ordinary chondrites tended to be

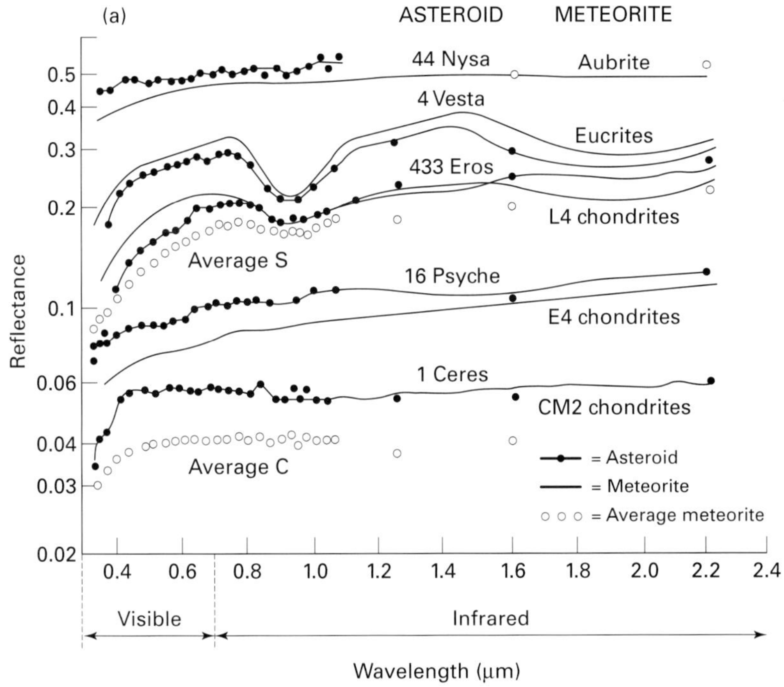

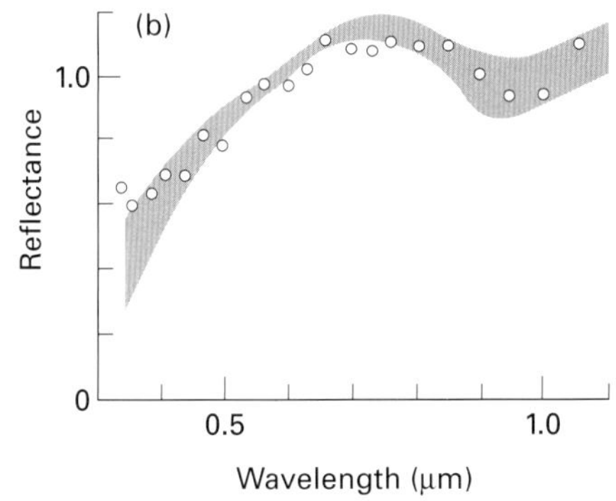

Fig. 11.9. (a) A comparison of asteroid and meteorite reflection spectra from the visible to the infrared. The solid lines show laboratory reflection spectra of several meteorite types, differing in mineralogy. These are compared to asteroid reflection spectra of surface minerals made with telescopes from Earth and designated by black dots on a thin line connecting them. Also noted are data points showing average S-type and C-type spectra (open circles) From these spectra, a close comparison seems to exist between the mineralogy of meteorites and asteroid surfaces. (Adapted from the work of Clark Chapman, Southwest Research Institute, Boulder, Colorado.) (b) The reflection spectrum of asteroid 1685 Toro shown here as open circles is compared to the reflectance spectra of several type L chondrites integrated here as a gray field. Note how the spectra of the two closely match each other especially with a prominent absorption band near 1 μm. Toro remains a good candidate as perhaps one of several parent bodies for the L chondrites. (Chapman, C.R. (1973). Minor planets and related objects X. Spectrophotometric study of the composition of 1685 Toro. *Astronomical Journal* **78**, 502.)

neutral in color (a flat, linear curve), not "reddened" (higher reflectance) like the metal in stony-irons. (No one knows why there is a different spectral signature in the metal of ordinary chondrites vs stony-iron metal.) But this still left the troubling problem of finding the parent body of the ordinary chondrites, the most common meteorites to fall to Earth.

Something of a compromise to the dilemma (or "S-type conundrum" as asteroid researchers refer to it) was reached in 1993. Recognizing that the great majority of S-type asteroids taken as a single type show substantial mineral diversity prompted Gaffey and his coworkers to divide the S-type into seven subtypes with varying ratios of olivine to pyroxene. Figure 11.11 shows a plot of the seven S-subtype fields, S(I)–S(VII). Major absorption bands due to a combination of olivine and pyroxene appear at 1 μm referred to as *Band I* and another major absorption band due to pyroxene appears at 2 μm, *Band II*. The vertical axis is centered around the center wavelength (1 μm) of Band I and the horizontal axis is the ratio of areas of Band II to Band I. Using Band II/Band I area ratios helps to minimize the effects of textural differences such as variations in grain size on the shape of the spectra. This diagnostic technique emphasizes spectral differences between olivine and pyroxene.[6] All the S-subtype asteroids plot within the areas noted somewhere along the curving line which was derived from spectra made in the laboratory of various mixtures of olivine and pyroxene. The olivine content increases from S(VII) to S(I). Pyroxene increases from S(I) to S(VII).

The ordinary chondrites were assigned to subtype S(IV) as the parent body with the other six S-subtypes assigned to various differentiated asteroids or to asteroids that have not yielded meteorite samples on Earth.[6] So the S-type asteroids appear to be a mix of chondritic and differentiated parent bodies. Of the 39 large S-type asteroids Gaffey studied, nearly 30% were S(IV) subtypes. Among them was the large

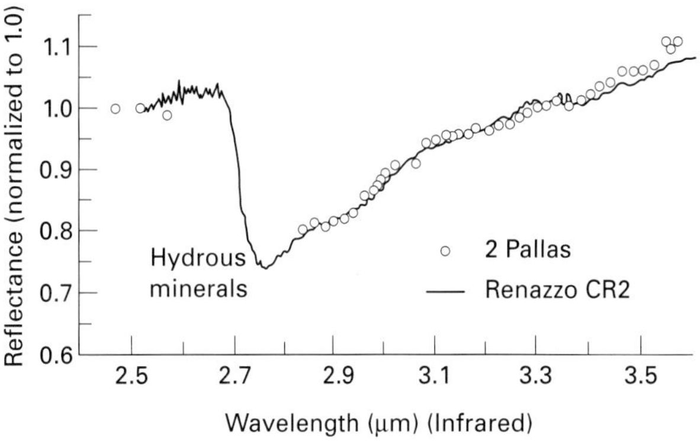

Fig. 11.10. Comparison of reflection spectra of the CR2 chondrite, Renazzo (solid curve), and the main belt asteroid, 2 Pallas (open circles). The match is almost perfect. (Adapted from Sato *et al.* (1997). *Meteoritics and Planetary Science* **32**, No. 4, 503–507.)[4]

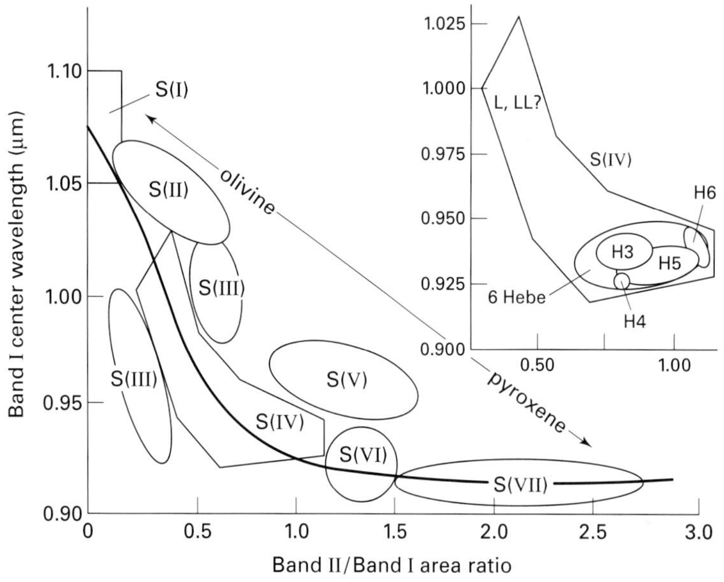

11.11. This graph shows the S-type asteroids subdivided into seven fields. The curve passing through the fields is derived from laboratory spectra of various mixtures of olivine and pyroxene. The upper left field is nearly pure olivine and the lower right field is nearly pure pyroxene with various mixtures between. Laboratory measurements of ordinary chondrite reflection spectra match best the S(IV) field. The other fields may represent some achondrites, stony-irons or possibly parent bodies not yet represented in meteorite collections. Specific absorption bands (I and II) are plotted to define the fields (see text). The lower part of the S(IV) field is occupied by the H chondrites of various petrographic types. The L and LL chondrites probably occupy other positions in the field not yet defined. The asteroid 6 Hebe plots within the H chondrite portion of the field. Variations in its plotted position represented here by an oval field is thought to denote variations in surface mineralogy as the asteroid rotates. There seems to be little doubt that 6 Hebe is an H chondrite parent body. (Adapted from the work of M. Gaffey, Rensselaer Polytechnic Institute 1993.)

asteroid 6 Hebe whose reflectance spectra placed it near the bottom of the S(IV) field where H chondrite compositions are found. We can assume that the L and LL chondrites will fall somewhere within the S(IV) field but this has yet to be determined.

A dynamical factor entered the picture in 1985 that strongly pointed to the S-type asteroids as the parent bodies of the ordinary chondrites – delivery of S-type asteroid fragments to Earth. In 1985, dynamicist J. Wisdom of MIT showed that the S-type asteroids most likely to deliver meteorite fragments to Earth are those found in the inner part of the asteroid belt at 2.5 AU. This is near the 3:1 Kirkwood Gap where resonance with Jupiter is particularly strong.[7] Asteroids within this gap take on chaotic motions that enlarge their orbital eccentricities and inclinations relative to the ecliptic and carries them out of the asteroid belt within the relatively short time of a million years or so. Some eventually become Earth-crossing. Significantly, S(IV) asteroids make up more than half of the S-type asteroids near the 3:1 resonance gap. Of these, asteroids 3 Juno, 6 Hebe, and 7 Iris are the leading candidates as the most likely ordinary chondrite parent bodies to supply meteorites to Earth.[8]

The great abundance of ordinary chondrite meteorites observed to fall on Earth, and the relative rariety of like bodies within the asteroid belt remains a dilemma. A number of suggestions have been made to explain the paradox. Of these, two in particular stand out as plausible, working either singly or in combination. (1) Ordinary chondrites come from a relatively few parent bodies that just happen to be well situated near the 3:1 escape zone. This does seem likely as we saw earlier. (2) Our interpretation of S-type spectra is misled due to *space weathering* of surface materials on the asteroids that changes the optical qualities of the common minerals found in ordinary chondrites.[9] Although space weathering on asteroids has yet to be well defined, there is a weathering effect upon minerals that have remained exposed for millions of years in the space environment. Bombardment by solar wind particles, micrometeorites and high-energy electromagnetic waves (ultraviolet, X-rays, cosmic rays) can and do change the spectral reflectance and albedo of surface minerals that could mislead us to conclude that S-type asteroid compositions are different from ordinary chondrites. The idea of space weathering on S-type asteroids can only be tested by close encounters with remote-sensing spacecraft. We will see in the next section how this was first accomplished during the close encounters with asteroids 951 Gaspra and 243 Ida.

M-type metal asteroids

All asteroid types possess some ambiguities in that their reflectance spectra and albedo can be assigned to more than one meteorite group. The M-type asteroids are no different. Their spectral reflectance shows a relatively flat curve rising gently in albedo with increasing wavelength (Fig. 11.9a). There are no deep absorption features due to silicate minerals. M-type asteroids appear to be primarily metal, perhaps the denuded core of a fragmented differentiated asteroid. The classic example is the main belt asteroid 16 Psyche. Metal-rich E chondrites show an almost identical nearly featureless spectrum. Recall that E chondrites are highly reduced meteorites with as much as 30% iron metal embedded in nearly pure enstatite. As enstatite produces no absorption features in its spectrum, an E chondritic asteroid would mimic an iron body. The majority of M-type asteroids seem to occupy the middle to the outer edge of the main belt, not a location where one would expect to find highly reduced asteroids such as E chondrites.

Physical characteristics of asteroids

For nearly two centuries astronomers observed asteroids as moving points of light. No telescope on Earth could resolve them into the extended sources they really are. Many asteroids varied in brightness over periods of a few hours suggesting an irregular-shaped body in rotation. Assuming that the albedos of the main belt asteroids were roughly the same, their apparent magnitudes were a crude indication of their relative sizes. This assumption, of course, is not true. A small, bright metallic or E-type asteroid could appear visually as bright as a much larger C-type asteroid of low albedo. Sometimes an orbiting asteroid would align itself with Earth and a distant star and would occult the star for a brief few seconds. With a knowledge of the distance of the asteroid from Earth as well as its orbital period, the asteroid's physical size could be calculated. But most asteroids are not spherical and a stellar occultation viewed from one location on Earth could only give one plane through the asteroid. Many observations from different positions on Earth were required to define the shape and size of the asteroid. The first stellar occultation by an asteroid (Vesta) was made visually from Sweden in 1958.

Instrumentation and techniques referred to today as *remote sensing* developed in the latter half of the twentieth century and began to define the physical characteristics of asteroids. This is done today primarily using three approaches: (1) reflectance spectrophotometry, which we saw has given researchers data allowing them to classify asteroids according to albedo and mineral composition through absorption bands in their reflectance spectra; (2) the laboratory analysis of meteorites and their comparison to asteroids; and (3) close spacecraft flyby or orbit of selected asteroids. The first two are Earth-based and limited by the great distances of the main belt asteroids. The third approach, remote sensing from spacecraft, has given us our first closeup look at main belt and near-Earth asteroids. Table 11.2 shows the asteroids that have been imaged by radar, the Hubble Space Telescope or by spacecraft to the year 2000.

Key questions awaited extended spacecraft visits to the asteroids, questions that could best be answered definitively by close scrutiny. Let us consider these briefly before we review some of the results of recent spacecraft rendezvous with main belt and near-Earth asteroids.

Asteroid	Date imaged	Image	Type	MB/NEA	Size (km)
1989PB	Aug., 1989	Radar	?	NEA	1, 1
4179 Tautatis	Dec. 8, 1992	Radar	?	NEA	2.5, 1.6
1620 Geographos	Aug. 30, 1994	Radar	?	NEA	5.1 × 1.8
951 Gaspra	Oct. 29, 1991	Galileo	S	MB	18 × 10.5 × 8.9
243 Ida	Aug. 28, 1993	Galileo	S	MB	56 × 15
Dactyl*	Aug. 28, 1993	Galileo	S	MB	1.6 × 1.2
4 Vesta	May, 1996	HST	V	MB	500
253 Mathilde	June 27, 1997	NEAR	C	MB	66 × 48 × 46
9969 Braille	June 28, 1999	Deep Space 1	V	NEA	2.2 × 1
433 Eros	Feb. 14, 2000	NEAR	S	NEA	33 × 13 × 13
216 Kleopatra	May 04, 2000	Radar	M	MB	217 × 94

Notes:
*Dactyl is a satellite of Ida.

Table 11.2. **Asteroid images**

This table chronologically lists the important developments in the history of asteroid imagery. Before 1989, physical properties of asteroids were studied through Earth-based telescopic observations. These included crude estimates of their shape by their variations in brightness as they rotated. Their sizes were roughly estimated when they occulted distant stars. All that changed when the first radar images were made of near-Earth asteroid 1989PB, showing two distinct lobes apparently orbiting around each other. Three years later near-Earth asteroid 4179 Toutatis came within 3.5 million kilometers of Earth and radar imaging showed a near contact binary asteroid. The first high-resolution images awaited the Galileo spacecraft's flyby of two main belt asteroids, 951 Gaspra and 243 Ida in 1991 and 1993, respectively. The Hubble Space Telescope's high-resolution image capabilities were demonstrated with the first surface details of the main belt asteroid 4 Vesta. 433 Eros was orbited by the NEAR spacecraft February 14, 2000, allowing details a few meters across to be imaged.

Space weathering

The number-one question in asteroid research today is: are the ordinary chondrites derived from the S-type asteroids? We saw that about 16% (subtype S(IV)) had reflection spectra near laboratory spectra of ordinary chondrites. But 16% does not seem to account for the vast numbers of ordinary chondrites that have reached Earth relative to other meteorite types. Alteration of the optical characteristics of surface minerals on most of the S-type asteroids due to *space weathering* provides a way out of this dilemma. Space weathering has taken place on the Moon's regolith. Studies of lunar soils subjected to solar wind irradiation and micrometeorite bombardment have shown, in a general way, what chemical and physical processes are at work on the lunar surface. The process operates on a microscopic scale. Apparently solar wind hydrogen (protons) bombards the surface minerals down to ~25 μm depth causing the reduction of Fe^{+2} to microscopic particles of metal.[10] At the same time, micrometeorites impacting the minerals generate an extremely fine iron metal that deposits on the surface regolith particles.[11] The effects of this iron deposition was to lower the albedo, diminish the spectral bands and decrease the overall reddening.

Interesting laboratory experiments with provocative results were made even as the Galileo spacecraft was approaching the asteroid, Gaspra, the first asteroid to be visited. L.V. Moroz powdered a sample of the L5 chondrite, Elenkova, and, using a 30–40 kHz pulsed laser in a vacuum, zapped 100 μm areas on the powdered material simulating the impacts of micrometeorites and the solar wind. He and his coworkers measured the reflectance spectra of the meteorite before and after the laser treatment. The results were more revealing than they had expected. The Elenkova spectrum had been converted from OC spectra into S-type spectra. Albedos were lowered, contrast in the spectral absorption bands was diminished and spectral continuum reddened to match the S-type spectra. Thus, rapid melting and recrystallization of surface minerals essentially converted OC characteristics into S-type characteristics. An additional observation in the Moroz experiment is that the alteration moved the Band I center wavelength of Fig. 11.11 to longer wavelengths. This represents an olivine enrichment which means that the OC could be represented not only by the subtype S(IV) asteroids but also by the more olivine-rich subtypes (S(I)–S(III)). The researchers argued that the effect is real. The rapid cooling of the tiny melts resulted in recrystallization of the olivine while the pyroxene, lacking the time to recrystallize, accumulated as a glassy mesostasis. Thus, the pyroxene band weakens as the olivine band strengthens.[12]

It remains now to look for the predicted effects of space weathering during the flyby and orbital missions, for only during such close encounters do spacecraft instruments have the spatial resolution to detect similarities and differences in surface mineralogy. If space weathering is actively changing the optical properties of the surface minerals, we would expect to see a heterogeneous composition across the surface. Newly exposed subsurface material in the form of ejecta blankets around impact craters or mass wasting of material down slope would show less or no space weathering compared to the overall surface.

Onion shell vs rubble pile asteroid models

Up to this point we have concentrated on surface characteristics. Close scrutiny of asteroid bodies should allow us to determine accurately the internal characteristics of asteroids. In earlier chapters I alluded to an internal model of an ordinary chondrite asteroid as a body of low porosity and high density. Its accretion of chondrules and dusty matrix material built up a sizeable undifferentiated body. The accretion process trapped short-lived radioisotopes which, during the first 10 million years, heated the interior. The center probably reached nearly 1000 K, metamorphosing the rock to petrographic Type 6. The radioisotopes were probably uniformly distributed throughout the body but the center retained the heat longer due to the poor thermal conductivity of silicate rock. Nearer the surface the buried rock suffered less thermal metamorphism building layers of progressively lower petrographic types (Fig. 11.12). Assuming that there was no disruption or fragmentation of the parent body we can imagine that the body had a low porosity of only a few percent and a density of over 3 g/cm^3. This is the *onion shell* model.

Calculations of the percentage volume of the layers representing each petrographic type match reasonably well the relative proportions of the various petrographic types in our collections. They tell us that probably >75% of the volume of the OC parent bodies taken from the center outward were metamorphosed to Types 5 and 6. Equilibrated ordinary chondrites (Types 5 and 6) are far more common than unequilibrated chondrites in our collections. The *Catalogue of Meteorites* (2000) shows 75% H5 and H6, 86% L5 and L6, and 82% LL5 and LL6. These equilibrated meteorites represent volumes deep in the interiors of their parent bodies where they can only be sampled by complete disruption. The onion shell model seems consistent with internal radioisotopic heating but inconsistent with impact heating. Impact heating would affect the least metamorphosed layers near the surface and could act to equilibrate these near surface rocks to higher metamorphic types. Regardless, the ordinary chondrites in our collections tell us that at least three OC parent bodies (H, L, LL) must have suffered repeated fragmentation to the core.

Many asteroid researchers do not believe that most asteroids are pristine, layered, high-density bodies defined by the onion shell model. Instead, they imagine them as great disrupted and reassembled piles of rocky rubble. Asteroids are likely to collide with sufficient energy to totally fragment them but not sufficient energy to cause the fragments to disperse on escape trajectories. If an asteroid is catastrophically disrupted to the core by a major collision *after* it has

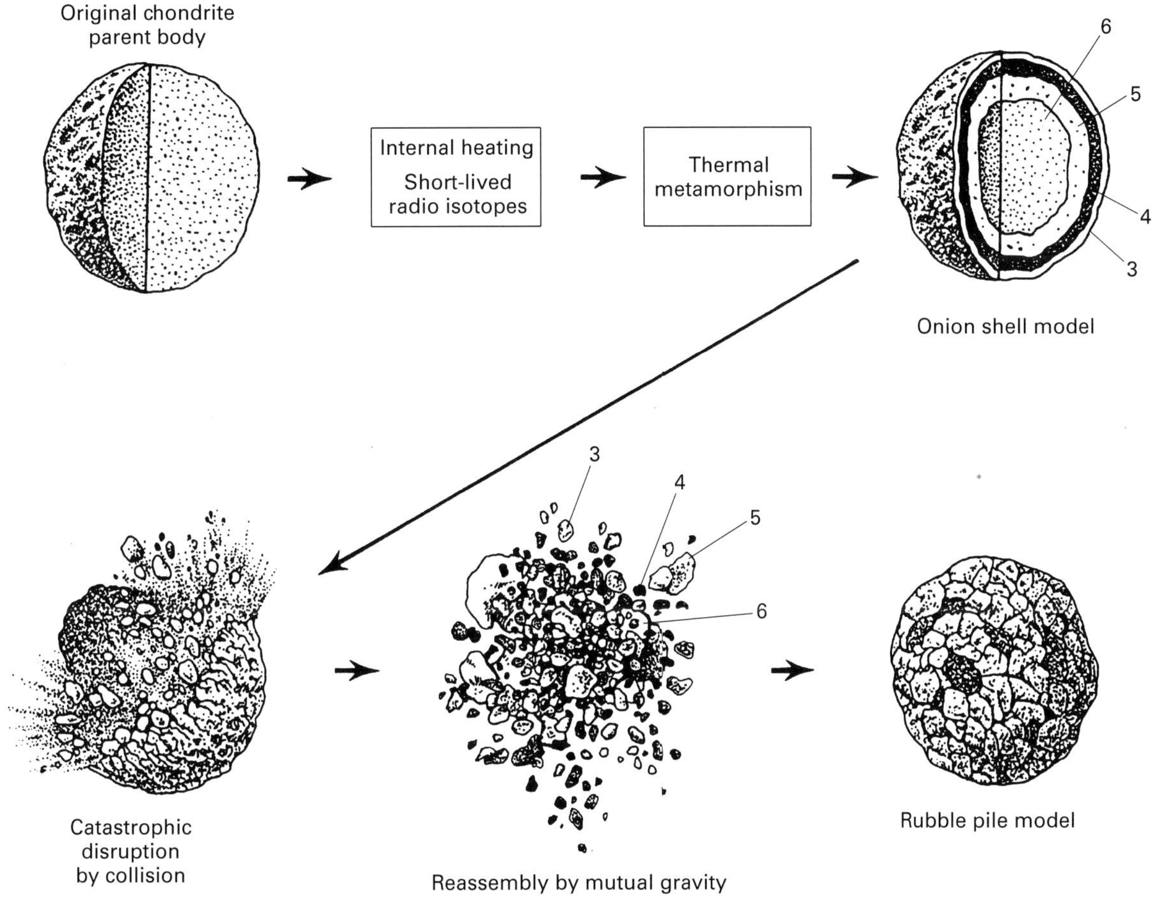

Fig. 11.12. Two models describe the interior of a chondritic asteroid parent body: onion shell and rubble pile. The onion shell model begins with an accreted primitive body internally heat by short-lived radioisotopes. The deep interior is thermally metamorphosed to petrographic Type 6 but surrounding layers receive less heating due to the slow conduction of heat through rock. This results in a layered structure with the least metamorphosed layers (Type 3) on the outside and progressively more metamorphosed layers in the deep interior. The rubble pile structure forms when the onion shell body is catastrophically disrupted by impact. The fragments gravitate together again, reassembling as a mix of petrographic types, into a rocky rubble pile.

metamorphosed into an onion shell structure, the pieces will probably not disperse but will reaccumulate together again into a gravitationally bound loose pile of rocks. Here the various petrographic types become mixed during reassembly (Fig. 11.12). Future, less energetic impacts will help to further fragment the pieces into smaller sizes building up a *megaregolith* and compacting the asteroid into a *rubble pile* body with a porosity of 25% or more and a density of perhaps <2.0 g/cm^3. Ordinary chondrites with genomict breccia textures probably bear witness to this rubble pile model.

One of the foundations of the onion shell model is the idea that the more deeply buried rock experienced the highest temperatures as well as the slowest cooling rate. One of the predictions of the model is that there should be an inverse relationship between cooling rates and petrographic type. To test this prediction required determining the cooling rates of several unbrecciated ordinary chondrites. G.J. Taylor and his colleagues at the Institute of Meteoritics, University of New Mexico, measured the sizes and nickel compositions of taenite grains in several H- and L-type ordinary chondrites. From this they determined the metallographic cooling rates based upon the methods of Wood (see Chapter 9). Their data showed that the cooling rates of H- and L-type chondrites *did not* correlate with petrographic type. This result implies either that the H- and L-type chondrites never had onion shell internal structures or that the H- and L-type parent bodies were disrupted and reassembled into a rubble pile while still above 500 °C, the temperature range in which cooling rates are recorded in metallic FeNi.[13]

Rubble pile models have been subjected to computer simulations which have revealed interesting characteristics. Since rubble piles initially are held together by mutual gravitation of their constituent parts, their ability to stay together is a function of their size (mass) and rotation period. Asteroid rotation periods range from 2.25 h (Icarus) to nearly 20 h. Some 83% rotate with periods between 4 and 16 h. Any rubble pile asteroid 200 m or more in diameter rotating faster than once every ~2.4 h (depending upon its bulk density) would very likely fly apart. Thus, we would not expect to find rubble pile asteroids with rotations ≤2.4 h. Another characteristic of an asteroid rubble pile is that it is actually more likely to survive a major impact. Computer simulations show that rubble piles readily absorb impact shock and only local damage is suffered. A crater may be formed but the asteroid will absorb the impact shock without disrupting. A solid, high-density asteroid is more likely to shatter into smaller pieces most of which would reassemble into a rubble pile.[14]

Asteroid 4 Vesta and HED meteorites

In 1970, T.B. McCord and his coworkers at the Institute of Geophysics and Planetology, University of Hawaii, published a short paper describing their work on the large main belt asteroid, 4 Vesta, in which they compared its reflection spectra to the Nuevo Laredo achondrite.[2] This meteorite is a eucrite, a member of the basaltic achondrites (HED) group. They were first to recognize the strikingly similar characteristics between Vesta and the eucrites. It was the first successful effort to relate a specific meteorite to an asteroid. What began as a modest two page report resulted in a major effort to connect Vesta with the HED meteorites.

Vesta is the third largest asteroid known, with a diameter of about 530 km and a rotation period of 5.3 h. As Vesta rotated, its reflection spectrum appeared to change, demonstrating that Vesta had a heterogeneous surface composition. The spectrum was eucritic with a strong pyroxene band near 1 μm, but as it rotated the spectrum changed with it to show a diogenitic or even an olivine spectrum. The interpretation was that Vesta was a differentiated asteroid and was heavily cratered. Lava flows of eucritic basaltic material covered the surface except where there were several large craters. The impacts that produced the craters excavated through the extrusive eucritic crustal layer to an intrusive plutonic layer or perhaps a plutonic body of diogenetic composition. The deepest parts of the craters, their centers, may have reached an olivine-rich mantle (Fig. 11.13).

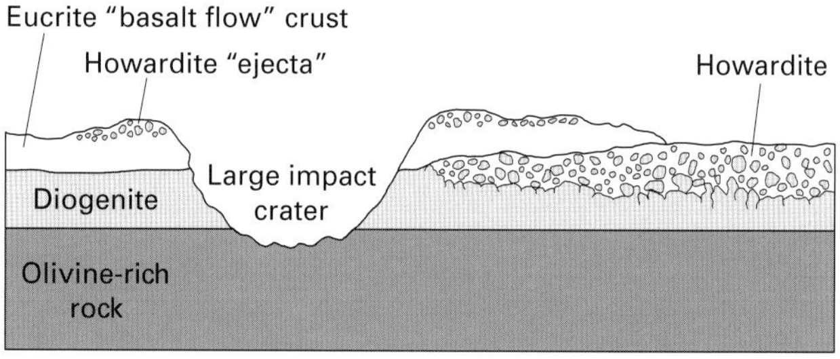

Fig. 11.13. Theoretical cross section of the crust and upper mantle of the differentiated asteroid, 4 Vesta. The crust is covered with eucritic rock of basaltic composition that has erupted onto the surface from fissures. Beneath the flows is a layer of plutonic rock with diogenetic composition. Impacts by meteorites penetrate the basaltic flows reaching the diogenite layer, excavating and mixing the two together. A regolith evolves by this gardening process that is welded together by other impacts to produce a third rock type, a howardite. This suite of HED rocks is thought to originate from 4 Vesta. There is at least one large crater on Vesta in which ultramafic mantle rock (olivine-rich) has been exposed but no ultramafic rock has reached Earth leading meteoriticists to believe that Vesta is still intact as a differentiated body.

There is little doubt now that the HED meteorites represent samples of the crust and the upper mantle of Vesta. Vesta, however, is compositionally unique among the asteroids. It is the only large main belt asteroid known that could be the source of the HED meteorites. But Vesta is in a stable orbit far from the 3:1 resonance "escape hatch". Dynamically speaking, the main body of Vesta itself could not be a contributor to the HED meteorites reaching Earth. This paradox was breached when in 1993, R.P. Binzel and S. Xu from MIT discovered that 20 small asteroids, all in the 5–10 km range and members of the Vesta family of asteroids, had spectra nearly identical to Vesta over a wavelength range between 0.4 and 1.0 μm and must therefore be chunks of crust and upper mantle fragmented off Vesta during a major impact event. These small asteroids all shared orbits similar to Vesta's but they were scattered between Vesta and a point near the 3:1 resonance at 2.5 AU forming a kind of continuous bridge of small V-type objects. These fragments from Vesta, called "Vestoids", are probably the sources of the HED meteorites. Small impacts among these fragments would break off meteorite-sized pieces that could easily be injected into the 3:1 resonance zone, which would change their orbits to highly eccentric Earth-crossing orbits within a few million years.[15]

To cap the evolving drama, researchers using the Hubble Space Telescope (HST) with its remarkable resolution imaged Vesta beginning in 1994. The first photographs showed Vesta with a somewhat flattened spheroidal shape. As the asteroid rotated, large circular areas in various shades of gray appeared. These were interpreted as impact craters that revealed varying mineralogies at different depths. Two years later, continuing with the HST, P.C. Thomas and coworkers at Cornell University discovered an enormous impact basin with a central rebound feature 12 km in height near the south pole of Vesta (Fig.11.14). The basin, much like the impact basins on the Moon, measures about 460 km across and 13 km deep. It is remarkable that Vesta survived an impact of this magnitude. Over a million cubic kilometers of rocky material was removed. Binzel's Vestoids are probably among the original fragments that escaped the asteroid. Three near-Earth asteroids with Vesta-like spectra were found in 1991 that are probably Vestoids that have made the jump from near-circular orbits in the main belt to Earth-crossing orbits. These are probably the source objects for the basaltic achondrites that continue to fall to Earth. Earth-crossing objects have short lifetimes relative to the age of the Solar System. They are destined to impact Earth or one of the other terrestrial planets within a few million years.[VI] It seems clear

[VI] HED meteorites are still making it to Earth. The latest, a diogenite called Bilanga, was observed to fall in Gomponsago, Burkina Faso, Africa, on October 27, 1999 (see Fig. 11.15).

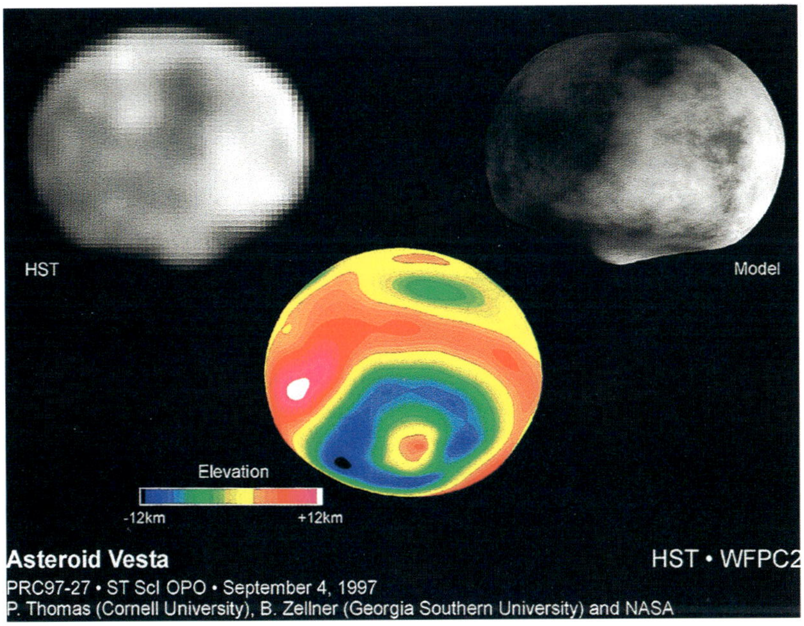

Fig. 11.14. Hubble Space Telescope image of the asteroid 4 Vesta taken in May 1996. (Upper left) The digital image appears asymmetric with many different elevations. The south pole appears to have a knob-like structure resting within a large impact area. (Lower center) This color-encoded elevation map of Vesta reveals an impact basin 460 km in diameter along with a central rebound peak. This may be the origin of the Vestoids and the HED meteorites. (Upper right) This three-dimensional model of Vesta was constructed from Hubble data. The impact basin and central uplift is clearly seen near the south pole. (Courtesy Ben Zellner, Georgia Southern University; Peter Thomas, Cornell University; and NASA.)

Fig. 11.15. This beautifully oriented meteorite, called Bilanga, is the most recent diogenite observed to fall. It was recovered at the time of fall on October 27, 1999. The specimen weighs 7.5 kg and measures 280 mm across the base of the cone. (Photo by Michael Casper.)

that there must be a reservoir of HED objects to continuously replenish the supply that are accelerated into Earth-crossing orbits. These could only come from Vesta fragments in the main belt.

Where are the mantle rocks?

When a chondritic body melts it differentiates into a concentric layered structure with an FeNi core, a thick olivine-rich mantle and a thin crust of pyroxene/plagioclase composition. Meteoriticists inventory their meteorites and attempt to fit the achondritic, stony-iron, and iron meteorites into an ideal differentiated parent body. The HED meteorites fit fairly well their concept of a crust of eucritic basalts, a diogenitic upper mantle rock, and a battered regolith composed of a mix of the two (howardites). The FeNi meteorites nicely fit the idea of a differentiated core body. The pallasites of roughly two-thirds olivine and one-third FeNi are thought to have formed at the core/mantle boundary. But where are the meteorites representing the mantles of these differentiated bodies? Meteorites from the mantle should be metal-free and olivine-rich, composed of as much as 80% olivine. (There are two known olivine-rich diogenites (one-third olivine) and a primitive achondrite, Divnoe, with 70 wt% olivine, and that's about it. The olivine-rich brachinites are not believed to have originated in a completely differentiated body; rather it is more likely that they originated from a partially differentiated body where metal and sulfide did not separate in the mantle.)

The paucity of mantle material could mean that very few chondritic asteroid parent bodies actually differentiated or were disrupted, but the number of irons representing different compositional groups and grouplets strongly suggest that at least 60 chondritic bodies differentiated. (Also, approximately 40 M-type asteroids are known and classified in the main belt.) The destruction of 60+ differentiated asteroids should have released and scattered large quantities of mantle material into the main belt. Some, like the HED meteorites, should have found their way to Earth.

There is a lack of direct evidence through meteorite samples that olivine-rich mantle rock has survived to the present. Likewise, there is no astronomical evidence that large differentiated bodies other than 4 Vesta have survived to the present in the main belt. Nor is there evidence of large numbers of mantle fragments currently present in the asteroid belt. The same can be said of basaltic crustal rock with the exception of HED material from Vestoids; Vesta's survival dictates that there can be few if any mantle rocks from Vesta. (The two known olivine-rich diogenites probably sample Vesta's mantle.)[16] It is possible that very early in the Solar System's history the crusts and mantles of numerous differentiated parent bodies were stripped away leaving behind a resilient tough FeNi core resistant to further fragmenting. Over the billions of years the mantle and crustal rock were broken into pieces by repeated impacts, pieces too small to be detected by current astronomical means. Cosmic-ray exposure age differences between irons (200–1000 Ma) and stones (~40 Ma) seem to bear this out. Thus, over the life time of the Solar System, we would expect only irons and stony irons to have survived to the present. The only reason that basaltic fragments from Vesta have survived is that they formed relatively recently. Vesta was impacted perhaps a billion years ago, creating fragments of basaltic rock that have been reaching Earth in very limited supply ever since.

If olivine-rich meteorites from asteroid mantles were reaching Earth we would expect to find them among the large cache of meteorites discovered in Antarctica over the past 30 years. Conditions in the Antarctic are more conducive to the survival of meteorites than in the temperate zones. Terrestrial ages of chondritic and achondritic Antarctic meteorites average about 140 000 years, although the record is over 2 Ma. Irons have shown terrestrial ages up to 5 Ma (see Chapter 3). Of the 20 000+ meteorites recovered in Antarctica so far, only two have shown the characteristics expected of mantle rock; both metal-free, olivine-rich diogenites.

It does not seem likely that out of all the differentiated asteroids that must have formed in the early Solar System that only one remains. We cannot rule out the possibility that space weathering has disguised these asteroids giving them the appearance of the numerous S or C-types, but the evidence from the world's meteorite collections seems to suggest that indeed olivine-rich asteroids, fragments of differentiated parent bodies, do not presently exist in the Solar System.[17]

Close encounters: flyby of asteroids 951 Gaspra and 243 Ida[18]

On its way to Jupiter, the Galileo spacecraft encountered the main belt S-type asteroid 951 Gaspra on October 29, 1991, the first asteroid to be visited by a spacecraft, passing it within 1600 km. Gaspra is a fragment sheared off of a much larger parent body. It is a small angular body only 18.2 × 10.5 × 8.9 km (Fig. 11.16). The asteroid shows a curious lack of large impact craters, only medium-sized craters that appear almost erased, smoothed over by ejecta from past impacts. Superimposed over this surface is a population of small, fresh-looking craters. This suggests that Gaspra's surface is relatively young, perhaps only 200 Ma. Its fragmented surface is not old enough to have accumulated large craters. Color images show only subtle differences across the asteroid but two distinct units appear in the infrared spectrum: a "blue" unit with a higher albedo and a deeper 1 μm absorption band, and a "red" unit with a slightly lower albedo and a shallower absorption band at 1 μm. Most of the asteroid's surface is covered with the red unit and only the ridges and fresh craters and ejecta show the blue unit. (The blue color is associated with ridges where down slope movement of regolith uncover fresh material. Fresh material is also seen

Fig. 11.16. This is the best color image of the asteroid 951 Gaspra taken by the Galileo spacecraft on October 29, 1991, from a distance of 5150 km. Gaspra, 18.2 km long, appears quite homogeneous over the entire visible surface and its color variations over the entire surface are likewise very subtle. They have been exaggerated here to clearly show them. The areas with subtle gray-blue tones appear to be fresh regolith exposed by landslides or impact. The fresher craters in particular show this color variation. These areas have a spectral absorption near 1.0 μm in the infrared. This is probably due to olivine. Redder areas are covered by an older regolith surface and tends to dominate low-lying areas. They show a lower albedo and a weaker absorption at 1.0 μm. This color variation is interpreted to be a consequence of space weathering. Gaspra is an S-type asteroid probably related to ordinary chondrite meteorites. (Courtesy NASA and JPL.)

inside the craters and in their ejecta. The true color of Gaspra is thought to be displayed by the blue units.) The subtle spectral differences between the blue and red units may be an effect of space weathering. A scan of the surface revealed a homogeneous but extremely olivine-rich composition. It may be an S-type body of perhaps subtype S(I) or S(II) or it may be a fragment of a differentiated body. The second choice seems less likely since scans across Gaspra's surface showed a homogeneous composition not expected of a differentiated body.

Asteroid 243 Ida was visited by the Galileo spacecraft two years later on August 28, 1993. Ida is roughly ellipsoidal in shape measuring $59.8 \times 25.4 \times 18.6$ km. Unlike Gaspra, there is nothing subtle about Ida's crater population. Ida has about ten times the craters of Gaspra, essentially saturated with large and small craters. Relative to Gaspra, it is an old surface, possibly 2 Ga. Ida is a member of the Koronis family of asteroids resulting from the breakup of an original parent body. It is possible that immediately after the impact disrupted the parent body some of the smaller Koronis family fragments impacted Ida, accelerating the cratering rate normally seen in the asteroid belt. Thus, Ida may not be as old as it appears from its cratering record.

Ida is covered by a much thicker regolith than Gaspra. It is thought to extend to a depth of 150 m. Large boulders scattered around the surface are witness to further fragmenting. To researchers, Ida appeared to be a good candidate for a rubble pile asteroid. The deep regolith can cover large boulders and fragmental material to form a megaregolith with considerable void space. If Ida is a rubble pile, its density should be much lower than a typical ordinary chondrite. One of the big surprises was the discovery of a 1.5 km diameter satellite orbiting some 45 km from Ida's surface (Fig. 11.17). Normally, a satellite's orbital period can be used to determine the mass of the body being orbited but from the spacecraft's perspective, the orbit was highly foreshortened (nearly edge-on) and did not allow an accurate mass to be found. Orbital constraints were recognized so that it was possible to estimate a reasonable mass. From the mass and known volume, Ida's bulk density, or at least its bulk density limits were determined. The density had to be between 2.0 and 3.1 g/cm^3. The upper value falls within the range of the ordinary chondrites. The lower value could also include ordinary chondrites if there was sufficient void space, assuming Ida is a rubble pile.

It was Ida's spectral variations that finally convinced the meteoritical world that space weathering did, indeed, exist. Just like on Gaspra, only many times more prominent, were the two color regions that divided the surface features into old and new. The old surface (red) appeared spectroscopically like a typical S-type asteroid, but the relatively newly excavated impact crater areas (blue) revealed spectra showing prominent deep absorption bands due to olivine and pyroxene. This was especially true of the ejecta from a large crater called Azzurra that has spread across a large portion of the surface. These spectra bear a remarkably close resemblance to ordinary chondrite spectra or subtype S(IV) asteroids. It seems clear that given sufficient time, space weathering changes the spectral characteristics of the surface of asteroid parent bodies of ordinary chondrites to mimic other S-type asteroids. This means that other S-subtypes, especially the olivine-rich subtypes may be disguised S(IV) ordinary chondrite asteroids.

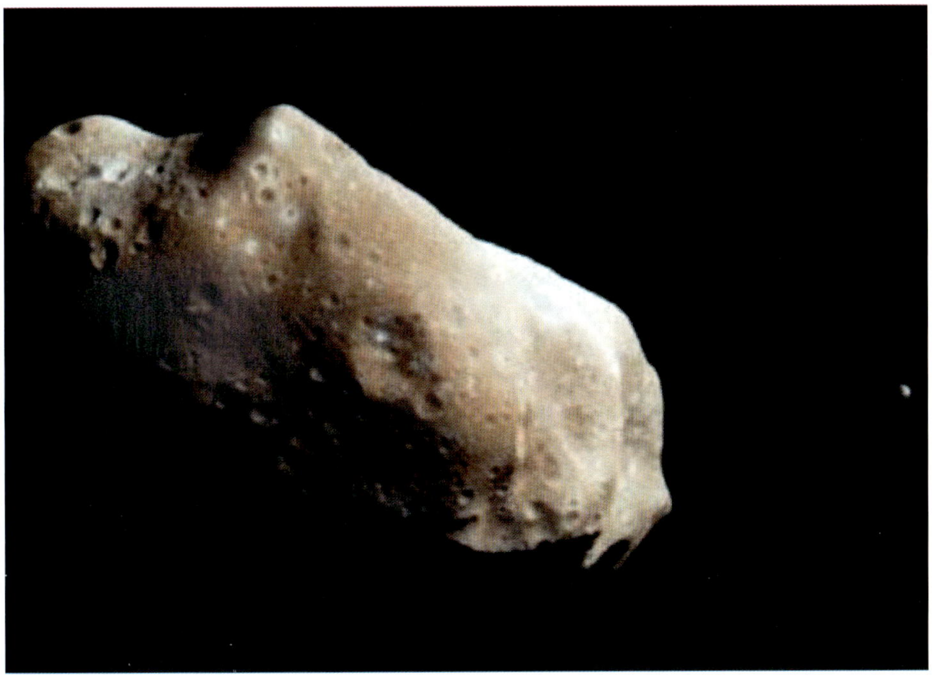

Fig. 11.17. This image of asteroid 243 Ida was taken on August 28, 1993, from a distance of 10950 km by the Galileo spacecraft on its way to Jupiter. This heavily cratered asteroid is 59.8 km long and is a fragment of a much larger parent body. Spectral studies of the surface of Ida have yielded strong evidence of space weathering that modifies the spectral characteristics of this S-type asteroid (see text). This wide angle view revealed for the first time a satellite of an asteroid. The satellite to the right of Ida, now called Dactyl, is approximately 48 km from Ida's surface in this view. Dactyl is a roughly spherical body 1.5 km in diameter and, like its parent, shows a cratered surface. (Courtesy JPL and NASA.)

Finale: NEAR encounter with 433 Eros

The first two close-up encounters with asteroids were fortuitous secondary targets of the Galileo mission. They happened to be near the planned trajectory of the Galileo spacecraft as it made its tortuous lengthy journey to Jupiter via two flybys of Earth and its Moon and one flyby of Venus, picking up gravitational energy (gravity assist) to accelerate the spacecraft to Jupiter.

There had not been any space mission with an asteroid as a primary target until the NEAR mission was conceived. NEAR, an acronym for *Near Earth Asteroid Rendezvous*, was the first mission in NASA's Discovery Program. NASA designated the Applied Physics Laboratory of Johns Hopkins University to manage the project. The NEAR spacecraft was the first to operate beyond the orbit of Mars on solar power alone, without the use of a thermonuclear power source. The mission called for one gravity assist through a flyby of Earth before heading out to the near-Earth asteroid, 433 Eros. The plan called for a flyby of one main belt asteroid, 253 Mathilde, during the first orbit around the Sun and before it encountered Earth (Fig. 11.18). After receiving a gravity assist from Earth on January 23, 1998, NEAR was scheduled for orbit insertion around Eros almost a year later, on January 10, 1999. But on December 20, only three days before the scheduled rendezvous, the mission nearly ended in disaster when its main engine misfired causing the orbit insertion to be aborted. Instead, the spacecraft performed a flyby on

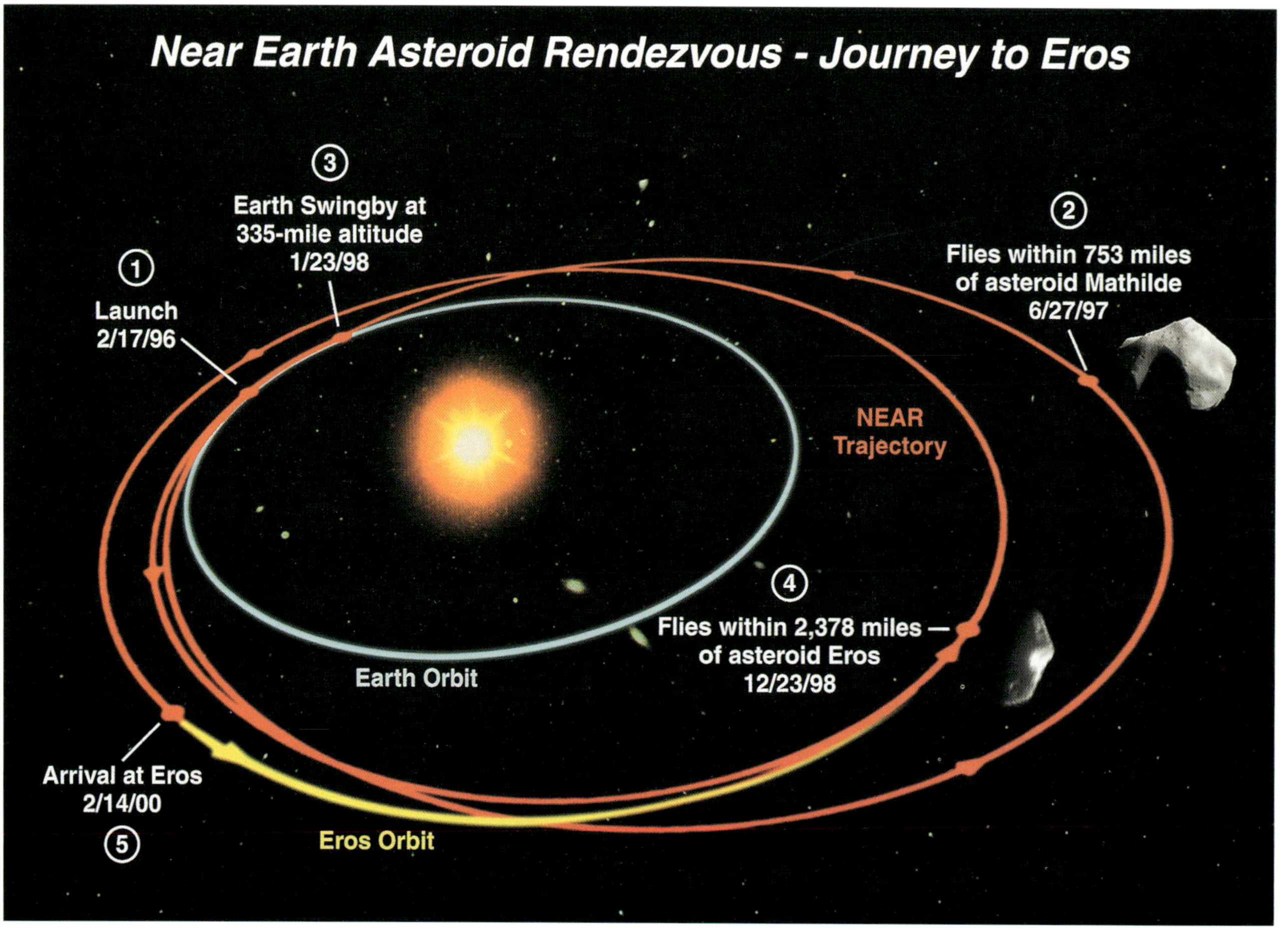

Fig. 11.18. Trajectory of the NEAR spacecraft to 433 Eros. To reach a trajectory that would intercept Eros' orbit, a gravity assist by Earth was necessary. On the first trajectory, NEAR encountered the C-type asteroid, 253 Mathilde. After a deep space maneuver, it returned to Earth's vicinity picking up gravitational energy that carried it to a rendezvous with Eros on December 23, 1998. A malfunction of its main engines shortly before encounter caused the orbital insertion to be aborted. An additional year would go by before NEAR again rendezvoused with Eros. This time an orbit insertion was successfully established on February 14, 2000, beginning a year-long study of this near-Earth asteroid. (Courtesy Applied Physics Laboratory, Johns Hopkins University, and NASA.)

Fig. 11.19. Global image of the C-type main belt asteroid, 253 Mathilde, during NEAR's encounter on June 27, 1997. Since the albedo of Mathilde's surface is only 4% the image necessarily had to be enhanced to better show the surface details. The enormous crater on the sunlit side measures about 20 km across. The apparent flat top is the rim of another large crater. (Courtesy APL, JHU and NASA.)

December 23, coming within 3830 km. A subsequent successful firing of the engine put NEAR on course to a second rendezvous with Eros. This unscheduled flyby turned out to be a good thing. It gave scientists an opportunity to see Eros relatively close-up and to make fundamental measurements that gave them valuable information, making the orbit insertion much less risky. On February 14, 2000, with a delay of over a year, scientists and engineers successfully placed NEAR into orbit around Eros marking the first time a spacecraft had achieved orbit around an asteroid. Also on that historic day NASA renamed the NEAR spacecraft the NEAR-Shoemaker spacecraft in honor of the American planetary science pioneer, Dr Eugene M. Shoemaker.

Flyby of 253 Mathilde

Earlier, on June 27, 1997, another historic first occurred when NEAR passed within 1212 km of the C-type asteroid 253 Mathilde. Mathilde was discovered over a century ago but not until 1995 did ground-based observations identify the asteroid as a C-type (Fig. 11.19). The asteroid measures $66 \times 48 \times 46$ km and has an albedo of 0.04, about half the reflectivity of charcoal. It is heavily cratered; five craters on the sunlit side had diameters larger than 20 km. Its surface is dark and colorless, showing no compositional variations over the 60% of its surface in sunlight at the time. Its appearance and low albedo are suggestive of a CM carbonaceous chondrite. Estimates of its volume made from photographs and its mass from gravitational effects on the spacecraft yielded a bulk density of only 1.3 g/cm^3. This is very low compared to a typical CM chondrite ($\sim$2.5 g/cm^3).[VII] This low density strongly suggests a porous rubble pile structure with an interior void space approaching 50%. The exceptionally large craters on Mathilde further suggest that it may have a rubble pile structure. The largest crater is 30 km across, approaching the size of Mathilde itself. Several others are in the 20 km range. If Mathilde was a dense OC-like body, there is a good chance that it would have disrupted. A rubble pile structure, however, is an excellent absorber of impact shock. Large craters would have formed as they actually did at the impact points but its rubble pile structure would have prevented its breakup. Mathilde's rotation period is 17.4 days, one of the longest known among the asteroids. (Only two other asteroids have slower rotations: 288 Glauke and 1220 Crocus.) It is possible that Mathilde's collision history may be responsible for slowing its rotation.

One characteristic surprised investigators. Mathilde was exceptionally bland in color and remarkably uniform in albedo. It seems to be homogeneous on the surface and throughout its interior. This is especially obvious noting the impact ejecta around small craters and on the interior walls of the largest craters all which show the same reflectivity and color as the general surface. This extreme homogeneity along with 1–3km diameter craters that saturate the surface speaks of a very ancient, primitive body.

NEAR-Shoemaker orbits 433 Eros

Meteoritical science took a giant step forward the day NEAR began orbiting Eros. At long last the opportunity had arrived to study a near-Earth S-type asteroid for a full year. Armed with a near-infrared spectrometer (NIS) that can measure reflected light at wavelengths between 0.8 and 2.7 μm, it was able to scan the surface with increasingly high spatial

[VII] Two other asteroid-like bodies are known to have exceptionally low densities. These are the moons of Mars, Phobos and Deimos, with densities of 1.9 and 1.8 g/cm^3, respectively. This, along with their low albedos ($\sim$0.04%), strongly suggest that they are captured C-type asteroids.

Fig. 11.20. These six color images of 433 Eros were made from a distance of 1800 km on February 14, 2000, during NEAR's final approach prior to orbit insertion. The images show Eros through one full rotation period (5h, 16 min.). The view point of the spacecraft is along Eros' north polar axis. The color shows approximately how Eros would appear to the unaided eye. Eros' largest crater, 5.5 km across and filled with shadow, can be seen in several of these images. (Courtesy APL, JHU, NASA.)

resolution as the orbit was gradually lowered from an initial spacecraft to surface distance of 350 km to a 50 km orbit through the year, revealing the distribution and abundance of surface minerals that for the first time provided a strong link between meteorites and asteroids. It also confirmed a space weathering phenomenon that measurements of both Gaspra and Ida had hinted. At the lowest altitudes (35 km) achieved by mid-2000 the NIS mapped the surface mineralogy at a resolution of 250 m. At the same time, an additional instrument, the X-ray/gamma-ray spectrometer determined the elemental composition of surface materials.

A multispectral imager (MSI) had the task of imaging Eros in visual and infrared light. By the end of the mission, it will have returned sufficient high-resolution images allowing an accurate map of the surface to be made with a resolution of 3 m. As this book is being written, the MSI is taking daily photographs that will be used to determine Eros' shape and size and its spin characteristics. Already (September 2000) NEAR has returned over 103 300 images. As NEAR approached Eros just prior to orbit insertion, a series of photographs was taken from which a motion picture was constructed, dramatically showing an end-over-end rotation around the narrowest part of the shoe-shaped asteroid (Fig. 11.20). From these images its rotation period of 5.27 h was confirmed.

Fig. 11.21. Two views of Eros seen from a distance of 355 km shows its two opposing hemispheres. The upper view is dominated by Eros' largest impact crater, 5.5 km across. The surface is covered by craters and the interior walls of the larger craters show a pattern of brightness variations, the result of slumping and mass movement. The lower view is dominated by a saddle-shaped depression, 10 km wide, in which the craters have been totally obliterated. Lineations of grooves and ridges can be seen around the rim of the depression. The total view is 31 km across and detail as small as 35 m can be seen. (Courtesy APL, JHU, and NASA.)

Overall global properties showed Eros to be a peanut-shaped mass with minimum and maximum diameters of 8.7 and 31.6 km respectively with a maximum dimension of 34 km. The largest crater, seen at a distance of several hundred kilometers, measured 5.5 km across and nearly diametrically opposed to the crater was a curious saddle-shaped depression 10 km across that showed a smooth surface with networks of lineations but with virtually no craters (Fig. 11.21.) As the resolution of the images increased with progressively lower orbits, wonderful details began to appear. Much of the surface

Fig. 11.22. Cratered terrain is the oldest surface on Eros. This image taken from an altitude of 30 km shows cratered terrain diametrically opposite the big "saddle" region near sunset time. Long shadows accentuate the craters and large, house-sized boulders cast long shadows on the surface. The scene is about 2.2 km across. (Courtesy APL, JHU, and NASA.)

is saturated with craters ranging from a few meters across to its largest crater (Fig.11.22). Overlapping craters show Eros to have a long impact history. Most are simple bowl-shaped craters showing various degrees of infilling and softening of their rims. Curious crater chains dot the surface, appearing similar to crater chains on the Moon. They may have been produced by multiple impacts from asteroids fragments traveling in tandem orbits. House-sized boulders are scattered here and there on the surface; ejecta from a past impact that failed to achieve escape velocity. It is possible that these blocks originated from the impact that created Eros' largest crater. Indeed, peering down into this 900 m-deep crater, sizeable boulders can be seen scattered on its flat floor. Among them is the largest boulder on Eros, 110 m across. Also inside the crater are signs of space weathering (Fig. 11.23). The walls show patches of brighter material that has been uncovered by overlying material slumping down its steep slopes. This bright material is relatively young compared to the darker material accumulating on the crater floor. On the side opposite the largest crater is the 10 km saddle-shaped depression. The depression is the antitheses of the opposing cratered terrain. It is completely devoid of craters, suggesting that the depression is relatively youthful compared to the cratered terrain; that is, either the craters that must have existed there earlier were removed by fragmentation or they have been completely covered over by a thick regolith. Abundant large boulders populate the terrain along with bright patches of newly exposed regolith (Fig. 11.24).

A laser range finder measures the distance between the spacecraft and the surface. This will ultimately result in a precise three-dimensional map of the surface as well as its overall shape and volume. Finally, a radio science experiment tracks the effects of Eros' gravity on the spacecraft. This leads to a determination of Eros' mass: 6.687×10^{15} kg. With its

Fig. 11.23. A false-color view inside Eros' largest impact crater, 5.5 km across, seen from an altitude of 50 km shows the floor littered with huge boulders, the largest over 100 m across, and covered with fine-grained material turned dark by chemical changes induced by micrometeorite impacts and solar wind particles. This regolith surface therefore shows effects of billions of years of space weathering. The bright areas on the walls of the crater are composed of relatively young material that has been less affected by space weathering. They were recently exhumed by landslides down the steep walls. The movement of old regolith material is especially evident on the opposing wall where alternate light and dark material can be seen. (Courtesy APL, JHU, and NASA.)

Fig. 11.24. A view from an altitude of 50 km shows the southern part of the "saddle" region. The craterless terrain within the depression contrasts sharply with cratered terrain. Boulders as small as 8 m in diameter stand out on the smooth regolith terrain. This surface must be relatively youthful since the craters that must have existed here either have been removed by major fragmentation or obliterated by fine-grained ejecta of later impacts. The large boulders probably originated as ejecta from these later impacts. Bright patches of younger regolith dot the surface especially around the boulders. (Courtesy APL, JHU, and NASA.)

volume and mass known, the density was calculated. The mean density of Eros has now been determined: 2.67 ± 0.1 g/cm^3. This is lower than expected for an OC-like body which is usually >3.0 g/cm^3. The low density suggests a porosity of ~20%, perhaps produced by an array of fractures, some of which reach the surface. The asteroid, however, appears to have a homogeneous internal structure (not a rubble pile) that is well consolidated much like monomict breccias in ordinary chondrites.

The most significant and long awaited comparison between ordinary chondritic meteorites and Eros is now being made. The infrared spectrometer detected two absorption features at the Band I and Band II positions (Fig. 11.11) produced by mixtures of olivine, orthopyroxene, and a "reddening agent", possibly iron metal, glass or an effect of space weathering.[19] The spectrum falls within the olivine-rich end of the S(IV) field (Fig. 11.25). The spectral properties seem to be consistent over the entire surface. Thus, Eros is a homogeneous body varying only in a minor way in mineral composition. This analysis combined with the density of 2.67 g/cm^3 and a porosity of <30% strongly suggests a close analog to an L or LL chondritic body that has been internally fractured. Thus, Eros is not a rubble pile but a well consolidated asteroid that has a history of impact cratering and fracturing but has been neither disrupted nor differentiated. Eros is a primitive body remaining essentially as it was when it was accreted in the first 10 million years of Solar System history.

The final approach

On February 12, 2001, almost one year to the day the NEAR spacecraft established orbit around 433 Eros, controllers at the Applied Physics Laboratory at Johns Hopkins commanded the spacecraft to begin a slow controlled descent to the surface. A series of four engine burns slowed the spacecraft from 32 km/h to about 8 km/h. As it approached the surface it provided an unprecedented series of images of the approaching surface. Figure 11.26a–d shows the four final images made before NEAR gently touched down on the surface. The final and closest image was made from an altitude of 120 m and shows part of a rock 4 m across. The finest detail is ~40 mm. The spacecraft landed with the speed of a brisk walk – and survived! Six hours after touch down controllers picked up telemetry signals indicating the spacecraft was healthy. For the next ten days, NEAR provided excellent data from the X-ray/gamma ray spectrometer that, when analyzed, should provide the most accurate data set on the surface composition yet obtained. The final wonders it recorded, close-up and personal, during this last telemetry from the surface are awaited with anticipation.

On a cold rocky fragment called 433 Eros, the NEAR spacecraft remains, asteroid-bound, in its final resting place.

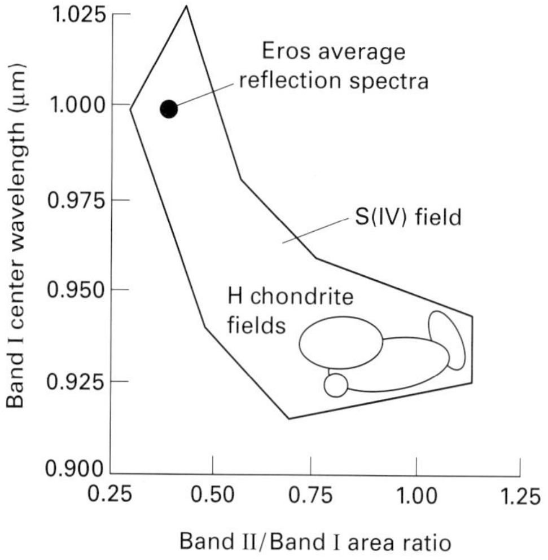

Fig. 11.25. The average NEAR infrared reflection spectrum data point plotted within the subtype S(IV) field. The plot lies in the olivine-rich end of the field suggesting a composition similar to L or LL chondrites for 433 Eros.[20]

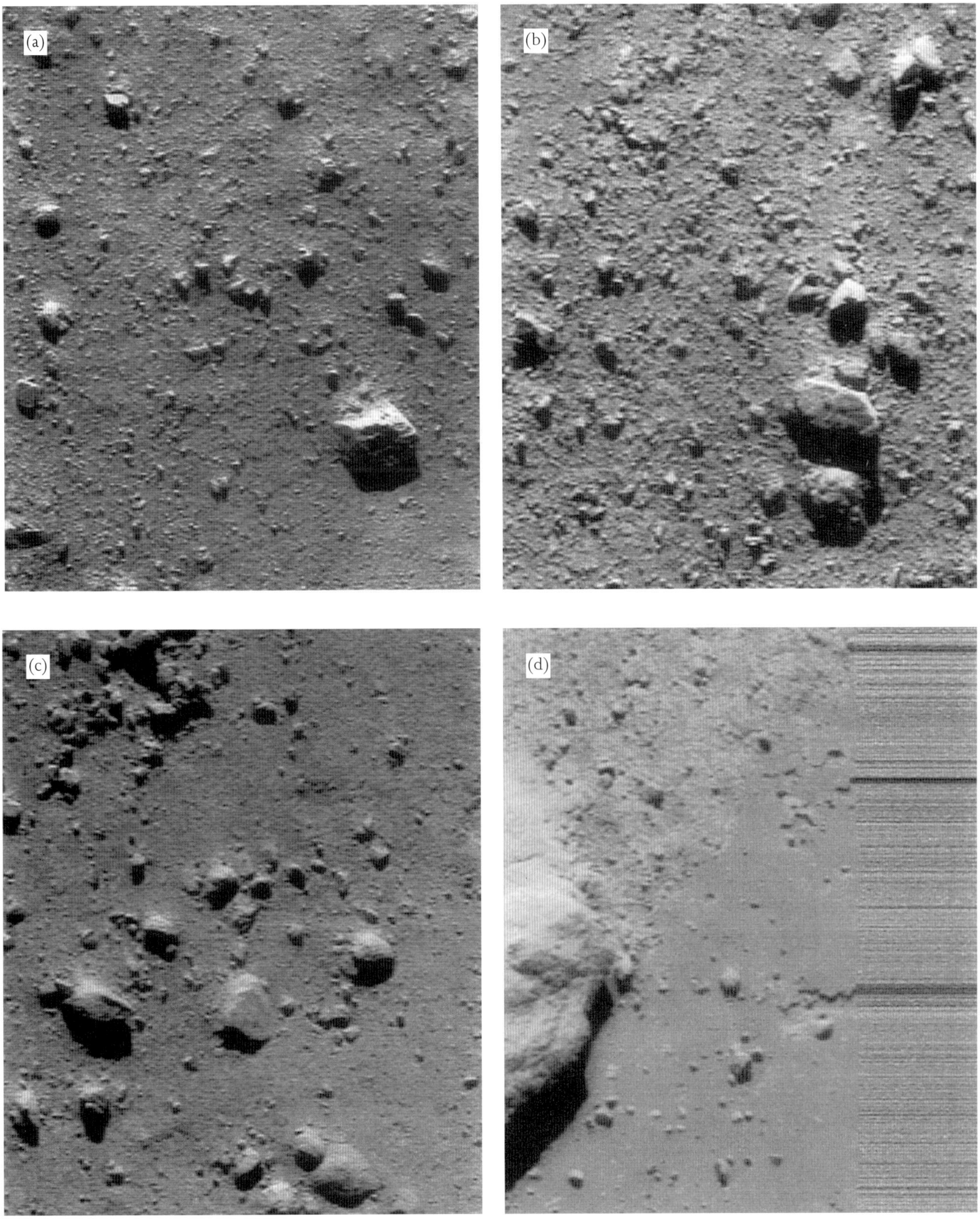

Fig. 11.26. These are the last four images taken by the NEAR-Shoemaker spacecraft as it slowly descended to the surface of 433 Eros on February 12, 2001. They are the highest resolution images ever taken of an asteroid surface. (a) Taken from an altitude of 1150 m, the image shows a scattered boulder field. The vertical field of view is 54 m and the large rock at the lower right is 7.4 m across. (b) From an altitude of 700 m, the image shows smaller boulders yet. The large oblong rock casting the black shadow is 7.4 m across. The smallest rock discernible is about 250 mm across. (c) The vertical dimensions of this image covers only 12 m, viewed from an altitude of 250 m. The cluster of small rocks in the upper left is about 1.4 m across. (d) This final image was taken from an altitude of 120 m (394 feet). The vertical field measures 6 m (20 feet) across. The large rock at the left edge is 4 m (12 feet) at the widest point on the edge of the picture. The finest detail discernable is only about 40 mm (<2 inches) across. Transmission was interrupted as NEAR made contact with the surface. Loss of signal is indicated by the horizontal lines on the right side of the picture.

Now silent, it remains functional and undamaged, a fitting monument to human ingenuity.

NEAR-Shoemaker has been a spectacular success, but did it solve the asteroid–meteorite problem? Are L and LL chondrites in our collection truly represented in 433 Eros? Remote sensing of Eros' surface has more than met all of the goals of the NEAR mission. It is satisfying to the astronomers who study asteroid reflection spectra with telescopes (the most remote of the remote sensing devices) that NEAR spectral studies have confirmed their own Earth-based observations. Yet doubts will remain, especially among the geochemists who would prefer, indeed insist upon, samples that can be studied in Earth-bound laboratories with all the sophisticated analytical tools at their disposal today. Thus, to the meteoriticist and geochemist, a more definitive answer must await a sample return mission.

References

1. Cunningham, C.J. (1988). Phase 1: The Visual Search, in *Introduction to Asteroids*, Willmann-Bell, Inc., Richmond, Virginia, pp. 3–10.
2. McCord, T.B., Adams, J.B. and Johnson, T.V. (1970). Asteroid Vesta: spectral reflectivity and compositional implications. *Science* **168**, 1445–1447.
3. McSween, Jr., H.Y. (1999). Chondrite parent bodies, *Meteorites and Their Parent Planets*, Cambridge University Press, Cambridge, p. 94.
4. Sato, K., Miyamoto, M. and Zolensky, M.E. (1997). Absorption bands near three micrometers in diffuse reflectance spectra of carbonaceous chondrites: comparison with asteroids. *Meteoritics and Planetary Science* **32**, No. 4, 503–507.
5. Gaffey, M.J. (1984). Rotational reflectance characteristics of asteroid 8 Flora: implications for the nature of the S-type asteroids and for the parent bodies of the ordinary chondrites. *Icarus* **60**, 83–114.
6. Gaffey, M.J., Bell, J.F., Brown, R.F., Burbine, T.H., Piatek, J.L., Reed, K. and Chaky, D.A. (1993). Mineralogical variations within the S-type asteroid class. *Icarus* **106**, 573.
7. Wisdom, J. (1985). Meteorites may follow a chaotic rout to Earth. *Nature* **315**, 731–733.
8. Gaffey, M.J. (1992). Narrowing the search for ordinary chondrites among the large S-type asteroids: identification and tests of three prime candidates (abstract). *Lunar and Planetary Science* **23**, 393–394.
9. Gaffey, M.J. (1995). The S(IV)-type asteroids as ordinary chondrite parent body candidates: implications for the completeness of the meteorite sample of asteroids (abstract). *Meteoritics* **30**, 507–508.
10. Allen, C.C., Morris, R.V. and McKay, D.S. (1995). Experimental space weathering of lunar soils (abstract). *Meteoritics* **30**, 479–480.
11. Kerridge, J.F. (1994). Production of superparamagnetic Fe^0 on the lunar surface (abstract). *Lunar and Planetary Science* **25**, 695–696.
12. Moroz, L.V., Fisenko, A.V., Semjonova, L.F., Pieters, C.M. and Korotaeva, N.N. (1996). Optical effects of regolith processes on S asteroids as simulated by laser shots on ordinary chondrite and other mafic materials. *Icarus* **122**, 366–382.
13. Taylor, G.J., Maggiore, P., Scott, E.R.D., Rubin, A.E. and Keil, K. (1987). Original structure, and fragmentation and reassembly histories of asteroids: evidence from meteorites. *Icarus* **69**, 1–13.
14. Asphaug, E., Ostro, S.J., Hudson, R.S., Scheeres, D.J. and Benz, W. (1998). Disruption of kilometre-sized asteroids by energetic collisions. *Nature* **398**, 437–440.
15. Binzel, R.P. and Xu, S. (1993). Chips off of asteroid 4 Vesta: evidence for the parent body of basaltic achondrite meteorites. *Science* **260**, 186–191.
16. Sack, R.O., Azeredo, W.J. and Lipschultz, M.E. (1991). Olivine diogenites: the mantle of the eucrite parent body. *Geochimica Cosmochimica Acta* **55**, 1111–1120.
17. Burbine, T.H., Meibom, A. and Binzel, R.P. (1996). Mantle material in the main belt: battered to bits? *Meteoritics and Planetary Science* **31**, 607–617.
18. Chapman, C.R. (1996). S-type asteroids, ordinary chondrites, and space weathering: the evidence from Galileo's fly-bys of Gaspra and Ida. *Meteoritics and Planetary Science* **31**, 699–725.
19. Veverka, J. *et al.* (2000). NEAR at Eros: imaging and spectral results. *Science* **289**, No. 5487, 2088–2097.

Additional reading

20. Meibom, A. and Clark, B.E. (1999). Evidence for the insignificance of ordinary chondritic material in the asteroid belt. *Meteoritics and Planetary Science* **34**, 7–24.
21. Yeomans, D.K. *et al.* (2000). Radio science results during the NEAR-Shoemaker spacecraft rendezvous with Eros. *Science* **34**, No. 5487, 2085–2088.
22. Zuber, M.T. *et al.* (2000). The shape of 433 Eros from the NEAR-Shoemaker lazer rangefinder. *Science* **289**, No. 5488, 2097–2101.
23. Trombka, J.I. *et al.* (2000). The elemental composition of asteroid 433 Eros: results of the NEAR-Shoemaker X-ray spectrometer. *Science* **289**, No. 5488, 2101–2105.

(a) The Aorounga impact structure. This is a heavily eroded impact crater located in the Sahara Desert of Northern Chad. Although little research has been done at the site, French scientists have collected rock samples that reveal the tell-tale signs of shock metamorphism. This is a peak-ring structure showing two distinct rings rising about 100 m above the surrounding terrain and separated by a shallow trough. The outer ring diameter is approximately 13 km. The crater once had a central uplift which has since eroded away. The structure is <350 million years old. Dark parallel streaks are sand deposits occupying shallow wind-cut valleys revealing the prevailing wind direction. The image was made from the Space Shuttle. (Courtesy Lunar and Planetary Institute, Houston and NASA.) (b) The Ouarkziz impact structure. This is a Space Shuttle image of a 4 km diameter impact crater located in the desert of northwest Algeria. It is located in folded and eroded sedimentary beds just right of center in this view. The single crater wall is almost perfectly circular with a slight opening to the south. Rocks from the vicinity collected some 30 years ago show planar deformation features, definitive evidence for a structure produced by impact. Estimates place the age of the crater at less than 70 million years. (Courtesy of Lunar and Planetary Institute, Houston and NASA.)

CHAPTER TWELVE

Terrestrial impact craters

Recognition that Earth bears scars of impacts by asteroids and comets was slow in coming. As we saw in Chapter 4, scientists in the late eighteenth century struggled to accept the simple, though to them profound, idea that rocks existed in space and that they could fall to Earth. Once this was understood and accepted, the origin of these rocks became the prime question. Chladni favored an interstellar origin far beyond the Solar System,[1] but prominent astronomers of the day claimed they had observed active volcanoes on the Moon, prompting the idea that meteorites were rocks expelled from lunar volcanoes into escape trajectories toward Earth. This idea persisted until it was discredited in the mid-nineteenth century.[2]

The discovery of the first asteroid in 1801 brought a glimmer of light to the problem of meteorite origins. Asteroids presented yet another possible solution. Many astronomers of the nineteenth century considered asteroids to be the broken remnants of a destroyed planet that once existed between the orbits of Mars and Jupiter. If this was true, then meteorites could be the remnants of this catastrophic breakup. Although there were a few scientists who promoted this hypothesis, the idea received surprisingly little attention probably because this would require an origin far beyond the Earth–Moon system, which suggested Chladni's interstellar hypothesis. Throughout the nineteenth century hundreds of asteroids were discovered, but they all seemed to be locked into stable orbits in a zone between 2 and 4 AU from the Sun. A link between meteorites and asteroids, if there was one, could not even be considered until a way was found to transport them or their fragments to the vicinity of Earth's orbit. The American astronomer Daniel Kirkwood cleared the way when he announced the discovery of gaps in the asteroid belt, unstable zones caused by Jovian gravitational perturbations (see Chapter 11). When 433 Eros was discovered to cross Mars' orbit in 1898, astronomers recognized that not all asteroids were gravitationally bound to the asteroid belt. Then the discovery in 1932 of the Amor and Apollo asteroids that either reached or crossed Earth's orbit left little doubt that there were asteroid parent bodies or their fragments out there that could collide with Earth. Every year fresh meteorites reached Earth from these Earth-crossing bodies. If relatively small meteoroids could reach Earth on a more or less regular basis, then the way was now clear to consider impacts with Earth of large crater-producing bodies. Yet few scientists in the first half of the twentieth century were bold enough to consider the question. They had good reasons for skepticism. The Moon's visible face was pockmarked with craters but in the minds of the astronomers up to the mid-twentieth century they remained volcanic constructs, not the result of impacts. Moreover, there was no other Solar System body known to have craters like the Moon. The Moon appeared unique. If the Moon's craters were indeed caused by impact, then Earth should have been impacted in like fashion. In fact, over all of its history, Earth must have been subjected to a larger number of impacts than the Moon. Yet, geologists could not point to any terrestrial structure they could positively identify as an impact crater.

There was an even more deeply rooted reason for geologists' resistance to the idea. It meant a return to catastrophism, an idea laid to rest by the Scottish gentleman farmer turned geologist, James Hutton, and Charles Lyell nearly two centuries earlier. Catastrophism began simply enough. Zoologists noted in the fossil record that many of the animals were unlike any species living at the time. Thus, these species must have died out. The eighteenth-century French zoologist Georges Cuvier (1769–1832) explained this extinction of life by a series of natural catastrophes, each catastrophe taking its toll of life forms after which new species would be created. Catastrophism expanded in the first half of the nineteenth century to include not only the fossil record but all of Earth's geologic record. This was a natural consequence of being able to observe such violent, uncontrolled natural phenomena as volcanic eruptions, earthquakes and floods on a time scale of a human life time. They could actually see these processes changing the landscape. Catastrophism also held religious overtones that could not be overlooked; namely, the Noachian flood recorded in Genesis. Predictably, clergy and some geologists began relating all catastrophes to the flood of Noah. It was, in essence, a compromise between science and religion. The idea of worldwide catastrophes orchestrating the fossil and geologic record persisted up to the time of Charles Darwin. Through this same period, James Hutton (1727–1797) developed a counter view to catastrophism. As a farmer he noted the slow cycle of weathering, erosion and deposition on his land. He viewed this cycle as persisting through enormous periods of time. He saw the origin of rocks through a variety of processes from the eruption and cooling of volcanic rocks through the steady accumulation of sediments that lithified by the action of gravity. He understood that the geologic processes he could observe operating then were the same processes that had operated in the distant past. Thus, he saw "the present as the key to the past". To him, there was no need for catastrophic events to explain Earth's evolving geology. Hutton published his observations and theories in his book, *Theory of the Earth, with Proofs and Illustrations*, in 1795. Thirty years later, the English geologist, Charles Lyell (1797–1875), wrote *Principles of Geology* in which he restated Hutton's views in more understandable form. All that was involved were observable Earth processes . . . and time. This philosophy, later called *uniformitarianism*, was hotly debated through the first half of the nineteenth century. Charles Darwin was profoundly influenced by Lyell's work and applied the ideas of uniformitarianism to his developing theory of evolution. By the time of the publication of the *Origin of Species* (1859) uniformitarianism had triumphed over catastrophism and became one of the founding principles of geology. From that point forward, geology, having its foundations firmly rooted in time, should never again be challenged by catastrophism. That Earth processes lead to gradual changes over time became an irrefutable axiom of geology. It was all a matter of time – millions of years of time. Impacts of asteroids or comets onto Earth through all of geologic time would be a major process sculpting Earth's surface in a matter of seconds, not millions of years, a direct violation of the principles of uniformitarianism that demanded slow, predictable events in small increments. The discovery and verification of impact craters on Earth would once again upset the rock-hard foundations of geology – and no one wanted to be an accessory to that!

An impact crater misjudged

Impact craters, if they existed on Earth, represented new geological territory, but the identifying criteria had not been defined. A good case in point involved the world's best preserved meteorite impact crater, Meteor Crater, in northern Arizona.

Sometime in the 1870s, cattlemen driving herds eastward must have come across the strange crater on the northern Arizona plains of the Colorado Plateau. They called it Coon Butte or Crater Mountain (Fig. 12.1). To them it was simply another old extinct volcano much like the hundreds of volcanic cones dotting the plains northeast of the town of Flagstaff, Arizona. But what seemed to set it apart from the other volcanoes was that it was surrounded by jagged pieces of nearly pure iron. (At least one sheep herder tried to sell a sample as silver. He didn't succeed.) In 1891, prospectors, thinking that they had come across a surface vein of iron ore, took samples to be assayed. One sample reached A.E. Foote, a mineral dealer in Philadelphia, who recognized the sample as an iron meteorite. Later the same year, Foote visited the crater, wrote a paper[3] on the meteorite-strewn field and mentioned the curious crater that did not appear to him to be volcanic. He described the crater in some detail mentioning the fragmented rock of the raised rim and how the horizontal layers of sandstone and limestone appeared to have been lifted and tilted outward clear around the perimeter. Today such a structure would have been highly suspect as an impact crater but with the absence of volcanic rocks in the area and having nothing in his experience to guide him, he could not explain this unusual structure. Amazingly, he did not attempt in his paper to relate the crater with the meteorites.

Foote's paper alerted Grover K. Gilbert, senior geologist of the United States Geological Survey. Gilbert was involved in the controversy over the origin of the Moon's craters. He had been studying photographs of the Moon in a futile attempt to prove them impact-related. If Coon Butte proved to be meteoritic then he would have a terrestrial equivalent with physical characteristics that may be applicable to his lunar impact crater hypothesis. In 1892, he sent a colleague, Willard D. Johnson, to Coon Butte to investigate. Johnson found the crater unremarkable and reported his findings to Gilbert. This was a crater that could be explained by well-understood geological processes, in this case, caused by a volcanic steam explosion. He had nothing at all to say about the meteoritic iron around the crater. (It is worth noting here that Johnson found no evidence of volcanic material in the immediate vicinity of Meteor Crater. The nearest volcanic constructs of any note are about 60 km to the northwest in the San Francisco Volcanic Field, home of about 400 cinder cones of Quaternary age in various states of degradation. A smaller field exists about 30 km to the south (Fig. 12.2).) The geology of the Colorado Plateau is nearly identical to the geology of the Grand Canyon, composed of generally undisturbed Paleozoic beds. These beds are highly disturbed within and around the crater.

Unsatisfied with Johnson's report, Gilbert decided to visit the crater himself later the same year (1892). During his investigation he made a number of incorrect deductions that

Fig. 12.1. This striking photo shows Meteor Crater in mid-winter after a fresh snowfall over the Colorado Plateau in northern Arizona. This is the world's first authenticated impact crater and the best preserved. Its rim-to-rim diameter is 1.2 km and its depth below the surrounding plain is 183 m with a 46 m high rim rising above the plain. The hummocky terrain below the outside rim are remnants of the ejecta blanket. The view is to the southeast. The road leads up the north rim to a group of buildings housing the Meteor Crater museum, an eating area and a tourist curio shop. (Photo courtesy of Alain Carion.)

Fig. 12.2. This is a view of Meteor Crater from the Space Shuttle at a distance of about 400 km. From this perspective the crater appears almost square in outline. A white ejecta blanket overlies the sloping outer rim. Canyon Diablo, winding its way north, comes within 3.2 km of the crater's west side. The iron meteorites are named after this geologic feature. To the south, about 40 km away, are Quaternary volcanic cinder cones, the closest volcanic constructs to the crater. (Courtesy the Lunar and Planetary Institute, Houston, and NASA.)

proved fatal to his eventual conclusions. From the apparent roundness of the crater he deduced that the meteoroid must have struck from a near vertical direction so that it had to lie beneath the center of the crater floor. He naturally expected the main projectile, being nearly pure iron, to be detectable from the surface with a magnetic compass. He tried to measure a disturbance of the magnetic field on the crater floor but failed. He expected that the projectile would create a hole equivalent to its own size. He estimated the volume of ejected rim material and found it equivalent to the volume of the hole. But if the projectile lay buried beneath the floor, he reasoned that it must be filling a large portion of the hole so that there should be a greater volume of material on the rim than it would take to fill the present crater. Unfortunately Gilbert did not understand the fundamentals of impact cratering to make correct judgements. The physics of impacting bodies would not be developed for another 50 years. For example, we know today that an object the size of the original iron meteorite (~56 million kilograms assuming a diameter of 24 m) would retain most of its initial cosmic velocity (~17 km/s) to impact. The energy released by the impact would instantly vaporize the body, causing a violent explosion that would create a crater dozens of times the size of the projectile, in this case approximately 50 times (1250 m). The shape of the crater does not dictate the incoming direction of the projectile until the angle of approach is nearly horizontal. The ensuing explosion would produce a symmetrical crater surrounded by a symmetrical ejecta blanket and rim. Based upon his observations, and without the benefit of hypervelocity impact science to help him, Gilbert's conclusions, published in 1896,[4] agreed with Johnson's. The crater was either a volcanic steam explosion or a limestone sink. (A field of limestone sinkholes does exist about 50 km southeast of the crater, formed within the Kaibab limestone of Permian age that covers much of the area around the crater.) And the meteorites? Either they arranged themselves by chance around the sink hole after it formed or, more plausibly, the meteorite impacted the surface fracturing a layer of rock that separated groundwater from deep hot rock causing a steam explosion that created the crater and scattered the surviving meteorites around it. So Gilbert's investigation was an opportunity lost.

Ten years later, a geologist and mining engineer from Philadelphia, Daniel Moreau Barringer, learned of Gilbert's survey and could not cast aside the role of the iron meteorites. Gilbert's considerable influence in the geological field had turned geologists away from the crater, accepting Gilbert's authoritative paper without a question. But not Barringer. Without even a survey of his own he was convinced of the tie between the crater and the meteorites. The crater, he insisted, was the result of an impact and the irons scattered around the crater were mere fragments of a yet-to-be-discovered main body that lay somewhere beneath the crater floor. Barringer's real interest in the crater was not as a scientific pursuit but as a commercial venture. He estimated that an iron–nickel body weighing between 5 and 15 million tons still existed centrally located beneath the floor. At $80.00 per ton for smelted iron, recovery of this body would be an important investment worthy of the effort and expense. The nickel alone (7 wt%) would be a find of incalculable importance since the United States had no nickel mines.

In 1903 Barringer applied for and received from the United States Government a lease on two square miles of land centered on the crater and he instructed his workmen to begin drilling immediately. Barringer and his associate Benjamin C. Tilghman visited the crater first in 1905 and came prepared to expand the exploratory drilling. Between 1903 and 1909 twenty eight holes had been dug in the floor to a maximum depth of 250 m. Here he encountered fractured rock showing shock features but no main meteoritic mass. Thinking that the meteorite came in at a low angle from the north and buried itself beneath the south rim, he began operations anew in 1920 with additional holes, this time drilled into the south rim. The deepest hole reached 412 m and jammed against something hard, which Barringer interpreted as the main mass. In 1923 a final hole was dug on the south rim outer slope in an attempt to reach the meteorite from outside the crater, but groundwater stopped operations at the 225 m level. After a quarter century of exploration in which some three dozen holes had been dug at a total expenditure of over $600 000, not one could be definitely shown to have encountered a solid iron body.[I] Operations ended in 1929, the year of Barringer's death.

Barringer had provided importance evidence that the crater was impact-related but 30 years would pass before Eugene M. Shoemaker, Edward T.C. Chao and B.M. Madsen provided the first indisputable evidence that finally labeled Meteor[II] Crater as the world's first proven impact crater – the discovery of the first natural occurrence of the high density polymorph of quartz, coesite.[5] This discovery and others shortly to follow finally dispelled the skepticism among geologists and almost immediately opened the door to further investigations of known circular structures scattered across the Earth.

[I] In 1929, F.R. Moulton, a well-known mathematician and astronomer from the University of Chicago was asked to advise Barringer on the problem of the "lost" meteorite. Several opinions were given in reports to Barringer. Moulton's final report was bleak indeed. He not only calculated that the mass may not be the hoped for 5–15 million tons but could be as little as 100 000 tons and as such was probably destroyed. This opinion, along with an earlier opinion by George P. Merrill, curator of minerals at the Smithsonian's Department of Geology that the meteorite may have been destroyed during the collision, was unwelcome news to Barringer's financiers but a milestone to the fledgling field of impact physics.

[II] The official name of the crater, *Meteor Crater*, given to the structure by the United States Board of Geographic Names in 1946 is an unfortunate misnomer. A meteor does not refer to the impacting body but to the light and other phenomena accompanying a meteoroid's passage through Earth's atmosphere. The proper name would have been the *Arizona Meteorite Crater*. Its second name, the *Barringer Meteorite Crater*, given to the crater by the Meteoritical Society is preferable as it honors the mining engineer Daniel Moreau Barringer who spent more than 25 years exploring the crater in an attempt to locate the massive iron. His explorations resulted in important contributions that helped to established the crater's impact origin. It was pointed out in Kathleen Mark's excellent book, *Meteorite Craters* (1987), University of Arizona Press, that the naming of the crater followed the official rules of nomenclature by naming geological features after the nearest post office. It seems that an official post office was established in 1905 five miles from the crater at "Meteor", Arizona. Although it was only temporary and no longer exists, the name has officially been Meteor Crater ever since.

Impact cratering in the Solar System

In the Solar System impact cratering is the single most important mechanism determining the surface characteristics of solid planetary bodies. Virtually every surface in the Solar System sufficiently solid to support impact features have them in abundance (Fig. 12.3a,b,c). Even the rock/ice satellites of the gaseous planets from Jupiter to Neptune (and probably Pluto) exhibit them though they have been severely modified by slumping and melting (Fig. 12.4). On the Moon, Mercury, and Mars the crater populations extend from circular pits a few meters wide to great impact basins hundreds of kilometers across, all attesting to a violent past history. Cratered terrain is considered the most ancient, dating back to the accretionary period of the planets. Virtually all of these planetary surfaces exhibit some reworking of the terrain either by volcanism, melting and flowing of ice-rock crust, or more recent impacts. Venus also has impact craters, its surface revealed for the first time through radar imaging from the Magellan/Venus space probe orbiting the planet in the early 1990s. Planet-wide ongoing volcanism that would continually renew the surface, removing impact craters in the process, may be occurring on Venus. Thus, taken together, Venus may harbor the youngest craters in the Solar System. Also, Venus' atmosphere has placed a constraint on the crater-forming mechanism. There are no craters smaller than about 3 km across, and there is a deficiency of craters less than 25 km in diameter. Venus' atmospheric density is 90 times Earth's, which destroys meteoroids in a size range that in a thinner Earth-like atmosphere would reach the surface to produce craters.

Without question, Earth and the Moon received saturation impacting of asteroidal and cometary bodies during the most active cratering period from 4.5 billion years ago to about 3.9 billion years ago. But throughout its geologic history

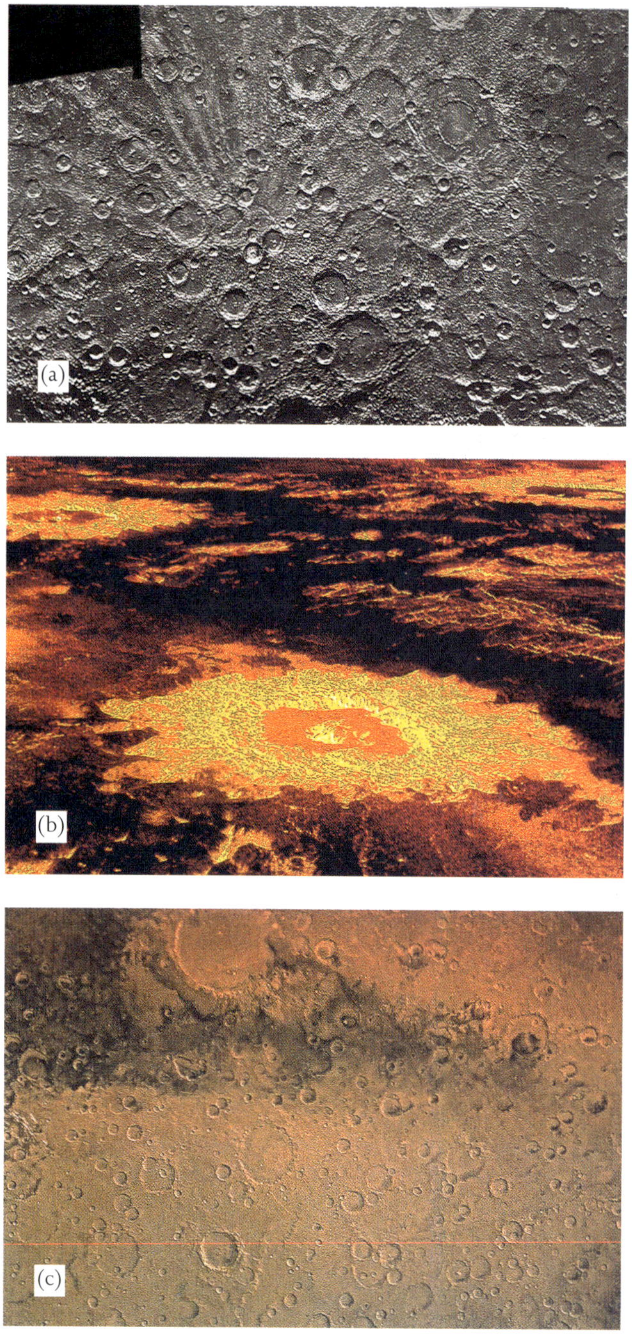

Fig. 12.3. (a) This is the cratered surface of the innermost planet, Mercury. The view is of the Bach quadrant in the southern hemisphere near the south pole. Virtually every crater type from small simple bowl shapes to complex craters with rebound peaks and peak-rings can be identified. The large crater in the upper right called Bach is 225 km in diameter and has peak-ring morphology. Another peak-ring crater is below center, the crater Bernini. It is a transitional form between rebound peak morphology to peak-ring morphology. White rays or ejecta point radially to a crater just to the left of center. This crater is relatively young since its ejecta overlies other craters in the area. (Courtesy NASA.) (b) The Magellan/Venus orbiter took this picture of three impact craters in the Lavinia region of Venus at latitude 27° south and 339° east. The spacecraft used radar imaging techniques to image the surface through Venus' thick cloud deck. The color is a computer simulation of what the surface colors may be like. The craters range in size from 37 km to 62 km. Each crater has a central rebound feature and a compact ejecta blanket. Ejecta does not travel very far, owing to Venus' dense atmosphere, 90 times the density of Earth's atmosphere. (Courtesy NASA.) (c) This is a portion of the cratered hemisphere on Mars, the oldest terrain on the planet. It is found south of Meridiani Terra at longitude 0° to 350° east and latitude 0° to 30° south. The largest crater is Schiaparelli, over 200 km in diameter. Like Mercury and Venus, every crater type is found in this view from simple to complex forms. The photograph was taken by the Viking Orbiter. (Courtesy NASA.)

Fig. 12.4. This is the saturnian moon, Dione. It is 1120 km in diameter and bears a surface saturated with impact craters. The largest crater is about 100 km in diameter. Many of the craters show well-developed central rebound peaks much like those found on the terrestrial planets. The craters are much more shallow, probably subject to slumping and deformation. Dione is half ice and half silicate rock and has the highest density (1.4 g/cm^3) of the saturnian moons. The higher percentage of silicates might promote greater internal radiogenic heating that in turn promotes more melting and gradual flattening of the crater rims. Crater counts indicate that Dione has undergone several periods of resurfacing. (Courtesy NASA.)

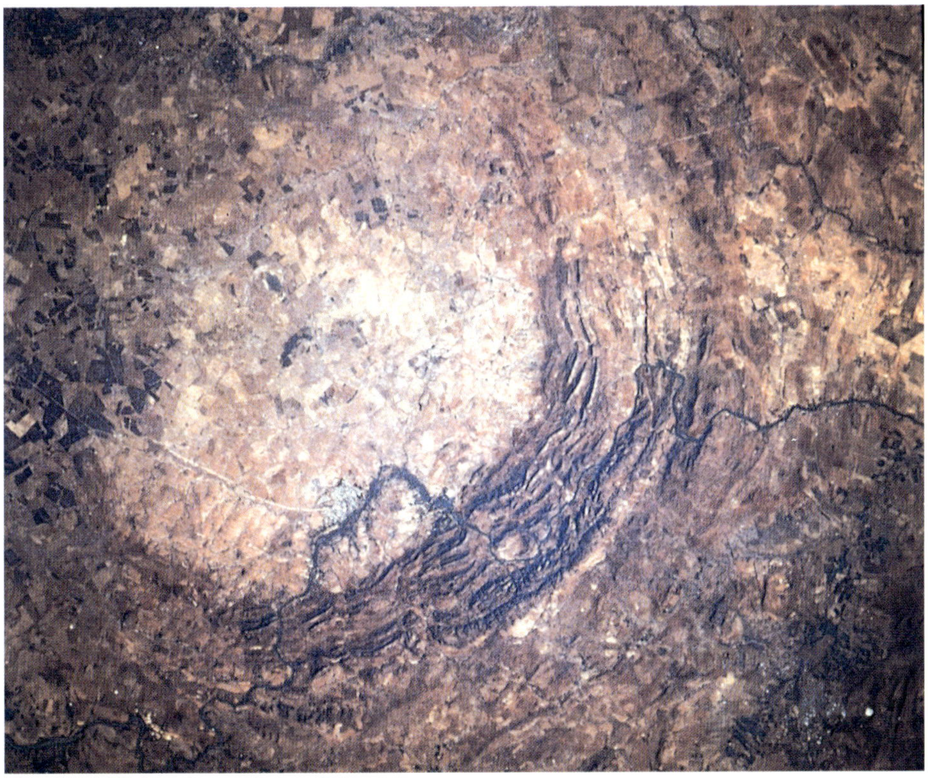

Fig. 12.5. The Vredefort structure in South Africa is one of the oldest terrestrial impact structures known, formed almost 2 billion years ago. The structure is highly eroded and partially buried by younger sedimentary rocks. Shock metamorphism effects have been found in rocks out to a radius of 70 km suggesting a diameter of 140 km. Also, shatter cones and high-pressure quartz have been found in this structure. This photo was made from the Space Shuttle. (Courtesy Lunar and Planetary Institute, Houston, and NASA.)

Earth, unlike the Moon, has been a highly active geologic environment. Weathering and erosion and plate tectonics have tended to modify, mask or even remove impact craters in time. After a few million years, erosional processes alone have obliterated small terrestrial craters in the 0.5–10 km diameter range so that compared to the Moon and other airless astronomical bodies the Earth is deficient of these craters. Thus, most small craters on Earth are also quite young. One of the oldest and largest recognized impact structures on Earth is the 140 km diameter Vredefort Ring located about 100 km southwest of Johannesburg, South Africa, with an age of 1.970 billion years. The structure is best seen from orbit. From that vantage point (Fig. 12.5) the structure appears highly reworked and clearly the southern half of the circular structure has been modified by geological processes, being replaced by new rock after the crater was formed. In all, over 150 impact structures have been identified over the past 40 years with 60% identified in the last 20 years of the twentieth century. Appendix G lists 158 impact craters[III] worldwide. About 30% of the structures are buried by post-impact sediments and four are entirely under water (Montagnais off Nova Scotia; Mjølnir in the western Barents Sea, north of Norway; Neugrund in the Baltic Sea off the coast of Estonia; and Tvären in the Baltic Sea, south of Stockholm.)[6] Of all the major planets capable of supporting impact craters, Earth appears, at least so far as our planet-wide investigations have presently shown, to have the smallest population.

[III] In Paul Hodge's book, *Meteorite Craters and Impact Structures of the Earth* (1994), Cambridge University Press, Cambridge, a distinction is made between meteorite craters and impact structures. Hodge points out that both are impact-related but the difference lies in the remaining structure. If the structure has retained a fresh unaltered and complete crater-like form, it is labeled a *meteorite impact crater*. The Barringer Crater is the "canonical" impact crater. On the other hand, if the structure is much older and highly eroded so that its impact origin is nearly unrecognized or even completely buried, it is labeled an *impact structure*. Most of the known impact features on Earth are best called impact structures.

Mechanism of impact cratering

A meteoroid or small asteroid striking Earth at a velocity between 10 and 30 km/s will pass through the atmosphere in a matter of seconds carrying a strong shock wave with it. A bow shock will develop on the forward end. This is a zone where atmospheric gases are compressed and heated between the shock wave and the surface of the meteorite. This heats the projectile's surface to the point of melting and vaporization and the air stream passing around the meteorite carries the ablated material into the train behind the meteoroid. At the lower end of its trajectory dynamic forces become quite large (~100 MPa at 15 km altitude) and usually fragment the meteoroid into numerous pieces still traveling more or less in the same direction. At impact they could produce a strewn field with numerous small craters confined to the scatter ellipse, some of which may be large enough to produce explosion craters like those in the Sikhote-Alin strewn field; or, if the breakup is near the ground and the cluster remains together they can produce a crater but with characteristics quite different from that of a single solid mass.

For our discussion of impact cratering we will assume that a large single projectile strikes the ground at hypervelocity. Impact geologists divide the impact process into three stages: *contact/compression*; *excavation*, and *modification*.[7]

Stage 1: contact/compression

The contact/compression stage begins when the projectile makes contact with the target rock. The projectile penetrates only 1 to 2 diameters before it is stopped (in competent rock), transferring all of its kinetic energy to the target rock. Two shock waves are generated: one into the target rock and the other into the meteorite's forward end (Fig. 12.6a). *Shock waves* are compressional waves that form whenever an impacting body's speed relative to the target rock in which the body is traveling exceeds the speed of sound propagation through the target rock (~3.5–5 km/s). The amplitude of the shock wave also exceeds the elastic limit or yield point of the target rock beyond which the rock can no longer return to its original physical state. Both the target rock and the meteorite suffer compressional shock pressures that reach several hundred gigapascals. These peak pressures occur in a hemispherical shock zone immediately involving the projectile and adjacent target rock. The expanding shock front moves rapidly through the target rock enlarging the hemispherical shell while increasing the radial distance from the initial contact point. The shock-wave pressures are reduced exponentially with increasing radius. Physicists have divided the expanding hemisphere into shock zones, each carrying shock pressures that are determined by each zone's distance from the contact point. For example, peak pressures can exceed 100 GPa at the contact point which produces melting and vaporization of the projectile and surrounding rock. Shock pressures then drop rapidly through the shock zones. Minerals in the target rock further from the contact zone are subject to reduced pressures (1–50 GPa) but still far exceed pressures experienced by "normal" violent terrestrial processes (explosive volcanic eruptions or earthquakes). These pressures cause distinctive shock metamorphism signatures in minerals composing the rock located in these zones (see Table 12.3).

This high-pressure state is released as the shock wave reaches the aft end of the meteorite and is reflected back through the meteorite as a *rarefaction* or *decompression* wave (fine dashed curve in Fig. 12.6b). This may spall off surviving fragments of the heated meteorite which may then be expelled during the excavation stage. (Meteorite fragments at Meteor Crater in which the Widmanstätten figures have been altered or even erased by severe heating (melting) have been found beneath the rim breccia as well as a kilometer or two beyond the rim on the surrounding plains.) As the decompression wave passes through the projectile aft to front end, it unloads the projectile, releasing it from the initial high pressures. This unloading results in the complete melting and/or vaporization of the projectile. The decompression wave continues on through the front of the projectile and into the target rock ending the compression stage. All of this occurs in <0.01 s for a projectile with a diameter of 100 m.

Stage 2: excavation

The excavation stage defines the morphology of the crater before modification by falling and slumping debris. Within the hemispherical envelope of expanding shock fronts some of the shock waves travel upward and contact the ground surface. These are immediately reflected downward as decompression waves. These waves exceed the strength of the target rock which shatters the rock and, at the same time, transfers some of the kinetic energy to the rock accelerating the rock outward at several kilometers per second. This begins the excavation process (Fig. 12.6c). As the decompression waves continues through the target rock, it moves the rock in a symmetrical fashion expanding the cavity as rock is thrown outward in a conical sheet or ejecta curtain.

Over the next several seconds the crater continues to grow as material is ejected. It reaches its final depth several seconds

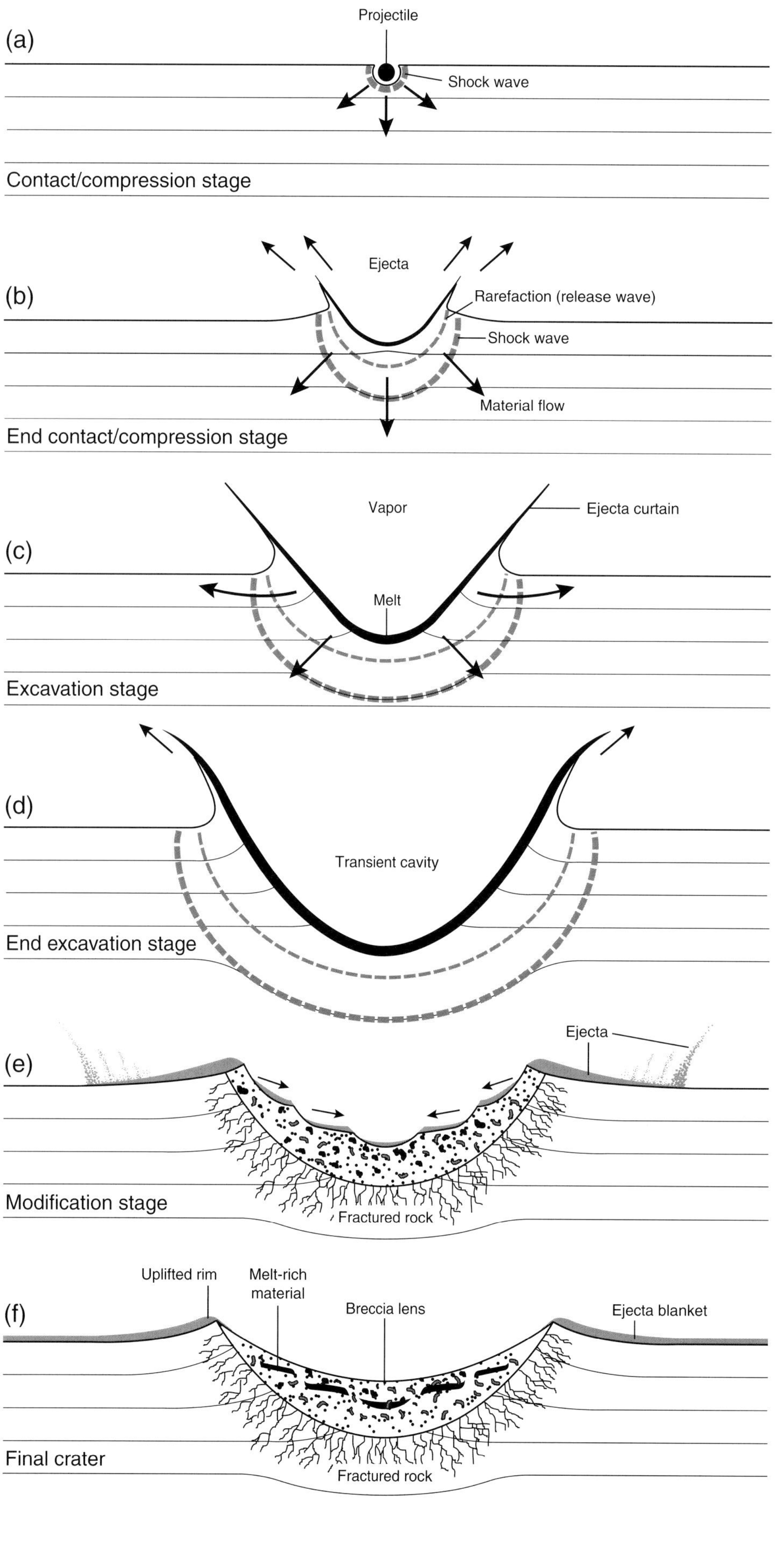

Fig. 12.6. This series of cross-section diagrams show the progressive development of a small, bowl-shaped simple impact structure in horizontally layered target rock: (a) Contact/compression stage: begins with initial penetration of projectile and outward radiation of shock waves through the target rock and projectile. (b) Start of excavation stage: continued expansion of shock wave into target rock; development of tensional wave (*rarefaction* or *release wave*) behind shock wave as the near-surface part of the original shock wave is reflected downward from the ground surface; interaction of rarefaction wave with ground surface accelerates near-surface material upward and outward. (c) Middle of the excavation stage: continued expansion of shock wave and rarefaction wave; development of melt lining in expanding transient cavity; well-developed outward ejecta flow (*ejecta curtain*) from the opening crater. (d) End of excavation stage: transient cavity reaches maximum extent to form melt-lined *transient crater*; near-surface ejecta curtain reaches maximum extent, and uplifted crater rim develops. (e) Start of modification stage: over-steepened walls of transient crater collapse back into cavity, accompanied by near-crater ejecta, to form deposits of mixed breccias (*breccia lens*) within the crater. (f) Final simple crater: a bowl-shaped depression, partially filled with complex breccias and bodies of impact melt. Times involved are a few seconds to form the transient crater (a)–(d), and minutes to hours for the final crater (e)–(f). Subsequent changes reflect the normal geological processes of erosion and infilling. (From *Traces of Catastrophe: A Handbook of Shock-Metamorphic Effects in Terrestrial Meteorite Impact Structures* (1998) by Bevan M. French. Lunar and Planetary Institute Contribution No. 954. Used with permission of the author and LPI.)

before it reaches its final diameter. The underlying target rock is much more resistant to movement than fragmental rubble on the surface. The edges of the enlarged crater are pushed up and folded backward into recumbent folds by the force of the explosion. At Meteor Crater, the succession of the rock layers on the rim were reversed by this process. Often pools of melted rock or *impact melt* line the floor of the crater. The shock wave that has penetrated the rock below the newly formed floor fragments the rock in place into an *autochthonous breccia sheet* (*auto-thō – nē – us*: forming in place where found). Mineral crystals in this crushed zone show substantial shock metamorphism on a microscopic scale. We will look at these shocked minerals later in this chapter.

Finally, when the shock and decompression waves have lost most of their energy through the excavation process, growth of the crater slows and then ceases. The excavation stage ends when the crater has reached its maximum size. Its diameter and depth are measured from the preexisting ground level to the crater's greatest depth below this level. The resulting bowl-shaped crater is wider than it is deep and a cross section through a diameter approximates a parabola. This is called the *transient crater* (Fig. 12.6d). This is the 'real' dimensions of the crater as a result of the dynamic forces involved. Formation of the transient crater ends the excavation stage and begins the modification stage.

Stage 3: modification

The modification stage changes the morphology of the transient crater (Fig. 12.6e,f). At this point the weakened shock waves no longer play a role in crater formation. Gravity now subjects the newly formed crater to modifications resulting in the final crater morphology. Subsequent collapse of the steep crater walls by down-slope movement enlarges the crater diameter by as much as 20% for small craters, and a paving of the crater floor by fall-back rock breccia and impact melt raises the floor level of the transient crater, forming a *breccia lens*. The rim and surrounding plain are recipients of fall back debris that builds the rim and forms a symmetrical ejecta blanket around the crater. The diameter and depth measured now from the rim gives the *final* crater dimensions. These are the dimensions most easily measured on photo images of extraterrestrial craters, but transient crater dimensions are usually determined in the field for terrestrial craters. In Table 12.1, both transient and rim to rim measurements are compared for a given projectile size.

Effects of changing parameters on crater size[IV]

When a large body with a mass of thousands of tons enters the atmosphere at hypervelocity (an average of around

Projectile diameter (m)*	Projectile density (g/cm³)	Impact angle (°)	Impact velocity (km/s)	Target rock density (g/cm³)	Rim–rim diameter (m)	Transient diameter (m)
100	**3.0**	90	15	3.0	2.74×10^3	1.76×10^3
100	**8.0**	90	15	3.0	4.02×10^3	2.43×10^3
50	8.0	90	15	3.0	2.21×10^3	1.42×10^3
100	8.0	90	15	3.0	4.21×10^3	2.43×10^3
100	8.0	90	**15**	3.0	4.02×10^3	2.43×10^3
100	8.0	90	**25**	3.0	5.24×10^3	3.05×10^3
100	8.0	**90**	15	3.0	4.02×10^3	2.43×10^3
100	8.0	**45**	15	3.0	3.51×10^3	2.17×10^3
100	8.0	90	15	3.0**(rock)**	4.02×10^3	2.43×10^3
100	8.0	90	15	3.0**(water)**	4.86×10^3	2.86×10^3

Note: *Assumes a sphere.

Table 12.1. **This table illustrates the various parameters used to predict impact crater sizes**

The parameters used to predict impact crater sizes head each column. To illustrate the importance and magnitude of these parameters, the parameter in question is varied while the others remain constant. The numerical magnitude of each varying parameter is noted in bold type. The resulting crater diameter in meters is listed in the right two columns giving both rim-to-rim diameters and transient crater diameters in meters. In the calculations we assume that the impacting bodies are spherical. (The numerical figures of Table 12.1 have been computed through a computer program developed by H. Jay Melosh and Ross A. Beyer of the Lunar and Planetary Laboratory, University of Arizona.)

[IV] For those who wish for a more mathematical treatment of impact cratering I suggest the authoritative and well written text by planetary geologist, H. Jay Melosh (1989) *Impact Cratering – A Geologic Process*, Oxford University Press, Inc., New York.

15 km/s for Earth) a series of events take place in rapid succession resulting in an impact crater of a size that depends in a complex way on many parameters: impact velocity; diameter (assuming a sphere here) and density of the projectile (from which the mass is calculated); impact angle; gravitational acceleration; target rock density; and target rock type. To illustrate the magnitude of importance of these parameters in the crater-forming process, let us vary the parameters one at a time to see how the resulting craters differ in size.

In Table 12.1 the columns include the above parameters. The pairs of values across show variations of a single parameter (bold type) with the others remaining constant. That parameter change will illustrate how the parameter affects both the final (rim-to-rim) diameter of the crater and the transient crater diameter. Note that the final diameter (rim-to-rim) is always larger than the transient diameter. In the examples in this section we will use the rim-to-rim diameter as it is more easily observed in the photographic examples given here. We begin with a change in density of the projectile. The upper line begins with a stony meteorite with a diameter of 100 m and a density of 3000 kg/m^3 (3.0 g/cm^3 – a typical chondritic density) impacting the Earth at a velocity of 15 km/s (the acceleration due to Earth's gravity, 9.8 m/s^2, is built into the program) with a trajectory angle of 90° to Earth's surface. The target rock (Earth's crustal rock) has a density of 3000 kg/m^3 and is competent (solid with low porosity). The result is a crater with a final diameter of 2.74×10^3 m.

Now let us keep all of the parameters the same except the composition of the meteorite. The second line shows the effects of changing from a typical chondritic projectile to an iron projectile so the density changes from 3000 kg/m^3 to 8000 kg/m^3 (8.0 g/cm^3). The diameter of the projectile has remained the same, thus the mass has increased substantially by a ratio of their densities 8:3 or 2.6 times. From the kinetic energy equation,

$$\text{K.E.} = \tfrac{1}{2}mv^2,$$

we see that the kinetic energy of the projectile increases directly as the mass increases. Thus, the crater diameter increases from 2.74×10^3 m to 4.02×10^3 m, an impressive 47%.

For an interesting variation of this compare two iron projectiles of different sizes but the same density. Clearwater Lakes impact structures in Canada or, on a much smaller scale, the Henbury craters in Australia, are examples of multiple impact structures formed by binary projectiles of the same composition and density, differing only in diameter (see Fig. 12.17). The second set of line pairs varies the diameter of two iron projectiles by a factor of two (50 m and 100 m – line pair number 2)) with all other parameters remaining constant. Keep in mind that the volume of a sphere changes as the *cube* of its radius. This will result in a mass difference between the two of a factor of 8. The result is two craters with final diameters of 2.21×10^3 and 4.02×10^3 m, nearly a doubling of the crater size. Thus, for a change in projectile diameter by a factor of two, the change in mass will be a factor of eight, which will change the crater diameter by approximately a factor of two.

Of all the parameters the mass and impact velocity are the most important. Referring again to the kinetic energy equation we are reminded that the kinetic energy of the projectile varies as the *square* of the velocity. Line pair number 3 compares two iron bodies of diameter 100 m impacting Earth at 15 and 25 km/s respectively, with all other parameters held constant. They produce crater diameters of 4.02×10^3 and 5.24×10^3 m respectively. The smaller crater is transitional between a simple and complex form. The additional 10 km/s changes the crater form to a complex crater. These forms are discussed in the next section.

Up to this point, we have considered impacts perpendicular to the target surface. Oblique impacts are another variable that must be taken into account when deriving a crater's final diameter. It seems at first self evident that a hypervelocity impact at an angle to the vertical will result in an elliptical crater. For the most part, this is not true. Even at very low impact angles down to 10° the shock wave generated during impact is still hemispherical so that the resulting explosion that opens the crater is symmetrical. Understandably, the shock wave produced at a low impact angle is weaker than that produced by a vertical impact angle since the absorption of energy of impact is less efficient for low-angle impacts. Thus, we expect and do find that the crater size relates to impact angle.

Let us see how the impact angle affects the diameter of the resultant crater. Once again we select two iron bodies maintaining all the parameters except the impact angle, which we change from 90° to 45° (line pair number 4). The resulting final crater diameter decreases from 4.02×10^3 m at 90° to 3.51×10^3 m at 45°, a change of about 14%.

Low oblique impact angles are often revealed in lunar craters by a nonsymmetrical ejecta blanket. These take on a butterfly or wedge shape appearance as material is scattered preferentially perpendicular to and along its direction of

[V] The Rio Cuarto impact structures in north central Argentina are an example of a low almost grazing incidence. First noticed by an airline pilot as he flew over them in 1990, the highly elliptical depressions, 10 in all, are aligned parallel to each other in a northeast–southwest direction. Together, they span a distance of about 30 km and range in size along their long axes from 250 m to 4.5 km. They appear to have been formed by a hypervelocity meteoroid approximately 150 m in diameter striking the Earth's surface at a glancing angle. Either a single projectile broke apart before impact producing multiple impacts from the fragments or the main mass may have fragmented on impact with the fragments continuing on down range, gouging out oblong shallow craters in the process. Impactite glass and two small chondritic meteorites have been found within the craters.[8]

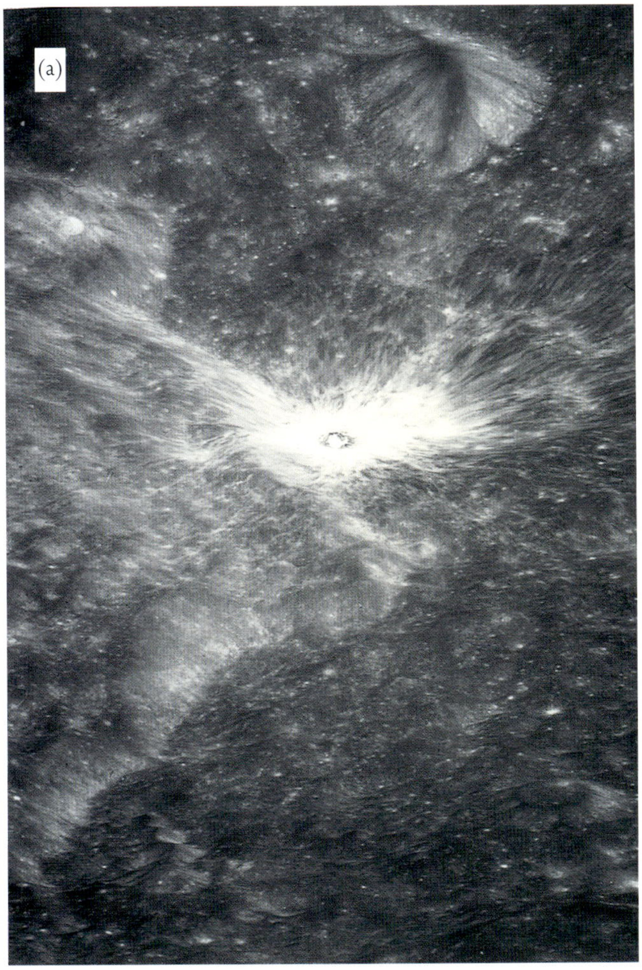

motion. There is a zone of avoidance behind the crater (Fig. 12.7a). At angles of less than 10° the crater becomes elliptical.[V] The motion of the projectile can be divided into vertical and horizontal components. The strong horizontal component simply means that there is less energy that goes into the target rock at impact because the projectile penetrates less deeply than in a 90° impact angle. A trough is gouged out of the target rock and material from the trough is ejected at right angles to the trajectory direction. Up range and down range along the trajectory direction remain clear of ejecta. If the trough length of the final crater is close to the actual diameter of the crater, the crater will be elliptical in shape (Fig. 12.7b). At a low angle of less than 10° the projectile may survive the compression and heating and actually remain solid, skipping across the surface leaving gouge marks in the target rock. This may be what we are seeing with the Rio Cuarto craters in north central Argentina.

The above calculations consider impact on competent continental rock with a density of 3000 kg/m^3. But Earth is about 70% covered with liquid water. Does that make a difference? Indeed it does. Referring again to Table 12.1, line pair number 5, all the parameters of both lines are the same. What has changed is the target material. The upper line considers competent rock; the lower line, liquid water. It may be surprising to see that an impact in water is more efficient at making craters than an impact on land. Note that a liquid water body produces, for a given projectile, a crater that is 20% larger. Moreover, the smaller crater is a transitional form

12.7. (a) Low-angle impacts of less than 10° from the horizontal produce not only elliptical craters but butterfly-shaped ejecta patterns as rock debris is squirted out of the forming crater at right angles to the direction of motion. Notice also a zone of avoidance at the forward and rear sides. This unnamed crater nestled in hummocky terrain on the Moon's far side shows the classic butterfly-shaped ejecta pattern. (Apollo 10 photo number 10-33-4868, NASA.) (b) This is the crater Messier found in Mare Fecunditatis on the Moon. The obviously elliptical crater has a long axis of 10 km. The origin of this crater is still controversial. Some believe its elliptical shape is the result of a low-angle impact from an impactor traveling west to east. Others believe it is actually two impacts occurring together producing two overlapping craters that fit neatly together. (Apollo 10 photo number 10-33-4906, NASA.)

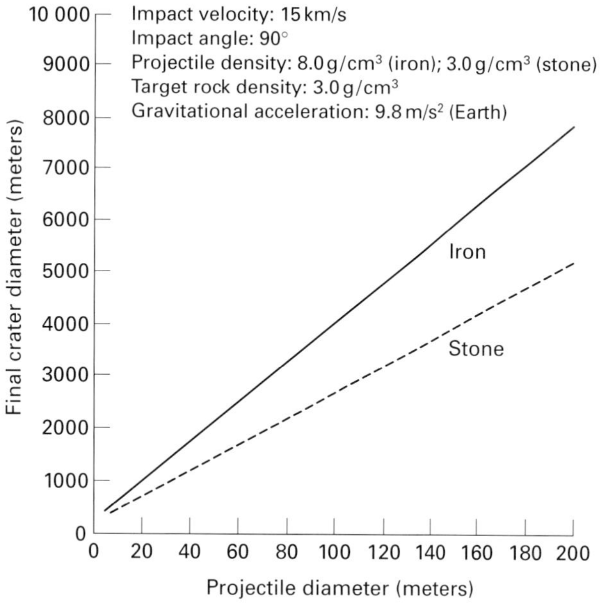

between simple and complex while the larger is a complex form. The energy of the impact in water is apparently more efficiently utilized in crater production than energy released in a land impact. Also the formation time in water is nearly double that in competent rock, 12.1 and 7.43 s respectively.

In summary, the most effective crater-producing projectile should be of iron composition, traveling about 30 km/s (near the maximum for Earth-crossing asteroids), and impacting at a 90° angle to Earth's surface. Figure 12.8 shows a graph of the change in crater diameter with a change in projectile size (assuming a sphere). Both stone and iron projectiles are compared.

Fig. 12.8. This graph relates the projectile size vs crater diameter for both stone and iron projectiles using the same parameters of velocity, impact angle, target rock density and Earth's gravitational constant. The graph shows that for a given projectile size, irons are more efficient crater producers than stones since irons are twice the mass of stones for a given projectile size and therefore carry more kinetic energy.

Morphology of impact craters

A brief look at the Moon with a modest telescope will demonstrate that not all craters are alike. There are two types, based upon their morphology: simple and complex. Of the complex craters we see secondary forms with central peaks, peak-rings and ring basins. This is a natural morphologic progression found on all the terrestrial planets including Earth. It is based upon the size of the projectile and the concomitant size of the crater. Basically, larger projectiles produce larger, more complex craters.

Simple craters

The most common crater type on the Moon and on Earth is simple craters. These are bowl-shaped structures with steeply raised rims and smooth interiors lacking any terracing or slumping (Fig. 12.9). The interior floor area is usually small with the smooth inner walls blending into the floor. The floor could be paved with impact melt. The rim is raised above the surrounding plain and is contiguous with an ejecta blanket arranged symmetrically around the crater. Simple craters are the deepest crater forms relative to their diameters. The ratio of depth to diameter for most are between 1:5 and 1:7. An excellent example of a simple crater is seen in Fig. 12.10 showing a remarkably circular bowl-shaped crater on Mars about 2550 m across surrounded by an extensive ejecta blanket.

Meteor Crater is the classic example and best preserved simple crater on Earth. Figure 12.11 shows a wide-angle photograph of its interior. Its rim stands about 45 m above the surrounding plain. In this view from the north rim looking south the floor is 173 m beneath the rim. The actual floor depth beneath the surrounding plain reaches 128 meters. Today the floor is overlain by lake-bottom deposits to a depth of about 60 m. Taking into account these deposits, the ratio of diameter to depth is very nearly 1:6. Beneath the floor extending down as deep as the shock wave reaches is usually a brecciated zone where the country rock has been fragmented in place, clear evidence of impact shock.

Another terrestrial impact crater of simple morphology is Monturaqui in the high Atacama Desert of Chile. This crater was recognized in 1962 from aerial photos and a few years later confirmed as an impact crater by field parties with the discovery of iron-bearing impactite glass around the crater, evidence of an iron projectile. The crater is only 0.46 km in diameter with a raised rim and depressed floor (Fig. 12.12).

Complex craters

Complex craters can take on a variety of morphologies. We look here at two common forms, those with central rebound peaks and those with peak ring structure. When the transition between simple and complex forms is made (see below) the first complex crater to appear differs remarkably from its simple counterpart. The crater is larger than the simple form; on Earth, usually larger than 4 km in diameter. It is no longer bowl-shaped but is much shallower, with a depth to diameter ratio of 1:10 to 1:20 (Fig. 12.13). We saw earlier that on impact, the meteorite and target rock beneath it are compressed, much of the impact energy being converted to heat sufficient to vaporize the meteorite and target rock causing a violent explosion and creating a hole many times larger than the diameter of the projectile. This is the initial or *transient* crater. Within seconds what was an initial deep crater floor rebounds as the compressional shock relaxes and the rock beneath the floor decompresses. The floor moves inward and upward causing faulting and fracturing of the rim into a series of slumping terraces as a central rebound mountain emerges. Shock waves from the explosion moving into the

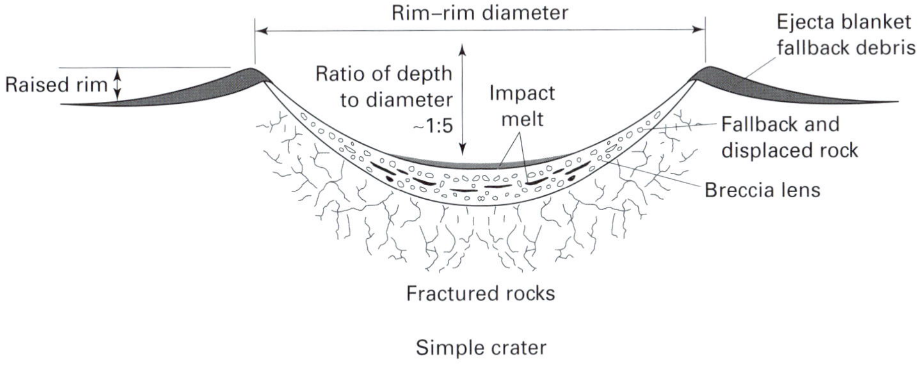

Fig. 12.9. This is a schematic cross section of a typical simple impact crater showing its steep bowl shape and a diameter to depth ratio of 1:5. The crater has been filled to approximately half its original depth with breccia and impact melt forming a breccia lens on top of which is the apparent or final crater floor. Beneath the breccia lens is fractured rock that was not ejected or displaced but remained in place (autochthonous breccia). Some of the fall back ejecta overlies the uplifted rim. Symmetrically around the crater is a hummocky deposit of ejecta.

Fig. 12.10. This is an unnamed simple crater on the surface of Mars. Notice its perfectly round steep bowl shape viewed from directly overhead. The crater is 2550 m in diameter rim to rim, about twice the size of Meteor Crater. Fine radial lines on the inside wall shows landsliding toward the small floor. Because of the light-scattering effects of Mars' atmosphere the shadows are not black as they are on the Moon. Thus, we can see into the shadows to the rubble-strewn floor. Ejecta around martian impact craters is different than that around lunar craters. Subsurface ice on Mars is melted briefly causing a slurry of rock-ice to move radially outward from the point of impact. On top of this flow are radial lines made by rock debris thrown from the crater. (Courtesy Lunar and Planetary Institute and NASA.)

rock beneath shatter the rock in place. The floor is composed of fragmented rock brought up from below by the explosion creating a breccia lens. If the target rock is igneous, it may melt forming impact melt pools that overlie the floor breccia, much of which is shocked if not melted by the impact. The 90 km diameter lunar crater Copernicus is the classic example of a complex crater with a central rebound feature and a faulted terraced rim (Fig.12.14).

The largest complex craters on Earth are a modification of the central rebound peak morphology. Instead of a single peak forming in the crater center, the central peak broadens out to become a ring of low lying mountains called peak-ring morphology (Fig. 12.15). The floor between the outer rim and the peak ring is usually depressed while the rebound floor in the center is raised or even slightly convex. If the crater is sufficiently large it may show a series of concentric peak rings evenly spaced on the floor. The famous Chicxulub crater in Yucatan, Mexico, is known to have at least four such rings with a possible fifth ring still being debated. Its diameter is 180 km or possibly larger. In addition a transitional form between the rebound peak and peak ring may have a cluster of low hills arranged in a rough circle enclosing a central group of hills. On Earth the outer rim of this crater type has often been eroded away leaving only the ring of mountains that mimic the rim of a much smaller crater. A good example of such an eroded structure is Gosses Bluff crater in Northern Territory, Australia (Fig. 12.16), where the peak ring, about 5 km across, is only the central rebound remnant of a much larger crater with a diameter of 22 km.

The transition between simple and complex craters is a function of the surface gravity of the planet. The greater the

Fig. 12.11. This is a view of the interior of Meteor Crater in northern Arizona. The view is looking due south to the opposite rim about 1275 m away. The rim stands about 45 m above the surrounding plane and the bowl-shaped cavity plunges 173 m to the floor. The floor is covered by about 60 m of lake-bottom deposits. (Photo by O. Richard Norton.)

Fig. 12.12. This is a view from the rim of the Monturaqui impact crater located in a remote part of the Atacama Desert in northern Chile near the Chile–Bolivia–Argentina border. The crater is among the hills of the Monturaqui mountains. It is a simple crater only 460 m in diameter. It was discovered in 1962 on aerial photos and visited by a field party confirming its meteoritic origin three years later. They found impactite with metal grains, shocked minerals and iron shale, evidence of an impact by an iron meteorite. The white material in the center is an evaporite, possibly gypsum. (Photo by Michael Cottingham.)

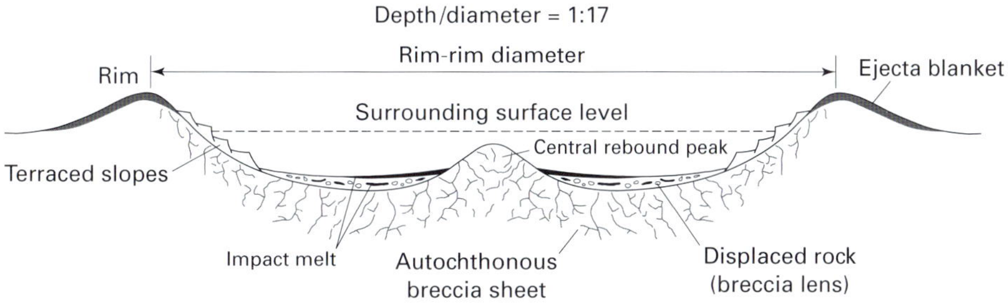

Fig. 12.13. This is a schematic cross section of a complex crater with central rebound peak morphology. The crater consists of a central uplift of deeper rock surrounded by a flat plain with a veneer of impact melted material mixed with allogenic breccias. The central peak is usually below the level of the plain. Inward movement along normal faults produces a stepped or multi-terraced rim. Ejecta material covers the rim and surrounding plain beyond. Beneath the crater floor is an autochthonous breccia sheet. The diameter of the final crater can be as much as twice the transient crater diameter and the depth to rim diameter can be as great as 1:20. On Earth craters from ~2 to 25 km may show central rebound peak morphology.

Fig. 12.14. This view looks down upon a northwest segment of the lunar crater Copernicus found in Oceanus Procellarum. The crater is 93 km in diameter and shows the classic signs of a complex crater. A group of hills in the center (right) are rebound features rising about 400 m above the floor but still within the confines of the crater rim. The walls show faulting and slumping into terraces. Lying on the terraces appear to be smooth impact melt material. The floor is smooth, probably covered with impact melt, and is saturated with small craters. The photograph was made by Lunar Orbiter V in 1967. The horizontal stripes are artifacts of the processing. (Courtesy NASA.)

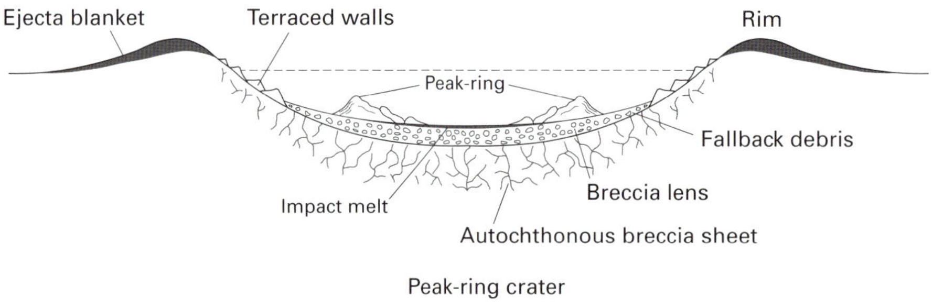

Fig. 12.15. The peak-ring morphology of complex craters is similar to central rebound peak morphology with a single important exception. Beyond about 25 km diameter, the central peak tends to be progressively replaced by a group of randomly placed hills or, in the most mature structures, a ring of hills. Terraced walls, a breccia lens composed of fall back debris and impact melt, and a zone beneath the breccia lens of rock fractured in place are all characteristics seen in central rebound peak morphology. On Earth, peak-ring morphology usually occurs with craters larger than about 25 km depending upon the characteristics of the target rock.

mass of the planet, the smaller is the diameter of the first-formed complex crater. The graph in Fig. 12.17 shows the terrestrial transition from simple crater to complex crater with central rebound peak to complex crater with peak-ring structure. The graph plots increasing projectile diameter against crater diameter for an iron meteorite with an impact velocity of 15 km/s, an impact angle of 90°, striking competent rock of 3000 kg/m^3. Transitional forms begin to occur with a projectile diameter of about 50 m and a crater diameter of about 2.21 km, approximately twice the size of Meteor Crater. A full complex crater form is reached when the projectile is about 105 m in diameter. The complex crater is 4.21 km in diameter and has a centrally located rebound peak. From this point on complex craters with central rebound features are formed until the projectile reaches about 618 m in diameter. The craters then change their morphology to a peak-ring structure. The transition crater diameter is 21.5 km. A marvelous example of this is the West Clearwater Lakes impact structure (Fig. 12.18). This 32 km diameter crater has a peak-ring morphology with the ring of mountains just clearing the surface of the lake. The East Clearwater impact structure is only 22 km diameter. It has a rebound floor hidden beneath the water but no peak-ring morphology. This is a classic example of the impact of a contact binary asteroid.

I mentioned earlier that the transition from simple to complex crater is a function of the surface gravity of the target body. To illustrate this, let us compare transitional forms for Earth and the Moon. Table 12.2 considers the same parameters used for the graph in Fig. 12.17. The single parameter change is the surface gravity of the Moon which is 0.16 of Earth's gravity. On Earth, we saw the transition crater diameter was between 2.21 and 4.21 km. For the same parameters on the Moon, the crater would be simple with a diameter of 5.66 km, about the middle of the simple crater range for the Moon. Only when the projectile reaches about 260 m in diameter will the crater become a complex form

Fig. 12.16. This image taken from the Space Shuttle shows Gosses Bluff in Northern Territory, Australia. The ring of hills surrounding a depressed center is not the outer rim of this crater. Rather, it is an inner mountain ring structure 5 km in diameter with the remainder of the uplift completely eroded to the surface. The true rim 22 km in diameter is also eroded to the ground. A terrestrial impact crater 22 km in diameter has just crossed the transition boundary from a central rebound peak to a peak-ring. This orbital image shows the circular extent of the crater surrounding the peak-ring. The rim material and part of the central uplift material was softer than the mountain ring so that after 142 million years, the crater has all but eroded away. (Courtesy Lunar and Planetary Institute, and NASA.)

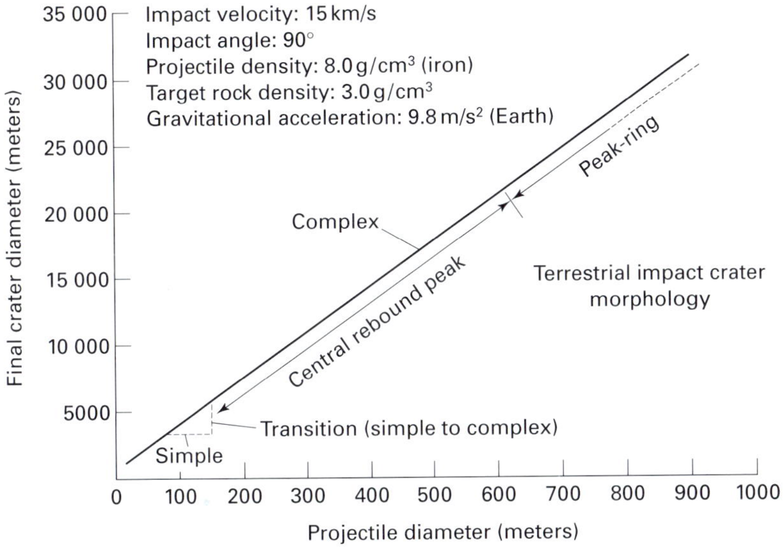

Fig. 12.17. This graph plots the impactor size against the crater size. It shows the terrestrial transition between a simple crater and the two types of complex craters. The change from simple to the first complex crater with central rebound peak occurs over a crater size between 2.21 and 4.21 km. Changing from a rebound peak morphology to a peak-ring morphology requires a much greater jump in size to ~22 km diameter.

Fig. 12.18. This image made aboard the Space Shuttle shows the Clearwater Lakes impact structures. The larger crater 32 km in diameter (west) harbors a ring of mountains inside the rim which just clears the level of the lake. The smaller lake (east) is on the transition line between a central rebound peak and a peak-ring at 22 km diameter. The central peak(s) is submerged below the lake. These structures formed at the same time as a contact binary asteroid broke up into two different masses immediately before impact. (Courtesy NASA.)

This chart gives the range over which transitional crater forms fall for the Earth and compares these ranges to the Moon. In both cases the variables are identical to those used earlier in Table 12.1. The projectile size in meters (bold) was increased until transitional crater forms appeared for both Earth and Moon. The data given were calculated for an iron projectile striking vertically at 15 km/s and impacting target rock with a density of 3000 kg/m^3 This chart shows that the transition from simple to complex craters is a function of the surface gravity of the impacted body.

Surface gravity	Transitional (S-CRP)*	Transitional (CRP-PR)**	Peak-ring
Earth = 1	2.21–4.21 km **(50–105 m)**	4.21–21.5 km **(105–618 m)**	>21.5 km **(>618 m)**
Moon = 0.16	12.1–20 km **(263–500 m)**	20–130 km **(500–3775 m)**	>130 km **(>3775 m)**

Notes:
*S-CRP = simple to central rebound peak.
**CRP-PR = central rebound peak to peak-ring.

Table 12.2. **Comparison of terrestrial and lunar crater transitional forms**

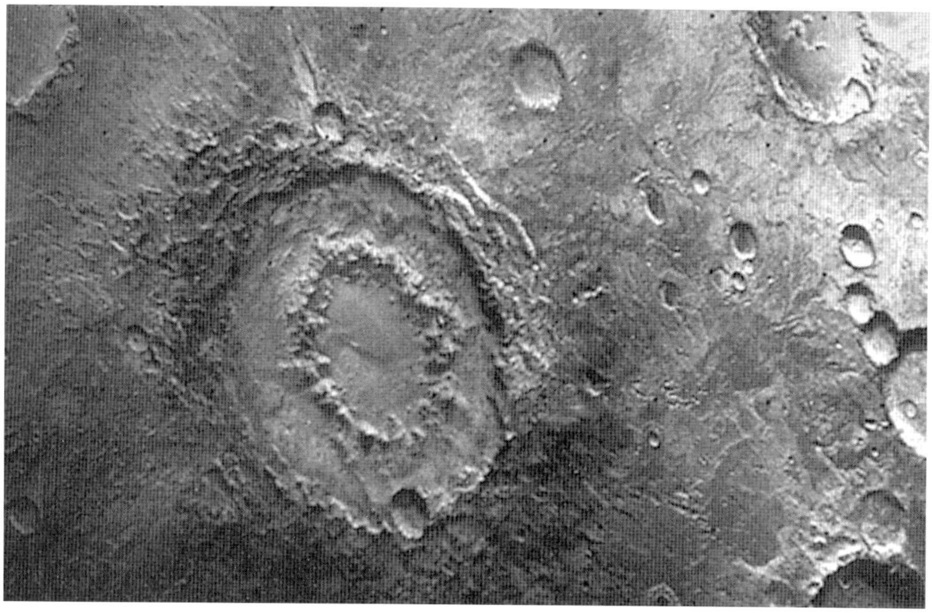

Fig. 12.19. Both Mercury and the Moon have peak-ring crater basins but compared to simple craters and complex craters with central rebound peaks they are very rare. Likewise, they are relatively rare on Mars. This photograph shows the 540 km diameter peak-ring basin, Lowell, with a well-developed circular ring of mountains enclosing an uplift in the center. A cloud partially obscures the central uplift. A massive ejecta blanket extends from the rim onto the plains. (Courtesy NASA.)

with central rebound mountain. The crater has reached about 12.1 km in diameter. This will vary somewhat depending upon the target rock density. A transition diameter of between 12 and 20 km is considered nominal for the Moon. The complex crater form with rebound peak continues to be produced until the projectile reaches about 3775 m diameter (2.3 miles). The resulting crater has reached peak-ring morphology at a diameter of about 130 km.

To summarize, the transition from simple to complex morphology is a function of surface gravity, which is related to planet size. The transition occurs at smaller crater diameters for larger planets. On Earth, the transition takes place between 2 and 4 km. On the Moon the first complex craters begin to make their appearance at about 12 to 20 km diameter but the best formed and more commonly occurring complex craters are over 50 km diameter. Mars, a planet half Earth's size and with only 1/3 the surface gravity, is intermediate with complex craters beginning to appear at 10 km and are well-formed at 30 km diameter (Fig. 12.19).

Criteria for the definitive recognition of terrestrial impact structures

Unlike the Moon, Mercury, and Mars, Earth has a dense atmosphere with dynamic weathering that, geologically speaking, makes relative short work of impact craters. The rocky brecciated rims and ejecta blankets are especially susceptible to rapid erosion. Those structures that have formed on the tough crystalline rock of the continental cratons are the best survivors of millions of years of weathering, retaining in many cases at least some of their tell-tale morphology. As a consequence, Earth is deficient in impact structures compared to the other terrestrial planets. There is every reason to believe that Earth has been impacted as much as any of the other terrestrial planets, but the early bombardment in the first half billion years of Solar System history, the scars of which are so prominent on the Moon, have been erased by erosion and plate tectonics on Earth. What remain today are either relatively recent structures formed over the past 50 million years (Cenozoic Era) or structures that have managed to survive from the early Paleozoic Era, 570 million years ago to the present. Only four structures have survived in recognizable form from the Precambrian Era (see Appendix G).

Over the past 20 years, 60% of the known impact structures have been discovered. We cannot always recognize crater forms from the ground. Satellite scrutiny has been invaluable in the recognition of especially very large impact structures that are easily overlooked or not recognized at all from the surface. But any structures suspected of impact origin must possess certain properties recognized by planetary impact scientists today as definitive. We look now at the indicators of impact.

Surface structures

These include all the morphological structures we previewed in the preceding section. If exposed on the surface, does the structure display a raised and overturned rim? A central rebound peak or peak ring and/or terraced walls? Are there remnants of a symmetrical brecciated ejecta blanket? Is there impact melt and/or shattered rock breccia on the floor? Is there a breccia lens beneath the floor? A positive response to any or all of these questions should raise suspicions of a possible impact structure. Since all of these structures are subject to weathering much of the original above-surface evidence may have been erased. The next step is to look for the definitive details.

Remnants of the meteorite

Perhaps the most obvious criterion is the presence of meteorite fragments in the crater's vicinity. (For craters formed by explosion, meteoritic material is almost never found in the crater.) We saw earlier that evidence of meteoritic material around craters is exceedingly rare. Meteoritic material in the form of fragments (Fig. 12.20) or as recondensed material trapped in impactites or in the soil have been found around only 10% of the 158 structures listed in Appendix G. It is all iron material. The single exception may be several stony meteorites found in the Rio Cuarto crater field in Argentina. These may not be related to the craters but may have coincidentally fallen into the craters after they

Fig. 12.20. These small iron meteorites were collected around the Henbury craters. They are fragments torn from the main mass(es) on impact. Almost all of the fragments have suffered mechanical distortions and many show alterations of their Widmanstätten structure due to severe heating. The craters, 13 in all, are only 10 000 years old. Their relative youth is not shared by the meteorites, which are heavily rusted and weathered. They are not likely to outlast the craters.

formed. It is unlikely that a multi-ton stony meteorite could survive atmospheric passage without major fragmentation whereupon it would probably not produce an impact (explosion) crater. Iron bodies are much more likely to survive with only minor fragmentation or spallation off the main mass and impact the surface intact and with sufficient energy to create an impact crater.

From the physics of impact it is understandable that little if any meteoritic material survives the crater-making process. The survival constraints are quite narrow. Iron meteorites over about 50 m diameter will be vaporized although there may be some fragments spalling off the main mass during the last few moments of atmospheric passage. This apparently was the case at Meteor Crater. Meteorites on the plains around the crater show undistorted Widmanstätten figures while those involved with the explosion that escaped vaporization were heated sufficiently to completely destroy their internal structure. Depending upon the parameters, it appears that meteorites producing impact craters with diameters above 1 to 1.5 km will not survive. Meteor Crater falls within this range. Only relatively young, small impact craters can be expected to have meteorite remnants. Thus, the presence of meteoritic material is evidence for an impact structure, but the converse is not true. Absence of meteoritic material is not evidence against an impact origin.

Shatter cones

The first criterion for the recognition of a meteorite crater was, of course, meteorite fragments arranged more or less symmetrically around the suspected crater. This was not very useful since, as we saw, very few craters are accompanied by pieces of the projectile. In 1905, shatter cones were first discovered in the Steinheim Basin in Germany by W. Branca and E. Graas (Fig. 12.21) but they were not thought to be associated with an impact structure. They labeled it a *cryptovolcanic* structure, a name they applied to a highly deformed and strongly brecciated circular structure believed to have been formed by powerful volcanic explosions although, in most cases, there seemed to be no evidence of volcanism. The term is still used to classify nonimpact structures of probable igneous origin. Through the next 40 years a growing number of structures labeled cryptovolcanic were identified, the term becoming a repository for any structure of an enigmatic or anomalous nature that could not be explained by conventional geophysics. Then, in 1947, R.S. Dietz working at the Kentland, Indiana, cryptovolcanic structure found shatter cones, the first diagnostic tool beyond meteoritic material that could be reliably applied as a criterion for identifying impact structures.[9]

Shatter cones are distinctive conical fragments of rock along which fractures have occurred. Divergent striations on the face of the cone fan out from an apex in a broad horse tail-like fashion (Fig. 12.22.) The apical angle is usually close to 90° although smaller acute angles may indicate higher shock pressures. Small parasitic half-cones cover the face of the master cone and their apices all point toward the tip of the master cone. They can range in size from only 1 cm measured along the cone axis to large structures greater than 12 m long, and can form in many rock types with the best morphology forming in fine-grained homogeneous rock such as limestone, dolomite, sandstone and quartzite. They are often found in swarms or nestled groups, cones in cones or mutually interfering cones and appear superficially like cone-in-cone structure typically found in sedimentary rock (Fig. 12.23a,b). They form immediately after the shock wave passes through the target rock and releases the rock under compression, fracturing the rock radially or apically. If the rock in the crater has not been folded or faulted after crater formation, the shatter cones are usually found in a preferred orientation, their apices pointing toward the crater center or point of impact. If the crater

Fig. 12.21. This shatter cone from the Steinheim Basin in southern Germany was found on the flanks of the central uplift. It is made of fine-grained hard limestone in which the shatter cone structure is especially well formed. Shatter cones were first found at Steinheim in 1905 but their significance was not understood for more than 50 years. The specimen is 97 mm in the vertical direction.

Fig. 12.22. This is an individual shatter cone in dolomite rock from the Wells Creek Basin, Tennessee, impact structure. Divergent striations surrounding the master cone extend from the apex of the cone (missing here) and fan out to a broad base or "horse tailing". Shatter cones were first reported from Wells Creek in the 1960s. The specimen is 113 mm along the apical axis.

center has rebounded the orientation may be tilted upward off the horizontal but they still remain radially oriented toward the impact point. In small complex craters shatter cones are often restricted to the central uplifts. For example, in the Sierra Madera impact structure, about 13 km in diameter, the shatter cones are found on the sides of the central uplift tilted upward parallel to the bedding planes, but in much larger structures like Sudbury, they can occur up to 35 km from the crater center. With Dietz's discovery of shatter cones at Kentland and later (1964) at Sudbury[10] (Fig. 12.24), he proposed the name *cryptoexplosion craters,* reserving cryptovolcanic to those structures that do not display impact evidence.

Shatter cones are important criteria for impact structures. For their formation, they require between 2 and 30 GPa of pressure. No known natural geologic phenomenon can produce them although they have been produced artificially in nuclear explosion craters.

Impactites

Impactites is a general name given to a variety of glassy to finely crystalline material created by fusion of target rock by heat generated during impact of a crater-producing meteorite. If the impactite material is a natural vitreous fused silica

Fig. 12.23. (a) Some of the best, most perfectly shaped shatter cones in the United States can be found in western Texas at the Sierra Madera impact structure about 30 km south of the town of Fort Stockton. This prize specimen shows a cluster of limestone shatter cones, some complete, others broken within a cluster about 110 mm in base dimension. All are pointing in the same direction which, *in situ*, would be toward the central uplift. (Photo by O. Richard Norton.) (b) Shatter cones have a superficial resemblance to a sedimentary formation call cone-in-cone structure. These curious structures are most often found in fibrous gypsum or limestone layers with their apices pointing downward. They form a series of concentric cones in clusters. The apical angles are between 30° and 60° with the cone axis perpendicular to the bedding planes. Unlike shatter cones, the sides of the cones do not show striations leading to the apex but instead are often marked by annular growth rings. This specimen is a typical example. Cone-in-cone structure is understandably often mistaken for an invertebrate fossil. The specimen is 11 cm in base dimension.

Fig. 12.24. The first distinctive shock metamorphism features discovered at the Sudbury structure in Ontario, Canada were shatter cones. This photograph shows large *in situ* shatter cones in the Mississagi quartzite found along the southern margin of the Sudbury Basin. The rock has been displaced from its original position by the impact. If returned to its former pre-impact position, the shatter cones would point toward the center of the basin. (Photograph by Robert S. Dietz.)

glass formed under high temperatures in a quartz sand environment, it is called *lechatelierite*. This term is also used for lightning-caused fulgurites as well as for glassy products of crater-producing impacts. Impactite in the form of a breccia set in a matrix of glass formed by impact shock is termed a *suevite*. Usually the breccia rock shows signs of shock metamorphism and there may be impact melt clasts. The term was first applied to impactite material found at the Ries Basin in Germany but is now commonly used for any brecciated glassy material found around impact structures (Fig. 12.25). Impactites often contain macroscopic iron–nickel globules or grains, remnants of the projectile (Fig. 12.26).

Impactite was found on the rims of the Wabar craters in Saudi Arabia (Fig. 12.27) and first described in 1933 by L.J. Spencer, keeper of minerals at the British Museum. Similar natural glass was found about the same time around the Henbury craters in central Australia.[11] The Wabar craters were formed by the impact of several iron meteorites, the largest estimated to be between 8.0 and 9.5 m in diameter (depending upon impact velocity) with an aggregate mass of >3500 tons. Eugene. M. Shoemaker and Jeffrey C. Wynn studied the Wabar craters in 1994 and 1995 and described several types of impactite glass they found.[12] Black glassy drop-like pellets cover the southeastern rim. They appear uniform in content and black in color. Under the microscope they are about 90% glass from the local sand and 10% iron–nickel. The iron–nickel is in a droplet form in a matrix of melted sand. It is the meteoritic iron that gives the pellets a black color, as seen in Fig. 12.27 (inset). A second kind of impactite glass formed as the shock wave expanded outside a hemispherical "melt" zone formed in the ground at the moment of impact. Around this zone, the expanding shock wave, now losing strength, compacted the sand/meteorite

Fig. 12.25. This is a sample of impactite from the Nordlingen Ries Basin in Heerhof, Bavaria, Germany. It is a breccia composed of large brecciated fragments of black glass and other crystalline rock fragments (white) both set in a fine clastic matrix. In this context it is referred to as *suevite*. Stishovite and coesite are found in suevites. The specimen is 102 mm × 75 mm. (Specimen furnished by Alain Carion.)

Fig. 12.26. Metal-impregnated impactite representing the remnants of the original iron impactor is found around a number of impact craters. This photograph shows the cut face of a vesicular impactite specimen collected at the Monturaqui Crater in Chile. The face reveals white grains of iron set in a glassy matrix. The specimen is about 18 mm wide. (Specimen furnished by Ronald N. Hartman.)

mix instead of melting it. Some of this compacted glass was caught in the subsequent explosion as the compressed sand/meteorite mass relaxed. This coated the impactite with a black patina skin from microscopic iron particles. Finally, a third type was immersed into molten glass at 10 000 °C near the locus of the explosion, turning the glass into a vesicular translucent yellow-white mass.

Impactite has been found around at least a third of the craters listed in Appendix G. It has become an important criterion for authentication. Meteor Crater seemed to be a likely

Fig. 12.27. Wabar Crater in the Empty Quarter of Saudi Arabia. The view is looking toward the rim of the largest of three craters. White and black glassy material is scattered over the southeast rim, produced by impact heating and fusion of the local quartz sand. (Photo by James P. Mandaville, Jr., courtesy of Aramco World.) (Inset) These black, glossy pellets, a form of impactite, are found on the southeast rim of Wabar Crater. They are about 90% glass and 10% iron–nickel. The metal has colored the glass black. Much smaller spheres of black glass can be seen adhering to the pellets. The small and large pellets are 3.5 and 8 mm across.

Fig. 12.28. These two specimens are examples of Libyan Desert Glass. They are yet another form of impactite glass but of an exceptionally pure silica glass. There is no impact crater associated with this glass. The specimens are 30 mm and 35 mm across.

crater for the formation of impactite as the explosion of the iron meteorite expelled masses of Coconino sandstone made of almost pure quartz sand. Huge boulders of this sandstone can be seen resting on the rim, lifted out of the crater by the force of the explosion. It therefore comes as no surprise that H.H. Nininger found impactities on the lower part of the northwest rim in 1954.[13] What is surprising is that the impactite "slag" was derived not from the Coconino but instead from the Kaibab limestone which normally lies on top of the Coconino except on the rim where its stratigraphy has been reversed. The limestone tested positive for nickel. Barringer had found lechatelierite derived from the sandstone during drilling operations but very little was found within the crater or on the rim slopes.[14]

Although impactites around a crater furnishes strong evidence for an impact origin, impactites have also been found in locations where craters are not known. An exceptionally pure, dense fused silica glass, Libyan Desert Glass, is found in the desert near Kufrah Oasis on the far eastern edge of Libya. This translucent yellow-green glass (Fig. 12.28) is located sufficiently near the B.P. impact structure to be suspect, but no association with the structure has been shown. For the present, Libyan Desert Glass has no known source and remains an enigma.

Impactites in the Wabar craters and Meteor Crater contain shocked minerals. If there was any doubt about the authenticity of these as impact craters, all that is immediately removed with the discovery of coesite.

Coesite and stishovite

Quartz is one of the most common minerals on Earth and quartz-rich rocks abound almost everywhere. Its wide distribution makes it ideal as a shock indicator for the recognition of impact structures. The most dense of the silica polymorphs is *coesite* and *stishovite*. Under high shock pressures and temperatures associated with the production of meteorite craters, quartz (density $= 2.65$ g/cm^3) is transformed into stishovite (density $= 4.35$ g/cm^3) at pressures of 15–40 GPa, and into coesite (density $= 2.93$ g/cm^3) formed at pressures of 30–50 GPa (Fig. 12.29). Coesite was first synthesized in 1953 by L. Coes[15] and stishovite in 1961. They were discovered in small quantities in 1960 at Meteor Crater.[5] This provided final conclusive proof of its impact origin. Volcanic

Fig. 12.29. Moderately shocked Coconino Sandstone from Meteor Crater, Arizona. The quartz shows multiple cleavage planes. The high-pressure silica phase, coesite, has formed in the interstices appearing black under plane-polarized light. The horizontal dimension of the photomicrograph is 3.2 mm. (Photo by Ted. E. Bunch.)

Mineral	Shock indicator	Pressure range (GPa)
Quartz	Planar fractures, *PDF	5–35
	Mosaicism	10–35
	Stishovite	15–40
	Coesite	30–50
	Lechatelierite	>50
Olivine	Planar fractures, PDF	5–45
	Mosaicism	10–35
	Ringwoodite	15–50
Plagioclase	Planar fractures, PDF	5–45
	Diaplectic glass (maskelynite)	30–45
	Jadeite	15–50
Pyroxene	PDF	30–45
	Majorite	15–50
	Mechanical twinning	5–40
Graphite	Cubic diamond	13
	Hexagonal diamond	70–140

Note: *Planar deformation features.

Table 12.3. **Mineral shock indicators**

This table lists the minerals used as shock indicators diagnostic for impact craters. The shock pressure range in GPa (gigapascals) is given for each.

Fig. 12.30. Mosaicism in a quartz crystals in impactite from the Gardnos impact structure, Norway. Notice the patchy extinction pattern across the grain showing shock metamorphism.

explosions are not known to produce pressures necessary to transform quartz into these dense polymorphs. Thus, these shock metamorphic features have never been found in volcanic rocks. Of all the criteria considered definitive for impact structures, the presence of coesite and stishovite (usually found together) are the most important.

Shocked minerals

Shocked minerals at impact structure sites stand with coesite and stishovite as the most definitive shock indicators. It is important to stress that the shock metamorphic effects described here do not occur in materials formed by any other known natural process. (Nuclear explosions have produced coesite within the crater.) They are uniquely characteristic of impact processes alone. All known impact structures possess at least some of these shock effects and some exhibit all of them. Typically, a large impact will produce shock pressures of 100 GPa or more within the target rock. This is far in excess of the pressures needed to produce the shock features listed in Table 12.3.

Of the various minerals displaying shock characteristics, quartz and feldspar are the most common minerals found at impact sites and show the clearest microscopic shock features. It is not uncommon to see undulatory extinction occurring in quartz that has been tectonically deformed. Thus, undulatory extinction cannot be used as an indicator of shock metamorphism. Mosaicism in quartz, however, requires a pressure of at least 10 GPa, far beyond the static pressure required to produce undulatory extinction. Mosaicism viewed under the polarizing microscope is distinctive, appearing as a patchy extinction pattern across the grain as the thin section is rotated (Fig. 12.30). Often shocked quartz crystals that show planar deformation features also exhibit mosaicism.

There are two types of planar features exhibited by shocked minerals. Planar fractures (PFs) are parallel sets of thin cleavage or shock lamellae in the crystal spaced about 20 μm apart and are often seen as several sets of parallel plates running in several directions (Fig. 12.31a,b). They are larger features and therefore easier to see under a light microscope than planar deformation features (PDFs). They usually form at lower pressures (~5–8 Gpa) than PDFs and may not be considered by themselves to be definitive of impact metamorphism unless other shock features are found around the impact structure. PDFs are finer features than PFs with parallel lamellae less than 2–3 μm in thickness and spaced between 2 and 10 μm apart (Fig. 12.32a,b). They are usually found as multiple sets of five or more orientations parallel to specific planes in the minerals' crystal lattice. They are composed of amorphous glass suggesting pressures of around 25 GPa. If the quartz has been altered by the post impact metamorphic processes the material in the planes may be converted back to quartz by recrystallization. In the process, arrays of small fluid inclusions, called *decorations*, may develop along the original PDF planes. These tiny inclusions are usually between 1 and 2 μm in diameter and preserve the original orientation (Fig. 12.32b) even though the original PDFs have vanished.

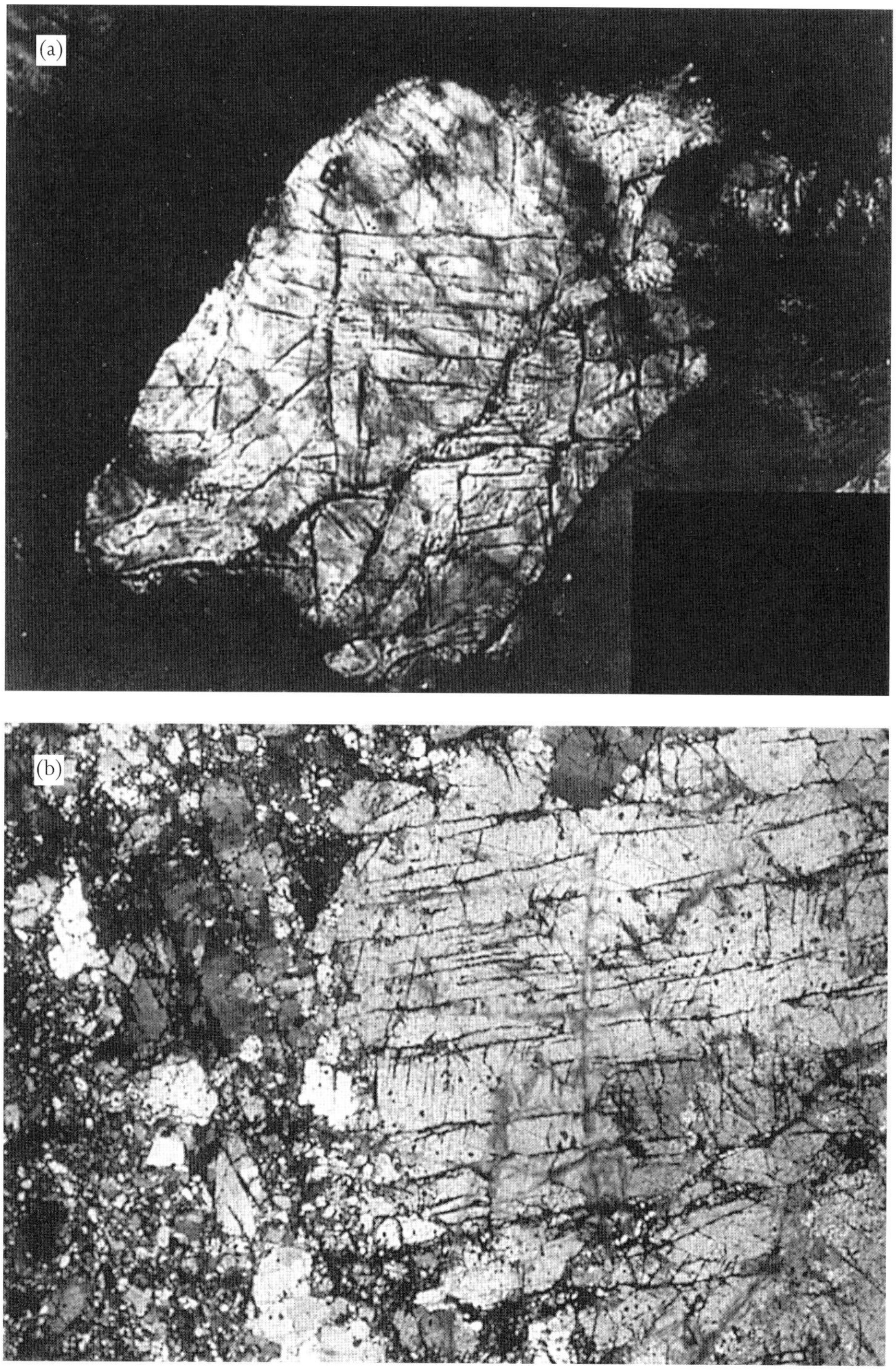

Fig. 12.31. (a) This quartz grain from moderately shocked Coconino Sandstone from the Barringer Meteorite Crater in Arizona shows mosaic extinction and multiple sets of cleavage cracks or planar fractures. View is in cross-polarized light. (Courtesy T.E. Bunch, Lunar and Planetary Institute, Houston.) (b) Intense fracturing of quartz in a coarse-grained metamorphosed orthoquartzite target rock from the Gardnos structure seen here in cross-polarized light. The large quartz grain (right) grades into a finer-grained recrystallized shear zone (left). The quartz grain is cut by numerous subparallel planar fractures (longer, dark, subhorizontal lines) and by much shorter planar features (short, dark, near-vertical lines) that originate along the fracture planes. These latter features may be relicts of actual PDFs. (Courtesy Dr Bevan M. French, Lunar and Planetary Institute, Houston.)

We saw earlier that quartz will be transformed to the high density forms, coesite or stishovite if shock pressures reach 30 GPa or greater. Quartz will melt at pressures greater than 50 GPa becoming the amorphous silica glass, lachatelierite. At the same shock pressures the vitrification of plagioclase will result in the transfiguration of plagioclase into maskelynite, a diaplectic glass, in which the external appearance of crystalline plagioclase remains (see Fig. 8.26).

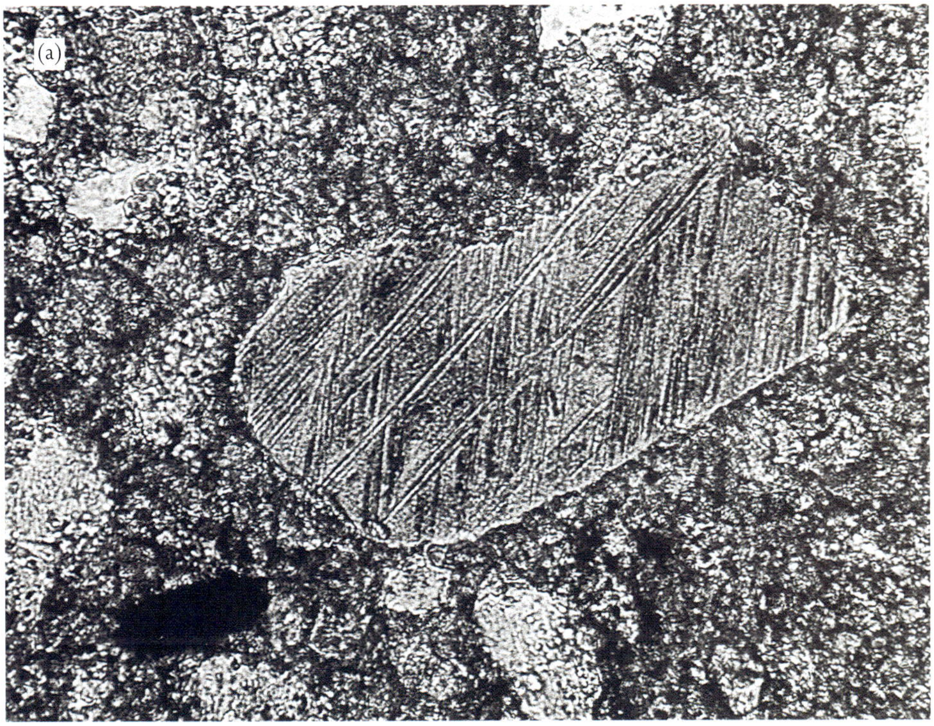

Fig. 12.32. (a) This is a shocked quartz grain containing multiple sets of planar deformation features (PDFs). The grain is included with rare sandstone fragments in a carbonate breccia dike that cuts the deformed basement rock at the Sierra Madera impact structure in Texas. The closely spaced PDFs give a distinctive darkened, yellowish appearance to the quartz grain. Viewed in plane-polarized light. (Courtesy Dr Bevan M. French, Lunar and Planetary Institute, Houston.) (b) A quartz grain from the Holleford, Canada, impact crater showing two sets of planar deformation features or PDFs. The material in the planes have been recrystallized back into quartz leaving behind strings of fluid inclusions marking the position the original PDFs. These are called *decorated* PDFs and are the result of post impact metamorphic annealing of the original quartz grain. Photographed in plane-polarized light. The horizontal dimension of the photomicrograph is 0.1 mm. (Photo courtesy Ted. E. Bunch.)

Catastrophic impacts on Earth

It has taken 100 years since the recognition of the first terrestrial impact crater to convince the majority of geologists that Earth has been subject to impact cratering over its entire geologic history to as recently as 50 000 years ago, or even less. Impact cratering is an important mechanism that has the potential to change the surface of the Earth. But equally important, impacts of asteroids and comets have also been a major element acting to change the course of evolution on Earth. And impacting is still going on.

The Tunguska event

On June 30,1908, at 7:17 a.m. in the remote forests of the Podkamennaya Tunguska River basin in central Siberia a huge fireball raced across the sky from the southeast. The object was probably a stony meteorite about 100 m in diameter. When it reached an altitude of about 8 km above the forest it exploded with the force of a 10 to 20 megaton bomb creating a searing heat that ignited the forests. Some 30 s later the shock wave from the explosion reached the surface, flattening the burning trees into a radial pattern. This was followed by a wind blast that extinguished the flames. This is the only time in recorded history that a meteorite has been seen to devastate the environment. An eyewitness in Vanovara, 40 km south of the explosion, gave this account:

> I was sitting on the porch of the house at the trading station of Vanovara at breakfast time and looking toward the north. I had just raised my axe to hoop a cask when suddenly in the north above the Tunguska River, the sky was split in two: high above the forest the whole northern part of the sky appeared to be covered with fire. At that moment, I felt great heat as if my shirt had caught fire; this heat came from the north side. I wanted to pull off my shirt and throw it away, but at that moment there was a bang in the sky, and a mighty crash was heard. I was thrown three sagenes [about 7 m] away from the porch and for a moment I lost consciousness. My wife ran out and carried me into the hut. The crash was followed by noise like stones falling from the sky, or guns firing. The Earth trembled, and when I lay on the ground I covered my head because I was afraid that stones might hit it. At the moment when the sky opened, a hot wind, as from a cannon, blew past the huts from the north.[16]

The Tunguska incident is still shrouded in mystery. There is little if any evidence of a projectile: no meteorites, no crater. Only a blown-down forest (Fig. 12.33). But there is little doubt that a meteorite had exploded over the Tunguska River region in 1908. Such an event is thought to occur about once a century. Smaller events occur much more often but usually harmlessly over the oceans of Earth, where there are no witnesses. But with today's military satellites equipped with

Fig. 12.33. Tunguska site of blown down trees. This photograph was made by Leonid Kulik, first to witness the devastated area, on April 13, 1927, nineteen years after the great event. The flash from the great explosion instantly set the trees aflame followed about 30 s later by an enormous shock wave that extinguished the flames but flattened the taiga over an area of 2200 km^2. The blast had the energy equivalent of a 20 megaton bomb.

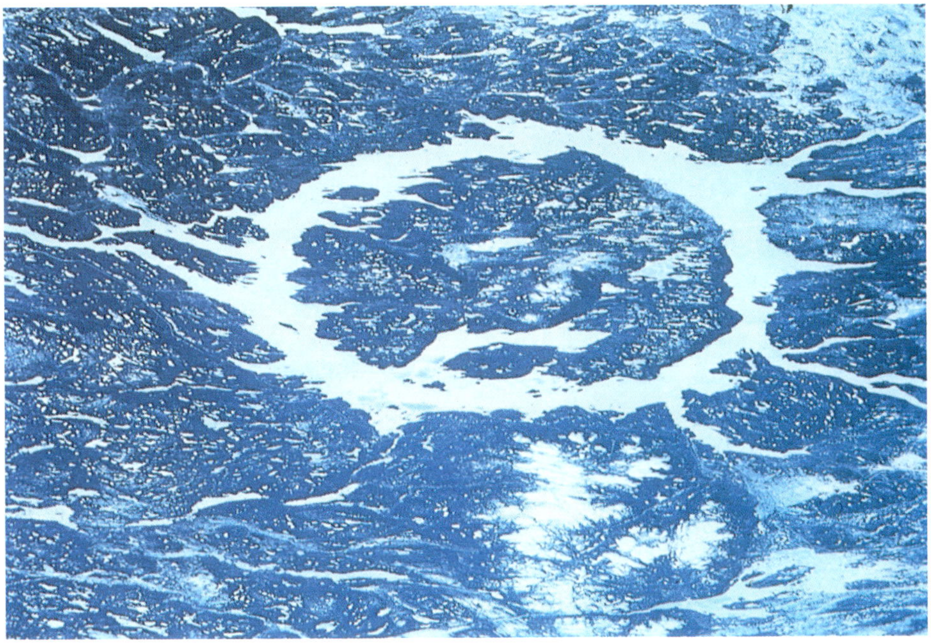

Fig. 12.34. Lake Manicouagan, viewed here from the Space Shuttle, is actually Canada's second largest impact structure. Once thought to be 64 km across, it has now been upgraded to about 100 km in diameter. The "O"-shaped lake is a curious feature formed by two semi-circular lakes, Lake Manicouagan and Lake Mushalagan that flow around the east and west sides respectively, and converge at the south point where they connect with a river that drains them. The two lakes together define the structure's central uplift. The uplift, though heavily eroded, peaks at 952 m above sea level. This is a winter view taken when the lakes and rivers were frozen. (Courtesy NASA.)

infrared sensing devices, many masses have been detected entering Earth's atmosphere. Over a period of 17 years (1975–1992) US Military satellites detected 136 explosions, some in the 20 kiloton range. (The Nagasaki atomic bomb had the yield of a 20 kiloton bomb.)

When catastrophism had its day

Terrestrial craters 100 km or greater in diameter have the potential to disrupt catastrophically the environment on Earth upon which all life depends. Perusing the list of known impact structures in Appendix G we find seven structures 100 km diameter or greater (Fig. 12.34). Others will undoubtedly be found in the future. Some may lie buried within the sediments of the ocean floor. That such impacts could disrupt the evolutionary flow of life on Earth was an idea few paleontologists or geologists took seriously. But in 1980 everything changed. A landmark paper[17] by University of California, Berkeley scientists Walter and Luis Alvarez and team members Frank Asaro and Helen Michel published in *Science* June 6, 1980 with the provocative title, "Extra-terrestrial cause for the Cretaceous–Tertiary extinction", sent shock waves around the world. It all began in the late 1970s when Walter Alvarez was searching for clues to the great extinction at the end of the Cretaceous Period 65 million years ago. He was studying rock layers outside the Italian town of Gubbio where the Cretaceous–Tertiary (K–T) boundary was especially well preserved. The layers of rock were a relatively uninterrupted sequence of limestone beds deposited in the Cretaceous seas as the calcareous shells of ancient microscopic animals called foraminifera (forams) rained down onto the floor of the sea, gradually compacting into hard limestone. Alvarez noticed a curious brown clay-like layer about 15 mm thick that interrupted the gray-white limestone and conveniently marked the K–T boundary (Fig. 12.35). Looking closely at the limestone immediately beneath the clay layer he could identify the fossilized shells of many forams commonly found in Late Cretaceous sediments, but the limestone immediately above the clay layer was all but devoid of foram shells (Fig. 12.36). The forams appeared to have suddenly become extinct for no apparent reason. Since the time of Darwin, evolution was thought to be a slow steady process in which lifeforms evolved and became extinct in a cycle of millions of years of time. But it appeared that the forams had died out over a matter of a few years at most. It was obvious to Alvarez that an extraordinary event had occurred in a wink of a geological eye immediately prior to the deposition of the clay layer. To determine the time it took to deposit the clay layer would tell him how long the event lasted that had stopped the limestone deposition and killed off the forams. What Alvarez needed was to find a standard from which to measure the rate of sedimentation. His father, Nobel Prize winning physicist Luis Alvarez, suggested that the rate of accumulation of meteoritic material, which was known to be uniform in time over the entire Earth, could be used as a standard. We saw earlier (Chapter 1) that roughly 110 metric tons of meteoritic and comet dust per day fall uniformly over the Earth. But how does one detect meteoritic dust mixed in with terrestrial material? Alvarez knew that the Earth's crust was depleted of noble metals whereas they were found in relative abundance in iron-bearing meteorites. An iron-rich asteroidal body contains platinum-group metals such as iridium and osmium in trace quantities but still far greater quantities than that found in Earth's crustal rock. Platinum group metals are siderophile elements that are attracted to iron. Most of these metals were carried to the Earth's core during differentiation leaving little behind in the crust.

Fig. 12.35. Dr Paul Whitsell points to the iridium-rich dark brown clay layer nestled between limestone beds marking the Cretaceous–Tertiary boundary. The white limestone beds beneath the clay layer mark the end of the Cretaceous Period while the tan limestone beds above the clay layer mark the beginning of the Tertiary Period. This location is about 2 km northeast of the Italian town of Gubbio in Umbria. (Photograph by Dorothy Sigler Norton.)

Noble metal concentrations in stony meteorites are between 1000 and 10 000 times that found in Earth's crustal rocks. Alvarez thought that the element iridium, though in low concentrations of about 1 p.p.m. (part per million) in meteorites, could be detected using newly devised neutron activation techniques used to determine trace elemental compositions. Frank Asaro, a specialist in neutron activation techniques at Berkeley irradiated a sample of Gubbio clay with neutrons causing the elements in the sample to emit gamma rays in amounts specific to each element. He detected the noble metal, iridium, in the clay sample but, to the team's surprise, in quantities far above the calculated amount expected if the rain of meteoritic dust was uniform in time and place. Asaro's measurements yielded 30 times more iridium than expected. The iridium spike suddenly became the most important issue. What was the cause?

After eliminating every conceivable source, the team came to the startling conclusion that the clay layer was the result of a meteorite impact of enormous dimension.[VI] Based upon

Fig. 12.36. These two photomicrographs show the tiny foraminifera found in Cretaceous sediments below but near the K–T boundary. The upper image shows the Tertiary Period with only a single species of foram. The lower image shows the Cretaceous Period with numerous forams. Most of the Cretaceous species died out at the end of the Cretaceous as a result of the Chicxulub impact. (Courtesy Alessandro Montaneri, Montanari Osservatorio Geologico di Coldigioco, Le Marche, Italy.)

[VI] Comets are usually considered along with asteroids as the possible K–T projectile. Scientists usually refer to the projectile as an asteroid or comet rather than an asteroid alone. Comet densities range from 0.5 to 1.2 times the density of water, leaving little room for silicate rock. If the iridium content in the silicate rock material of comets is similar to that in stony asteroids, a comet about twice the size of the 10 km asteroid would be needed to provide the same amount of iridium.

Fig. 12.37. This striking painting depicts the K–T impact a fraction of a second after the 12 × 5 km asteroid fragment contacts the floor of the shallow sea on the Yucatan Peninsula. Long compression cracks develop in the crater-marked asteroid as molten material is left behind in its atmospheric wake. The atmosphere has been partially pushed aside producing a cylinder of darker blue along the asteroid's atmospheric path. The shock wave preceding the asteroid has radially parted the dense cloud deck. The water surface around the point of impact is flattened by the preceding shock wave and a wave 1 km high rapidly expands away from the impact point traveling at several kilometers per second flooding the coast land. (Artwork by Dorothy Sigler Norton, 2000.)

their iridium measurements they concluded that an asteroid or comet about 10 km in diameter (the actual projectile was probably a fragment of irregular shape much like the asteroid Gaspra or the nucleus of Comet Halley) had struck Earth 65 million years ago releasing the energy of between 4×10^8 and 4×10^9 megatons of TNT.

If indeed a 10 km diameter asteroid/comet had impacted Earth, the explosion would have sent hundreds of cubic kilometers of Earth and projectile rock into and above the atmosphere. Much of this material was in the form of fine dust laden with concentrations of platinum-group metals that gradually settled out over the entire Earth adding a layer of clay-like material to the existing stratigraphic sequence, a boundary clay marking the end of the Mesozoic Era and the extinction of 65% of the world's species, including the dinosaurs.

The Berkeley team suggested a test of the theory. Geologists must search worldwide for the tell-tale iridium spike at the K–T boundary wherever these layers are preserved. This challenge was at first reluctantly received by skeptical geologists but those who chose to join the search were rewarded. Astonished, there were iridium spikes everywhere they looked. In Texas the boundary layer contained 43 times more iridium than surrounding rock; in Montana, 127 times; and in Haiti, 300 times more iridium. Over one hundred sites have been examined and all of them show excesses of iridium. Catastrophism had indeed returned adding this time an extraterrestrial event of unimaginable violence (Fig. 12.37).

The Chicxulub crater

The iridium spike was sobering evidence of a cataclysmic event that must have occurred at the K–T boundary, but it was still circumstantial evidence. Certainly an impact of this magnitude would have created a crater over 200 km in diameter. Where does such a crater hide? Was it covered by kilometers of limestone beneath the deepest sea? Was its fate to be subducted into the Earth's mantle? Had erosion taken its toll and worn it down to a peneplane surface with its remnants carried to the sea? The crater was an essential element necessary to convince the world that a celestial hammer had dealt the final *coup de gras* to the dinosaurs.

While Walter Alvarez was at Gubbio taking samples, Glen Penfield, an American geophysicist on loan to Pemex,

Mexico's national oil company, was making a magnetic survey of the Gulf waters north of the Yucatan Peninsula looking for thick limestone beds where oil might be trapped. Daily, Penfield would fly a gridwork pattern over the Gulf using a magnetometer to define the sediments. After each flight he would piece together strip charts showing sediments on the Gulf floor. It was during this activity that a magnetic disturbance began to show up in the shallow waters north of the Yucatan Peninsula. Before long the charts pictured a semicircular feature which seemed to Penfield to be foreign to the natural Gulf floor. The structure was buried in more than a kilometer of Cenozoic limestone sediments. Using borrowed gravity survey maps made by Pemex in the 1960s that extended the survey onto the Yucatan Peninsula, he saw immediately that, when connected together, the two maps made a complete circular structure that was half on land and half under water. The object was over 160 km in diameter centered on the coast near the small village of Chicxulub. Penfield had serendipitously found the long sought impact structure. In 1990, core samples made just outside the outer rim in the 1950s were located, stored at the University of New Orleans. In these samples Glen Penfield and Alan Hildebrand, then a graduate student at the University of Arizona, found shocked quartz, a definitive sign of an impact structure.[18]

In the last decade of the twentieth century, Virgil Sharpton and coworkers at the Lunar and Planetary Institute in Houston performed gravity anomalies studies at the Yucatan site (Fig. 12.38) resulting in the discovery of what appears to be a series of concentric rings surrounding a central uplift placing the crater in the category of a multiring impact basin. Figure 12.39 is an interpretation of the gravity anomaly data showing the presence of at least four rings. There is some evidence of an additional ring outside the fourth. If this is the true outer rim of the structure it would extend the diameter to approximately 300 km. Estimates of the crater's true diameter remains the subject of vigorous debate.

Asteroid or comet?

An impact structure of this magnitude along with the amount of iridium found at the K–T boundary places constraints on the projectile size and mass. Three different objects are considered possible projectile candidates: near-Earth asteroids; short-period comets; and long-period comets. The average impact velocity of near-Earth asteroids is about 20 km/s. That of short-period and long-period comets is 30 and ~55 km/s, respectively. Of the three, only the long-period comets have the necessary kinetic energy to produce a multiring basin of this size. The worldwide iridium abundance at the K–T boundary dictates the amount

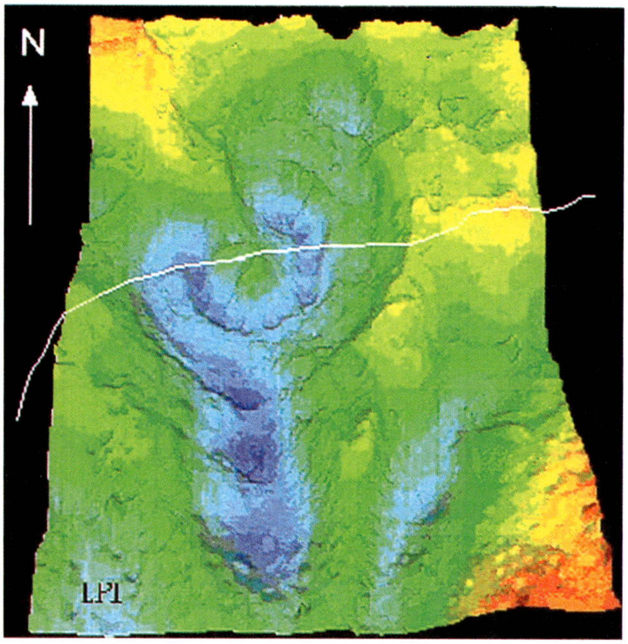

Fig. 12.38. The Chicxulub impact structure is seen here in three dimensions, produced by gravity field intensity scans across the structure bringing out gravity anomalies. The map is color coded spectrally with deep blue representing the lowest gravity values and the red representing the highest values. The Chicxulub basin is the green and blue circular area centered on the town of Chicxulub on the Yucatan Peninsula. The shape of the peninsula is superimposed over the map. The crater, buried in more than 1 km of limestone sediments, shows a raised center (green circular area) centered on the peninsula surrounded by a trough (blue) and a series of concentric rings (green), best seen in Fig. 12.39. (Plot by V.L. Sharpton, Lunar and Planetary Institute, Houston, Texas, and University of Alaska, Fairbanks.)

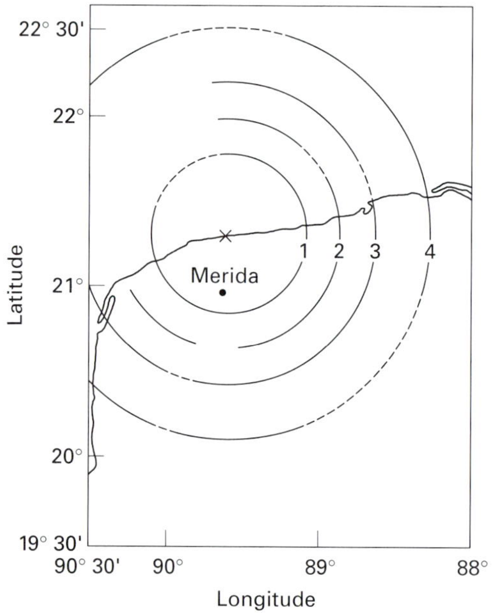

Fig. 12.39. This map interprets the gravity anomaly data showing the position of four discrete rings giving the Chicxulub crater the appearance of a multiringed impact basin commonly seen on the terrestrial planets and the Moon. There is evidence of yet a fifth ring outside the fourth ring that could extend the crater's diameter to about 300 km. (Map by V.L. Sharpton, Lunar and Planetary Institute, Houston, Texas, and University of Alaska, Fairbanks.)

of chondritic material (which carries the noble metals) needed in the comet projectile's mass. Long-period comets from 10–21 km in diameter containing a minimum of 50% frozen gases and volatile nonchondritic material (hydrocarbons), could have sufficient chondritic material (≤50%) to make a sphere 7.6 km in diameter that could carry the total iridium found worldwide at the K–T boundary. This projectile traveling at an impact speed of ~55 km/s would produce a transient crater of 130 km ±30 km diameter.[19] Subsequent modification of the transient crater by uplifting of the floor and widening of the basin flanks by collapse could have extended the final crater diameter to ~300 km.

There is still considerable debate about the relationship between the K–T impact and the presumed concomitant extinction of species. The naysayers argue that all major extinctions, including the K–T extinction, are marked beforehand by long-term declines of many species including the dinosaurs for reasons that still elude paleontologists. The extinction was selective with the more environmentally sensitive species found in the tropics and subtropics already in decline before the end of the Cretaceous. Other more-tolerant species continued on through the K–T boundary to the Paleocene Epoch apparently undisturbed by the environmental consequences of the impact. Sea-level changes worldwide are given as a probable reason for extinctions of environmentally intolerant species. The causative agent is often cited as the massive volcanic Deccan Trap eruptions in India which polluted the atmosphere with volcanic dust creating a greenhouse effect that raised the temperature by 3–4 °C and melted the ice caps, raising the sea level. Supporters of the theory of extinction by asteroid impact point to the strong evidence (the iridium anomaly) collected over the past 20 years that suggests the impact was a *final* causative agent leading to mass extinctions of species already in decline.

The Permian–Triassic extinction

The search goes on for other large impact craters that may have been involved with other known mass extinctions. The Permian–Triassic extinction 250 million years ago is especially interesting since life very nearly became extinct on the planet at that time. About 90% of all marine species and 70% of land vertebrates perished. Recently a team of scientists lead by Luann Becker of the University of Washington found new evidence that the Permian extinction was triggered by a collision with an asteroid or comet. The impact occurred when all of the continental landmasses were joined into the supercontinent, Pangaea. Although the team does not know the location of the crater today, if indeed a crater still exists, their search in the Permian–Triassic sedimentary beds in China and Japan have revealed large complex carbon molecules containing 60–200 carbon atoms linked in the form of a geodesic called fullerenes or Buckyballs (buckminsterfullerenes, named for the geodesic dome inventor, Buckminster Fuller). These structures are not native to Earth but are thought to have formed in carbon-rich stars. The team found noble gases trapped inside the geodesic molecular structures. The gases, argon and helium, reveal ratios of isotopes normally not present in the Solar System. (Fullerenes are found in some carbonaceous chondrites with similar noble gas ratios.) These fullerenes were probably delivered to Earth on a long-period comet that spent most of its time in the Oort Cloud of comets, halfway to the nearest star, an ideal place to gather fullerenes released by carbon stars. Estimates of the comet size necessary to scatter fullerenes around the Earth in quantities found in the sediments are between 6 and 12 km diameter, close to the size of the projectile at the K–T boundary. The evidence shows that the extinction occurred very rapidly on a geologic scale, in less than 100 000 years and possibly as little as 8000 years. Like the Cretaceous Period, many species were in decline at the end of the Permian. The P–T impact, though not considered the sole causal agent of the extinctions, is thought to have triggered events such as climate and sea-level changes that ultimately led to the great extinction. With this research, fullerenes have become the new tracer of extinction impacts. Iridium was found in the P–T sediments but at much lower levels than at the K–T boundary, suggesting differences in composition between the two projectiles.[20]

References

1. Marvin, U.B. (1996). Ernst Florens Friedrich Chladni (1756–1827) and the origins of modern meteorite research. *Meteoritics and Planetary Science* **31**, 545–588.
2. Gould, B.A. (1859). On the Supposed Lunar Origin of Meteorites. *Proceedings of the American Association for the Advancement of Science* **13**, 181–187.
3. Foote, A.E. (1891). A new locality for meteoric iron, with a preliminary notice of the discovery of diamonds in the iron. *American Journal of Science* (3rd series) **42**, 413–417.
4. Gilbert, G.K. (1896). The origin of hypotheses, illustrated by the discussion of a topographic problem. *Science* (new series) **3**, 1–13.
5. Chao, E.T.C., Shoemaker, E.M. and Madsen, B.M. (1960). First natural occurrence of coesite. *Science* **132**, 220–222.
6. Grieve, R.A.F. (1997). Target earth: evidence for large-scale impact events. *Annals of The New York Academy of Sciences* **882**, 319–352.
7. French, B.M. (1998). *Traces of Catastrophe: A Handbook of Shock-Metamorphic Effects in Terrestrial Meteorite Impact Structures*. LPI Contribution No. 954, Lunar and Planetary Institute, Houston, 120 pp.
8. Schultz, P. and Lianza R. (1992). Recent grazing impacts on the Earth recorded in the Rio Cuarto crater field, Argentina. *Nature* **355**, 234–237.
9. Dietz, R.S. (1947). Meteorite impact suggested by orientation of shatter cones at the Kentland, Indiana, disturbance. *Science* **105**, 42–43.
10. Dietz, R.S. (1964). Sudbury structure as an astrobleme. *Journal of Geology* **72**, 412–434.
11. Spencer, L.J. (1933). Meteoritic iron and silica glass from the meteorite craters of Henbury, Central Australia; and Wabar, Arabia. *Mineralogy Magazine* **23**, 387–404.
12. Shoemaker, E.M. and Wynn, J.C. (1998). The day the sands caught fire. *Scientific American*, Nov. 1998 issue.
13. Nininger, H.H. (1954). Impactite slag at Barringer Crater. *American Journal of Science* **252**, 277–290.
14. Nininger, H.H. (1956). Arizona's Meteorite Crater. *American Meteorite Laboratory*, pp. 117–134.
15. Coes, L. (1953). A new dense crystalline silica. *Science* **118**, 131.
16. Krinov, E.L. (1966). *Giant meteorites*, Pergamon Press, New York, p. 147.
17. Alvarez, L., Alvarez, W., Asaro, F. and Michel, H.V. (1980). Extraterrestrial cause for the Cretaceous–Tertiary extinction. *Science* **208**, 1095–1107.
18. Hildebrand, A.R. *et. al.* (1991). Chicxulub crater: a possible Cretaceous/Tertiary boundary impact crater on the Yucatan peninsula, Mexico. *Geology* **19**, 867–871.
19. Sharpton, V.L. and Marìn, L.E. (1997). The Cretaceous–Tertiary impact crater and the cosmic projectile that produced it. *Annals of the New York Academy of Sciences* **822**, 353–380.
20. Becker, L., Poreda, R., Hunt, A. and Rampino, M. (2001). Impact event at the Permian–Triassic boundary: evidence from extraterrestrial noble gases in fullerenes. *Science* **291**, 1530–1533.

Appendix A Classification of meteorites (A.E. Rubin, 2000)

CHONDRITES	
Carbonaceous chondrites	
CI	aqueously altered; chondrule-free; volatile-rich
CM	aqueously altered; minichondrule-bearing
CR	aqueously altered; primitive-chondrule-bearing; metal-bearing
CO	minichondrule-bearing; metal-bearing
CV	large chondrule-bearing; abundant CAIs; partially aqueously altered
CK	large chondrule-bearing; darkened silicates
CH	microchondrule-bearing; metal-rich; volatile-poor
Ungrouped	(e.g., Coolidge; LEW 85332)
Ordinary chondrites	
H	high total iron
L	low total iron
LL	low total iron; low metallic iron
"HH"	(Chondritic silicates in Netschaevo IIE – an iron)
R chondrites	
R	highly oxidized; $\delta^{17}O$-rich
Enstatite chondrites	
EH	high total iron; highly reduced; minichondrule-bearing
EL	lower total iron; highly reduced; moderately sized chondrules
Ungrouped	(e.g., LEW 87223)
IAB/IIICD silicates	subchondritic composition; chondrule-free; planetary-gas-bearing
Ungrouped chondrites	(e.g., Deakin 001)
PRIMITIVE ACHONDRITES	
Acapulcoites	chondritic abundances of plagioclase and troilite; medium-grained
Lodranites	subchondritic abundances of plagioclase and troilite; coarse-grained
Winonaites	IAB-silicate-related
Ungrouped	(e.g., Divnoe)
DIFFERENTIATED METEORITES	
Asteroidal achondrites	
Eucrites	basalts
Diogenites	orthopyroxenites
Howardites	brecciated mixture of basalts and orthopyroxenites
Angrites	fassaitic–pyroxene-bearing basalts
Aubrites	enstatite achondrites
Ureilites	olivine–pyroxene–carbonaceous matrix-bearing
Brachinites	equigranular; olivine–clinopyroxene–orthopyroxene-bearing
Martian meteorites	
Shergottites	basalts and lherzolites
Nakhlites	cumulus–augite-bearing pyroxenites
Chassigny	dunite
ALH 85001	orthopyroxenite
Lunar meteorites	
Mare basalts	
Impact breccias	anorthositic- and mare-dominated regolith and fragmented breccias
Stony irons	
Pallasites	metal plus olivine; core/mantle boundary samples
	main-group pallasites
	Eagle-Station-trio pallasites
Mesosiderites	metal plus basaltic, gabbroic and orthopyroxenitic silicates
Ungrouped	(e.g., Enon; Mt. Egerton)

Irons

Magmatic groups	(IC, IIAB, IIC, IID, IIF, IIIAB, IIIE, IIIF, IVA, IVB)
Nonmagmatic groups	(IAB/IIICD, IIE)
Ungrouped	(e.g., Britstown; Denver City; Guin; Sombrerete)

Printed with permission from A.E. Rubin and *Meteoritics and Planetary Science* (1997) **32**, 232.

Appendix B Formation ages of selected meteorites

Meteorite	Type	Method	Age	Source
Chondrites				
Tieschitz	H3	Rb/Sr	4.53 ± 0.06	Minster and Allègre (1979)
Menow	H4	Ar/Ar	4.48 ± 0.06	Turner *et al.* (1978)
Ochansk	H4	Ar/Ar	4.48 ± 0.06	Turner *et al.* (1978)
Richardton	H5	Rb/Sr	4.39 ± 0.03	Evensen *et al.* (1978)
Guarena	H6	Rb/Sr	4.46 ± 0.08	Wasserburg *et al.* (1969)
Butsura	H6	Ar/Ar	4.48 ± 0.06	Turner *et al.* (1978)
ALH 77015	L3	Ar/Ar	4.51 ± 0.10	Kanoeka (1980)
Tennasilm	L4	Pb/Pb	4.55 ± 0.13	Unruh (1982)
Bjurböle	L4	Ar/Ar	4.51 ± 0.08	Turner (1969)
Bruderheim	L6	Pb/Pb	4.53 ± 0.04	Unruh (1982)
Peace River	L6	Rb/Sr	4.46 ± 0.03	Gray *et al.* (1973)
Parnallee	LL3	Rb/Sr	4.53 ± 0.04	Hamilton *et al.* (1979)
Soko-Banja	LL4	Rb/Sr	4.45 ± 0.02	Minster and Allègre (1981)
Krähenberg	LL5	Rb/Sr	4.60 ± 0.14	Kemp and Muller (1969)
Olivenza	LL5	Rb/Sr	4.53 ± 0.16	Sanz and Wasserburg (1969)
Jelica	LL6	Rb/Sr	4.42 ± 0.04	Minster and Allègre (1981)
Indarch	EH4	Rb/Sr	4.46 ± 0.08	Gopolan and Wetherill (1970)
Abee	EH4	Ar/Ar	4.52 ± 0.03	Bogard *et al.* (1983)
St. Sauvein	EH5	Rb/Sr	4.46 ± 0.05	Minster *et al.* (1979)
Allende	CV3	Ar/Ar	4.55 ± 0.03	Jessberger *et al.* (1980)
Differentiated				
Juvinas	Euc	Rb/Sr	4.50 ± 0.07	Allègre *et al.* (1975)
Stannern	Euc	Sm/Nd	4.48 ± 0.07	Lugmair and Scheinin (1975)
Stannern	Euc	Rb/Sr	3.20 ± 0.05	Birck and Allègre (1978)
Ibitira	Euc	Rb/Sr	4.42 ± 0.25	Birck and Allègre (1978)
Kapoeta	How	Rb/Sr	4.44 ± 0.12	Papanastassiou *et al.* (1974)
Y-7308	How	Rb/Sr	4.48 ± 0.06	Kanoeka (1981)
Norton Co.	Aub	Rb/Sr	4.39 ± 0.04	Minster and Allègre (1976)
Angra dos Reis	Ang	Sm/Nd	4.56 ± 0.04	Jacobsen and Wasserburg (1984)
Shergotty	SNC	Rb/Sr	0.36 ± 0.16	Jagoutz and Wänke (1986)
Nakhla	SNC	Rb/Sr	1.34 ± 0.02	Papanastassiou *et al.* (1974)
Mundrabille	IAB	Ar/Ar	4.57 ± 0.06	Niemeyer (1979)
Pitts	IAB	Ar/Ar	4.54 ± 0.06	Niemeyer (1979)
Colomera	IIE	Rb/Sr	4.51 ± 0.04	Sanz *et al.* (1970)
Kodaikanal	IIE	Rb/Sr	3.70 ± 0.1	Burnett and Wasserburg (1967)
Weeberoo	IIE	Rb/Sr	4.39 ± 0.07	Evensen *et al.* (1979)
Estherville	Mes	Rb/Sr	4.54 ± 0.20	Brouxel and Tatsumoto (1991)
Vaca Muerta	Mes	Pb/Pb	4.54 ± 0.15	Ireland and Wlotzka (1992)

References

Allègre, C.J., Birck, J.L., Fourcade, S. and Semet, M.P. (1975). Rubidium-87/strontium-87 age of Juvinas basaltic achondrite and early igneous activity in the solar system. *Science* **187**, 436–438.

Birck, J.L and Allègre, C.J. (1978). Chronology and chemical history of the parent body of basaltic achondrites studied by the ^{87}Rb–^{87}Sr method. *Earth and Planetary Science Letters* **39**, 37–51.

Bogard, D.D., Unruh, D.M. and Tatsumoto, X. (1983). ^{40}Ar–^{39}Ar and U–Th–Pb dating of separated clasts from the Abee E4 chondrite. *Earth and Planetary Science Letters* **62**, 132–146.

Brouxel, M. and Tatsumoto, M. (1991). The Estherville mesosiderite: U–Pb, Rb–Sr, and Sm–Nd isotopic study of a polymict breccia. *Geochimica Cosmochimica Acta* **55**, 1121–1133.

Burnett, D.S. and Wasserburg, G.J. (1967). ^{87}Rb–^{87}Sr ages of silicate inclusions in iron meteorites. *Earth and Planetary Science Letters* **2**, 397–408.

Evensen, N.M. Carter, S.R., Hamilton, P.J., O'nions, R.K. and Ridley, W.I. (1978). A combined chemical–petrological study of separated chondrules from the Richardton meteorite. *Earth and Planetary Science Letters* **42**, 223–236.

Evensen, N.M., Hamilton, P.J., Harlow, G.E., Klimentidis, R., O'nions, R.K. and Prinz, M. (1979). Silicate inclusions in Weekeroo Station: planetary differentiates in an iron meteorite. *Lunar and Planetary Science* **X**, 376–378.

Gopolan, K. and Wetherill, G.W. (1970). Rubidium–strontium studies on enstatite chondrites: whole-rock and mineral isochrons. *Journal of Geophysical Research* **75**, 3457–3467.

Gray, C.M., Papanastassiou, D.A. and Wasserburg, G.J. (1973). The identification of early condensates from the solar nebula. *Icarus* **20**. 213–239.

Hamilton, P.J., Evensen, N.M. and O'nions, R.K. (1979). Chronology and chemistry of Parnallee (LL3) chondrules. *Lunar and Planetary Science* **X**, 494–496.

Ireland, T.R. and Wlotzka, F. (1992). The oldest zircons in the Solar System. *Earth and Planetary Science Letters* **109**, 1–10.

Jacobsen, S.B. and Wasserburg, G.J. (1984). Sm–Nd isotopic evolution of chondrites and achondrites, II. *Earth and Planetary Science Letters* **67**, 137–150.

Jagoutz, E. and Wänke, H. (1986). Sr and Nd isotopic systematics of Shergotty meteorites. *Geochimica Cosmochimica Acta* **50**, 939–953.

Jessberger, E.K., Dominik, K.B., Staudacher, T. and Herzog, G.F. (1980). ^{40}Ar–^{39}Ar ages of Allende. *Icarus* **42**, 380–405.

Kanoeka, I. (1980). ^{40}Ar–^{39}Ar ages for L and LL chondrites from Allan Hills, Antarctica: ALHA 77015, 77214, and 77304. *Memoirs of the National Institute of Polar Research*, Special Issue No. 17, 177–88.

Kanoeka, I. (1981). ^{40}Ar–^{39}Ar ages of Antarctic meteorites: Y74191, Y75258, Y7308, Y74450, and ALH 765. *Memoirs of the National Institute of Polar Research*, Special Issue No. 20, 250–263.

Kemp, W. and Muller, O. (1969). The stony meteorite Krähenberg. Its chemical composition and the Rb–Sr age of the light and dark portions. *Meteorite Research*, P.M. Millman (ed.), Dordrecht: D. Reidel, pp. 418–428.

Lugmair, G.W. and Scheinin, N.B. (1975). Sm–Nd systematics of the Stannern meteorite [abstract]. *Meteoritics* **10**, 447–448.

Minster, J.F. and Allègre, C.J. (1976). ^{87}Rb–^{87}Sr constraints on the primitive chronology of meteorites (abstract). *Meteoritics* **11**, 336–337.

Minster, J.F. and Allègre, C.J. (1979). ^{87}Rb–^{87}Sr chronology of H chondrites: constraints and speculations on the early evolution of their parent body. *Earth and Planetary Science Letters* **42**, 333–347.

Minster, J.F. and Allègre, C.J. (1981). ^{87}Rb–^{87}Sr dating of LL chondrites. *Earth and Planetary Science Letters* **56**, 89–106.

Minster, J.F., Ricard, L.P. and Allègre, C.J. (1979). ^{87}Rb–^{87}Sr chronology of enstatite meteorites. *Earth and Planetary Science Letters* **44**, 420–440.

Niemeyer, S. (1979). ^{40}Ar–^{39}Ar dating of inclusions from IAB iron meteorites. *Geochimica Cosmochimica Acta* **43**, 1829–1840.

Papanastassiou, D.A. and Wasserburg, G.J. (1974). Evidence for late formation and young metamorphism in the achondrite Nakhla. *Geophysical Research Letters* **1**, 23–26.

Sanz, H.G. and Wasserburg, G.J. (1969) Determination of an internal ^{87}Rb–^{87}Sr isochron for the Olivenza chondrite. *Earth and Planetary Science Letters* **6**, 335–345.

Sanz, H.G., Burnett, D.S. and Wasserburg, G.J. (1970). A precise ^{87}Rb–^{87}Sr age and initial ^{87}Sr/^{86}Sr for the Colomera iron meteorite. *Geochimica Cosmochimica Acta* **34**, 1227–1239.

Turner, G., Enright, M.C. and Cadogan, P.H. (1978). The early history of chondrite parent bodies inferred from ^{40}Ar–^{39}Ar ages. *Proceedings of the Ninth Lunar and Planetary Science Conference*, pp. 989–1025.

Turner, G. (1969). Thermal histories of meteorites by the ^{39}Ar–^{40}Ar method. *Meteorite Research*, P.M. Millman (ed.), Dordrecht: D. Reidel, pp. 407–417.

Unruh, D.M. (1982). The U-Th-Pb age of equilibrated L chondrites and a solution to the excess radiogenic Pb problem in chondrites. *Earth and Planetary Science Letters* **58**, 75–94.

Wasserburg, G.J., Papanastassiou, D.A. and Sanz, H.G. (1969). Initial strontium for a chondrite and the determination of a metamorphism or formation age. *Earth and Planetary Science Letters* **7**, 33–43.

Appendix C Minerals in meteorites

About 275 minerals are known to exist in meteorites. This is only ~7% of the minerals known to have formed in the terrestrial environment. Sixty of these minerals have been listed here in alphabetical order. They have been selected for their importance as primary and accessory minerals and nearly all of them appear in the text. They also appear in the index under "minerals" for convenient reference. The name of each mineral is followed by its formula, mineral class and crystal system.

Akaganèite [β-FeO(OH,Cl)]; hydroxide
This is a major corrosion product in the terrestrial weathering of FeNi in all meteorites. Akaganèite is the major carrier of chlorine indigenous to the environment, but not in meteorites. The low-nickel iron, kamacite, is converted directly to akaganèite within the meteorite.

Albite [$NaAlSi_3O_8$]; tectosilicate; triclinic
Albite is the Na end member of the plagioclase solid solution series. It is very rare in meteorites. Minor amounts are found in SNC meteorites.

Anorthite [$CaAl_2Si_2O_8$]; tectosilicate; triclinic
Anorthite is the calcium end member of the plagioclase solid solution series. It is a common accessory mineral in chondrites and achondrites. It is a major mineral in eucrites and an accessory mineral in angrites. Also found in refractory inclusions in C chondrites.

Augite [$(Ca,Na)(Mg,Fe,Al)(Si,Al)_2O_6$]; inosilicate; monoclinic
A calcium-rich clinopyroxene found in some achondrites. Accessory amounts in eucrites and nakhlites and a major pyroxene in shergottites.

Awaruite [(Ni_3Fe)]; metal; isometric
A nickel-rich iron metal similar to taenite found in minor amounts in CV chondrites and in small amounts in CK and R chondrites.

Bronzite [$(Mg,Fe)SiO_3$]; inosilicate; orthorhombic
An orthopyroxene in the solid solution series between magnesium-rich enstatite and iron-rich ferrosilite. In ordinary chondrites its composition is usually about $Fs_{(10-20)}$.

Bytownite [$(Na,Ca)Al_2Si_2O_8$]; tectosilicate; triclinic
A calcium-rich member of the plagioclase series with a composition $Ab_{(10-30)}$ $An_{(70-90)}$. Is often found in eucrites along with anorthite and in very small amounts in angrites.

Calcite [$CaCO_3$]
Rare in meteorites. Usually found along veins in CI chondrites. Often found associated with magnetite.

Chromite [$FeCr_2O_4$]; isometric; oxide
Found in many meteorite groups. Is the dominant oxide in ordinary chondrites. Often found as small, black and opaque euhedral and subhedral grains in chondrules.

Clinoenstatite [$MgSiO_3$]; inosilicate; monoclinic
Primarily a meteoritic pyroxene mineral. It is the end member of the monoclinic pyroxene series $MgSiO_3$–$FeSiO_3$, clinoenstatite–clinoferrosilite. The "clino" in the name refers to the crystal system. Most enstatite on Earth is found in the orthorhombic crystal system. Clinoenstatite may be recognized under the microscope by its low birefringence and polysynthetic twinning. Common in ordinary chondrites.

Clinopyroxene [$(Ca,Mg,Fe)SiO_3$] (see **Clinoenstatite**)

Coesite [SiO_2]; tectosilicate; monoclinic
A very dense polymorph of quartz produced by high shock pressures on quartz sandstone material. Generally thought to be the product of crater-forming meteorite impacts. Coesite was first found around Meteor Crater in Arizona.

Cohenite [$(Fe,Ni,Co)_3C$]; carbide; orthorhombic
Iron–nickel carbide found as an accessory mineral primarily in coarse octahedrite iron meteorites. Also found as a minor mineral in Type 3 ordinary chondrites. Oxidizes a bronze color and is often associated with schreibersite. It can be distinguished from schreibersite under a petrographic microscope (see text).

Copper [Cu]; native metal; isometric
Found widely in trace amounts in ordinary chondrites and iron meteorites. Trace amounts also found in some CV (Allende). It is usually found as tiny inclusions in FeNi and troilite.

Diamond [C]; native element; isometric
A polymorph of graphite produced by shock pressure during impact either in space or on Earth. Found in some iron meteorites with graphite nodules and in the carbonaceous matrix in ureilites. Also found in CM chondrites as interstellar diamonds.

Diopside [$CaMgSi_2O_6$]; inosilicate; monoclinic
A calcium and magnesium-rich clinopyroxene and is an end member of a solid solution series with hedenbergite ($CaFeSi_2O_6$) as the iron end member. Occurs in E chondrites, aubrites and mesosiderites. Also found in small quantities in refractory inclusions in CM chondrites.

Enstatite [$MgSiO_3$]; inosilicate; orthorhombic
Enstatite is the magnesium-rich end member of the enstatite-ferrosilite ($FeSiO_3$) solid solution series of the orthopyroxenes. It is the major mineral in all ordinary, carbonaceous and R chondrites. Enstatite is also the primary mineral in enstatite chondrites and achondrites and the basaltic achondrites.

Epsomite [$MgSO_4.7H_2O$]; hydrous sulfate
Hydrated magnesium sulfate (gypsum). Found lining microscopic veins in CI chondrites. Also in matrices of CM chondrites.

Fassaite [$Ca(Mg,Ti,Al)(Al,Si)_2O_6$]; inosilicate; monoclinic
Fassaite is a Ca-rich clinopyroxene sometimes referred to as a Al–Ti diopside or a variety of augite. It is the primary mineral in the Angra dos Reis angrite (93 vol% fassaite). Also a minor phase in some CV chondrules and refractory inclusions.

Fayalite [Fe_2SiO_4]; nesosilicate; orthorhombic
The iron end member of the olivine complete solid solution series in the system, forsterite (Mg_2SiO_4)–fayalite (Fe_2SiO_4). Fayalite content is diagnostic for ordinary chondrite petrographic types. It is a major phase in all chondrites except E chondrites. Usually meteoritic olivine is no greater than Fa_{30}. R chondrites are highly oxidized so their FeO is high, Fa_{38-41}.

Feldspathoids
These are silicates that are chemically similar to feldspars. The primary chemical difference between feldspathoids and feldspars is the amount of SiO_2. Feldspathoids contain about two-thirds the silica as the feldspars. The most common meteoritic feldspathoids are sodalite, nepheline and melilite**.** All three are found in chondrules and refractory inclusions in CV chondrites.

Ferrosilite [$FeSiO_3$]; neosilicate; orthorhombic
The iron-end member of the orthopyroxene solid solution series. See **Enstatite**.

Forsterite [Mg_2SiO_4]; neosilicate; orthorhombic
The magnesium-end member of the olivine solid solution series. See **Fayalite**.

Goethite [α-FeO(OH)]; hydroxide; orthorhombic
A major secondary mineral; the product of terrestrial weathering of FeNi in meteorites. See Chapter 3.

Graphite [C]; native element; hexagonal
A common accessory mineral in iron meteorites, ordinary chondrites and ureilites. Occurs as nodules often associated with troilite. May be the site of diamond and lonsdaleite in IA irons and in ureilites. Also found in CI and CM chondrites and some E chondrites.

Hypersthene [(Mg,Fe) SiO_3]; inosilicate; orthorhombic
An orthopyroxene of the solid solution series from enstatite to ferrosilite. It is more iron-rich than either enstatite or bronzite with a Fs_{20-30}. Hypersthene is a major phase in diogenites appearing a light green or brown. It is also common in L-group ordinary chondrites.

Ilmenite [$FeTiO_3$]; oxide; rhombohedral
A black, opaque, slightly magnetic mineral; the principal ore of titanium. Occurs as a common accessory mineral in terrestrial igneous rocks, achondrites, lunar mare basalts and martian basalts.

Kamacite [FeNi]; metal; isometric
Low nickel–iron metal alloy, containing between 4 and 7.5 wt% nickel. The principal metal in iron and stony-iron meteorites and accessory metal in ordinary chondrites and minor metal in some achondrites. In the FeNi phase diagram kamacite is the alpha-iron.

Lawrencite [$(Fe,Ni)Cl_2$]; chloride
A brown or green iron-nickel mineral once thought to be an abundant accessory mineral in iron and stony-iron meteorites involved in the terrestrial weathering of these meteorites. It has been discredited (see Chapter 3).

Lonsdaleite [C]; element; hexagonal
A hexagonal polymorph of diamond. Occurs in ureilites and IAB irons produced by shock metamorphism of graphite on the parent body. Lonsdaleite has been artificially produced in the laboratory.

Magnetite [Fe_3O_4]; oxide; isometric
Opaque, black, strongly magnetic iron oxide commonly found in the matrices of carbonaceous chondrites and in small amounts in ordinary chondrites and some achondrites. A common mineral in the fusion crusts of stony meteorites and forms a black coating on terrestrially weathered iron meteorites.

Maskelynite a glass
A diaplectic glass, not a true mineral. Has the composition of plagioclase that has been transformed into a glass by shock metamorphism. Most commonly found in shocked plagioclase-bearing shergottites and ordinary chondrites. Presence of maskelynite glass is diagnostic for shocked meteorites suffering impact pressures of around 30 GPa.

Melilite [$(Ca,Na)_2(Al,Mg)(SiAl)_2O_7$] sorosilicate; tetragonal
Melilite is a group of minerals whose composition varies within a solid solution between gehlenite [$Ca_2Al(Si,Al)_2O_7$] and akermanite [$Ca_2MgSi_2O_7$]. Found in refractory inclusions in CV chondrites. Gehlenite is the major phase of melilite found in large chondrules in the Allende CV3 with a smaller component of akermanite.

Nepheline [$(Na,K)AlSiO_4$]; tectosilicate; hexagonal (see **Feldspathoids**)

Oldhamite [CaS]; sulfide; isometric
A dark brown, nearly opaque meteoritic mineral found in minor amounts in highly reduced meteorites like aubrites and in all petrographic types of E chondrites.

Oligoclase [$Ab_{90}An_1$–$Ab_{70}An_{30}$]; tectosilicate; triclinic
A sodium-rich plagioclase involving the plagioclase feldspar solid solution series between albite and anorthite. It is found in highly equilibrated ordinary chondrites.

Olivine [$(Mg,Fe)_2SiO_4$]; neosilicate; orthorhombic
A series of minerals formed by a complete solid solution from forsterite (Mg_2SiO_4) and fayalite (Fe_2SiO_4). Magnesium-rich olivines are much more common in meteorites than iron-rich olivines. A major mineral in all chondrites, pallasites, and some achondrites. Rare in E chondrites and aubrites.

Orthoclase (K-feldspar) [$KAlSi_3O_8$]; tectosilicate; monoclinic
A very rare phase in meteorites. Found in accessory amounts in eucrites and as a mesostasis in nakhlites.

Orthopyroxene [$(Mg,Fe)SiO_3$]; inosilicate; orthorhombic
A group of Ca-poor pyroxene minerals involving a solid solution between end members enstatite and orthoferrosilite. Orthopyroxenes are a major phase in nearly all meteorites. Most phases of the group are magnesium-rich with the orthoferrosilite composition usually not beyond Fs_{30}.

Pentlandite [$(Fe,Ni)_9S_8$]; sulfide; isometric
Resembles pyrrhotite in bronze color but is not magnetic until heated. Often associated with troilite in meteorites. Found in accessory amounts in the matrix and chondrules of CO, CV, CK, and CR chondrites. Small amounts in R chondrites and in pallasites.

Perovskite [$(CaTi)O_3$]; oxide; isometric
A high temperature calcium–titanium oxide found in refractory inclusions (CAIs) in carbonaceous chondrites.

Phyllosilicates
A large class of hydroxyl-bearing silicate minerals usually occurring in stacked flat sheets. They comprise four large groups including the serpentine group, smectite (clay) mineral group, mica group and chlorite group. Of the four, the first two are most important in meteorites. They occur as a result of aqueous alteration of meteoritic minerals. They are found most commonly in carbonaceous chondrites (see **Serpentine** and **Smectite Group**).

Pigeonite [$(Fe,Mg,Ca)SiO_3$]; inosilicate; monoclinic
A Ca-poor clinopyroxene with a 5–15 mole% $CaSiO_3$. It is a major phase in eucrites and a cumulate mineral along with augite and orthopyroxene in shergottites. Iron-rich olivine is rimmed by pigeonite in nakhlites.

Plagioclase [$Ab_{100}An_0$–Ab_0An_{100}]; tectosilicate; triclinic
An important class of silicate minerals with compositions between albite ($NaAlSi_3O_8$) and anorthite ($CaAl_2Si_2O_8$). A sodic plagioclase feldspar mineral occurring as microcrystalline to

coarse-grained crystals in ordinary chondrites and as calcic feldspar grains in HED achondrites. Calcic plagioclase is replaced by maskelynite in shocked shergottites.

Plessite [FeNi]; metal

A fine-grained intergrowth of kamacite and taenite commonly present in octahedrites and some chondrites.

Pyrrhotite [$Fe_{1-x}S$]; sulfide; hexagonal

A magnetic iron sulfide found in meteorites that are deficient in iron with respect to sulfur. It is similar in appearance to troilite in meteorites. An accessory mineral in CM chondrites.

Quartz [SiO_2]; tectosilicate; hexagonal

Extremely rare in meteorites. Found in small quantities in eucrites and other calcium-rich achondrites and in the highly reduced E chondrites.

Ringwoodite [$(Fe,Mg,)_2SiO_4$]; oxide; isometric

An olivine with a spinel structure; a cubic dimorph of olivine. First found in shock veins in an ordinary chondrite in 1969. A high-pressure mineral in which magnesium-rich olivine is converted to ringwoodite at pressures of about 150 kbar. An indicator of impact shock in meteorites.

Schreibersite [$(FeNi)_3P$]; phosphide; tetragonal

An iron–nickel phosphide common as an accessory mineral in iron and stony-iron meteorites. Often oriented parallel to Neumann lines in kamacite plates. Silvery-white when fresh and tarnishes to bronze. Often found surrounding troilite nodules. A true extraterrestrial mineral not found on Earth except in meteorites.

Serpentine [$Mg_3Si_2O_5(OH)_4$]; phyllosilicate; monoclinic–orthorhombic

A group of hydrous minerals produced by the aqueous alteration of the magnesium silicates, olivine and pyroxene, in meteorites. Abundant in the matrices of CI and CM chondrites, usually fine-grained and mixed with organic matter.

Silicon Carbide [SiC]

Occurs as interstellar dust grains in CM chondrites (Murchison) and E chondrites.

Smectite Group

A group of clay-like minerals with complex compositions including montmorillonite and saponite clays. These have been found in CM chondrites and SNC meteorites.

Sodalite [$Na_8(AlSiO_4)_6 Cl_2$]; tectosilicate; isometric (see **Feldspathoids**)

Spinel [$MgAl_2O_4$]; oxide; isometric

A group of minerals with the general formula AB_2O_4 where A represents Mg, Fe (ferrous), zinc and manganese in any combination and B represents Al, Fe (ferric), and Cr. It occurs in meteorites as small, usually opaque octahedrons. Present in small amounts in chondrules, aggregates and refractory inclusions in CV chondrites. Chromite ($FeCr_2O_4$) is the dominant oxide in ordinary chondrites.

Stishovite [SiO_2]; tectosilicate; tetragonal

High-pressure extremely dense polymorph of quartz produced by meteorite impact into quartz-bearing rock. It is usually associated with coesite and forms at static pressures of over 100 kbar. Its occurrence is diagnostic for terrestrial impact craters.

Taenite [FeNi]; metal; isometric

A γ-phase iron–nickel alloy with variable nickel from 27% to 65% in iron meteorites. It occurs in iron meteorites as thin lamellae bordering kamacite plates or as intergrowths with kamacite to form plessite.

Troilite [FeS]; sulfide; hexagonal

A bronze-colored iron sulfide occurring as an accessory mineral in nearly all meteorites. It is found as nodules in iron meteorites and is often associated with graphite nodules. In chondritic meteorites it is usually found as small blebs or grains in both chondrules and matrix averaging about 6 wt%. It differs from pyrrhotite by lacking an iron deficiency and is not magnetic.

Whitlockite [$Ca_9MgH(PO_4)_7$]; phosphate

An important phosphate mineral in ordinary chondrites, R chondrites and CV chondrites. Also known as merrillite.

Wollastonite [$CaSiO_3$]; inosilicate; triclinic

End member in the pyroxene composition system $CaSiO_3$–$MgSiO_3$–$FeSiO_3$. Frequently the composition of a pyroxene is stated in terms of molecular percentages of the above three end members: Wo (wollastonite), En (enstatite), Fs (ferrosilite).

Reference

Rubin, A.E. (1997). Mineralogy of meteorite groups. *Meteoritics and Planetary Science* **32**, 231–247.

Appendix D Preparing and etching iron meteorites

Witnessing the etching of an iron meteorite is a fascinating experience. It is analogous to watching a picture appear on a piece of black and white print paper during chemical development. And like the printing process, etching a meteorite to bring out the delicate Widmanstätten and Neumann structures relies on judgement and experience. Thus, etching a meteorite is as much an artistic endeavor as a carefully controlled scientific procedure. The etching process is the icing on the cake, that is, it is the final process which culminates a lengthy and carefully executed preparation of the specimen. Etching an iron meteorite without paying the price in time and effort to prepare the specimen correctly is a losing proposition. Let us go through the steps that, if followed carefully, will result in a fine museum-quality specimen.

Selecting and treating the meteorite specimen before cutting

If choice of the specimen is an option, it is always best to choose a specimen that is free of rust. A rusted and corroded exterior usually is an indicator that the specimen has been invaded by corrosive substances in the interior also, especially if there are deep cracks evident on the surface. A rusting meteorite should be cleaned with a wire brush attached to a grinder, beginning with a coarse brush and ending with a fine one. Safety glasses should always be used during the cleaning operation as it is a sure thing that some of the metal bristles will come loose to become projectiles endangering your eyes. Caution should be taken to clean irons until they appear uniformly black. If bright metal appears during this cleaning operation you have over-cleaned the specimen. You are removing the outer black magnetite coating which should be preserved. There are always cavities in iron meteorites that brushes cannot reach. These can be cleaned using a rust removal chemical. If such a chemical is used, it must be neutralized after use.

Iron meteorites found beneath the surface in desert environments are often coated with a calcium carbonate layer called *caliche*. This can be easily dissolved with a 10% solution of nitric, hydrochloric or sulphuric acid. The meteorite can be emersed in the acid. This should be done before the rust is removed as the rust provides a level of protection against the acid. When the caliche has been removed, the meteorite should be washed thoroughly then neutralized with ammonium hydroxide for 5–10 min, washed again, and thoroughly dried.

Cutting the meteorite

Preparing an iron meteorite for the etching process begins with cutting the meteorite to expose the largest surface area possible. It is desirable for gross specimens to be cut in half, and the cut face prepared for etching, since this affords examination of the gross specimen's exterior morphology as well as its etched interior. These are scientifically and commercially more valuable than a slab. A cut slab is easier to prepare and etch since it will lie flat, making the etching process and subsequent microscopic examination easier to perform. If a cut slab is selected, both sides should be cut parallel to each other and cut reasonably flat, which helps to control the distribution of the etching solution. An irregular mass is much more difficult to handle during the etching process. Slabs cut off a main mass are often cut too thin so that the specimen does not remain flat and often suffers from a "turned down edge", to use an optician's term. I recommend a slab/thickness ratio of about 25:1, where the slab's longest dimension is considered. Thus, a slab of 150 mm in its longest dimension should have a thickness of 6 mm. Keep in mind that cutting thinner slices involve more waste per slab than cutting thicker slabs. A diamond blade should be used with a water-soluble cutting solution and a wetting agent. Since iron meteorites are extremely hard, a blade of sufficient thickness should be used to prevent the blade from flexing during the cutting operations (approximately 0.625 inches for a 12-inch blade). Ideally, the diamond blade should have a diameter at least three times the maximum dimension of the specimen being cut. The blade should be allowed to cut at its own speed. Forcing a blade to cut faster will only result in distorting the blade, an expensive mistake for blades that cost over US$125 each! I have found the best way to cut a large specimen is to equip the cutting table with weights hanging over the edge of the moving support system that applies an even force to the blade, letting the diamond saw cut at its own pace using gravity alone. When the meteorite is all but cut through, great care should be taken to hand guide the final few minutes of cut. If the final cut is hastened, it may leave a raised edge that will take hours to grind down to conform to flatness with the rest of the face.

Preparing the cut face

Before surface preparation begins the meteorite should be checked to see if there is any rust evident. If you are preparing a surface that has been cut years earlier and rust has appeared, the rust must be removed not only on the surface but within the fine cracks between kamacite plates or around troilite and graphite nodules. There are excellent rust removers and inhibitors available specifically for iron meteorites.[1]

After cutting, the face to be etched should be ground flat. This is not unlike the preparation of optical flats in an optical shop. Many preparators choose to use sanding belts to grind away the grooves made by the diamond blade. (A good cut fed at the rate of about an inch per hour should show no saw marks.) Since there is usually no cooling agent furnished with such grinders, care must be taken here to keep the meteorite cool by frequent dipping into cold water.

[1] I am indebted to Mr Bill Mason, president of Uncommon Conglomerates of St. Paul, Minnesota, for many personal conversations regarding the use of corrosion removers and inhibitors for iron and stony-iron meteorites. This company has available kits that include the chemicals necessary for cleaning and preserving iron meteorites.

Temperatures can reach several hundred degrees Celsius at the contact face. If allowed to get too hot (over ~400 °C), the meteorite will begin to lose its structure at the ground surface and additional grinding will be necessary to get to a deeper unaffected depth. Belt sanding, if not carefully done, could easily turn down the edges and additional sanding time will be necessary to bring the slab to a flat-to-the-edge surface. To check for flatness, a straight-edge such as a steel ruler should be placed across the face. The specimen face is flat enough to proceed to fine grinding only if light cannot be seen through the meteorite/straight-edge contact.

An excellent surface to use for the following grinding operations is a thick piece of plain parallel glass. The glass should be at least 4 mm thick and beveled on the edges. It should be no less than 600 mm long by 300 mm wide, allowing long strokes to be applied. The glass should be supported by a stout flat work table. Several grades of silicon carbide or aluminum oxide must be progressively used to arrive at a semi-polished state. (Some workers prefer to use wet/dry Carborundum paper rather than loose Carborundum. The glass would then act only as a backing rather than a grinding tool.) Beginning with 150 grade grit, the meteorite is ground until all pits and scratches from the previous grade have been removed. Use a bright nonfrosted incandescent light to illuminate the meteorite face with special emphasis on a zone about 4–5 mm from the edge. At this time you will be able to see if the slab has a turned edge. If so, pit marks from the previous grade will be evident along the edge seen under a 10× magnifier. Use of a felt-tipped pen to scribe lines across the slab from center to edge will reveal the flatness of the specimen after a few strokes. Residual marks will show where the low spots are located. The next grade should not be used until the slab is flat and the pits removed. This applies to all the grades up to the final grade. To insure that the slab is being ground evenly, you should constantly rotate the specimen a few degrees per stroke especially during the final grades. You must pass through grades 150, 220, 400, and 600. Some preparators even go to 1000 as a final finish. This gives the surface a semi-polished look on which the filament from the lamp can be imaged. Care must be taken to wash off the previous grade thoroughly before proceeding to the next, otherwise scratches from residual coarser grades will mar the final surface. At the final grinding stages, cleanliness is essential.

Etching the specimen with nitric acid

Once you are satisfied that all the pits and scratches have been removed from the ground surface and that the surface is flat to the edge, you are ready for the etching process. Prepare a 10% solution of HNO_3 by mixing 10 ml of concentrated nitric acid with 90 ml of ethanol (ethyl alcohol) making a solution called Nitol by H.H. Nininger. Ethanol is used instead of water because a water/acid solution reacts with troilite inclusions to produce brown stains that can only be removed by regrinding the specimen. You can use a water/nitric acid mix only of you know that troilite is not present. After the etching solution is prepared, have on hand a shallow tray in which you have placed household ammonia. This will act as a neutralizing agent to neutralize residual amounts of acid on and in the meteorite. This should be used full strength. The etching process works because the solubility of kamacite and adjacent taenite to nitric acid differs substantially. Kamacite is more readily dissolved while the taenite resists dissolution. The result is that the kamacite dissolves to a minuscule level below the resistant taenite border. This effects the luster of the two minerals, the kamacite appears a dull gray while the taenite appears a bright silver (or black depending upon the lighting direction.)

Use latex gloves to prevent oils from your hands leaving fingerprints on the specimen. The gloves will also protect your hands from the acid solution when you are handling the specimen during the etching process. Place the specimen in a porcelain shallow dish making certain that the slab is level. If an irregular end piece is being etched, use clay to stick the irregular end to the dish and to level it. With a soft artist's paint brush apply the etching solution, making certain the solution is quickly spread over the entire surface. Continually move the solution over the surface recharging your brush with fresh solution every few seconds. At a 10% acid concentration, the Widmanstätten structure should begin to appear in a minute or two. The process will proceed faster if the etching solution and meteorite has been warmed to about 100 °F. From this point onward, it is a matter of judgement as to the depth of the etch. Continue to apply the etching solution making certain that it is evenly spread across the slab. Pick up the slab and examine the sheen. The kamacite plates will be distinct and have a satin luster while the narrow taenite band bordering the kamacite, if you can see it at all, should appear silvery and bright in direct illumination. As you tilt the specimen you should see the kamacite alternate from dark to light gray when viewed from different angles. Look for delicate Neumann lines across the kamacite plates. They will look like scratches but, unlike scratches, the lines will be straight and parallel. At this point, the etching procedure is nearing completion. At any point in the etching process you can rinse the specimen with water, stopping the reaction momentarily to examine the progress. If the process is then continued, be certain the meteorite is dried thoroughly before applying the etchant. Refer to the various photographs in Chapter 9 to help you judge the progress. If you have prepared both sides for etching, only one side should be etched at a time. After one side is completed it should be washed with water, neutralized with ammonia and thoroughly dried. Place a protective layer of duct tape over the etched side to prevent acid solution from contacting the the face and proceed as above with the reverse side.

Once the meteorite has been etched and rinsed with water, it should be immersed in ammonia, a weak base, to neutralize the remaining acid. This will take about 5 min. The slab is then washed thoroughly in luke-warm water using a soft brush to scrub the faces and sides. Dry as completely as possible and then immerse the slab in 100% ethanol. Ethanol is a potent drying agent that will absorb any residual water remaining on and in the meteorite. Leave it in this bath covered for about an hour. Finally, place the slab in an oven set at the minimum level (about 120 °F) and leave for an additional hour before removing.

Etching with ferric chloride

Etching with ferric chloride usually brings terror to the hearts of the seasoned meteorite collector and museum curator. This is because it was thought that chlorine in the form of ferrous chloride resided within many iron meteorites that reacted with the meteorite to produce an ugly yellow-green goo (ferric chloride) exuding between the kamacite plates and eventually destroying the display surface if not the entire meteorite (see Chapter 3). Now we know that chlorine is not indigenous to iron meteorites but is picked up from the environment during the weathering process.

We can apply ferric chloride to the prepared surface of an iron meteorite to etch the meteorite. If ferric chloride is applied in the presence of water an instant reaction occurs with the byproducts being a hydroxide of iron, hydrochloric acid (HCl) and ferric chloride again. It is not the ferric chloride that does the etching. Rather, it is the HCl byproduct. The acid in this case is stronger than the nitric acid etchant normally used. It produces a deep etch in a much shorter time, often less than a minute, especially if the solution is heated to 100 °F. Interestingly, the HCl does not seem to react with troilite inclusions to produce ugly brown stains. Just how permanent the results may be remains to be seen over the years. I have experimented with ferric chloride, comparing it to a nitric acid/alcohol etch, and have found it comparable or better. It seems to produce Widmanstätten figures with substantially more contrast and distinctiveness. Figure 9.25 shows a Nantan iron meteorite etched with ferric chloride. Note that the troilite nodules are clear of stains. The etch is deep and the contrast excellent. Many kamacite plates show distinct Neumann lines. The ferric chloride reaction is very rapid, often too rapid to prevent over-etching. Signs of over-etching are a darkening of the etched face and a drop in contrast between the kamacite plates. Ferric chloride in liquid form is the etchant of choice when etching printed circuit boards. It is readily available at most electronics stores.

Application of a protective coating

After the specimen has cooled to room temperature, you can either elect to coat the surface with a protective sealant finish or leave it uncoated. If left without a sealant, the specimen should be placed in a glass-covered box (a Riker© specimen box) with the finished face held tightly against the glass. This will keep air from the surface, reducing the likelihood of rust forming during humid conditions. Water-absorbing silica gel can be added to the box to maintain a dry environment. If a protective coating is wanted, a coating of polyurethane from a spray can is commonly used to seal the surface. This coating does not produce a complete barrier to the atmosphere and usually a meteorite prone to rusting will eventually rust beneath the coating. A better sealant is available that is applied with a brush.[I] This is a three-component epoxy mixture that contains corrosion inhibitors providing a solid monomolecular protective layer bonded to the meteorite surface. This coating seals the surface, preventing moisture transmission through the coating to the surface. This material was originally developed for the military as a preservative to be used on stored armored military equipment. Once applied, the meteorite should be quickly covered with a lid to prevent dust from accumulating on the still-wet surface.

Appendix E Testing a meteorite for nickel

Throughout this book, we have seen that iron alloyed with nickel is quite common in meteorites, so much so that for all but a very few stony achondrites, the presence of nickel is an important diagnostic characteristic that, if shown to be present, almost certainly identifies the specimen as a meteorite. A common qualitative test for the presence of nickel can be made in a high school chemistry lab.

The test requires only a tiny piece of material, far less than 1 g. All that is necessary is to file a few scrapings of the specimen onto a white paper towel. These are then transfered to a wide-mouthed flask. Slowly add about 10 ml of concentrated HCl and heat until the grains are dissolved. Then add a few drops of HNO_3 to oxidize the iron (we are assuming that this is meteoritic iron alloyed with nickel). This will caused the iron to precipitate out of solution as ferric hydroxide. The ferric hydroxide can be decanted with a fine filter leaving a clear solution. (If you wish, the iron can be made to remain in solution by adding a few drops of citric acid.) The solution should now be neutralized with NH_4OH (ammonium hydroxide) until it is slightly basic. Check that it is basic with litmus paper. The solution should be clear of precipitate. If not, filter the precipitate. With the filtrate now basic, add a few drops of *dimethyl glyoxime*. If nickel is present, the solution will turn a bright cherry red. This test is very sensitive and will detect nickel in amounts of $\ll 1\%$. Since nickel always occurs in amounts of more than 4% in meteoritic iron, the test is more than adequately sensitive.

A similar test can be made by using the same procedure described above, but adding ammonium sulphide to the clear filtrate. If nickel is present, a black precipitate of nickel sulphide will appear in the clear fluid.[1]

The test also can be made *in situ* on iron flakes in chondritic meteorites or directly on the surface of an iron slab without removing any material. Simply place a small drop of warmed concentrated HCl onto the grain. Allow it to remain for several minutes. Remove the acid with a pipette and place it in a small test tube. Add a few drops of ammonium hydroxide to the acid. Then filter out the precipitate and pour the filtrate on a white paper towel. Add a drop of dimethyl glyoxime to the wetted paper and the characteristic pink color will appear if nickel is present.

[1] Farrington, O.C. (1915). *Meteorites: Their Structure, Composition, and Terrestrial Relations*. Published by the author, p. 3.

Appendix F Meteorite museum collections and selected research facilities

This appendix lists the important collections of the world, both for public display and for meteoritical research. Certainly one of the best ways to learn to recognize meteorites is to study them in public museums where the best examples are usually displayed. Public access to meteorite collections was therefore an important criterion in this selection. Besides public display, there are many facilities that maintain laboratories to study meteorites. These may or may not have accompanying display areas. The more important of these have been included in this compilation. Many of the research facilities are not open to the public but may grant interested parties permission to visit. If you have a specimen you believe may be a meteorite, these laboratories will often provide analytical services. If the specimen is a meteorite, you may wish to consider donating it to the laboratory or give the laboratory a sample from the specimen (50–100 g) for further study. This is traditionally considered payment for the analytical services rendered the finder.

Argentina

Argentine Natural History Museum, Avenida Angel Gallardo 470, C1405DJR Buenos Aires, Argentina.

Australia

Australian Museum, 6 College Street, Sydney, New South Wales 2000, Australia. Displays meteorites, tektites and impact material.

Melbourne Museum of the Museum of Victoria, Carlton Gardens, Carlton. PO BOX 666E, Melbourne, Victoria 3053, Australia. Features 307 specimens representing 166 meteorite falls and finds, and more than 3000 tektites.

Research School of Earth Sciences, Australian National University, Canberra, Australian Capital Territory 2600, Australia.

South Australian Museum, North Terrace, Adelaide, South Australia 5000, Australia. Collection contains 150 meteorites and tektites.

Western Australian Museum, Francis Street, Perth, Western Australia 6000, Australia.

Austria

Institute of Geochemistry, University of Vienna, Althanstrasse 14, A1090 Vienna, Austria.

Natural History Museum, Burgring 7, Vienna 1014, Austria. Over 1700 specimens, comprising the fourth largest collection in the world.

Brazil

National Museum, Quinta da Boa Vista, São Cristóvão, Brazil. Exhibit includes the Bendegó meteorite, 5360 kg, the largest found in the southern hemisphere.

Canada

Earth and Atmospheric Sciences, University of Alberta, Edmonton, Alberta T6G 2E3, Canada. Houses more than 1000 specimens from more than 130 locations.

National Meteorite Collection, Museum of the Geological Survey of Canada, 601 Booth Street, Ottawa, Ontario K1A OE8, Canada.

Royal Ontario Museum, 100 Queen's Park, Toronto, Ontario M5S 2C6, Canada.

China

Institute of Geology, Chinese Academy of Sciences, Beijing 100029, People's Republic of China.

Czech Republic

National Museum, Václavské námesti 68, 11579, Prague 1, Czech Republic.

Denmark

Geological Museum, University of Copenhagen, Øster Voldgade 5–7, DK-1350 Copenhagen K, Denmark. Collection includes main masses of four Danish meteorites and the notable iron called Agpalilik (sixth largest in the world) and other Cape York irons. Altogether about 500 meteorites.

Technical University of Denmark, Department of Metallurgy, Building 204, DK-2800 Lyngby, Denmark.

Egypt

Egyptian Geological Museum, Atar El Nabi, Miser El Kadima, Cairo, Egypt. Collection includes Nakhla specimens, but none on display.

France

Museum of the School of Mines of Paris, boulevard St. Michel 60, 75006 Paris, France. Contains over 300 meteorite samples.

National Museum of Natural History, 61 rue Buffon, 75005 Paris, France. Approximately 2000 meteorites in the collection. Contains 344g of the Chassigny SNC achondrite.

Rathaus, Ensisheim, Alsace, France. Main mass of the oldest existing observed fall in the western hemisphere (1492) preserved here.

Germany

Institute of Mineralogy, Petrology and Geochemistry, Albert Ludwigs University of Freiburg, Albertstrasse 23b, D-79104 Freiburg, Germany.

Institute of Planetology, Münster University, Wilhelm-Klemm-Str.10, D48149 Münster, Germany.

Max Planck Institute for Chemistry, Cosmochemistry Department, Joh.-Joachim-Becher-Weg 27, D-55128 Mainz, Germany. A collection of nearly 1250 meteorite specimens maintained for research purposes.

Museum of Natural Science, Humboldt University, Invalidenstrasse 43, D-10115 Berlin, Germany. Contains about 3800 meteorites from more than 1287 locations dating back to1781.

India

Geological Survey of India, 27 Jawaharlal Nehru Road, Calcutta 700016, India. Maintains a collection of nearly 700 meteorites housed in the Central Petrological Laboratories in Calcutta, and a meteorite gallery in the Indian Museum in Calcutta.

National Geophysical Research Institute, Uppal Road, Hyderabad 500 007, India.

Physical Research Laboratory, Ahmedabad 380009, India.

Italy

Civic Museum of Natural History, Corso Venezia 55, I-20121 Milan, Italy.

Mineralogy Museum of the University of Rome, Piazzale Aldo Moro 5, 00185 Rome, Italy.

Vatican Collection, Vatican Observatory Museum, Papal Palace at Castel Gandolfo, Vatican, Italy. Not open to the public. Contains ~1000 specimens.

Japan

National Institute of Polar Research, 9–10, Kaga 1-chome, Itabashi-ku, Tokyo 173-8515, Japan. The largest collection of meteorites in the world. Not open to the public.

Mexico

Centro Civico Constitution, Culiacan, Sinaloa, Mexico. Home of the 22 ton Bacubirito meteorite.

Institute of Geology, Universidad Nacional Autónoma de Mexico, Ciudad Universitaria, Mexico D.F. 04510. Specimens of Toluca include main mass, 400 kg.

School of Mines, Mexico City, Mexico. Contains El Morito, an 11 ton oriented iron and the two Chupaderos irons that together weigh 21 tons.

Namibia

Grootfontein, Namibia. The Hoba iron meteorite, largest in the world at approximately 60 tons, remains where it was found on a farm 12 miles west of Grootfontein.

New Zealand

Canterbury Museum, Christchurch, New Zealand. Contains a 485 kg Canyon Diablo iron.

Russia

Fersman Mineralogical Museum, Russian Academy of Sciences, Leninskiy prospect 18-2, Moscow 117071, Russia. Features more than 447 falls, about 158 from the former Soviet Union, and includes a piece of the Pallas iron, as well as the largest Sikhote-Alin piece, 1745 kg.

Vernadsky Institute of Geochemistry and Analytical Chemistry, Russian Academy of Sciences, U1. Kosigina 19, Moscow 117975, Russia. About 1500 specimens, 300 on display, representing ~100 falls. Also impactites and tektites.

Slovakia

Astronomical Institute, Slovak Academy of Sciences, Dúbravská 9, 842 28 Bratislava, Slovakia.

Slovak National Museum, Vajanského nábrezie 2, 814 36 Bratislava, Slovakia.

South Africa

Department of Geochemistry, University of Cape Town, Rondebosch, CP, South Africa.

South African Museum, 25 Queen Victoria Street, Cape Town, South Africa. Contains three large Gibeon irons and other meteorites.

Spain

National Museum of Natural Sciences, José Gutiérrez Abascal, 2, 28006 Madrid, Spain. Collection contains 230 specimens, 18% from Spain.

Sweden

Swedish Museum of Natural History, Frescativägen 40, SE 10405 Stockholm, Sweden. Collection contains ~1000 specimens, 307 falls represented.

Switzerland

Institute for Isotope Geology and Mineral Resources, ETH-Zurich, Soneggstrasse 5, CH-8092 Zürich, Switzerland.

Natural History Museum, Bernastrasse 15, CH-3005 Bern, Switzerland. Has 300 specimens, many from Switzerland.

Ukraine

Natural History Museum of the National Academy of Sciences of Ukraine, B. Hmelnitskogo 15, Kiev-30, 252 650, Ukraine.

State Scientific Centre of Environmental Radiogeochemistry, National Academy of Sciences of Ukraine, Palladina 34, Kiev-142, 252 180, Ukraine.

United Kingdom

The Natural History Museum, Cromwell Road, London SW7 5BD, United Kingdom. Around 2000 specimens. The fifth largest collection in the world.

Planetary Sciences Research Institute, Open University, Milton Keynes MK7 6AA, United Kingdom.

United States of America

American Museum of Natural History, Central Park West at 79th Street, New York, New York 10024. A highlight is the Willamette Meteorite, the largest iron found in the United States and one of the largest in the world.

California Desert Information Center, 831 Barstow Road, Barstow, California 92311. Home of the Old Woman iron meteorite, 2753 kg, found in nearby mountains.

Center for Meteorite Studies, Arizona State University, Tempe, Arizona 85281. More than 1200 falls. A small museum open to the public.

Cosmochemistry Group, Department of Chemistry and Biochemistry, University of Arkansas, Fayetteville, Arkansas 72701.

Division of Geological and Planetary Sciences, California Institute of Technology, Pasadena, California 91124.

The Field Museum of Natural History, Roosevelt Road at Lake Shore Drive, Chicago, Illinois 60605. Over 1300 falls represented in the collection, many on display.

Griffith Observatory and Planetarium, 2800 East Observatory Road, Los Angeles, California 90027.

Institute of Geophysics and Planetary Physics, University of California, Los Angeles, California 90095.

Institute of Meteoritics, University of New Mexico, Albuquerque, New Mexico 87106. A very nice museum with many meteorites exhibited, including Norton County main mass.

Laboratory for Planetary Studies, Space Sciences Building, Cornell University, Ithaca, New York 14853.

Lunar and Planetary Laboratory, University of Arizona, Tucson, Arizona 85721.

Meteor Crater, 20 miles west of Winslow, Arizona, Exit 233 on Interstate 40. Museum features an extensive display of meteorites and cratering phenomena.

Mineralogical Museum, Harvard University, 24 Oxford Street, Cambridge, Massachusetts 02138

NASA/Johnson Space Center, Houston, Texas 77058.

National Museum of Natural History, Smithsonian Institution, Constitution Avenue at 10th Street, Washington, D.C. 20560. The 3rd largest collection in the world. Highlights include the Goose Lake (1169.5 kg) and Tucson Ring (688 kg) irons.

Oscar E. Monnig Collection, Dept. of Geology, Texas Christian University, Ft. Worth, Texas 76129.

Peabody Museum of Natural History, Yale University, 170 Whitney Avenue, New Haven, Connecticut 06520. The oldest collection in the United States, dating from 1807 and containing pieces from the first recorded fall in the New World, the Weston, Connecticut stones.

Planetary Geosciences Division, University of Hawaii, Honolulu, Hawaii 96822.

Appendix G Known terrestrial impact craters

This is a list of impact craters with features that strongly support their impact origin. The list contains 158 craters ranging in size from explosion pits 10 m in diameter to 200 km diameter impact structures with raised rims and central uplifts. Most of the craters do not show evidence of the impactor, that is, meteorites, meteoritic shale or metallic spherules. Seventeen are associated with meteoritic material all of which are irons. No known meteorite crater has been proven associated with stony meteorites. This may be because stony meteorites are more subject to atmospheric disintegration before impact. Or if surviving impact, remnant fragments may have succumbed to terrestrial weathering so they did not survive the age of their craters. Craters with associated meteorites are relatively youthful. All except Wolfe Creek are under 50 000 years old and are listed under the "Age" column as recent.

The evidence supporting an impact origin for the craters is summarized under the "Evidence" column. This is denoted (a)–(i) as follows: (a) exposed on the surface; (b) central rebound peak; (c) peak-ring structure; (d) raised rim; (e) shatter cones; (f) shattered rock breccia or breccia lens; (g) impact melt or impactite; (h) shock metamorphism; and (i) associated meteorites, iron oxide, or iron spherules. If iron meteorites are associated with the crater, their classification is given in parenthesis.

Question marks mean either that definitive evidence is lacking or that the evidence remains in question. Suspected meteorite impact craters are currently under study around the world and new craters are being added at an increasing pace to the list each year. This list should therefore be considered transitional. There is no doubt that new evidence will continue to strengthen the authenticity of these craters as well as add new ones in the near future.

Name	Latitude	Longitude	Size (km)	Evidence	Age (Ma)
Acraman, Australia	32°01′S	135°27′E	160	a,e,h	570
Ames, Oklahoma, USA	36°15′N	98°10′W	16	f,g,h	470 ± 30
Amguid, Algeria	26°05′N	04°23′E	0.450	a,d,h	0.10
Aorounga, Chad, Africa	19°06′N	10°15′E	17	a,h	200
Aouelloul, Mauritania	20°15′N	12°41′W	0.390	a,g,i (FeNi in glass)	3.10 ± 0.30
Araguainha Dome, Brazil	16°46′S	52°59′W	40	a,b,d,e,f,h	249 ± 19
Avak, Alaska, USA	71°15′N	156°38′W	12	f,h	100 ± 5
Azuara, Spain	41°10′N	00°55′W	30	a,h	130
B.P. Structure, Libya	25°19′N	24°20′E	2.80	a,h	120
Barringer, Arizona, USA	35°02′N	111°01′W	1.186	a,d,e,f,g,h,i (IAB)	0.049
Beaverhead, Montana, USA	44°36′N	113°00′W	60	a,e,h	600
Bee Bluff, Texas, USA	29°02′N	99°51′W	2.4	?	40
Beyenchime-Salaatin, Russia	71°50′N	123°30′E	8	a,h	65
Bigach, Kazakhstan	48°30′N	82°00′E	7	a	6.00 ± 3.00
Boltysh, Ukraine	48°45′N	32°10′E	25	b,g,h	88.00 ± 3.00
Bosumtwi, Ghana	06°32′N	01°25′W	10.5	a,g,h	1.30 ± 0.2
Boxhole, N. Terr, Australia	22°37′S	135°12′E	0.17	a,i (IIIAB)	0.03
Brent, Ontario, Canada	46°05′N	78°29′W	3.8	a,e,f,g	450 ± 30
Campo Del Cielo, Argentina	27°38′S	61°42′W	0.05	a, i (IAB)	recent
Carswell Lake, Sask., Canada	58°27′N	109°30′W	39	a,e,f,h	115 ± 10
Charlevoix, Quebec, Canada	47°32′N	70°18′W	54	a,b,d,e,h	357 ± 15
Chassenon, Haut-Vienne, France	45°50′N	00°56′E	10	a,e,f,g	150–170
Chesapeake Bay, Virginia, USA	37°15′N	76°05′W	85	b,d,f,h	35.5 ± 0.6
Chicxulub, Yucatan, Mexico	21°20′N	89°30′W	170	b,c,f,g,h	64.98 ± 0.05
Chiyli, Kazakhstan	49°10′N	57°51′E	5.5	a	46 ± 7

Name	*Latitude*	*Longitude*	*Size (km)*	*Evidence*	*Age (Ma)*
Clearwater East, Quebec, Canada	56°05′N	74°07′W	22	a,g	290 ± 20
Clearwater West, Quebec, Canada	56°13′N	74°30′W	32	a,c,g	290 ± 20
Connolly Basin, Australia	23°32′S	124°45′E	9	a	60
Crooked Creek, Missouri, USA	37°50′N	91°23′W	7	a,e,f,h	320 ± 80
Dalgaranga, West Australia	27°45′S	117°05′E	0.021	a,i (MES)	0.03
Decaturville, Missouri, USA	37°54′N	92°43′W	6	a,e,f	300
Deep Bay, Sask., Canada	56°24′N	102°59′W	13	a,f	100 ± 50
Dellen, Sweden	61°55′N	16°39′E	15	a,h	110 ± 2.7
Des Plaines, Illinois, USA	42°03′N	87°52′W	8	a,e	280
Dobele, Latvia	56°35′N	23°15′E	4.5	?	300 ± 35
Eagle Butte, Alberta, Canada	49°42′N	110°35′W	19	?	65
El'Gygtgyn, Russia	67°30′N	172°05′E	18	a,h	3.5 ± 0.5
Flynn Creek, Tennessee, USA	36°17′N	85°40′W	3.55	a,e,f	360 ± 20
Frombork, Poland	54°20′N	19°41′E	0.1	a,i (magnetic spherules)	?
Gardnos, Norway	60°39′N	09°00′E	5	a,b,f,h	500 ± 10
Glasford, Illinois, USA	40°36′N	89°47′W	4	e,f	430
Glover Bluff, Wisconsin, USA	43°58′N	89°32′W	3	a,e	500
Goat Paddock, West Australia	18°20′S	126°40′E	5.1	a,e,f	50
Gosses Bluff, N. Terr., Australia	23°50′S	132°19′E	22	a,e	142.5 ± 0.5
Gow Lake, Canada	56°27′N	104°29′W	4	a,b,d,e,f	250
Goyder, N. Terr., Australia	13°29′S	135°02′E	3	a,g,h	>136
Granby, Sweden	58°15′N	15°33′E	3	f	470
Gusev, Russia	48°21′N	40°14′E	3.5	f	65
Gweni-Fada, Chad, Africa	17°25′N	21°45′E	14	a	<345
Haughton Dome, NW Terr., Canada	75°22′N	89°41′W	24	b,d,e,h	21.5 ± 1.0
Haviland, Kansas, USA	37°35′N	99°10′W	0.015	a,i (PAL)	recent
Henbury, N. Terr., Australia	24°35′S	133°09′E	0.157	a,g,i (IIIAB)	0.01
Holleford, Ontario, Canada	44°28′N	76°38′W	2.35	f,h	550 ± 100
Ile Rouleau, Quebec, Canada	50°41′N	73°53′W	4	a,e,f	300
Ilumetsä, Estonia	57°58′N	25°25′E	0.08	g,h	recent
Ilyinets, Ukraine	49°06′N	29°12′E	4.5	f,g,h	395 ± 5
Janisjärvi, Russia	61°34′N	30°33′E	14	a,e,f,g,h	698 ± 22
Kaalijarvi, Estonia	58°24′N	22°40′E	0.11	a,e,f,i (IAB)	recent
Kalkkop, South Africa	32°43′S	24°34′E	0.64	a,h	<1.8
Kaluga, Russia	54°30′N	36°15′E	15	f,g	380 ± 10
Kamensk, Russia	48°20′N	40°15′E	25	b	65 ± 2
Kara, Russia	69°05′N	64°18′E	65	b,d,f,g,h	73 ± 3
Kara-Kul, Tajikistan	39°01′N	73°27′E	52	a	25
Kardla, Estonia	58°59′N	22°40′E	4	f,h	455
Karla, Russia	54°54′N	48°00′E	12	a,f	10
Kelly West, N. Terr., Australia	19°56′S	133°57′E	10	a,e,g	550
Kentland, Indiana, USA	40°45′N	87°24′W	13	a,e,f,h	300
Kgagodi Basin, Botswana	22°29′S	27°35′E	3.5	f,g,h	180

Name	Latitude	Longitude	Size (km)	Evidence	Age (Ma)
Kursk, Russia	51°40′N	36°00′E	5.5	b,d,f,g	250 ± 80
Lac Couture, Quebec, Canada	60°08′N	75°20′W	8	a,g	430 ± 25
Lac La Moinerie, Canada	57°26′N	66°37′W	8	a,g	400 ± 50
Lake Paasselkä, Finland	62°12′N	29°23′E	3.5	f,h	<70
Lappajarvi, Finland	63°09′N	23°42′E	17	a,f	77.30 ± 0.4
Lawn Hill, Queensland, Aus.	18°40′S	138°39′E	18	a,e,f	515
Liverpool, N. Terr., Australia	12°24′S	134°03′E	1.6	a,h	150 ± 70
Lockne, Sweden	63°00′N	14°48′E	7	?	540 ± 10
Logancha, Russia	65°30′N	95°48′E	20	a,b,e	25 ± 20
Logoysk, Belarus	54°12′N	27°48′E	17	f,g,h	40 ± 5
Lonar, India	19°59′N	76°31′E	1.8	a,d,f,g	0.052 ± 0.01
Lumparn, Finland	60°05′N	20°04′E	9	f,g	1000
Macha, Russia	59°59′N	118°00′E	0.3	a,i (metallic spherules)	0.007
Manicouagan, Quebec, Canada	51°23′N	68°42′W	100	a,b,c,e,g	212 ± 1
Manson, Iowa	42°35′N	94°31′W	35	b,d,f,g	65.7 ± 1
Marquez Dome, Texas, USA	31°17′N	96°18′W	22	e,h	58 ± 2
Middlesboro, Kentucky, USA	36°37′N	83°44′W	6	a,b,e	300
Mien, Sweden	56°25′N	14°52′E	9	a,g,h	121 ± 2.3
Misarai, Lithuania	54°00′N	24°36′E	5	f,g,h	395 ± 145
Mishina Gora, Russia	58°40′N	28°00′E	4	a,f,h	360
Mistastin Lake, Labrador, Canada	55°53′N	63°18′W	28	a,b,d,g	38 ± 4
Mjølnir, Norway	73°48′N	29°40′E	40	a,d,h	143 ± 20
Montagnais, Nova Scotia, Canada	42°53′N	64°13′W	45	?	50.5 ± 0.76
Monturaqui, Chile	23°56′S	68°17′W	0.46	a,g,i (IAB)	1
Morasko, Poznan, Poland	52°29′N	16°54′E	0.01	a,i (IIICD)	0.01
Morokweng, South Africa	26°28′S	23°32′E	70	b	145 ± 2
Mount Darwin, Tasmania, Australia	42°15′S	145°36′E	1	a,g	0.70 ± 0.01
Neugrund, Estonia (Baltic Sea)	59°12′N	23°19′E	7	f,h	540
Newporte, North Dakota, USA	48°58′N	101°58′W	3.2	a,h	<500
New Quebec, Quebec, Canada (see Pingualuit)					
Nicholson Lake, NW Terr., Canada	62°40′N	102°41′W	12.5	a,e,f,h	<450
Nördlinger Ries, Bayern, Germany	48°53′N	10°37′E	21 × 24	a,b,d,e,f,g,h	14.6 ± 0.1
Oasis, Libya	24°35′N	24°24′E	11.5	a,e,f	120
Obolon, Ukraine	49°30′N	32°55′E	15	a,b,h	215 ± 25
Odessa, Texas, USA	31°45′N	102°29′W	0.168	a,i (IAB)	0.05
Ouarkziz, Algeria	29°00′N	07°33′W	3.5	a	70
Piccaninny, Western Australia	17°32′S	128°25′E	7	a	360
Pilot Lake, NW Terr., Canada	60°17′N	111°01′W	5.8	a,h	445 ± 2
Pingualuit	61°17′N	73°40′W	3.44	a,d,g,h	1.4 ± 0.1
Popigai, Siberia	71°30′N	111°00′E	100	f,g,h	35 ± 5
Presqúile, Quebec, Canada	49°43′N	78°48′W	12	b,e	500
Pretoria Salt Pan, South Africa	25°24′S	28°05′E	1.13	a,g,h	0.2
Puchezh-Katunki, Russia	57°06′N	43°35′E	80	b,h	220 ± 10

Name	Latitude	Longitude	Size (km)	Evidence	Age (Ma)
Ramgarh, India	25°20′N	76°37′E	3	d,f,h	?
Ragozinka, Russia	58°18′N	62°00′E	9	b,d,e,f,g,	55 ± 3
Red Wing Creek, N. Dakota, USA	47°36′N	103°33′W	9	b,d,e,h	200 ± 25
Riachao Ring, Brazil	07°43′S	46°39′W	4.5	a	200
Rio Cuarto, Argentina	30°52′S	64°14′W	4.5	a,d,g,h	0.1
Rochechouart, France	45°50′N	00°56′E	23	a,f,g,h	186 ± 8
Roter Kamm, Namibia	27°46′S	16°18′E	2.5	a,f,h	5 ± 0.3
Rotmistrovka, Ukraine	40°00′N	32°00′E	2.7	f,g,h	140 ± 20
Saaksjärvi, Finland	61°24′N	22°24′E	5	e,h	514 ± 12
Saint Martin, Canada	51°47′N	98°32′W	40	a,g	220 ± 32
Serpent Mound, Ohio, USA	39°02′N	83°24′W	6.4	a,e,f,h	320
Serra da Cangalha, Brazil	08°05′S	46°52′W	12	a	300
Shoemaker (Teague Ring), W. Australia	25°52′S	120°53′E	30	a	1685 ± 5
Shunak, Kazakhstan	47°12′N	72°42′E	3.1	a,f,h	12 ± 5
Sierra Madera, Texas, USA	30°36′N	102°55′W	13	a,e,f	100
Sikhote Alin, Siberia	46°07′N	134°40′E	0.027	a,i (IIAB)	recent
Siljan, Sweden	61°02′N	14°52′E	55	a,b,d,e,h	368 ± 1.1
Slate Islands, Ontario, Canada	48°40′N	87°00′W	30	a,b,d,e,h	350
Sobolev, Russia	46°18′N	138°52′E	0.053	a	recent
Spider, Western Australia	16°44′S	126°05′E	13	a	570
Steen River, Alberta, Canada	59°31′N	117°37′W	25	?	95 ± 7
Steinheim, Germany	48°40′N	10°04′E	3.8	a,e	14.8 ± 0.70
Sterlitamak, Russia	53°40′N	55°59′E	0.0094	a,i (IIIAB)	recent
Strangways, N. Terr., Australia	15°12′S	133°35′E	25	a,d,f,h	470
Sudbury, Ontario, Canada	46°36′N	81°11′W	200	a,e,f,g,h	1850 ± 3
Suvasvesi North, Finland	62°41′N	28°06′E	4	a,h	250
Tabun-Khara-Obo, Mongolia	44°06′N	109°36′E	1.3	a	3
Talemzane, Algeria	33°19′N	04°02′E	1.75	a,d,e,h	0.5–3
Tenoumer, Mauritania	22°55′N	10°24′W	1.9	a,h	2.5 ± 0.5
Ternovka, Ukraine	48°01′N	33°05′E	12	e,h	280 ± 10
Tin Bider, Algeria	27°36′N	05°07′E	6	a	70
Tokrauskaya, Kazakhstan	47°44′N	75°29′E	220–250	e,f,g	?
Tookoonooka, Queensland, Australia	27°00′S	143°00′E	55	b,f,h	128 ± 5
Tvären, Sweden	58°46′N	17°25′E	2	?	455
Upheaval Dome, Utah, USA	38°26′N	109°54′W	5	a,b,e,f,h	65
Ust-Kara, Russia	69°18′N	65°18′E	25	g	73 ± 3
Vargeao Dome, Brazil	26°50′S	52°07′W	12	a	70
Veevers, Western Australia	22°58′S	125°22′E	0.08	a,i (IIAB)	1
Vepriaj, Lithuania	55°06′N	24°36′E	8	e,f	160 ± 30
Vredefort, South Africa	27°00′S	27°30′E	140	a,e,f,h	1970 ± 100
Wabar, Saudi Arabia	21°30′N	50°28′E	0.097	a,e,g,i (IIIAB)	0.01
Wanapitei Lake, Ontario, Canada	46°45′N	80°45′W	7.5	a	37 ± 2
Wells Creek, Tennessee, USA	36°23′N	87°40′W	14	a,e,f	200 ± 100

Name	Latitude	Longitude	Size (km)	Evidence	Age (Ma)
West Hawk Lake, Manitoba, Canada	49°46′N	95°11′W	3.15	a	100 ± 50
Wetumpka, Alabama	32°31′N	86°11′W	6.5	b(?),d,g,h	81.5 ± 1.5
Wolfe Creek, Western Australia	19°18′S	127°46′E	0.875	a,b,d,i (IIIAB)	0.3
Zapadnaya, Ukraine	49°44′N	29°00′E	4	b	115 ± 10
Zeleny Gai, Ukraine	48°42′N	32°54′E	2.5	h	120 ± 20
Zhamanshin, Kazakhstan	48°24′N	60°58′E	13.5	a,g	0.9 ± 0.1

Appendix H Summary of meteorites by classification (compiled by Bernd Pauli, June 2001)

This is a compilation of all known meteorites worldwide as of June 2001. The summary has been divided into three categories: worldwide; Antarctica (USA); and Antarctica (Japan). The worldwide category is tabulated into falls and finds and includes meteorites from the Sahara Desert. The Antarctica meteorites are divided into finds by American and Japanese field parties working on opposite sides of the continent over the past 30 years. Details on more than 22 000 of these meteorites may be found in: *The Catalogue of Meteorites*, Fifth Edition (2000), by Monica M. Grady, Head of the Petrology and Meteoritics Division in the Department of Mineralogy at The Natural History Museum, London, published by Cambridge University Press, Cambridge, UK.

H-group

	Worldwide			Antarctica (USA)	Antarctica (Japan)	
Class	**Fall**	**Find**	**Total 1**	**Total 2**	**Total 3**	**∑ Total (1–3)**
H	017	027	044	000	022	066
H3	007	042	049	025	064	138
H3–4	002	005	007	000	000	007
H3/4	000	002	002	004	003	009
H3–5	002	008	010	002	000	012
H3–6	002	009	011	000	000	011
H3.2	000	001	001	001	000	002
H3.3	000	000	000	001	000	001
H3.4	000	001	001	002	000	003
H3.5	000	001	001	006	001	008
H3.6	000	003	003	007	002	012
H3.7	000	012	012	009	002	023
H3.8	000	013	013	011	000	024
H3.8–4	000	001	001	000	000	001
H3.8–5	000	001	001	000	000	001
H3.8/4	000	002	002	000	000	002
H3.9	000	009	009	004	001	014
H3.9–4	000	001	001	000	000	001
H3.9–5	000	002	002	000	000	002
H3.9/4	000	004	004	001	000	005
H3.9–6	000	002	002	001	000	003
H4	057	337	394	331	723	1448
H4–5	001	022	023	010	003	036
H4/5	000	045	045	005	013	063
H4–6	000	007	007	001	000	008
H5	138	742	880	2322	361	3563
H5–6	004	028	032	010	000	042
H5/6	000	049	049	006	003	058

H-group (cont.)

Class	Fall	Worldwide Find	Total 1	Antarctica (USA) Total 2	Antarctica (Japan) Total 3	$\sum$ Total (1–3)
H6	092	385	477	1287	268	2032
H6–7	000	000	000	001	000	001
H7	000	000	000	001	003	004
H(L)3	000	001	001	000	000	001
H(L)6	000	001	001	000	000	001
H/L	000	000	000	000	001	001
H/L3.9	001	000	001	000	000	001
H/L4	000	001	001	000	001	002
Total	**323**	**1764**	**2087**	**4048**	**1471**	**7606**

L-group

Class	Fall	Worldwide Find	Total 1	Antarctica (USA) Total 2	Antarctica (Japan) Total 3	$\sum$ Total (1–3)
L	011	031	042	000	009	051
L(H)3	000	001	001	000	000	001
L(L)3	000	001	001	000	000	001
L(LL)3	000	002	002	003	000	005
L(LL)3–5	000	001	001	000	000	001
L(LL)3.6	000	001	001	000	000	001
L(LL?)3	000	001	001	004	000	005
L–LL6	001	000	001	000	000	001
L/LL3	000	003	003	000	001	004
L/LL3.8	000	001	001	000	000	001
L/LL4	002	003	005	000	000	005
L/LL4–5	000	001	001	000	000	001
L/LL5	002	001	003	000	000	003
L/LL5–6	000	001	001	000	000	001
L/LL5/6	000	005	005	000	000	005
L/LL6	002	009	011	000	000	011
L3.0	000	000	000	001	000	001
L3.0/3.7	000	000	000	001	000	001
L3.0/3.9	000	000	000	001	000	001
L3	004	019	023	074	051	148
L3–4	001	001	002	000	000	002
L3–5	001	006	007	000	000	007

L-group (cont.)						
	Worldwide			Antarctica (USA)	Antarctica (Japan)	
Class	Fall	Find	Total 1	Total 2	Total 3	$\sum$ Total (1–3)
L3–6	001	004	005	000	000	005
L3.1	000	000	000	002	001	003
L3.2	000	001	001	004	000	005
L3.3	000	000	000	006	000	006
L3.2/3.5	000	000	000	001	000	001
L3.2/3.6	000	000	000	001	000	001
L3.2/3.7	000	000	000	001	000	001
L3.2/3.8	000	001	001	000	000	001
L3.4	000	002	002	042	000	044
L3.4/3.7	000	000	000	001	000	001
L3.5	000	001	001	027	001	029
L3.5/3.7	000	000	000	001	000	001
L3.5/3.8	000	000	000	001	000	001
L3.5/3.9	000	000	000	001	000	001
L3.5/4	000	000	000	001	001	002
L3.6	000	001	001	018	007	026
L3.6/4	000	000	000	001	000	001
L3.7	002	007	009	012	002	023
L3.7–4	000	000	000	001	000	001
L3.7/3.9	000	000	000	001	000	001
L3.7/4	000	000	000	001	000	001
L3.8	000	008	008	010	001	019
L3.9	000	003	003	001	000	004
L3.9–6	000	001	001	000	000	001
L3.9/4	000	001	001	000	000	001
L3/4	001	003	004	000	002	006
L4	022	130	152	174	138	464
L4–5	000	002	002	000	000	002
L4–6	000	001	001	000	000	001
L4/5	000	022	022	000	002	024
L5	065	298	363	923	082	1368
L5–6	000	018	018	000	000	018
L5/6	000	032	032	000	007	039
L6	256	857	1113	2786	544	4443
L6–br	000	000	000	002	000	002
L6/7	000	004	004	002	002	008
L6/LL6	000	001	001	000	000	001
L7	000	001	001	013	001	015
Total	**371**	**1487**	**1858**	**4118**	**852**	**6828**

LL-group

Class	Worldwide			Antarctica (USA)	Antarctica (Japan)	
	Fall	Find	Total 1	Total 2	Total 3	∑ Total (1–3)
LL	001	003	004	000	019	023
LL(L)3	000	005	005	000	000	005
LL(L?)3	000	000	000	002	000	002
LL(L)3–6	000	001	001	000	000	001
LL(L)3.2	000	001	001	000	000	001
LL/L4	000	001	001	000	000	001
LL3	004	012	016	003	014	033
LL3–6	002	002	004	000	000	004
LL3.0	000	000	000	000	003	003
LL3.1	001	000	001	000	002	003
LL3.1/3.5	000	000	000	001	000	001
LL3.2/3.4	000	000	000	002	000	002
LL3.2–3.4	000	001	001	000	000	001
LL3.2/3.5	000	000	000	001	000	001
LL3.2/3.7	001	000	001	000	000	001
LL3.3	001	001	002	001	000	003
L3.3/3.5	000	000	000	001	000	001
L3.3/3.6	000	000	000	001	000	001
L3.3/3.7	000	000	000	001	000	001
LL3.4	001	001	002	006	000	008
LL3/4	000	001	001	000	000	001
LL3.4/3.5	000	001	001	000	000	001
LL3.5	000	002	002	005	000	007
LL3.6	001	000	001	006	000	007
LL3.7	000	002	002	004	001	007
LL3.8	000	001	001	003	000	004
LL3.8–6	000	001	001	000	000	001
LL3.9	001	003	004	000	000	004
LL4	006	020	026	022	024	072
LL4–5	000	004	004	000	000	004
LL4–6	001	011	012	000	002	014
LL4/5	000	001	001	000	000	001
LL4/6	000	001	001	000	000	001
LL5	014	033	047	889	016	952
LL5–6	000	023	023	000	000	023
LL5–7	000	001	001	000	000	001
LL5/6	000	005	005	000	004	009
LL6	038	127	165	237	069	471
LL6–br	000	000	000	001	000	001

LL-group (cont.)

Class	Fall	Worldwide Find	Total 1	Antarctica (USA) Total 2	Antarctica (Japan) Total 3	∑ Total (1–3)
LL7	000	005	005	003	002	010
LL7(?)	000	000	000	001	000	001
Sahara 97009*	000	001	001	000	000	001
Sahara 97039*	000	001	001	000	000	001
Sahara 97042*	000	001	001	000	000	001
Total	**072**	**273**	**345**	**1190**	**156**	**1691**

Source: * Sahara 97009, Sahara 97039, and Sahara 97042 have fayalite contents near the top of the LL range, but have O isotopes and bulk composition that are distinct from LL (Sexton, A.S. *et al.* (1998). Anomalous chondrites from the Sahara. *Meteorics and Planetary Science*, **33**, Suppl., A143).

Other ordinary chondrites

Class	Fall	Worldwide Find	Total 1	Antarctica (USA) Total 2	Antarctica (Japan) Total 3	∑ Total (1–3)
Classified but not grouped	000	001	001	000	000	001
Grouped but not classified	031	032	063	019	000	082
No group/class	001	000	001	000	000	001
Total	**032**	**033**	**065**	**019**	**000**	**084**
T.O.C.	***798***	***3557***	***4355***	***9375***	***2479***	***16209***

Carbonaceous chondrites

Class	Fall	Worldwide Find	Total 1	Antarctica (USA) Total 2	Antarctica (Japan) Total 3	∑ Total (1–3)
C? or C	001	000	001	002	002	005
C1	000	000	000	000	003	003
C2	001	001	002	096	000	098
C3	000	006	006	000	000	006
C4	000	002	002	001	002	005
C5	000	000	000	000	002	002
C6	000	000	000	000	003	003
CH	000	005	005	006	000	011
CH3	000	000	000	001	000	001
CI	005	000	005	000	000	005
CK	000	000	000	000	001	001
CK3	001	005	006	000	000	006

Carbonaceous chondrites (cont.)

	Worldwide			Antarctica (USA)	Antarctica (Japan)	
Class	Fall	Find	Total 1	Total 2	Total 3	∑ Total (1–3)
CK4	002	007	009	013	001	023
CK4–5	000	000	000	001	000	001
CK4/5	000	003	003	000	000	003
CK5	000	002	002	047	002	051
CK5–6	000	000	000	001	000	001
CK6	000	000	000	001	000	001
CM	000	000	000	001	000	001
CM1	000	001	001	002	000	003
CM2	017	007	024	057	069	150
CO	000	002	002	000	000	002
CO3	005	058	063	011	017	091
CO3.3	000	000	000	000	007	007
CO3.3/3.4	000	001	001	000	000	001
CO3.5	000	000	000	000	001	001
CR	000	002	002	000	000	002
CR2	002	013	015	056	007	078
CR3.8	000	001	001	000	000	001
CV	000	001	001	000	001	002
CV2	000	001	001	000	000	001
CV3	009	020	029	020	005	054
CV3 anom.	000	000	000	001	000	001
C2 UNGR	000	000	000	005	000	005
C5/6–Ungr.	000	001	001	000	000	001
Total	**043**	**139**	**182**	**322**	**123**	**627**

Enstatite chondrites

	Worldwide			Antarctica (USA)	Antarctica (Japan)	
Class	Fall	Find	Total 1	Total 2	Total 3	∑ Total (1–3)
E	001	000	001	000	001	002
E?	001	000	001	000	000	001
E anom	000	000	000	001	000	001
E3	000	000	000	000	013	013
E3 anom	000	000	000	006	000	006
E4	000	001	001	003	001	005
E4–5	000	001	001	000	000	001
E4/5	000	001	001	000	000	001
E5	000	000	000	001	007	008

Enstatite chondrites (cont.)

	Worldwide			Antarctica (USA)	Antarctica (Japan)	
Class	Fall	Find	Total 1	Total 2	Total 3	∑ Total (1–3)
E6	000	002	002	001	002	005
E7	000	001	001	001	000	002
Total	**002**	**006**	**008**	**013**	**024**	**045**
EH	000	001	001	000	000	001
EH3	003	052	055	040	003	098
EH4	003	001	004	003	004	011
EH4–5	000	000	000	007	000	007
EH5	002	000	002	002	001	005
EH6	000	000	000	000	002	002
EH–EL	000	000	000	000	001	001
Total	**008**	**054**	**062**	**052**	**011**	**125**
EL	001	000	001	000	000	001
EL anom	000	001	001	000	000	001
EL3	000	000	000	010	000	010
EL4	000	001	001	000	000	001
EL4–5	000	001	001	000	000	001
EL5	000	000	000	002	000	002
EL6	006	006	012	011	000	023
EL6/7	000	001	001	000	000	001
Total	**007**	**010**	**017**	**023**	**000**	**040**
Total E	***017***	***071***	***088***	***088***	***035***	***211***

Other chondrites

	Worldwide			Antarctica (USA)	Antarctica (Japan)	
Class	Fall	Find	Total 1	Total 2	Total 3	∑ Total (1–3)
Kakangari Rumuruti	001	001	002	001	000	003
R	000	000	000	005	003	008
R3	000	000	000	001	000	001
R3.7	000	001	001	000	000	001
R3.8	000	001	001	000	001	002
R3.9	000	001	001	000	000	001
R3–4	000	001	001	000	000	001
R3–5	000	002	002	000	000	002
R3–6	001	003	004	001	000	005

Other chondrites (cont.)

Class	Worldwide Fall	Worldwide Find	Worldwide Total 1	Antarctica (USA) Total 2	Antarctica (Japan) Total 3	∑ Total (1–3)
R4	000	004	004	000	000	004
R5	000	001	001	000	000	001
Ungrouped chondrites						
CB (Bencubbin)	001	003	004	002	000	006
GRO 95551	000	000	000	001	000	001
Impact melt	000	002	002	001	000	003
CHANOM	000	002	002	000	000	002
DAV 92308	000	000	000	002	000	002
LEW 87241 (fusion crust)						
LEW 88432 (H chon. metal)	000	000	000	001	000	001
UNCL	010	005	015	000	000	015
Terrestrial dolomite	000	000	000	002	001	003
Total other chondrites	**013**	**027**	**040**	**017**	**005**	**062**
Total all chondrites	***871***	***3794***	***4665***	***9802***	***2642***	***17109***

Achondrites

Class	Worldwide Fall	Worldwide Find	Worldwide Total 1	Antarctica (USA) Total 2	Antarctica (Japan) Total 3	∑ Total (1–3)
ACAP	001	004	005	007	002	014
ACAP-LOD	000	000	000	004	000	004
ALOD	001	002	003	004	004	011
ANGR	001	002	003	002	001	006
AUB	009	002	011	035	001	047
BRACH	000	006	006	004	001(?)	011
ADIO	011	001	012	018	068	098
AEUC	027	039	066	115	065	246
AHOW	019	020	039	043	026	108
UNIQUE	000	000	000	001	000	001
LUN	000	012	012	008	007	027
SNC	004	015	019	005	001	025
AURE	005	040	045	052	012	109
AWIN	001	005	006	001	005	012
Total	**079**	**148**	**227**	**299**	**193**	**719**

Ungrouped Yamato meteorites: 0006 specimens (stones)
Unclassified Yamato meteorites: 5653 specimens
Unclassified US-Ant. meteorites: 0002 specimens

Stony-irons and irons

	Worldwide			Antarctica (USA)	Antarctica (Japan)	
Class	Fall	Find	Total 1	Total 2	Total 3	∑ Total (1–3)
Stony-irons						
Mesosiderites	008	029	037	026	004	067
MESANOM	000	002	002	000	000	002
Pallasites	005	040	045	005	002	052
Total MES/PAL	**013**	**071**	**084**	**031**	**006**	**121**
Irons						
I	000	005	005	000	000	005
IA?	000	001	001	000	000	001
IAB	005	102	107	015	001	123
IAB-ANOM	000	005	005	004	000	009
IC	000	009	009	000	000	009
IC-ANOM	000	002	002	000	000	002
IIAB	006	071	077	020	004	101
IIAB-ANOM	000	000	000	001	000	001
IIC	000	008	008	000	000	008
IID	003	013	016	000	000	016
IID-ANOM	000	001	001	000	000	001
IIE	001	015	016	000	001(?)*	017
IIE-ANOM	000	002	002	001	000	003
IIF	001	004	005	000	000	005
IIIAB	008	214	222	006	002	230
IIIAB-ANOM	001	003	004	000	000	004
IIICD	001	035	036	001	000	037
IIICD-ANOM	002	003	005	000	000	005
IIIE	000	013	013	000	000	013
IIIF	000	008	008	000	000	008
IVA	004	055	059	001	000	060
IVA-ANOM	000	005	005	000	000	005
IVB	000	013	013	000	000	013
Ataxite UNGR	000	030	030	000	000	030
IRANOM	005	047	052	001	002	055
Hexahedrites, no class	001	003	004	000	000	004
Twannberg	000	001	001	000	000	001
Octahedrites, no class	007	054	061	001	002	064
No group/no class	002	034	036	015	000	051
Metal frags.	000	000	000	002	000	002
Ungrouped	000	002	002	000	000	002
Total irons	**047**	**758**	**805**	**068**	**012**	**885**

Source: *M. Ebihara *et al.* (1996). Yamato 791093, an anomalous IIE iron (abs. *Meteoritics*, **31**, Suppl. 1996, p. A041).

Grand total of Bernd Pauli meteorite database as of June 08, 2001

Class	Worldwide Fall	Worldwide Find	Worldwide Total 1	Antarctica (USA) Total 2	Antarctica (Japan) Total 3	$\sum$ Total (1–3)
Stones	0950	3942	4892	10101	2835	17828
Stony-irons	0013	0071	0084	00031	0006	00121
Irons	0047	0758	0805	00068	0012	00885
Unknown+	0012	0009	0021	00000	0000	00021
Doubtful	0128	0024	0152	00000	0000	00152
Doubtful and no fall/find info	0000	0018	0018	00000	0000	00018
Pseudometeorites	0008	0024	0032	00000	0000	00032
Pseudo (no date)	0000	0033	0033	00000	0000	00033
Ungr. Yamato	0000	0000	0000	00000	0006	00006
Uncl. Yamato	0000	0000	0000	00000	5651	05651
Uncl. US-Ant.	0000	0000	0000	00002	0000	00002
GRAND TOTAL	**1158**	**4879**	**6037**	**10202**	**8510**	**24749**

Note: +stone/group/class = no information.

Discrepancies:

08510	Ant (Japan) total records in database	08510	total records in statistical survey
10218	Ant (USA) total records in database	10202	total records in statistical survey
06078	Worldwide without Antarctica	06037	total records in statistical survey
24806		24749	

	Antarctica (USA)	Antartica (USA)	Worldwide
24806	10218	stones 10114	6078
−24749	−10202	irons 00072	−6037
Δ = 57 records	Δ = 16 records	St-i 00031	Δ = 41 records
	But: ⇒	10217	
		Δ = 1 record	

24806
−24764
Δ = 42 records (worldwide Δ = 41 records + Ant USA = Δ1 record)

The (Δ = 41 records) worldwide difference may be due to counting the "obscure" meteorites repeatedly within different categories (pseudomets., unclassed, ungrouped). Another possibility might be that I overlooked some bizarre classifications [example: LL(L)3–6 or H(L)6, etc.].

Glossary

ablation. The removal of material from a meteoroid through heating and vaporization as it passes through Earth's atmosphere.

acapulcoite. A primitive achondrite in which only partial melting and differentiation has taken place on the parent body. It has chondritic composition with some chondritic textures surviving. Acapulcoites are related to lodranites.

achondrite. A meteorite whose parent body has gone through melting and differentiation. These meteorites have crystallized from a magma. Achondrites include all stony meteorite types except the ordinary chondrites, carbonaceous chondrites and enstatite chondrites.

agglutinate. A characteristic of the surface of a planetary or subplanetary body in which small clasts of the regolith have been bonded together by impact-melted glass.

albedo. The percentage reflectivity of incoming incident light reflecting off the surface of an astronomical body.

allochthonous breccia. A brecciated rock unit that has been transported from its original place of formation. In impact craters, it is the fragmental material ejected during the excavation stage and usually forms a rim deposit and a symmetrical ejecta blanket around the crater.

Amor asteroid. An asteroid whose perihelion distance lies just outside Earth's orbit, defined as between 1.017 and 1.3 AU from the Sun.

amphoterite. An ordinary chondrite with low metal and low total iron. It is an obsolete term replaced by LL chondrite.

angrite. An achondritic meteorite composed primarily of the calcium–aluminum–titanium-rich pyroxene, fassaite. Accessory minerals include calcium-rich olivine and anorthite.

anhedral. Said of an individual mineral crystal usually of igneous origin that has failed to develop bounding crystal faces expressing its internal crystal structure.

anisotropic. A mineral crystal whose physical properties vary with crystallographic direction. All crystals except those belonging to the isometric system are anisotropic. See *isotropic*.

aphelion. In an elliptical orbit about the Sun, it is the furthest distance between a planet and the Sun.

Apollo asteroid. A class of near-Earth asteroids defined as having a mean distance from the Sun greater than 1.0 AU and a perihelion of greater than 1.017 AU. Apollo asteroids are Earth-orbit crossers and candidates as meteorite producers.

asteroid. A rocky or metallic body of subplanetary size usually but not necessarily confined to the asteroid belt.

asteroid belt. A zone between the orbits of Mars and Jupiter, between 2.0 and 4.0 AU from the Sun where the main belt asteroids are located.

Astronomical Unit (AU). The mean distance between the Earth and Sun; a standard unit of distance equal to 1.496×10^8 km.

ataxite. An iron meteorite composed of almost pure taenite with a nickel content of greater than 16 wt% and showing no macroscopic structure.

Aten asteroids. A class of near-Earth asteroid with a mean distance of less than 1.0 AU and an aphelion of greater than 0.983 AU. These asteroids are entirely within the confines of Earth's orbit.

aubrite. A meteorite formed by igneous processes composed primarily of the pyroxene, enstatite. It is highly reduced with little elemental iron; also called an enstatite achondrite.

autochthonous breccia. A breccia that has formed in the place where it was originally found before brecciation. In impact craters, it is the entire brecciated unit that has fragmented due to shock waves passing through the unit but that stays in place, i.e. not being ejected during crater formation. The rock unit is usually found below the floor of the crater and in its walls.

basalt. A common fine-grained, mafic volcanic igneous rock usually erupted onto Earth's surface from vents or fissures. The mineral content is primarily plagioclase and pyroxene.

basaltic achondrite. Achondrites that are members of the HED class of meteorites. They have textures and compositions similar to terrestrial basalts and are believed to originate on the asteroid 4 Vesta.

body-centered. A type of crystal lattice in which an atom is centered within the lattice and surrounded equidistant by eight other atoms. See *face-centered*.

bow shock. A boundary in front of a moving body in which there is a buildup of pressure as the body passes through a resisting medium at hypervelocity.

brachinite. A rare, primitive achondrite composed almost entirely of equigranular olivine.

breccia. A rock made up of angular clasts of previous generations of rock cemented together by fine grained matrix material. A breccia is a common textural feature of stony meteorites. See *genomict*; *monomict*; *polymict* breccia.

breccia lens. A lens-shaped unit of allochthonous breccia rock forming the floor of an impact crater. It is composed of ejected and fall back fragmental rock and impact-melted rock covering the floor of a transient crater.

brustseite. A German term referring to the vertex end of an oriented meteorite.

CAI. Highly refractory inclusions rich in calcium, aluminum and titanium. They are thought to be among the first minerals to condense out of the cooling solar nebula and are commonly found in C chondrites, especially CV3 and CM2 chondrites.

carbonaceous or C chondrites. A clan of chondritic meteorites with near solar compositions. They contain several wt% carbon and show evidence of aqueous alteration. They are the most highly oxidized of the chondrites as well as the most primitive.

Catastrophism. The hypothesis that worldwide violent, short-lived geologic events have acted in the past to shape Earth's crust. At the same time Earth's biosphere is affected by the sudden extinction of animal and plant species and the equally sudden rise of new species. Early ideas of Catastrophism (eighteenth century) were based upon the Biblical account of the Noachian worldwide flood. Catastrophism was rejected in the nineteenth century and replaced by James Hutton's doctrine of uniformitarianism, but returned in the guise of extraterrestrial impacts by asteroidal or cometary bodies in the late twentieth century. See *Uniformitarianism*.

Centaurs. These are asteroidal bodies orbiting in a zone between Saturn and Neptune. They may be cometary bodies that have depleted their volatiles.

chalcophile elements. These are elements that have an affinity for sulfur rather than metals or silicates and form sulfides in meteorites.

chassignite. A very rare achondritic meteorite similar to terrestrial dunite, made mostly of olivine. It is thought to be of Martian origin.

chondrite. A primitive stony meteorite containing chondritic texture, i.e. containing chondrules and has a composition, less volatiles, similar to the Sun. See *chondrules*.

chondrules. Spherical to subspherical millimeter-sized bodies of igneous origin found in ordinary chondrites, carbonaceous chondrites (except CI), enstatite chondrites and R chondrites.

commensurate orbits. Asteroid orbits in the main asteroid belt whose orbital periods are simple fractions of Jupiter's orbital period.

complex crater. An impact crater displaying complex morphology: shallow depth to diameter, flat floor, central rebound peak, concentric peak rings, and faulted and terraced walls. On Earth, complex craters begin with a diameter of about 4 km. Transition between simple and complex craters varies inversely with the gravitational acceleration (mass) of the impacted body, being different for different planetary bodies. See *simple crater*.

condensation sequence. A sequence of condensation of solid mineral grains from nebular gas in response to gradually decreasing temperatures.

corona. The extremely hot atmosphere of the Sun extending to several millions of kilometers.

cosmic dust. A general term for microscopic particles produced by comets as they loose their volatiles and trapped dust; or dust produced by collisions among asteroids; or dust shed by massive red giant stars. See *interplanetary dust particles* (IDPs).

cosmic-ray exposure age. The period of time in which a meteorite fragment has been exposed to cosmic rays while in space. This is the time between an impact on the parent body which produces the meteorite fragment and the fragment's arrival on Earth.

cosmic velocity. This is the velocity of an orbiting meteoroid with respect to the Sun.

cryptocrystalline texture. Texture of a meteorite or constituents of a meteorite (chondrules) in which the individual crystals are too small to be seen with the aid of a light microscope.

cryptoexplosion structures. A circular structure showing signs of an intense, sudden explosive release of energy but without evidence of volcanic or tectonic activity. They often exhibit intense structural deformation, shock metamorphism and shatter cones. It is a term suggested by Robert Dietz in 1959 to distinguish these structures from cryptovolcanic structures that display evidence of volcanic activity but no signs of impact. Many cryptoexplosion structures have since been proven to be impact-related.

cumulate. An igneous rock made up of relatively large crystals that settled out of a magma by gravity and accumulated on the floor of a magma chamber.

daughter nuclide. A radioactive or stable nuclide produced through the decay of a radioactive parent nuclide.

devitrification. The conversion of a glass to a crystalline texture while in the solid state.

differentiation. A process in which a homogenous planetary or subplanetary body melts and gravitationally separates into layers of different density and composition; the body usually separates into a core, mantle and crust.

diogenite. An achondritic meteorite composed of magnesium-rich orthopyroxene cumulate. It may represent the upper mantle or lower crust of the asteroid 4 Vesta. It is related to howardites and eucrites.

distribution ellipse. An elliptical area usually covering several square kilometers over which meteorites of a multiple fall tend to fall, with the more massive meteorites distributed on the far end of the ellipse. See *strewn field*.

E chondrites. Enstatite chondrites; a highly reduced chondritic meteorite composed of the magnesium-rich orthopyroxene, enstatite, and iron–nickel metal.

eccentricity. Referring to an elliptical orbit, it is the degree to which an orbit is elongated. It is a ratio of the distance between the foci and the major axis of the ellipse.

ecliptic. The plane of the Solar System defined as a projection of the Earth's orbit against the sky. It is the apparent yearly path of the Sun. Positions of Solar System objects are measured with respect to the ecliptic as a reference plane.

ejecta blanket. A blanket of allochthonous rock debris ejected from a forming impact crater out onto the surrounding terrain. It usually completely encircles the crater but may exhibit a ray pattern.

entry velocity. The velocity of a meteoroid at the beginning of the visible trail of a fireball. The initial velocity of a meteoroid at the top of the Earth's atmosphere.

equant. A crystal in an igneous or sedimentary rock that possesses the same dimensions in all directions. Synonym: *equidimensional*.

equilibrium. The state in which mineral phases in a rock are stable; they do not undergo further changes in time as long as conditions remain the same.

eucrite. The most common achondritic meteorite type. It is igneous in origin and is similar in composition and texture to terrestrial basalts. Thought to represent a lava flow on the asteroid 4 Vesta.

euhedral. Mineral crystals fully bounded by well-formed typical crystal faces.

exsolution. Unmixing. A process by which a homogeneous primary crystalline phase separates into two or more secondary crystalline phases without change in the bulk composition. This occurs at subsolidus temperatures.

face-centered. A type of crystal lattice in which atoms occupy the centers of all crystal faces so that every atom is surrounded by twelve neighboring atoms in the lattice.

fall. An observed fall of a meteorite that is subsequently recovered. See *find*.

find. A meteorite that was found but not observed to fall. See *fall*.

formation interval. The length of time between the origin of the Solar System and the formation of a meteorite or parts thereof. This is usually within an interval of 10 million years.

fractional crystallization. A process in which, at specific temperatures, minerals crystallize out of a magma so that there is no longer reaction between the crystals and the original liquid, thereby changing the composition of the magma. Synonym: *fractionation*.

fusion crust. A dark glassy layer, usually black or dark brown that forms during melting of the exterior of a meteoroid passing through Earth's atmosphere. It is made up of glass and magnetite.

gardening. Reworking and mixing of a regolith surface by impacts from meteoroids.

gas-retention age. The age of a meteorite calculated from the amount of retained gaseous daughter isotopes. Gas retention of a meteorite begins with cooling and solidification of its parent body but the radiogenic clock is reset during collision events in which the accumulated gas escapes. Gas retention ages record impact events in a meteorite's history.

gegenschein. A faint, diffuse glowing region situated on the ecliptic plane opposite the Sun, produced by sunlight reflecting or back scattering off interplanetary dust particles.

geocentric velocity. The velocity of a meteoroid with respect to Earth.

genomict breccia. A brecciated meteorite in which the individual clasts are compositionally of the same group but have differing petrographic characteristics.

glass. A solid material that has no crystal structure. Glasses occur during very rapid cooling of a melt in which there is no time for the constituent atoms to arrange themselves into an orderly atomic lattice. See *isotropic*.

H chondrite. A group of chondritic stony meteorites belonging to the ordinary chondrite clan. They have the highest total iron of the ordinary chondrites.

half-life. The interval of time in which half of the remaining atoms of a radiogenic isotope in a given sample have decayed to daughter isotopes. This is a fundamental quantity needed to calculate radiometric ages of meteorites.

heliocentric. Sun-centered; pertains to heliocentric velocities of planetary and subplanetary bodies in orbit around the Sun.

hexahedrite. A class of iron meteorites containing primarily low nickel (~5%) kamacite.

howardite. A brecciated achondrite containing lithic clasts of eucrites and diogenites: a polymict breccia.

hypervelocity impacts. This is the velocity at which an impacting body produces compressional forces that exceed the static compressive strength of the target rock. This varies with the composition and structural competency of the target rock material. Most explosive crater-forming events on Earth occur when the impacting body exceeds 5 km/s.

idiomorphic. A texture of an igneous rock in which the crystals are completely bounded by crystal faces. Synonym: *automorphic*; *euhedral*.

igneous rock. A rock that has solidified and crystallized from a melt.

impact craters. Craters produced by the impact of asteroids or comets. When fresh, they display surface expressions such as depressed floors, raised and faulted rims and central uplifts. See *impact structures*.

impact melt. Melted and recrystallized rock material produced by the heat of impact of a meteorite. Impact melt occurs as sill or dike-like bodies intruded into the breccia lens of the resulting crater or ejected into the surrounding ejecta blanket.

impact structures. These are impact craters but distinguished from them by their lack of obvious surface morphology. They are usually very old craters in which most or all above surface impact crater morphology has eroded away.

impactite. A general term for vesicular, glassy material formed from melted target rock produced upon impact of a crater-forming meteorite.

interplanetary dust particles (IDPs). Micron-sized dust particles usually of chondritic composition that is ubiquitous along the Solar System plane and is thought to originate from comets and/or debris from asteroid fragmentation.

interstellar grains. Submicron-sized solid grains thought to be ejected by red giant stars. The main constituents are carbon (diamonds), silicon carbide, and graphite.

ion. An atom that has gained an electric charge either by loss or gain of one or more electrons.

iron meteorites. Meteorites composed of almost pure iron with accessory amounts of nickel and cobalt. They are thought to be remnants of core bodies from differentiated asteroids that have been completely disrupted.

isotope. One or more atoms of the same element with the same number of protons but with differing numbers of neutrons. Their atomic numbers are the same but their atomic masses differ. Most isotopes are stable but some are radioactive.

isotropic. A crystal whose physical properties do not vary with crystallographic direction. As a result, isotropic minerals go to extinction under crossed Polaroids in any optical rotation angle. Isometric minerals are isotropic.

Kirkwood gaps. Gaps in the asteroid belt produced by orbital resonance between the asteroids and Jupiter. This resonance causes instabilities in those orbits in the gap positions that eventually carries the occupying asteroids and asteroid fragments (meteoroids) out of the main belt and into Earth-crossing orbits. See *orbital period*.

Kuiper belt. A disk-shaped region lying along the ecliptic plane outside the orbit of Neptune thought to contain billions of cometary bodies. It is probably the source for short period comets, comets with orbital periods less than 200 years.

L chondrite. A group within the ordinary chondrite clan containing metal and combined iron in amounts intermediate between the H and LL chondrites.

LL chondrite. A group within the ordinary chondrite clan that contains the lowest amounts of metal and combined iron.

lamellae. Thin plates or layers of a mineral often compressed between other layers of similar or different composition. Thin taenite lamellae between kamacite plates in the Widmanstätten structure of an octahedrite iron meteorite is an example.

lechatelierite. An amorphous fused silica glass produced by melting of high silica target material such as quartz sand by heat generated during impact of a large crater-producing meteorite.

lithophile elements. Elements that have an affinity for silica or tend to concentrate in the silicate phase. Examples: Si, Ca, Al, Si, Na, K and rare earth elements.

lodranite. A subclass of the primitive achondrites class showing only partial melting and differentiation. Related to acapulcoites.

mafic minerals. Silicate minerals that contain primarily magnesium and /or iron. These ferromagnesian minerals form mafic igneous rock.

magma. Molten rock materials containing dissolved gases and mineral crystals. Through fractional crystallization processes minerals crystallize out of the magma and interlock to form igneous rock.

main sequence. A line on the HR diagram plotting star temperatures vs stellar absolute magnitudes. All stable hydrogen-burning stars occupy the main sequence until their hydrogen fuel is exhausted, whereupon they evolve off the main sequence as their primary characteristics (mass and chemical composition) change.

mass fractionation. A natural chemical and physical process that causes ratios of isotopes of a single element to change in a manner that is dependent on the differences in their masses. For example, mass fractionation of oxygen isotopes causes their isotopic ratios to change subject to the chemical and physical conditions during their formation.

matrix. Fine-grained material between inclusions and chondrules in chondrites. This material usually has the same composition as the chondrules, primarily magnesium-rich olivine and pyroxene.

mesosiderite. A class of stony-iron meteorites consisting of a mixture of iron–nickel and broken rock fragments of magnesium-rich silicate minerals. The rock fragments are similar in composition to eucrites and diogenites.

mesostasis. The last material to solidify from a melt. It is usually found as interstitial fine-grained material or glass between crystalline minerals in an igneous rock.

metal. A general term for uncombined metal in meteorites, usually iron alloyed with nickel to form kamacite and/or taenite depending upon the nickel content.

metastable. A mineral that is stable with respect to the conditions in which it exists but is capable of reacting with its medium with the production of energy if sufficiently disturbed.

meteor. The light produced when a meteoroid enters Earth's atmosphere and heats to incandescence due to friction with atmospheric atoms.

meteor shower. A predictable rain of cometary dust particles that enters Earth's atmosphere annually from a specific direction and is usually associated with specific comets.

meteorite. A meteoroid from space that enters Earth's atmosphere and lands on the surface intact and recoverable. It is probably a fragment off an asteroid parent body.

micrometeoroid. A meteoroid of submillimeter size that probably came from comets or disintegrated asteroids.

monomict breccia. A brecciated meteorite composed of angular fragments and matrix all of like composition.

mosaicism. A characteristic of a mineral crystal seen microscopically under crossed polarizers in which extinction is not uniform but checkered into a mosaic pattern due to small irregularities within the crystal. This occurs when the crystal has been distorted by shock metamorphism.

Mosaicism is an indicator of shock effects produced by an impacting body.

nakhlite. A rare class of achondritic meteorite composed of the clinopyroxene, augite, and olivine. It is a Martian meteorite of which only four are known. They are related to shergottites and chassignites.

Neumann lines. Sets of parallel-running lines frequently seen across kamacite plates in octahedrites or across hexahedrites. They appear when the meteorite is etched with weak acid. They represent mechanical twinning boundaries produced by impact shock.

noble metals. Precious and/or rare metals such as platinum, gold, iridium, osmium. Also called the platinum group of metals.

nucleosynthesis. The synthesis of elements within stars of solar mass or greater by various thermonuclear processes during the life of the star; or formed instantaneously by the explosion of supernovae.

octahedrite. An iron meteorite of intermediate nickel content composed of low-nickel kamacite bounded by high-nickel taenite arranged in plates on the faces of an octahedron. Acid etching reveals the Widmanstätten structure.

onion shell model. A model of a hypothetical chondritic parent body that has been heated concentrically and differentially to the point of thermal metamorphism from the center to the surface, resulting in concentric layers of different petrographic grades. See *rubble pile model.*

Oort cloud of comets. A cloud of comets thought to exist in a spherical zone between 20 000 and 150 000 AU away from, and centered on, the Sun, from which are derived the long-period comets. First hypothesized by astronomer Jan Oort in 1950.

ophitic texture. A texture found in some terrestrial igneous rocks and eucritic meteorites in which lath-shaped plagioclase crystals are partially or completely included in pyroxene crystals, typically augite.

orbital period. The period of time taken by an orbiting body to make one complete circuit of the Sun relative to the stars; the sidereal period.

ordinary chondrite. A common clan of stony meteorites characterized by a primitive homogeneous texture in which igneous chondrules and iron–nickel grains are uniformly distributed within a matrix of similar composition. The clan is divided into H, L, and LL chemical groups based upon their total iron.

oriented meteorite. Meteorites that show a preferred orientation of fall, which produces a conical shape for the meteorites as they ablate in the atmosphere. About 5% of all meteorites show orientation.

oxidation. A chemical change in which there is an increase in the oxidation number of an atom as a result of losing electrons. The oxidation number defines the combining capacity of an element or compound.

paired meteorites. Meteorites that have fallen simultaneously some distance from each other but that are found through analysis to be fragments of the same mass and are therefore considered the same meteorite.

pallasite. A type of differentiated stony-iron meteorite characterized by an immiscible combination of olivine and iron–nickel metal. The metal forms a continuous network enclosing olivine nodules.

parent body. Astronomical bodies of subplanetary to planetary size that fragment upon collision to produce meteorites. Synonyms: *planetesimal; asteroid.*

penetration hole. The hole made by an impacting meteorite that does not explode upon impact. Varies in depth depending upon ground surface characteristics but usually has the diameter of the impacting meteorite.

perihelion. In an elliptical orbit around the Sun, it is the nearest distance between the orbiting body and the Sun. See *aphelion.*

perturbation. A small gravitational force on an orbiting body exerted by another, more massive body, often resulting in a gradual change of the orbit.

petrographic type. A scale used to denote the texture of chondritic meteorites; denotes increasing metamorphism in chondrites. Synonym: *petrologic type.*

phase diagram. A diagram illustrating the stability of minerals at certain temperatures, pressures, and compositions. Phase diagrams for metal alloys in iron meteorites show continuing recrystallization in the solid state.

phenocryst. The most conspicuous mineral grains in a rock with porphyritic texture.

piezoglypts. See *regmaglypts.*

planar deformation features (PDFs). Multiple sets of parallel planes 2–3 μm apart oriented along specific crystal lattice planes. These are most commonly and best observed in shocked quartz grains although they are also seen in feldspar crystals. PDFs are the most definitive criteria indicating shock metamorphism by a crater-reproducing meteorite.

poikilitic texture. A rock texture in which small euhedral mineral grains are scattered without a common orientation within a larger typically anhedral grain of different composition. For chondritic meteorites, poikilitic texture is often seen with small olivine grains enclosed in orthopyroxene grains.

polymict breccia. A rock made up of angular fragments or clasts from other rocks of different compositions.

polymorph. A mineral that has two or more crystal forms. An example is the element carbon: graphite, a hexagonal form; diamond, an isometric form.

porosity. A percentage of the bulk volume of a rock occupied by pore space.

primary characteristics. The original physical characteristics of a primitive meteorite showing no signs of change through subsequent thermal or aqueous alteration.

prograde orbit. A counterclockwise orbit of a body as seen from the Sun. See *retrograde orbit.*

racemic proportions. Equal proportions of left- and right-handed optical isomers. This pertains to an abiotic mixture of amino acids in carbonaceous meteorites. In contrast, amino acids in biologic life forms on Earth are composed primarily of left-handed stereoisomers.

radiant point. A point in the sky in a specific constellation from which meteors of a meteor shower seem to diverge; an illusion of perspective.

radiogenic isotope. An isotope of an element produced through radioactive decay of the parent isotope.

reflectance spectra. Absorption spectra produced as light from the Sun is partially absorbed by mineral crystals on the surface of an asteroid and then reflected back into space.

refractory phase. Element or mineral that is stable at high temperatures. The first phases to condense out of the solar nebula at high temperatures. Example: *CAIs*.

regmaglypts. Well-defined subspherical, elongated or polygonal depressions or indentations on the surface of meteorites. They are ablation features produced during the melting phase of a meteorite's atmospheric passage. Their shape depends upon the atmospheric orientation of the meteorite. Synonym: *piezoglypts*.

regolith. A layer of unconsolidated rocky debris found on the surface of asteroids, the Moon and other solid celestial bodies of subplanetary size. It is produced through a history of meteoroid and micrometeoroid impacts. See *gardening*.

regolith breccia. A chondritic brecciated meteorite made of consolidated lithified regolith material. Such meteorites have dark-light texture and represent material from beneath the surface as well as on the surface of its parent body.

retardation point. The point in a meteoroid's path through Earth's atmosphere where its cosmic velocity has dropped to zero and the meteoroid falls freely by Earth's gravity alone.

retrograde orbit. An orbit in which the orbiting body travels in a direction opposite the direction of the planets; clockwise seen from the Sun. A few of the moons of the outer planets as well as many comets travel in retrograde orbits.

rubble pile model. A hypothetical model of a chondritic parent body in which the petrographic types are distributed randomly throughout the body. This distribution is caused by a history of catastrophic impacts that break apart the entire parent body which then randomly reassembles the fragments into a rubble pile.

Rumuruti (R) chondrites. A small clan of meteorites similar to ordinary chondrites but is much more oxidized; little if any metal exists, the metal being incorporated into the minerals.

secondary characteristics. Characteristics resulting from thermal metamorphism, partial melting and aqueous alteration reducing the parent body's primary characteristics to secondary characteristics. They define the physical and chemical history of the meteorite's parent body after its origin but before breakup.

semimajor axis. In an elliptical orbit, it is half the major axis: perihelion + aphelion/2. It is the mean distance between a planet and the Sun.

shatter cones. Cone-shaped fragments of rock with apical angles of near 90° found in and around impact craters. They are formed by fracturing of target rock by shock waves generated during impact of a crater-producing meteorite. They are best formed in fine-grained rock such as limestone or sandstone and can range in size from a few centimeters to several meters.

shergottite. One of the SNC martian meteorites. Basaltic in composition with pyroxene (pigeonite and augite) plagioclase, and maskelynite as major components.

shock metamorphism. Permanent physical, chemical, mineralogic, and morphologic changes observed in terrestrial rocks and minerals and meteorites as a result of a transient high-pressure shock wave passing through the host body due to hypervelocity impacts.

siderophile elements. The geochemical class of elements with an affinity for the metallic phase rather than silicate or sulfur phases. Examples: *Fe, Ni, Co, Cu, P, platinum group metals*.

silicated iron. An iron meteorite that contains inclusions of silicate minerals. Many inclusions show chemical relationships (oxygen isotopic ratios) with ordinary chondrites and stony achondrites.

simple crater. A small bowl-shaped impact crater with a raised rim and surrounding ejecta blanket. Unlike complex craters, they have no central uplifted area. On Earth they are between 1 and 4 km in diameter with a depth to diameter ratio averaging about 1:6. Meteor Crater in northern Arizona is the classic example.

SNC meteorites. An acronym for shergottites, nakhlites, and chassignites. All three are rare achondritic meteorites with young isotopic ages and are thought to have originated on Mars.

snow line. A point 2.5 AU from the Sun where it is thought water-ice first condensed out of the solar nebula.

solar nebula. The disk-shaped cloud of gas and dust that gave birth to the Sun and Solar System 4.6 billion years ago.

solid solution. Substitution of ions, usually metals, of similar atomic size, producing variations in the composition of minerals; usually forms a solid solution series of related minerals. Examples: *Na plagioclase–Ca plagioclase; forsterite–fayalite*.

space weathering. Changes in the spectral properties of surface minerals on asteroidal bodies due to impacts by solar wind particles and micrometeorites. Space weathering may disguise the true spectral characteristics of the asteroid.

spectrophotometer. An instrument that measures the reflection spectra and albedo of the surface materials of an asteroid.

sporadic meteor. An unpredictable, isolated meteor not associated with a periodic meteor shower.

stones. A common term for meteorites with stony compositions.

stony-iron meteorites. A class of meteorites that contain both silicate minerals and iron–nickel in approximately equal proportions. Examples: *pallasites; mesosiderites.*

strewn field. An elliptical field defining the distribution of meteorite specimens from a multiple meteorite fall.

subhedral. An individual mineral crystal in an igneous rock that is partly bounded by its own crystal faces and partly bounded by surfaces formed against preexisting crystals. See *euhedral; anhedral.*

subsolidus temperature. The temperature at which a chemical system is below its melting point but which can continue to react in the solid state.

suevite. A form of impactite composed of a glassy material and brecciated rock fragments showing signs of shock metamorphism. Suevite is often found around suspected meteorite impact craters. Nickel–iron particles from the original impactor are sometimes found trapped within the glassy mass.

supernova. A star of several solar masses that explodes due to instabilities in its core. The explosion results in the creation of new elements that are dispersed into interstellar space, eventually to intermingle with nebular dust and gas.

terminal velocity. The velocity of a freely falling meteoroid due to Earth's gravity after its cosmic velocity has been reduced to near zero in the atmosphere; usually between 320 and 640 km/h. This often marks the end of the visible trail of a fireball.

terrestrial age. The period of time a meteorite has resided on Earth, determined by the decay of radionuclides produced by cosmic rays while in space.

terrestrial planets. The four Earth-like planets nearest the Sun: Mercury, Venus, Earth, and Mars. All are high-density planets; differentiated with iron cores, iron-rich mantles and silica-rich crusts.

thermal metamorphism. Changes in the chemical and physical characteristics due to internal heating of the parent body, probably by the decay of aluminum-26. Melting temperatures are not reached so that all changes are in the solid state. This is responsible for the various petrographic types of ordinary and carbonaceous chondrites.

tertiary characteristics. Characteristics produced by fragmentation of a meteorite's parent body, shock metamorphism, atmospheric ablation, impact and terrestrial weathering.

thermoluminescence. An emission of light when a meteorite is heated. It is the release of energy produced by ionizing radiation (cosmic rays) which was stored and locked in the mineral crystals while in space. It is a measure of the meteorite's history of irradiation.

thermonuclear reactions. Fusion reactions taking place in stars in which hydrogen is fused to form helium and energy. Later reaction after hydrogen becomes depleted results in the production of heavier elements up to carbon and oxygen in solar mass stars and iron in the more massive stars.

transient crater. The initial impact crater resulting from the compression and excavation of target and surrounding rock but before modification by fall back debris, faulting and slumping of the crater rim.

Uniformitarianism. The axiom that geological processes occurring today that modify the Earth are the same processes that occurred in the past, and that past geological events can be explained by natural laws and forces acting today. The rate of change is slow but not necessarily uniform. Originally put forth by James Hutton in 1788 and expanded upon by geologist Charles Lyell in 1830. This guiding principle has been seriously challenged by twentieth-century geologists who have discovered that catastrophic impacts of extraterrestrial bodies over all of geologic time have acted over a short period of time to modify Earth's environment and the evolution of life on Earth. See *Catastrophism.*

ureilite. A rare type of achondrite meteorite consisting of pyroxene grains (pigeonite) and olivine set in a carbon-rich matrix. Diamonds were first found in this meteorite type.

volatile elements. Elements that are the last to condense out of a cooling gas. Volatile elements condense from a gas or evaporate from a solid at low temperatures relative to refractory elements. Volatiles are the first materials to be lost when a meteorite is heated.

Widmanstätten structure. An intergrowth of plates of kamacite within taenite that grow on the octahedral faces of octahedrite meteorites.

winonaites. A rare class of primitive achondrites that have been partially melted and differentiated. They are associated with IAB irons.

xenolith. An inclusion of a foreign rock in an igneous host rock that is not chemically related to the host rock.

zodiacal light. Light from the Sun scattered by interplanetary dust particles along the ecliptic plane and between Earth and the Sun.

General index

Page numbers in bold type refer to illustrations.

Meteorite index

Below are all of the meteorites specifically discussed in this book, many of which are illustrated. Entry page numbers in bold type are either color or black and white images. A 'ts' next to the bold page number indicates the image is a photomicrograph of a thin section of the meteorite. In the entries the name of the meteorite is followed by its meteorite class or clan designation: chondrite, achondrite, stony-iron, or iron. This is followed by the group designation in parentheses. These designations are often abbreviated in the journals as they are here. If you are not familiar with these abbreviations, they are listed at the end of the index.

Meteorite classification abbreviations

Acapulcoite ACAP
Angrite ANGR
Anomalous ANOM
Ataxite D
Aubrite AUB
Brachinite BRACH
Carbonaceous C
Diogenite DIO
Enstatite E
Eucrite EUC
Hexahedrite H
Howardite HOW
Lodranite LOD
Lunar LUN
Mesosiderite MES
Octahedrite O
Ordinary Chondrite OC
Pallasite PAL
Rumurutiite R
Martian SNC*
Unclassified UNCL
Ungrouped UNGR
Ureilite URE
Winonaite WIN

*Shergottite, Nakhlite, Chassigny